Drug Discovery for Leishmaniasis

Drug Discovery Series

Titles in the Series:

1: Metabolism, Pharmacokinetics and Toxicity of Functional Groups
2: Emerging Drugs and Targets for Alzheimer's Disease; Volume 1
3: Emerging Drugs and Targets for Alzheimer's Disease; Volume 2
4: Accounts in Drug Discovery
5: New Frontiers in Chemical Biology
6: Animal Models for Neurodegenerative Disease
7: Neurodegeneration
8: G Protein-Coupled Receptors
9: Pharmaceutical Process Development
10: Extracellular and Intracellular Signaling
11: New Synthetic Technologies in Medicinal Chemistry
12: New Horizons in Predictive Toxicology
13: Drug Design Strategies: Quantitative Approaches
14: Neglected Diseases and Drug Discovery
15: Biomedical Imaging
16: Pharmaceutical Salts and Cocrystals
17: Polyamine Drug Discovery
18: Proteinases as Drug Targets
19: Kinase Drug Discovery
20: Drug Design Strategies: Computational Techniques and Applications
21: Designing Multi-Target Drugs
22: Nanostructured Biomaterials for Overcoming Biological Barriers
23: Physico-Chemical and Computational Approaches to Drug Discovery
24: Biomarkers for Traumatic Brain Injury
25: Drug Discovery from Natural Products
26: Anti-Inflammatory Drug Discovery
27: New Therapeutic Strategies for Type 2 Diabetes: Small Molecules
28: Drug Discovery for Psychiatric Disorders
29: Organic Chemistry of Drug Degradation
30: Computational Approaches to Nuclear Receptors
31: Traditional Chinese Medicine
32: Successful Strategies for the Discovery of Antiviral Drugs

33: Comprehensive Biomarker Discovery and Validation for Clinical Application
34: Emerging Drugs and Targets for Parkinson's Disease
35: Pain Therapeutics; Current and Future Treatment Paradigms
36: Biotherapeutics: Recent Developments using Chemical and Molecular Biology
37: Inhibitors of Molecular Chaperones as Therapeutic Agents
38: Orphan Drugs and Rare Diseases
39: Ion Channel Drug Discovery
40: Macrocycles in Drug Discovery
41: Human-based Systems for Translational Research
42: Venoms to Drugs: Venom as a Source for the Development of Human Therapeutics
43: Carbohydrates in Drug Design and Discovery
44: Drug Discovery for Schizophrenia
45: Cardiovascular and Metabolic Disease: Scientific Discoveries and New Therapies
46: Green Chemistry Strategies for Drug Discovery
47: Fragment-Based Drug Discovery
48: Epigenetics for Drug Discovery
49: New Horizons in Predictive Drug Metabolism and Pharmacokinetics
50: Privileged Scaffolds in Medicinal Chemistry: Design, Synthesis, Evaluation
51: Nanomedicines: Design, Delivery and Detection
52: Synthetic Methods in Drug Discovery: Volume 1
53: Synthetic Methods in Drug Discovery: Volume 2
54: Drug Transporters: Role and Importance in ADME and Drug Development
55: Drug Transporters: Recent Advances and Emerging Technologies
56: Allosterism in Drug Discovery
57: Anti-aging Drugs: From Basic Research to Clinical Practice
58: Antibiotic Drug Discovery: New Targets and Molecular Entities
59: Peptide-based Drug Discovery: Challenges and New Therapeutics
60: Drug Discovery for Leishmaniasis

How to obtain future titles on publication:
A standing order plan is available for this series. A standing order will bring delivery of each new volume immediately on publication.

For further information please contact:
Book Sales Department, Royal Society of Chemistry, Thomas Graham House, Science Park, Milton Road, Cambridge, CB4 0WF, UK
Telephone: +44 (0)1223 420066, Fax: +44 (0)1223 420247,
Email: booksales@rsc.org
Visit our website at www.rsc.org/books

Drug Discovery for Leishmaniasis

Edited by

Luis Rivas
Centro de Investigaciones Biológicas, Madrid, Spain
Email: luis.rivas@cib.csic.es

and

Carmen Gil
Centro de Investigaciones Biológicas, Madrid, Spain
Email: carmen.gil@csic.es

ROYAL SOCIETY OF CHEMISTRY

THE QUEEN'S AWARDS FOR ENTERPRISE: INTERNATIONAL TRADE 2013

Drug Discovery Series No. 60

Print ISBN: 978-1-78262-889-7
PDF eISBN: 978-1-78801-017-7
EPUB eISBN: 978-1-78801-258-4
ISSN: 2041-3203

A catalogue record for this book is available from the British Library

The Royal Society of Chemistry is a charity, registered in England and Wales, Number 207890, and a company incorporated in England by Royal Charter (Registered No. RC000524), registered office: Burlington House, Piccadilly, London W1J 0BA, UK, Telephone: +44 (0) 207 4378 6556.

For further information see our web site at www.rsc.org

Printed in the United Kingdom by CPI Group (UK) Ltd, Croydon, CR0 4YY, UK

Foreword

This book has been produced to provide to the reader with an integral and updated approach to the chemotherapy of leishmaniasis. Aside from the partial success of reservoir and vector control, as well as promising advances in vaccine development against human and canine leishmaniasis, chemotherapy is nowadays the workhorse to fight this disease.

To fulfill this goal, the book is structured as 19 chapters grouped under three major topics: general aspects of leishmaniasis, the role of medicinal chemistry in *Leishmania* pharmacology and the description of some appealing targets to define new therapeutic candidates.

The first topic provides an actual assessment of leishmaniasis and its chemotherapy. J. Alvar and B. Arana describe the current situation of leishmaniasis and its impact on human health (Chapter 1). The next two chapters tackle the current strategies on chemotherapy, including new candidates under clinical trial (C. E. Mowbray), and how miltefosine was implemented as a first-line leishmanicidal drug (F. Alves *et al.*).

The second thematic block focuses on medicinal chemistry efforts in the quest for new leishmanicidal drugs. To this end, this part starts with the description of the different drug assays used to test leishmanicidal drugs (V. Yardley and M. Konordiu), and the process of high-throughput screening (J. Martin *et al.*) to define new candidates from chemical libraries. The growing importance of "omics" (C. Fernández-Prada *et al.*) and *in silico* technologies (C. Roca *et al.*) for uncovering new drug candidates ensues. The next three chapters describe the current scaffolds for dug development obtained either from chemical synthesis (A. Martínez and C. Gil), or from natural products (J. H. G. Lago and A. G. Tempone), as well as the renaissance of organo-metallic compounds which overcome previous pharmacological hurdles of current antimonials (B. Rodrigues do Prado *et al.*). This part finishes with a description of the different approaches for

Drug Discovery Series No. 60
Drug Discovery for Leishmaniasis
Edited by Luis Rivas and Carmen Gil

Published by the Royal Society of Chemistry, www.rsc.org

drug delivery (K. C. Carter and A. B. Mullen), essential for yielding an improved chemotherapeutical outcome.

The final section of the book describes the differential targets that are or can be exploited in the development of new leishmanicidal drugs. These targets include the nucleic acids and their associated biological processes (B. Norris-Mullins and M. A. Morales), the phospholipid matrix (L. Rivas *et al.*), nutrient transporters (D. Zilberstein), carbon (H. Acosta *et al.*) and redox metabolism (H. Castro *et al.*). Chapter 17 (R. M. Reguera *et al.*), describes a representative example of a drug development programme based on a defined target: the topoisomerases. Finally, two important aspects with a relevant influence on the final result of leishmanicidal therapies are drug resistance and the interphase parasite-macrophage, which are described in the chapters written by A. Mondelaers *et al.* and A. Descoteaux respectively.

In all, this highly relevant and timely book emphasizes not only the great progresses achieved so far but also the crucial need for future discovery of new drug candidates and future development of new leishmanicidal drugs.

In view of the high cost and the long period of time needed for developing a new drug, the book conveys a clear and urgent message to health authorities around the world: it is an absolute priority to make every effort to protect the drugs currently in use, to prevent, by all available means, the appearance of resistance (*e.g.* strict use of the drugs at the right dose and the right time, use of combination therapy when relevant and feasible, *etc.*) and to put in place a strong pharmaco-vigilance program at national level to allow early detection of resistance.

This useful book deserves to be made widely available and adopted and used worldwide.

Philippe Desjeux
Former chief of leishmaniasis control at WHO HQ and former leishmaniasis coordinator at TDR, Geneva, Switzerland. Former senior program officer at iOWH and PATH, grantees of the B & M Gates Foundation.

Contents

I. Appraisal of Leishmaniasis Chemotherapy, Current Status and Pipeline Strategies

Chapter 1 Leishmaniasis, Impact and Therapeutic Needs 3
Jorge Alvar and Byron Arana

1.1 The Natural History of Leishmaniasis 3
1.1.1 Post-kala-azar Dermal Leishmaniasis (PKDL) 10
1.1.2 *Leishmania*–HIV Co-infection 10
1.1.3 Asymptomatic Carriers 11
1.1.4 Outbreaks 11
1.2 Control Measures 12
1.2.1 Diagnostics and Biomarkers 12
1.2.2 Disease Control Strategies by Region 14
1.3 Existing Treatment Options, Recent Advances and Unmet Needs 16
1.3.1 Visceral Leishmaniasis 16
1.3.2 Cutaneous Leishmaniasis 18
1.3.3 Drug Access 19
1.4 Conclusion. Development of an Elimination Tool 20
References 21

Chapter 2 Anti-leishmanial Drug Discovery: Past, Present and Future Perspectives 24
Charles E. Mowbray

2.1 Introduction 24

Drug Discovery Series No. 60
Drug Discovery for Leishmaniasis
Edited by Luis Rivas and Carmen Gil

Published by the Royal Society of Chemistry, www.rsc.org

2.2 Current Drugs for Leishmaniasis 25
2.2.1 Lessons Learned from the Properties of Current Drugs for Leishmaniasis 27
2.3 Medicinal Chemistry Challenges and Strategies to Discover New Anti-leishmanial Drugs 27
2.3.1 Phenotypic Drug Discovery Approaches 28
2.4 Future Perspectives 34
Acknowledgements 34
References 35

Chapter 3 From Bench to Bedside: Development and Optimization of Clinical Therapies for Visceral Leishmaniasis 37
Fabiana Alves, Jean-Yves Gillon, Byron Arana and Thomas P. C. Dorlo

3.1 Introduction 37
3.2 The Role of Preclinical Models in Preclinical to Clinical Translation 38
3.2.1 Strategies for *in vitro* Screening 38
3.2.2 Preclinical Animal Models for Visceral Leishmaniasis 39
3.2.3 Issues in Interpretation of Preclinical Models 40
3.2.4 How can Pre-clinical Data be Translated to Facilitate Clinical Development? 41
3.3 Challenges and Opportunities to Optimize Therapies for Leishmaniasis: The Case of Miltefosine 42
3.3.1 Discovery and Development History of Miltefosine for Leishmaniasis 42
3.3.2 Changing Patterns in Clinical Efficacy 43
3.3.3 Dose Optimization of Miltefosine for Future Combination Strategies 46
3.3.4 Translational Approach to Manage Miltefosine's Main Safety Concern: Teratogenicity 49
3.4 Final Remarks 50
Acknowledgements 50
References 51

II. Methodologies and Medicinal Chemistry Strategies to Discover and Develop New Treatments

Chapter 4 Drug Assay Methodology in Leishmaniasis: From the Microplate to Image Analysis 57
Vanessa Yardley and Markella Koniordou

4.1 Introduction 57
4.2 Drug Assay Methods 60
4.3 Models of Infection 60
4.3.1 *In Vitro* 60
4.3.2 *Ex Vivo* 63
4.3.3 *In Vivo* 63
4.4 Detection 65
4.4.1 Conventional Methods 66
4.4.2 Transgenic *Leishmania* and Related Methods 68
4.5 Technological Innovations and Developments 70
4.6 Summary 71
4.6.1 Optimism for the Future 72
Acknowledgements 72
References 73

Chapter 5 The Pursuit of Novel Anti-leishmanial Agents by High-throughput Screening (HTS) of Chemical Libraries 77
Julio Martin, Juan Cantizani and Imanol Peña

5.1 New Chemical Entities (NCEs) are an Unmet Need for Therapeutics of Kinetoplastid Diseases 77
5.2 The Challenge of Choosing the Right Screening Tactics and Assays 79
5.2.1 The Screening Approach: Target *vs.* Phenotypic Screening 80
5.2.2 The Parasite 83
5.2.3 The Host Cell 84
5.2.4 The Screening Assay Format 85
5.2.5 The Analysis of Screening Data 85
5.2.6 Hit Triage: Progression Cascade and Selection Criteria 87
5.3 Case Studies of HTS Campaigns for Anti-Leishmanials 93
5.4 Conclusions 97
References 97

Chapter 6 Omics and Their Impact on the Development of Chemotherapy Against *Leishmania* **101**
Christopher Fernández-Prada, Isabel M. Vincent, Élodie Gazanion and Rubens L. Monte-Neto

6.1 Introduction 101
6.2 Genomics 103
6.2.1 Classical Genomic Strategies in *Leishmania* 103
6.2.2 Novel Gain-of-function Genomic Approaches 104
6.2.3 Current Limitations and Future Development of Genomic Approaches 107
6.3 Transcriptomics 108
6.3.1 Transcriptomic Toolkit for Target Elucidation in *Leishmania* 108
6.3.2 Recent Advances in and Perspectives on Transcriptomics for Studying Chemotherapy Against *Leishmania* 111
6.4 Proteomics 112
6.4.1 Proteomics Methods 112
6.4.2 Proteomics in the Discovery of Modes-of-action and Drug-resistance Mechanisms in *Leishmania* 113
6.4.3 The Future of Proteomics in Drug Discovery 115
6.5 Metabolomics 115
6.5.1 Metabolomics Methods 116
6.5.2 Metabolomics and Anti-leishmanial Drugs 117
6.5.3 Resistance and Synergy 118
6.5.4 The Future of Metabolomics in *Leishmania* Drug Development 119
6.6 Concluding Remarks 120
Acknowledgements 120
References 121

Chapter 7 *In silico* Tools for Target Identification and Drug Molecular Docking in *Leishmania* **130**
Carlos Roca, Víctor Sebastián-Pérez and Nuria E. Campillo

7.1 Introduction 130
7.2 *Leishmania* Targets Identification 131
7.3 Anti-leishmanial Drug Development 133
7.3.1 Ligand-based Drug Design 134
7.3.2 Structure-based Drug Design 137
7.3.3 Drug Repurposing 144

7.4 Target Databases 144
7.5 Concluding Remarks and Outlook 146
Acknowledgements 147
References 147

Chapter 8 Medicinal Chemistry Strategies to Discover New Leishmanicidal Drugs 153
Ana Martinez and Carmen Gil

8.1 Introduction 153
8.2 Functional Scaffolds for Drug Discovery Against *Leishmania* 155
8.2.1 Quinones 155
8.2.2 Nitroaromatic Compounds 157
8.2.3 Nitrogen Heterocycles 158
8.3 Targets for Leishmaniasis Drug Discovery 161
8.3.1 Enzymes of Polyamine Biosynthesis 162
8.3.2 Enzymes of Thiol Metabolism 164
8.3.3 Enzymes of Folate Metabolism 165
8.3.4 Sterol Biosynthesis Pathway 167
8.3.5 *Leishmania* Proteases 168
8.3.6 *Leishmania* Kinases 170
8.3.7 *Leishmania* Phosphodiesterases 171
8.4 Conclusions 172
Acknowledgements 172
References 172

Chapter 9 Natural Products as a Source of New Drugs Against *Leishmania* 179
João Henrique G. Lago and Andre G. Tempone

9.1 *Leishmania*—A Challenging Parasite 179
9.2 The Interest in Natural Products as Scaffolds: Using a Rational Approach for Drug Discovery 181
9.2.1 Major Topics During Biological (*In Vitro* and *In Vivo*) Tests of Natural Products 182
9.3 From a Crude Extract to a Pure Compound—Aspects of Bioactive Compounds Isolation and Chemical Identification 184
9.4 The Anti-leishmanial Potential of Natural Products in Animal Models 185
9.4.1 Alkaloids 185
9.4.2 Terpenoids 187

9.4.3 Saponins 187
9.4.4 Phenylpropanoids 188
9.4.5 Flavonoids 189
9.4.6 Lignoid 191
9.4.7 Naphthoquinones 191
9.4.8 Iridoid 192
9.5 Conclusions 196
Acknowledgements 196
References 196

Chapter 10 Organometallic Compounds in Chemotherapy Against *Leishmania* **199**
Bruno Rodrigues do Prado, Arshad Islam, Frédéric Frézard and Cynthia Demicheli

10.1 Introduction 199
10.2 History and Clinical Use of Antimony-based Drugs 201
10.3 Organoantimonial Compounds 203
10.4 Organobismuth Compounds 208
10.5 Organotin(IV) Compounds 213
10.6 Organotellurium(IV) Compounds 215
10.7 Iron- and Ruthenium-based Organometallic Compounds 215
10.8 Palladium, Rhodium and Iridium-based Organometallic Compounds 218
10.9 Conclusion 219
Acknowledgements 219
References 219

Chapter 11 New Avenues for Drug Delivery in *Leishmania:* Using Treatment of Visceral Leishmaniasis with Amphotericin B as an Exemplar **224**
Katharine C. Carter and Alexander B. Mullen

11.1 Introduction 224
11.2 Drug Delivery Systems (DDS) 225
11.3 Parenteral Administration 227
11.4 Oral Administration 228
11.5 Pulmonary Administration 231
11.6 Summary 231
References 232

III. The Quest for Achille's Heel of Leishmania. Singular Targets as New Avenues for Drug Development

Chapter 12 Addressing the Molecular Biology of *Leishmania* for Drug Development **237**
Brianna Norris-Mullins and Miguel A. Morales

12.1 Introduction 237
12.2 The *Leishmania* Genome 238
12.3 The *Leishmania* Transcriptome 238
12.3.1 Polycistronic Transcription and *Trans*-splicing Mechanisms 239
12.3.2 Spliceosome 241
12.3.3 Mitochondrial mRNA Editing 241
12.4 Post-transcriptional Regulation of Gene Expression 243
12.4.1 3′ UTR Control and mRNA Degradation 243
12.5 Perspectives 243
References 244

Chapter 13 The Physical Matrix of the Plasma Membrane as a Target: The Charm of Drugs with Low Specificity **248**
Luis Rivas, Montserrat Nácher-Vázquez and David Andreu

13.1 Outline for an Antimicrobial Peptide-based Chemotherapy against Leishmaniasis 248
13.2 General Appraisal of Peptide-based Therapies 249
13.3 Natural History of AMPs 249
13.4 Mechanism of Action of Antimicrobial Peptides 250
13.4.1 Molecular Characteristics of AMPs 250
13.4.2 Antimicrobial Peptide–Membrane Interaction 250
13.4.3 Models for Antimicrobial Peptide–Membrane Interaction 251
13.4.4 The Plasma Membrane of *Leishmania* as a Target for Antimicrobial Peptides 252
13.5 Natural AMPs as Leishmanicidal Agents 254
13.6 Assessment of Plasma Membrane Permeabilisation by AMPs in *Leishmania* 260
13.7 Intracellular Targets 262
13.8 Induction of Programmed Cell Death of *Leishmania* by Antimicrobial Peptides 262

13.9 Structure–Activity Relationship of Leishmanicidal Antimicrobial Peptides 263
13.10 Antimicrobial Peptides in Animal Models of Leishmaniasis 268
13.11 Immunomodulation by Leishmanicidal AMPs 268
13.12 The Outlook for AMP-based Therapies for Leishmaniasis 269
13.12.1 Intrinsic Resistance of *Leishmania* to Antimicrobial Peptides 269
13.12.2 Role of Enviromental Conditions for AMPs on CL Lesions 269
13.12.3 Synergism and AMPs for Leishmaniasis 270
13.12.4 Antimicrobial Peptide Production. Scale–up Production or *In situ* Expression 270
13.13 Conclusions 271
Acknowledgements 272
References 272

Chapter 14 Nutrient Transport and Sensing as Pharmacological Targets for Leishmaniasis 282
Dan Zilberstein

14.1 Introduction 282
14.2 The Metabolic Environment Inside the Vector and Host 283
14.2.1 The Insect Vector 283
14.2.2 Development Inside the Host 284
14.3 Amino Acid Transport and Homeostasis 285
14.3.1 The Amino Acid Pool in *Leishmania* Cells 285
14.3.2 LdAAP24: A New Role for Permease N-termini in Transport Regulation 286
14.3.3 A *Leishmania*-Specific Gene Encodes the Arginine Transporter AAP3 Family 287
14.3.4 Lysine Transport 288
14.3.5 Transporters of Other Amino Acids 288
14.4 Glucose Transport 288
14.5 Purine, Pyrimidine and Nucleoside Transport 290
14.6 Nutrient Sensing Has a Role in Virulence 290
14.6.1 Intracellular Sensing: Response to Purine Starvation 290
14.6.2 Extracellular Sensing: Arginine-deprivation Response 291

14.6.3 Proteins Involved in Sensing are Localized in the Flagella Membrane 291

14.7 Can Nutrient Transporters Deliver Drugs into *Leishmania* Cells? 292

14.7.1 Amino Acid Transporters 292

14.7.2 Pentamidine Transport in *Leishmania* 292

14.8 Concluding Remarks 293

Acknowledgements 293

References 293

Chapter 15 Carbon Metabolism as a Drug Target in *Leishmania* **297**

Héctor Acosta, Meng Yuan, Ana J. Cáceres, Wilfredo A. Quiñones, Juan Luis Concepción and Paul A. M. Michels

15.1 Introduction 297

15.2 Carbon Metabolic Network of *Leishmania* 300

15.3 Drug Targets of Carbon Metabolism in *Leishmania* 303

15.4 Functional and Structural Properties of Carbon-metabolic Enzymes that are Potential Drug Targets in *Leishmania* 304

15.5 Drug Discovery 308

15.6 Conclusions 311

Acknowledgements 312

References 312

Chapter 16 The Redox Metabolism and Oxidative Stress in *Leishmania* as a Crossroads for the Lethal Effect of Drugs **316**

Helena Castro, Margarida Duarte and Ana M. Tomás

16.1 Reactive Oxygen and Nitrogen Species and *Leishmania* 316

16.1.1 Generation of ROS and RNS within *Leishmania* 317

16.1.2 Exogenous Sources of Oxidative and Nitrosative Stress to *Leishmania* 320

16.2 Elimination of ROS and RNS 321

16.2.1 Superoxide Dismutases 321

16.2.2 Class I Haem Peroxidases 322

16.2.3 The Trypanothione Redox System 323

16.3 Conclusions 335

Acknowledgements 336

References 336

Chapter 17 DNA Topoisomerases as Promising Targets for *Leishmania* Chemotherapy **348**
Rosa M[a] Reguera, José M. Escudero-Martínez, Bárbara Domínguez-Asenjo, Camino Gutiérrez-Corbo and Rafael Balaña-Fouce

17.1 Introduction 348
17.2 Supercoiled *vs.* Relaxed DNA. The Need for DNA Topoisomerases 350
17.3 DNA Topoisomerases in *Leishmania* 352
17.3.1 *Leishmania infantum* DNA Topoisomerase IB 353
17.3.2 Inhibitors of DNA Topoisomerase IB 356
17.4 Other DNA Topoisomerases in *Leishmania* 361
17.4.1 Type II DNA Topoisomerases 361
17.4.2 Type IA DNA Topoisomerases 362
17.4.3 Inhibitors of DNA Topoisomerase II 362
17.5 Conclusions 363
Acknowledgements 364
References 364

Chapter 18 Molecular Basis of Drug Resistance in *Leishmania* **371**
Annelies Mondelaers, Sarah Hendrickx, Guy Caljon and Louis Maes

18.1 Emerging Drug Resistance 371
18.2 Methods to Study Resistance 372
18.3 Molecular Mechanisms of Resistance and Current Limitations 372
18.3.1 Antimonials 373
18.3.2 Miltefosine 376
18.3.3 Paromomycin 378
18.3.4 Amphotericin B 379
18.4 Conclusions 379
Acknowledgements 379
References 380

Chapter 19 The Macrophage–Parasite Interface as a Chemotherapeutic Target in Leishmaniasis **387**
Albert Descoteaux

19.1 Introduction 387
19.2 Phagocytosis and Phagolysosome Biogenesis 389

19.3 The Membrane Fusion Machinery and the Phagosome 389
19.4 *Leishmania* and Phagosome Remodelling 390
19.5 The Parasitophorous Vacuole as a Potential Target for Drug Development 391
19.6 Conclusions 393
References 393

Subject Index **396**

I. Appraisal of Leishmaniasis Chemotherapy, Current Status and Pipeline Strategies

CHAPTER 1

Leishmaniasis, Impact and Therapeutic Needs

JORGE ALVAR* AND BYRON ARANA*

Drugs for Neglected Diseases *initiative* (DND*i*), Switzerland
*Email: jalvar@dndi.org; barana@dndi.org

1.1 The Natural History of Leishmaniasis

Leishmaniasis is a group of diseases caused by a series of species of *Leishmania*, protozoa transmitted by the bite of infected female sand flies. The parasite in its flagellated form, called a promastigote, evades the action of complement in the bloodstream by entering mononuclear phagocytic cells (macrophages) as a result of recognition of surface ligands on the parasite by receptors in the macrophage membrane. Once inside, the parasite multiplies within a parasitophorus vacuole in a new form without a flagellum, called amastigote, by evading the oxygen cascade and other mechanisms activated by the host cell. The macrophage finally ruptures and multiple amastigotes enter new macrophages (Figure 1.1). There are two clinical forms of leishmaniasis in humans, cutaneous leishmaniasis (CL) (Box 1.1) and visceral leishmaniasis (VL) (Box 1.2). The former causes skin ulcers which normally heal spontaneously, although they may leave scars, causing stigma. The latter (also called *kala-azar*) affects vital organs such as the spleen, liver and bone marrow, and results in death if untreated.

In brief, cutaneous leishmaniasis is mediated by a Th1 response whereas kala-azar (visceral leishmaniasis) is mediated by a Th2 response.[1]

Drug Discovery Series No. 60
Drug Discovery for Leishmaniasis
Edited by Luis Rivas and Carmen Gil

Published by the Royal Society of Chemistry, www.rsc.org

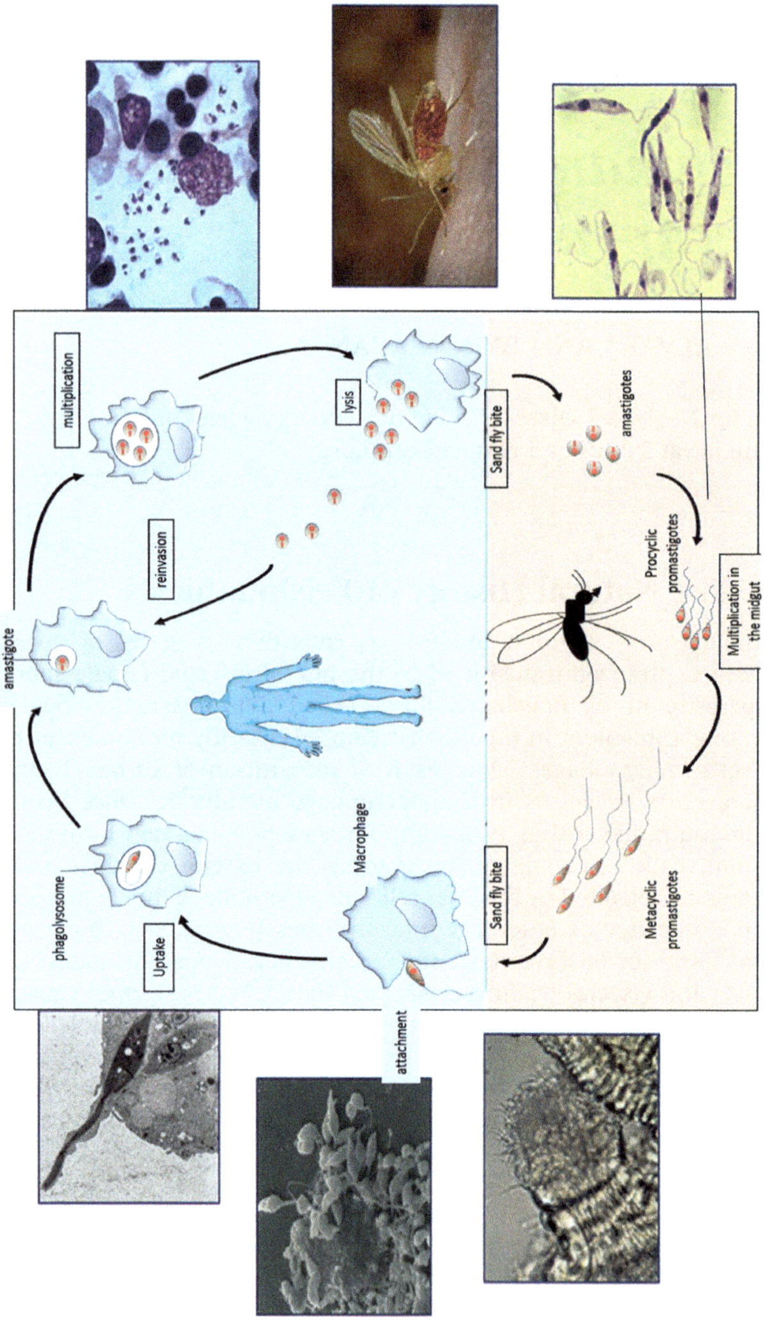

Figure 1.1 Cycle of *Leishmania* transmission.

Box 1.1 Cutaneous leishmaniasis. Epidemiological considerations

- Cutaneous leishmaniasis is endemic in 98 countries.
- 0.6 to 1.2 million new cases occur annually, which means that there is a new case every 30 seconds.
- Cutaneous leishmaniasis is not a single disease, different clinical features and courses of illness are observed.
- Although not a fatal disease, lesions can cause disfiguration, mutilation and carry a lot of stigma, especially in women.
- 75% of all cases are reported from only ten countries: Afghanistan, Algeria, Brazil, Colombia, Costa Rica, Ethiopia, Iran, Sudan, Peru and Syria.
- More than ten *Leishmania* species, dozens of reservoirs and hundreds of different vectors have been identified in the transmission cycle.
- CL is not considered a priority and treatment is not provided free of cost in many countries.
- Traditional vector control methods are not effective or practical for at-risk populations.
- No rational measures for the control of reservoir hosts are available in the New World.
- Control is unlikely to be achieved by a single intervention. A combination of case management strategies, integrated vector control and animal reservoir control, if relevant, is required and should be tailored to each context.
- Current treatment recommendations are based on weak evidence, the number of treatments options is limited and associated with high toxicity.

In CL, T-lymphocytes initiate the cellular response, causing Th1 subpopulations to release lymphokines, including gamma interferon ($IFN-\gamma$) and interleukin-2 (IL-2). These lymphokines attract macrophages to kill the parasite, and activate oxygen-dependent mechanisms and those dealing with the fusion of lysosomes with the parasitophorus vacuole. Meanwhile, the *Leishmania* parasite is able to neutralize this lymphokine response by releasing inhibitors of lysosome hydrolases, and of reactive oxygen species and nitrogen intermediates. Thus the parasite survives and multiplies inside the macrophage until the infection is controlled. In contrast, in VL the $CD4^+$ T-lymphocytes stimulate Th2 subpopulations to produce IL-4, IL-5 and IL-10, meaning that macrophages are not activated and the disease progresses and invades vital organs such as the spleen, liver and bone marrow. In these cases, the cooperation of T and B cells favours development of the humoral response, with production of specific antibodies against

Box 1.2 Visceral leishmaniasis. Epidemiological considerations

- Visceral leishmaniasis is a fatal disease: patients need life-saving treatment.
- 150 000–300 000 cases per year, with around 90% focused in a few countries (India, Sudan, South Sudan, Brazil, Bangladesh, Ethiopia) although a sharp decrease is noticeable in Asia.
- India is an important endemic country: historically, nearly 60% of all cases worldwide occur in the state of Bihar. Around 20 000 cases are estimated in the East Africa region, which is prone to devastating outbreaks.
- VL is poverty-related and leads to economic devastation of already poor households; this is partly due to a lack of cheap, effective treatments at the lowest health care level.
- 50–70% of patients are children.
- In anthroponotic areas of transmission (South Asia, East Africa), case detection and treatment of VL and post-kala-azar-dermal leishmaniasis (PKDL) is one strategy to control the disease.
- There are important geographical features with regards to efficacy of rapid diagnosis tests and treatments in VL.
- In addition, response to drug treatment has been shown to vary between and within regions: higher doses (of AmBisome® and paromomycin) are required to achieve efficacy in East African and Latin American as opposed to South Asian foci. Geographical variation has also been shown within East Africa. To date neither regional variability in pharmacokinetic profiles in humans nor drug sensitivities of parasites have been shown to be responsible.

Leishmania, but this has little effect since the parasite is hidden inside the macrophage.[2]

Around 20 species of *Leishmania* cause CL but only two cause kala-azar. From an epidemiological point of view, the majority of species follow a zoonotic cycle (infected animal reservoir–sand fly–human), whereas *Leishmania donovani* and *Leishmania tropica* follow an anthroponotic cycle (infected human reservoir–sand fly–human); this difference is critical in terms of control.[3]

Disease progress after the primary CL lesion depends on the infecting *Leishmania* species; patients can develop a mucocutaneous form (MCL; species belonging to the subgenus *Viannia* in the New World), relapse [recidivans leishmaniasis (RL) due to *L. tropica*], or develop diffuse cutaneous leishmaniasis (DCL; *Leishmania amazonensis*). In the case of *L. donovani*, a percentage of visceral leishmaniasis patients treated develop a skin condition named post-kala-azar-dermal leishmaniasis (PKDL), a condition not seen in visceral leishmaniasis, due to *Leishmania infantum* (Figure 1.2).

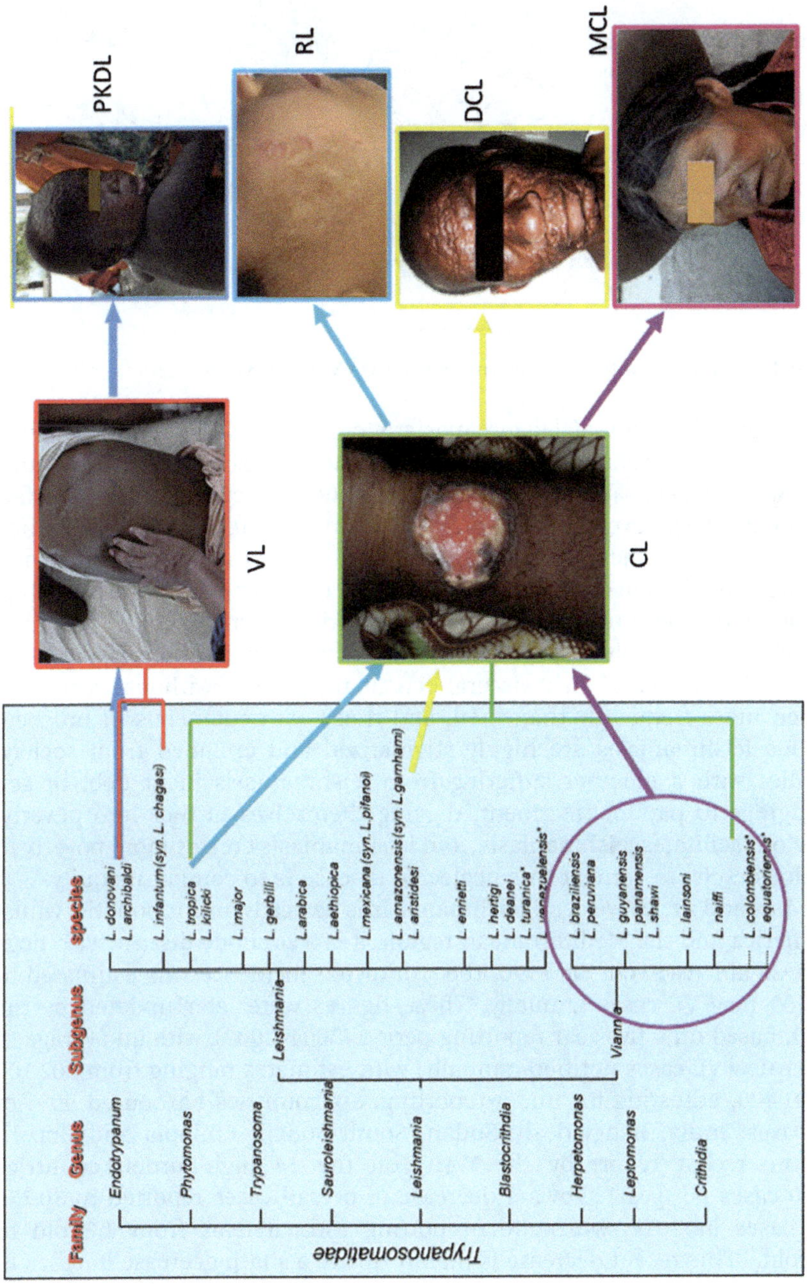

Figure 1.2 *Leishmania* species and forms of leishmaniasis.

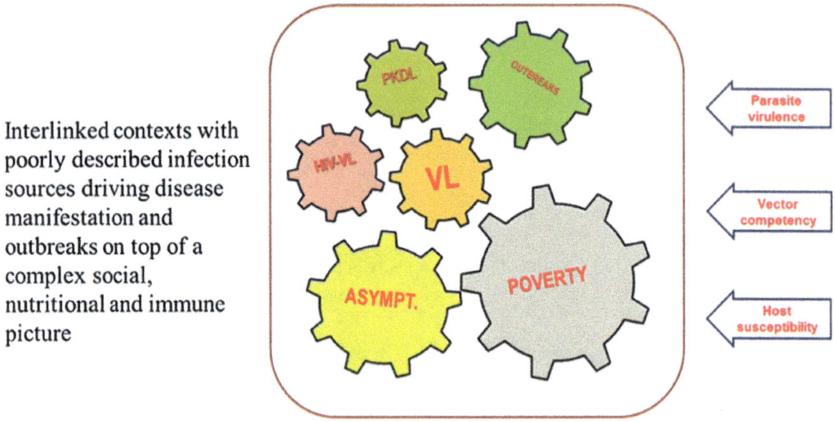

Figure 1.3 Global context of leishmaniasis. (ASYMPT., Asymptomatic.)

The natural history of leishmaniasis is determined by various elements that allow transmission, very frequently in the context of poverty, in which parasite virulence, sand fly competence and host susceptibility are key factors (Figure 1.3). Poverty is not an abstract concept but the context in which the disease flourishes, whose various components can be measured and weighted, all of them contributing to a greater or lesser extent: poor housing, deteriorated environment, low income, gender discrimination, illiteracy, lack of access to health care, malnutrition, displacement and war, *etc.* More than 20 000 patients with visceral leishmaniasis die without treatment, women more frequently than men, and those with cutaneous or mucocutaneous leishmaniasis are highly stigmatized and excluded from society. Families with a member suffering from leishmaniasis incur debt or sell belongings to pay for treatment, digging themselves further into poverty. Oblivion facilitates leishmaniasis, and leishmaniasis creates more poverty in a vicious cycle; to combat this neglected disease is to combat inequity.[4]

In Asia and Africa, visceral leishmaniasis is basically anthroponotic, whilst in America and the Mediterranean region, it is a zoonotic disease with dogs as the main reservoir. In 1990, the worldwide incidence was estimated at 500 000 new VL cases annually. These figures were later updated by the WHO, based on a five-year reporting period (2004–2008), with an average of 58 221 new VL cases notified annually, with estimates ranging from 202 100 to 391 400, adjusting for under-reporting. Six countries harboured 90% of the cases: India, Bangladesh, Sudan, South Sudan, Ethiopia and Brazil.[5] A more recent report by the WHO on the 14 high-burden countries (>100 cases per year) shows a decrease of overall cases reported to 30 758 new cases in 2014, with under-reporting estimated as from 1.2-fold to fourfold.[6] This recent decrease is mainly due to a sharp decrease in cases in the South East Asia region, from approximately 50 000 reported cases in 2006 and 2007 to 10 311 cases in 2014. This reduction can be attributed partly to the successful elimination campaign, a possible naturally fluctuating trend of incidence and improvements in living conditions of the local population. Currently, the region with the highest burden worldwide is eastern Africa, with most of the cases observed in Ethiopia, Kenya, Somalia, Sudan, South Sudan and Uganda. The six countries that currently represent 91% of the overall VL burden are India, Ethiopia, Sudan, South Sudan, Somalia and Brazil.

Cutaneous leishmaniasis is reported to be endemic in 98 countries, but 75% of all cases are reported from only ten of these: Afghanistan, Algeria, Brazil, Colombia, Costa Rica, Ethiopia, Iran, Sudan, Peru and Syria. The WHO estimates an incidence of around 0.6–1.2 million new cases per year, although just a fraction of all these numbers are officially reported for a variety of reasons.[5] Unlike the incidence of most other neglected tropical diseases, the incidence of CL is increasing, especially in the last 3–5 years due to migration or displacement of large populations due to crises in the Middle East and North Africa. According to recently published data from the WHO, 12 of the countries with the highest CL prevalence worldwide officially reported 153 027 cases in 2014. In these 12 high-burden countries, the number of cases tripled from over 50 000 in 1998 to over 150 000 in 2005. It was noted, however, by at least one of the 12 countries included in the 2014 report, that there was an underreporting factor of between 2.8 and 4.6.[6]

The real numbers might be even higher if we take into consideration the fact that reporting is problematic in several places, such as refugee camps or in countries where the health care infrastructure has collapsed. In Syria, for example, where millions of people have been internally or externally displaced and many CL cases are occurring among people living in refugee camps. Recent foci have also been described in several countries, including Burkina Faso, Iran, Spain, Argentina and Brazil.[7,8]

The global mean age-standardized disability-adjusted life years (DALYs; a measure used to calculate the burden of a disease) due to CL has also increased, from 0.29 per 100 000 inhabitants in 1990, to 0.58 in 2013. Nine countries were found to have significantly higher DALYs (per 100 000 inhabitants) compared with the mean: Afghanistan, Sudan, Syria, Yemen, Iraq, Burkina Faso, Bolivia, Haiti and Peru.[9]

In epidemiological terms, regardless of whether the parasite reservoir is human or a non-human sylvatic or domestic mammal, the animal has to fulfil a series of requisites to be considered a primary or secondary reservoir, and this is critical for the control programs: suffering from highly enzootic, chronic infection and proximity to the competent sand fly vector and human. In the case of anthroponotic transmission, prompt diagnosis and treatment are key not only to curing the patient but also to interrupting transmission. In the case of zoonotic transmission, the objective is to reduce the reservoir burden. In both cases, reduction of contact between sand flies, parasite reservoirs and humans will ultimately lead to such a low level of transmission that the cycle stops. In the absence of control programs, the epidemiological incidence of visceral leishmaniasis is recurrent, peaking every 10 years or so, probably when herd immunity is lost. If R_0 is greater than 1 in the animal or human reservoir, the disease will re-emerge. Although not properly understood and therefore modelled, there are other specific situations in which infected humans may play a role in transmission, basically PKDL, HIV–VL co-infection, and—possibly—asymptomatic carriers.

1.1.1 Post-kala-azar Dermal Leishmaniasis (PKDL)

PKDL is a skin condition that usually develops after treatment of visceral leishmaniasis. PKDL manifests clinically as macular, papular or nodular lesions. PKDL occurs in 1–40% of VL cases, depending on the geographical area. The drug used in the treatment of primary visceral leishmaniasis may be a factor in its incidence, but this is a hypothesis that has not yet been proven. Patients treated with sodium stibogluconate (SSG) seem to be the most prone to developing PKDL, but it occurs at unknown rates after any other medicinal treatment for visceral leishmaniasis. Studies are needed to define PKDL incidence in patients treated with recently introduced new treatment modalities. PKDL patients may infect sandflies, and play a role of pivotal importance in maintaining transmission during inter-epidemic periods.[10] The magnitude of this risk is yet to be estimated *via* standardized infectivity studies.

1.1.2 *Leishmania*–HIV Co-infection

Leishmania–HIV co-infection is a growing phenomenon, propelled by the spread of the HIV pandemic to rural areas and the emergence of visceral leishmaniasis in suburban areas. Both HIV and *Leishmania* not only contribute separately to the impairment of the immune response by targeting the same cells (macrophages), but also exert a synergistic deleterious effect on the host cells, increasing both virus replication and parasite multiplication, and favouring progression of the disease into AIDS. HIV–VL co-infected patients have a higher risk of treatment failure, a higher risk of relapse (especially in those with baseline CD4 counts of less than 100 μl^{-1} or who do not have an increase in CD4 count during follow-up), and higher rates of mortality.[11] Anti-retroviral therapy and anti-*Leishmania* secondary prophylaxis should be initiated in this population as soon as possible.[12] Furthermore, the higher number of parasites in peripheral blood in these patients may increase the chances of transmission *via* sand flies, or when sharing contaminated syringes and could contribute to the spread of drug-resistant strains, especially in anthroponotic visceral leishmaniasis. A total of 35 countries have reported cases of co-infection, with Ethiopia having by far the highest prevalence of HIV–VL worldwide (15–30% of VL cases in certain areas).[13]

1.1.3 Asymptomatic Carriers

The role of asymptomatic carriers in transmission is poorly understood. It has already been determined that large numbers of individuals in endemic areas are infected with *Leishmania* but do not develop any signs or symptoms of visceral leishmaniasis. The reported ratio of asymptomatic infections to clinical cases of visceral leishmaniasis varies widely, for example, from 4 : 1 in Kenya to 50 : 1 in Spain. Moreover, in 8 out of 316 seropositive carriers in Bihar (2.5%), parasite presence was shown by PCR, meanwhile in Iran PCR positivity was seen in 12.5% of 802 direct agglutination test (DAT)-positives.[14,15]

A crucial question is whether asymptomatic carriers can infect sand flies and if so, the control measures to be used. At the blood bank of a teaching hospital in Spain, 122 bags from asymptomatic donors were screened. Living parasites were isolated from the buffy coat of 3 of the 30 blood bags that were shown to contain seropositive samples.[16] However, these findings have not been reproduced. Another teaching hospital in Spain tested 483 blood donors and found that 70 (14.5%) samples were positive in a cell proliferation assay (CPA) after *in vitro* stimulation of peripheral blood mononuclear cells (PBMC) with soluble *Leishmania* antigen (SLA). Although CPA positivity shows immune reactivity to the parasite, its presence in the blood was transient, as parasites were not detected by quantitative PCR (qPCR) in any case.[17] Moreover, indirect xenodiagnosis was performed in 21 volunteers with negative results.[18] In canine leishmaniasis, it has been shown using xenodiagnosis that up to 50% of naturally-infected asymptomatic dogs can transmit *Leishmania* to sand flies.[19] Extrapolating these results in canine leishmaniasis to asymptomatic humans may be too risky, given what is currently known, and it is better not to speculate on the target product profile (TPP) for a drug to be used in this group of subjects; more research is needed before embarking on drug development. Today, the only good reason to put an asymptomatic carrier onto preventive chemotherapy with conventional medicines is if he or she has to be immunosuppressed for medical indications, *i.e.* transplantation.[20]

1.1.4 Outbreaks

Temporary movement due to conflict and famine may contribute to the creation of outbreaks caused by *L. donovani* and eventually to the establishment of new epidemic foci of anthroponotic VL as non-immune populations migrate into endemic areas and, on their return, import the parasite into previously non-endemic areas (as happened in Libo Kemkem, Ethiopia). Concomitant factors influencing transmission within displaced populations may include a high prevalence of malnutrition and a lack of access to diagnostic and treatment facilities, as well as vector barriers.[21] Migrant populations exposed to loss of assets and subsequent famine may face extremely high mortality if they become infected, as in South Sudan where it is estimated that 100 000 people (one third of the Nuer population) died in the outbreak in the 1990s [case fatality rate (CFR) 35%].[22] On the other hand, urban zoonotic leishmaniasis due to *L. infantum* has been identified as an increasing problem in Latin American countries (Belo Horizonte, Asunción, Posadas) and also in Europe (Tblisi, Madrid).[23–25]

Zoonotic VL (ZVL) is influenced by human activities, primarily migration from rural to urban areas, deforestation, environmental changes and changes in the ecology of vectors and reservoirs. Brazil accounts for more than 90% of all ZVL cases reported worldwide, and has registered a marked increase in the incidence of ZVL in the last 40 years. Migrants have settled in shanty towns built at the edges of main cities and close to forested areas. This may have established a link between the urban and sylvatic cycles, bringing humans closer to vector breeding sites and facilitating vector adjustment to peri-domestic habitats, with stray dogs playing a key role in closing the gap between the sylvatic and domestic cycles. In areas of poor sanitation, vectors can take advantage of the creation of new breeding sites resulting from the accumulation of organic waste generated by humans and domestic animals; the presence of dogs in households may amplify the disease cycle.[26]

As in VL, immigrants or displaced populations contribute to the appearance of new foci of CL or the occurrence of CL outbreaks. This situation is particularly worrying in the Middle East where immigrants, travellers or people traveling from this region may import the disease to non-endemic countries where the parasite may be able to form new cycles of transmission in foci where the vector may find a favourable habitat and where appropriate zoonotic reservoirs are present.[8,27] In Israel, an expansion of *Leishmania major* infections over large areas in the southern part of the country and the occurrence of *L. tropica* outbreaks in the northern part of the country have been recently reported, following new construction in populated areas.[28]

As part of national control programs specific studies are currently being conducted to cover the main epidemiological gaps.

1.2 Control Measures

1.2.1 Diagnostics and Biomarkers

Clinical findings are poorly specific for VL; additional serological testing (rk39 rapid diagnostic test) is recommended in the Indian Sub-continent, and serology followed by direct demonstration of parasites is recommended in East Africa. Rapid serologic diagnostic tests (RDTs) based on the rK39 antigen have shown excellent sensitivity for symptomatic infection in the Indian subcontinent, but less so in Africa.[29] Tests based on a synthetic recombinant of various protein antigens, rK28, have shown similar results in Sudan, with good performance in India. However, there is a need for a non-invasive test of cure based on antigen detection for active disease (*e.g.* parasite DNA by PCR or loop mediated isothermal amplification (LAMP)) which turns negative upon cure, to avoid the false-positives that serology yields because of antibody persistence, and to predict the disease outcome (relapse). Our partners are examining neopterin–PCR, IgG1 or cytokine release assays as markers of likely relapse in the context of our on-going clinical trials.

There was a remarkably high rate of PCR-positive results when asymptomatic individuals were tested in endemic areas (up to 45% of asymptomatic individuals in studies in India). Many of these PCR-positive asymptomatic individuals are negative by rapid tests based in rK39 (up to 38%), and they have lower levels of parasitaemia than symptomatic individuals (as low as less than 0.2 parasites ml^{-1}).[14] Studies on the value of IL-2 as a marker of infection, but not of the disease, are progressing.[17]

For PKDL, PCR (or another molecular method), appears to be the most effective diagnostic approach, considering the different forms and severity that may lead to low parasite presence in lesions.[30] In HIV–*Leishmania* co-infected patients, serology is less sensitive than in immunocompetent VL patients. For example, rK39 has shown 87% sensitivity in symptomatic VL patients that are HIV-negative and 77% in similar cohorts that are HIV-positive. Conversely, parasite burden rises with HIV infection: HIV-positive VL patients may have 10 to 10 000 parasites ml^{-1}. As cure is less predictable in HIV-infected patients, a test of cure might be helpful in HIV-prevalent populations. In HIV-co-infected patients, diagnostic evidence of infection would increase the probability of drug efficacy.[31]

Identifying markers of disease progression from asymptomatic towards VL, or from VL towards PKDL would be extremely useful for predicting the infection and disease outcome. In addition, markers that could correlate the infectivity of asymptomatic carriers and PKDL forms are urgently needed for modelling transmission and implementing control measures. Finally, biomarkers that could identify the chances of developing resistance would be extremely helpful in monitoring patients and mapping drug resistance.[32]

Diagnosis of CL is based on clinical features, epidemiological data and laboratory testing. Abundant diagnostic methods have been described, with large variations in sensitivity and specificity. Traditionally, diagnostic tests for CL have been grouped as direct tests (visual observation of the *Leishmania* parasite) which include parasitological examination by microscopy, histopathology and parasite culture, or indirect tests, which include serology and molecular diagnostics. The selection of the diagnostic test employed often depends on the available infrastructure and resources of the diagnostic facility rather than on diagnostic accuracy.[33,34]

The direct identification of amastigotes in Giemsa-stained lesion smears of biopsies, scrapings or impression smears is the most used method in routine situations and to confirm diagnosis in subjects participating in clinical trials. Culturing parasites is a tedious and time-consuming method that is usually done in reference centres or laboratories.

Although several serological tests have been evaluated, only a few are commercially available and probably none are used on a routine basis due to low sensitivity, caused either by the poor humoral response triggered by the infection or by the difficulties of developing a test with good sensitivity for all *Leishmania* species causing the disease. Serological tests are mainly based on enzyme-linked immunosorbent assay (ELISA), western blot, lateral flow assay or direct agglutination.[35]

Molecular diagnostic tests are assumed to have better sensitivity and specificity than traditional diagnostic methods and allow the use of less invasive sampling for diagnosis. Molecular tests also allow the identification of *Leishmania* species infecting the patient, an important variable when analysing efficacy data in clinical trials or when choosing the best treatment in infected patients. PCR, either as a single test, in a nested format, or as a quantitative assay, has been widely tested using different protocols and sources of clinical material. Unfortunately, almost every laboratory uses it is own methodologies and set of primers, making the adoption of a standardized protocol impossible at this stage.

For CL, clinical features, mainly re-epithelization of an ulcerated lesion or flattening of a non-ulcerated lesion, together with absence of inflammatory signs in the borders of the lesion, are used as parameters to define clinical cure. The presence of parasites in healed lesions has been reported, and its relation to clinical relapse has not been clearly determined, hence most clinical trials have outcome criteria that define cure based purely on clinical observations.[36]

Nevertheless, parasite recrudescence is a rare and slow-developing event which is difficult to assess or predict, and markers that correlate with treatment effect allowing the prediction of long-term clinical outcome are needed to better assess treatment efficacy or to determine risk of relapse.

1.2.2 Disease Control Strategies by Region

1.2.2.1 South Asia

A regional Kala-azar elimination program was signed by the governments of India, Nepal and Bangladesh in 2005, which lapsed in September 2015.[37] The Memorandum of Understanding (MoU) between the three countries was renewed in 2015 adding Thailand and Bhutan. The strategy consists of indoor residual spraying (IRS), case detection (rK39) and drug treatment (currently with miltefosine), surveillance, social mobilization and operational research. Distribution of bed nets has not been shown to have a huge beneficial effect although there are differing opinions. However, modelling has also shown that treatment of cases alone is insufficient to control the disease due to the presence of PKDL and asymptomatic carriers whose role in infectivity is yet to be properly determined. The elimination program has nevertheless resulted in a sharp decrease in figures in the Indian subcontinent having reduced mortality in 94%, incidence decreased 79% and the global target of elimination was achieved in 87% of the endemic districts.[6,38]

1.2.2.2 East Africa

Unlike South Asia, there is no regional elimination program. Kenya is the first country to have drafted a national strategy. Currently case detection and treatment form the core activities, and are dependent on NGOs or WHO support. Case detection is *via* serology (DAT or rK39) and parasitology.[3] The recommended treatment is the sodium stibogluconate plus paromomycin (SSG and PM) combination. South Sudan faces recurrent outbreaks, whereas the number of cases has been stable in Sudan and Ethiopia. The situation in South Sudan has returned to normality after the outbreak that started in 2014 and affected 6000 patients.

1.2.2.3 Latin America

As this is a zoonotic transmission area, control of the reservoir host forms the mainstay of control, including sacrifice of infected dogs, use of insecticide-impregnated collars and vaccines, all of them with controversial results. Humans are treated with conventional courses of Glucantime or liposomal amphotericin B.[39]

1.2.2.4 Outbreak Response

Recurrent massive epidemics are the result of displaced non-immune populations suffering from malnutrition, poor living conditions and the collapse of health services. Mapping risks (social, environmental and climatic) is one of the most efficient tools for predicting outbreaks in anthroponotic VL. Preparedness and a rapid response can efficiently reduce the impact of the epidemics, as was proven during the outbreak in South Sudan of 2009–2011 with 25 000 VL patients treated and a CFR of less than 5%, much lower than the 35% CFR during the epidemic in early 1990s.[40]

In zoonotic VL, a better understanding of the man-made environmental changes that often strengthen the sylvatic and domestic cycles can help in controlling urban outbreaks that normally cause social and political alarm. A good example of this is the 2010–2015 outbreak in Madrid, which affected more than 600 patients.[41]

There are no vaccines or chemoprophylaxis available to protect against CL, hence most control programs primarily focus on finding and treating cases. This is of particular importance in anthroponotic CL settings where person-to-person transmission can be reduced by early identification and treatment of active patients. Improved housing and environmental sanitation might also be beneficial in this situation. Individual protection, the use of insect repellent, impregnated clothes or bed nets, has been recommended for people deployed for short periods in endemic areas, but their effectiveness for permanent residents has not been measured. Evidence in the literature as to the efficacy of any animal reservoir intervention programmes in areas of zoonotic CL is scant and mixed.

Current treatments of CL are poorly justified and have sub-optimal effectiveness. Treatment options include topical or systemic regimens. Recommendations vary depending on *Leishmania* species, geographical region and clinical presentation.[3]

1.3 Existing Treatment Options, Recent Advances and Unmet Needs

1.3.1 Visceral Leishmaniasis

Until recently, antimonial monotherapy for 20–30 days has been the mainstay of treatment. However, in the last 15 years, liposomal amphotericin B, followed by paromomycin and miltefosine have been developed or made available for use. These treatments suffered from being too long, toxic or expensive. Therefore, additional work by the Drugs for Neglected Diseases initiative (DND*i*) and others has been conducted to optimize current drug regimens. A summary of current VL treatments is shown in Table 1.1.

SSG & PM for 17 days has now been shown to be as safe and effective as the standard 30-day treatment with SSG in East Africa. This regimen has been recommended as a first-line treatment for the region by a recent WHO expert committee.[3] A pharmacovigilance study enrolling 3000 patients from four countries showed an efficacy rate of 92%, and a mortality rate of 0.9.[42]

Phase III trials have also been completed in 2010 for three double-drug combinations (miltefosine and paromomycin, AmBisome® and miltefosine, AmBisome® and paromomycin) and single-dose AmBisome® in India—all of which show efficacy rates of above 95%.[43] Although more information is needed, especially from India, some progress on feasibility and field effectiveness has been made recently in South Asia. In Nepal, miltefosine is the first-line drug in the elimination program and after seven years of use, it showed an initial cure rate of 95.8% although the relapse rate at six and twelve months is 10.8% and 20.0%, respectively.[44] The results of pharmacokinetic (PK) studies have recently indicated that Nepalese children are under-exposed to miltefosine, which could explain the relapses. The same has been shown in African children. In Bangladesh, on the other hand, AmBisome® 10 $mg\,kg^{-1}$ single infusion showed a cure rate of 97.7% at six months follow up in a feasibility–efficacy study enrolling 300 patients treated by local doctors in sub-district hospitals (upazillas).[43,45] In India, an observational retrospective study followed patients treated with multiple doses of AmBisome® (20 $mg\,kg^{-1}$). Among the immunocompetent patients (984), 827 (84.0%) were traced in order to determine their long-term outcomes. Of these, 20 patients (2.4%) had relapsed: 0.3% (2 out of 767) at 6 months follow-up, 3.7% (14 out of 383) at 12 months and 2.4% (4 out of 164) at 15 months. The mean $\pm$ SD time-to-relapse was 9.6 $\pm$ 3.0 months.[46] A cohort event monitoring study is planned to start in 2017 in India under the KalaCORE consortium.

Table 1.1 Summary of current VL treatments (based on WHO expert committee, 2010). The drug price is considered here, but indirect costs are not included.

Regimen[a]	Manufacturer	Cost US$	Efficacy[b]	Comments
AmBisome® IV 10 mg kg^{-1} single dose	Gilead	126	95% (P3)	India only, poor efficacy in Africa; requires cold chain
AmBisome® 20 mg kg^{-1} over four doses	Gilead	252	98% (P4)	India only, poor efficacy in Africa; requires cold chain
Amphotericin B 1 mg kg^{-1} over 30 days	Multiple	20	95% (P3)	Toxicity is main barrier to use
Miltefosine (MF) 100 mg day^{-1} oral for 28 days	Paladin	65–150	94% (P3) 82% (P4)	Low compliance and teratogenicity
Paromomycin base (PM) 11 mg kg^{-1} day^{-1} IM for 21 days	Gland Pharma	15	94% (P3)	India only, poor efficacy in Africa
SSG 20 mg kg^{-1} day^{-1} for 30 days	Albert David	55.8	65–95% (P3)	Some toxicity; resistance in Bihar, but not elsewhere
Glucantime 20 mg kg^{-1} day^{-1} for 30 days	Sanofi-Aventis	59.3	90% +	Some toxicity; mainly used in Latin America
AmBisome® 5 mg kg^{-1} + MF 100 mg kg^{-1} day^{-1} for 7 days	Gilead Paladin	88.2–109.5	97.5% (P3)	India only; requires cold chain; teratogenicity of MF
AmBisome® 5 mg kg^{-1} + PM 11 mg kg^{-1} day^{-1} for 7 days	Gilead Gland Pharma	79	97.5% (P3)	India only; requires cold chain
MF 100 mg kg^{-1} + PM 11 mg kg^{-1} both for 10 days	Paladin Gland Pharma	30.2–60.7	98.5% (P3)	India only; teratogenicity of MF; effectiveness?
SSG 20 mg kg^{-1} + PM 11 mg kg^{-1} both given IM for 17 days	Albert David Gland Pharma	44	91% (P3)	Africa only; used in a wide field setting by MSF

[a]Assuming 35 kg patient.
[b]The phase of the study for which the percentage of efficacy was shown is given in brackets.

In Africa, due to the recent warning on cardiac toxicity for the conventional amphotericin B–SSG combination due to the SSG component, effort needs to be focused on adjusting the correct AmBisome®–miltefosine (MF) dosage to reach over 90% efficacy (LEAP 0211). All combination treatments should be evaluated for additional toxicities before clinical studies. AmBisome® alters the distribution of amphotericin B, and since the amphotericin B released into plasma is minimal, side effects are much less frequent than for amphotericin deoxycholate. In addition, it has been shown in *L. donovani*-infected macrophages that it is possible to induce resistance to all drugs used in combination, although it is much easier to induce resistance with MF–PM than with the other combinations. In regional meetings with VL experts and investigators, the consensus is that the ideal treatment for East Africa would be a combination of oral drugs that is efficacious, safe and affordable.

To date, treatment options for PKDL and HIV–VL remain limited and unsatisfactory, requiring long and often repeated courses of drug treatment, including with antimonials. A proof of concept study aiming at the elimination of ZVL with existing tools would determine if it is possible to eliminate this form of the disease in given countries or areas.

On the diagnostic side, better tools, including antigen-based tests, are required that can evaluate not just diagnosis but outcome and cure in a non-invasive manner. Standardized pharmacodynamic markers that can quantitatively evaluate drug response to treatment are also required.

1.3.2 Cutaneous Leishmaniasis

Given the clinical spectrum of the disease (Figure 1.4), the ideal treatment for CL should be based on the clinical presentation: local therapies for patients with small and few lesions located in anatomical areas amenable to topical treatment; systemic oral drugs for subjects suffering from numerous or large lesions, subjects with lesions which are potentially disfiguring, disabling or located in areas which make local therapy impossible (face, joints, toes, fingers, eyelids, lips, ears). Subjects with leishmaniasis recidivans, diffuse CL or PKDL should benefit from treatment with anti-leishmanial drugs in combination with an immune response modifier (IRM) to accelerate and enhance a Th1-type immune response.

The therapeutic options available to treat CL are limited, and the current evidence for recommendations for treatment is generally weak. Pentavalent antimonials (Sb^V), either meglumine antimoniate or sodium stibogluconate, at doses of 20 mg kg^{-1} body weight for 20 days, are the most recommended and used drugs worldwide. Antimonials have been the drugs of choice for more than six decades despite their high cardiac, hepatic and renal toxicity. Treatment compliance is very low due to the development of a high incidence of local adverse events and low tolerance. Antimonials are also contraindicated in pregnant and breastfeeding women, in patients with cardiac or hepatic problems and in immunosuppressed patients.

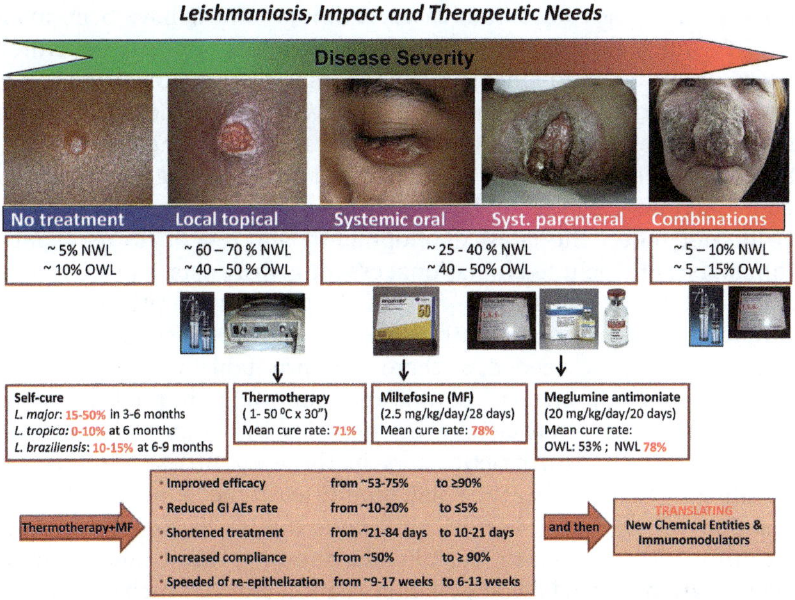

Figure 1.4 Cutaneous leishmaniasis, spectrum of skin involvement and response to current therapeutic methods. (GI AEs: Gastrointestinal Adverse Effects; NWL: New World Leishmaniasis; OWL: Old World Leishmaniasis)

WHO recommendations are based on the causative species, geographical area and the clinical features of the disease. Recommendations vary from no treatment to topical or systemic approaches.[3]

Local therapies (thermotherapy, cryotherapy, paromomycin ointment, local infiltration with antimonials) are good options with less systemic toxicity, and are recommended for patients with infections due to *L. mexicana* or *L. major*, or for patients with small and few lesions, while systemic treatments (miltefosine, antimonials, pentamidine, or amphotericin B formulations) are recommended for more complicated cases, in non-responders to topical treatments or in immunosuppressed patients. Intramuscular pentamidine is mainly used for *Leishmania guyanensis* infections. Miltefosine has been shown to be effective in infections due to *Leishmania panamensis* but less effective in infections due to *Leishmania braziliensis* or *L. tropica*. Amphotericin B formulations, either deoxicolate or liposomal, have been used in a relatively limited number of cases and mainly for the treatment of complicated cases, such as mucosal, diffuse or disseminated CL that do not respond to miltefosine or antimonials.[3]

1.3.3 Drug Access

Access to medicines for the treatment of visceral, cutaneous and mucocutaneous leishmaniases is very problematic in the poverty-stricken countries that have the highest burden of cases. Although efforts have been made by the WHO, medical non-governmental organizations and manufacturers to improve access to medicines for leishmaniasis, problems persist. In the context of weak health systems in many *Leishmania*-endemic countries, a number of factors contribute to lack of access to medicines.[47]

The medicines are relatively expensive and are often unaffordable, despite price reductions negotiated by the WHO. National treatment protocols often do not reflect the latest developments, and few or no medicines for leishmaniasis are included in national essential drug lists, a process that can take several years. Medicines for which preferential prices have been negotiated in low-income countries are not always registered by the companies where they are needed, as there is no profitable market. If medicines are not registered, special permission for their import is required. Registration is also often lacking in countries with very few cases, with a direct effect on medical practitioners in these countries, who experience great difficulty in obtaining the small quantities of medicines needed *ad hoc*. The supply of medicines is not continuous due to bad program management, from forecasting to sustainable funding. Most anti-leishmanial medicines are produced by only a single manufacturer, which implies a total dependence, including pricing policy or shortages. Problems with quality, low production capacity and lack of an adequate forecast of needs (resulting in long lead times for drug orders) regularly cause stock ruptures in endemic countries. There are no central buffer stocks that can be accessed in such cases, and there is no platform for indicating drug needs; therefore, the quantities needed globally cannot be estimated and drug production cannot be appropriately planned.

Overall monitoring of access to anti-leishmanial medicines should be strengthened, with pricing, registration and global needs taken into account. Moreover, the regulation of drug policies and quality assurance should be also strengthened at all levels. Uncontrolled access to medicines (*e.g.* over-the-counter availability of miltefosine) could lead to misuse, suboptimal treatment and, in the long term, drug resistance. The use of counterfeit medicines (toxic batches of antimonials, fake miltefosine) has led to several avoidable deaths in the past.

1.4 Conclusion. Development of an Elimination Tool

In the absence of a preventive vaccine, future short-course oral new chemical entities (NCE), with an excellent safety profile, could potentially be used as a tool for VL elimination. Thus, in addition to treating clinical cases of VL and PKDL, the NCE would ideally be administered to asymptomatic cases. The concept of preventive chemotherapy could be incorporated into the DND*i* strategy for VL. Development of such a tool would depend on the safety profile (currently none of the existing drugs have a sufficiently good profile to justify use for this indication) and the genetic barrier of parasites developing resistance to the NCE (see Chapter 18).

Even though some progress has been made in developing new treatments for CL, it seems that the currently available tools probably represent almost the entire therapeutic arsenal for the coming years; however, there may be some opportunities to improve current treatment options and available tools. Options include combining local with systemic treatments, *i.e.* one application of thermotherapy plus a short course of miltefosine for non-complicated CL cases, or perhaps optimizing the use of liposomal amphotericin B using a higher dosage but lower number of doses in patients with complicated CL.

References

1. I. Müller and P. Kropf, in *Immunity to Parasitic Infection*, ed. T. J. Lamb, Wiley-Blackwell, 2012, p. 153–164.
2. P. Kaye and P. Scott, *Nat. Rev. Microbiol.*, 2011, **9**, 604.
3. WHO technical report series; no. 949. Control of the leishmaniasis: report of a meeting of the WHO Expert Committee on the Control of Leishmaniases, Geneva, 2010.
4. J. Alvar, S. Yactayo and C. Bern, *Trends Parasitol.*, 2006, **22**, 552.
5. J. Alvar, I. D. Velez, C. Bern, M. Herrero, P. Desjeux, J. Cano, J. Jannin, M. den Boer and W. H. O. L. C. Team, *PLoS One*, 2012, **7**, e35671.
6. WHO, Weekly epidemiological record Relevé épidémiologique hebdomadaire, *Wkly Epidemiol. Rec.*, 2016, **91**, 285–296, www.who.int/wer.
7. I. M. Mosleh, E. Geith, L. Natsheh, M. Abdul-Dayem and N. Abotteen, *Trop. Med. Int. Health*, 2008, **13**, 855.
8. A. Alawieh, U. Musharrafieh, A. Jaber, A. Berry, N. Ghosn and A. R. Bizri, *Int. J. Infect. Dis.*, 2014, **29**, 115.
9. C. Karimkhani, V. Wanga, L. E. Coffeng, P. Naghavi, R. P. Dellavalle and M. Naghavi, *Lancet Infect. Dis.*, 2016, **16**, 584.
10. E. E. Zijlstra, A. M. Musa, E. A. Khalil, I. M. el-Hassan and A. M. el-Hassan, *Lancet Infect. Dis.*, 2003, **3**, 87.
11. J. Alvar, P. Aparicio, A. Aseffa, M. Den Boer, C. Canavate, J. P. Dedet, L. Gradoni, R. Ter Horst, R. Lopez-Velez and J. Moreno, *Clin. Microbiol. Rev.*, 2008, **21**, 334.
12. E. Diro, K. Ritmeijer, M. Boelaert, F. Alves, R. Mohammed, C. Abongomera, R. Ravinetto, M. De Crop, H. Fikre, C. Adera, R. Colebunders, H. van Loen, J. Menten, L. Lynen, A. Hailu and J. van Griensven, *PLoS Neglected Trop. Dis.*, 2015, **9**, e0004087.
13. J. Alvar, C. Cañavate, B. Gutierrez-Solar, M. Jimenez, F. Laguna, R. Lopez-Velez, R. Molina and J. Moreno, *Clin. Microbiol. Rev.*, 1997, **10**, 298.
14. R. K. Topno, V. N. Das, A. Ranjan, K. Pandey, D. Singh, N. Kumar, N. A. Siddiqui, V. P. Singh, S. Kesari, N. Kumar, S. Bimal, A. J. Kumar, C. Meena, R. Kumar and P. Das, *Am. J. Trop. Med. Hyg.*, 2010, **83**, 502.
15. M. Fakhar, M. H. Motazedian, G. R. Hatam, Q. Asgari, M. Kalantari and M. Mohebali, *Ann. Trop. Med. Parasitol.*, 2008, **102**, 577.

16. C. Riera, R. Fisa, M. Udina, M. Gallego and M. Portus, *Trans. R. Soc. Trop. Med. Hyg.*, 2004, **98**, 102.
17. A. V. Ibarra-Meneses, E. Carrillo, C. Sanchez, J. Garcia-Martinez, D. Lopez Lacomba, J. V. San Martin, F. Alves, J. Alvar and J. Moreno, *Clin. Microbiol. Infect.*, 2016, **22**, 739.e1.
18. R. Molina, personal communication.
19. R. Molina, C. Amela, J. Nieto, M. San-Andres, F. Gonzalez, J. A. Castillo, J. Lucientes and J. Alvar, *Trans. R. Soc. Trop. Med. Hyg.*, 1994, **88**, 491.
20. E. Carrillo, N. Carrasco-Anton, F. Lopez-Medrano, E. Salto, L. Fernandez, J. V. San Martin, J. Alvar, J. M. Aguado and J. Moreno, *PLoS Neglected Trop. Dis.*, 2015, **9**, e0004179.
21. J. Aagaard-Hansen, N. Nombela and J. Alvar, *Trop. Med. Int. Health*, 2010, **15**, 1281.
22. J. Seaman, A. J. Mercer and E. Sondorp, *Int. J. Epidemiol.*, 1996, **25**, 862.
23. M. O. Harhay, P. L. Olliaro, D. L. Costa and C. H. Costa, *Trends Parasitol.*, 2011, **27**, 403.
24. G. Babuadze, J. Alvar, D. Argaw, H. P. de Koning, M. Iosava, M. Kekelidze, N. Tsertsvadze, D. Tsereteli, G. Chakhunashvili, T. Mamatsashvili, N. Beria, I. Kalandadze, M. Ejov and P. Imnadze, *PLoS Neglected Trop. Dis.*, 2014, **8**, e2725.
25. E. Carrillo, J. Moreno and I. Cruz, *Trends Parasitol.*, 2013, **29**, 579.
26. P. Desjeux, *Comp. Immunol. Microbiol. Infect. Dis.*, 2004, **27**, 305.
27. S. L. Sharara and S. S. Kanj, *PLoS Pathog.*, 2014, **10**, e1004438.
28. D. Gandacu, Y. Glazer, E. Anis, I. Karakis, B. Warshavsky, P. Slater and I. Grotto, *Emerging Infect. Dis.*, 2014, **20**, 1605.
29. F. Chappuis, S. Rijal, A. Soto, J. Menten and M. Boelaert, *Br. Med. J.*, 2006, **333**, 723.
30. S. Verma, R. Kumar, G. K. Katara, L. C. Singh, N. S. Negi, V. Ramesh and P. Salotra, *PLoS One*, 2010, **5**, e10107.
31. B. Monge-Maillo, F. F. Norman, I. Cruz, J. Alvar and R. Lopez-Velez, *PLoS Neglected Trop. Dis.*, 2014, **8**, e3021.
32. F. Chappuis, S. Sundar, A. Hailu, H. Ghalib, S. Rijal, R. W. Peeling, J. Alvar and M. Boelaert, *Nat. Rev. Microbiol.*, 2007, **5**, 873.
33. H. J. de Vries, S. H. Reedijk and H. D. Schallig, *Am. J. Clin. Dermatol.*, 2015, **16**, 99.
34. R. J. Robinson, S. Agudelo, C. Muskus, J. F. Alzate, C. Berberich, D. C. Barker and I. D. Velez, *Trans. R. Soc. Trop. Med. Hyg.*, 2002, **96**(Suppl. 1), S169.
35. A. Masmoudi, W. Hariz, S. Marrekchi, M. Amouri and H. Turki, *J. Dermatol. Case Rep.*, 2013, **7**, 31.
36. P. Olliaro, M. Vaillant, B. Arana, M. Grogl, F. Modabber, A. Magill, O. Lapujade, P. Buffet and J. Alvar, *PLoS Neglected Trop. Dis.*, 2013, **7**, e2130.
37. Regional strategic framework for elimination of kala-azar from the South-East Asia region (2005-2015), ed. New Delhi: Regional Office for South-East Asia, SEA-VBC-85 (Rev-1), World Health Organization, 2005.

38. P. L. Olliaro, T. A. Shamsuzzaman, B. Marasini, A. C. Dhariwal, A. Be-Nazir, D. Mondal, M. R. Banjara, P. Das, S. Sundar, S. Rijal, B. Arana, J. Alvar, D. Argaw, R. W. Peeling, A. Kroeger and G. Matlashewski, *PLoS Neglected Trop. Dis.*, 2017, **11**, e0005190.
39. L. Solano-Gallego, G. Miro, A. Koutinas, L. Cardoso, M. G. Pennisi, L. Ferrer, P. Bourdeau, G. Oliva and G. Baneth, *Parasites Vectors*, 2011, **4**, 86.
40. A. Abubakar, J. A. Ruiz-Postigo, J. Pita, M. Lado, R. Ben-Ismail, D. Argaw and J. Alvar, *PLoS Neglected Trop. Dis.*, 2014, **8**, e2720.
41. A. Arce, A. Estirado, M. Ordobas, S. Sevilla, N. Garcia, L. Moratilla, S. de la Fuente, A. M. Martinez, A. M. Perez, E. Aranguez, A. Iriso, O. Sevillano, J. Bernal and F. Vilas, *Euro Surveill.*, 2013, **18**, 20546.
42. R. Kimutai, A. M. Musa, S. Njoroge, R. Omollo, F. Alves, A. Hailu, E. A. Khalil, E. Diro, P. Soipei, B. Musa, K. Salman, K. Ritmeijer, F. Chappuis, J. Rashid, R. Mohammed, A. Jameneh, E. Makonnen, J. Olobo, L. Okello, P. Sagaki, N. Strub, S. Ellis, J. Alvar, M. Balasegaram, E. Alirol and M. Wasunna, *Clin. Drug Investig.*, 2017, **37**, 259.
43. S. Sundar, J. Chakravarty, D. Agarwal, M. Rai and H. W. Murray, *N. Engl. J. Med.*, 2010, **362**, 504.
44. S. Rijal, B. Ostyn, S. Uranw, K. Rai, N. R. Bhattarai, T. P. Dorlo, J. H. Beijnen, M. Vanaerschot, S. Decuypere, S. S. Dhakal, M. L. Das, P. Karki, R. Singh, M. Boelaert and J. C. Dujardin, *Clin. Infect. Dis.*, 2013, **56**, 1530.
45. D. Mondal, J. Alvar, M. G. Hasnain, M. S. Hossain, D. Ghosh, M. M. Huda, S. G. Nabi, S. Sundar, G. Matlashewski and B. Arana, *Lancet Glob. Health*, 2014, **2**, e51.
46. S. Burza, P. K. Sinha, R. Mahajan, M. A. Lima, G. Mitra, N. Verma, M. Balasegaram and P. Das, *PLoS Neglected Trop. Dis.*, 2014, **8**, e2603.
47. M. den Boer, D. Argaw, J. Jannin and J. Alvar, *Clin. Microbiol. Infect.*, 2011, **17**, 1471.

CHAPTER 2

Anti-leishmanial Drug Discovery: Past, Present and Future Perspectives

CHARLES E. MOWBRAY

Drugs for Neglected Diseases *initiative*, 15 Chemin Louis Dunant, 1202 Geneva, Switzerland
Email: cmowbray@dndi.org

2.1 Introduction

The drugs currently used for treating leishmaniasis suffer from limitations in efficacy and safety and are not well adapted to the needs of patients. Combining these drugs has led to improvements in efficacy and safety in some geographical regions, but many patients are still in need of more effective, safer and more convenient treatments.[1,2]

The existing drugs for leishmaniasis were re-purposed from other therapeutic indications and are far from optimal, relying largely on painful intravenous and intramuscular injections, with the exception of miltefosine, which is orally active. In recent years, orally acting new chemical entities have been designed and selected for development for treating visceral leishmaniasis, and also have potential for the cutaneous form of the disease. These new drug classes have been discovered using phenotypic drug discovery methods and offer great promise for developing new treatments, but their mechanisms of action are often not well understood. Efforts to de-convolute the mechanisms of action of these candidates and newer

Drug Discovery Series No. 60
Drug Discovery for Leishmaniasis
Edited by Luis Rivas and Carmen Gil

Published by the Royal Society of Chemistry, www.rsc.org

target-based drug discovery approaches should open the door for discovery of further drug classes and candidate molecules.

This chapter describes the evolution of drug discovery approaches for leishmaniasis, explores the properties of emerging drug candidates and considers the potential for these efforts to deliver new treatments to patients.

2.2 Current Drugs for Leishmaniasis

A small number of drugs are currently available for the treatment of leishmaniasis in its visceral and cutaneous forms, and for patients with HIV co-infection. Examination of the chemical structures and selected properties of the most commonly used of these drugs shown in Table 2.1 immediately reveals their limitations. Sodium stibogluconate and meglumine antimoniate are complexes of antimony, a toxic heavy metal, which were discovered more than 70 years ago. These complexes are very polar, do not readily cross the biological membranes required for absorption of orally acting medicines and must be administered by repeated, painful intramuscular injections over 30 days. Amphotericin B is a complex natural product originally identified as an antifungal agent and successfully re-purposed for the treatment of leishmaniasis. The high molecular weight and polar nature of Amphotericin B prevent its oral absorption and so it is delivered as a liposomal formulation by single or repeated intravenous infusion. Although an effective anti-leishmanial drug, Amphotericin B can also cause significant toxicity to human cells. Formulation of Amphotericin B as Ambisome® in liposomes, which are selectively phagocytosed by macrophages, helps to target the drug to infected cells whilst sparing other cell types and reducing toxicity, but even this optimized form can cause serious side effects and also requires a cold-chain to reach the patient. Paromomycin is an aminoglycoside antibiotic repurposed for the treatment of leishmaniasis. The polar nature of paromomycin excludes oral delivery and requires administration by painful intramuscular injections, and this treatment can result in serious side effects including reversible nephrotoxicity and hepatotoxicity and irreversible deafness. The last drug in Table 2.1, miltefosine is a re-purposed anticancer drug with an unusual chemical structure, however, it is orally active, avoiding the need for injections or infusions. Unfortunately, the advantage of oral delivery of miltefosine is somewhat offset by the gastrointestinal disturbances and risk of teratogenicity reported for this drug. Finally, although not shown in the table, it should be noted that pentamidine (administered by intramuscular injections or short intravenous infusions) has limited application for treatment of cutaneous leishmaniasis in the New World, and also provides secondary prophylaxis for HIV–visceral leishmaniasis (VL) co-infected patients to reduce the risk of post-treatment relapse.

In attempts to overcome the limitations of this small set of anti-leishmanial drugs, researchers have explored the potential of combination therapies to increase efficacy, reduce toxicity, improve toleration and shorten

Table 2.1 The chemical structures and selected properties of existing anti-leishmanial drugs.

Drug name	Meglumine antimoniate	Sodium stibogluconate (SSG)	Liposomal amphotericin B	Paromomycin sulphate	Miltefosine
Chemical Structure of free acid or base	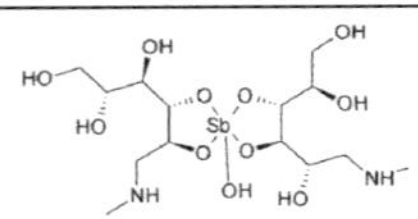	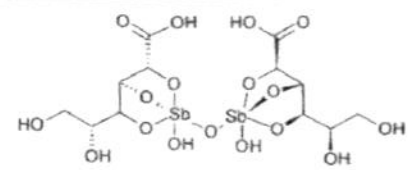	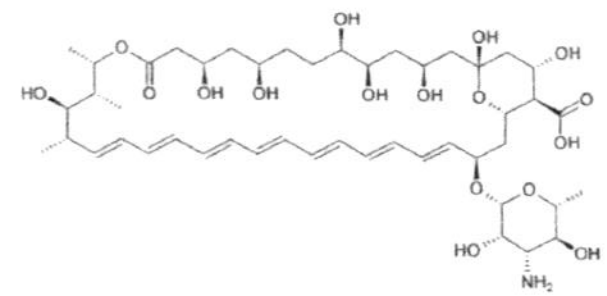	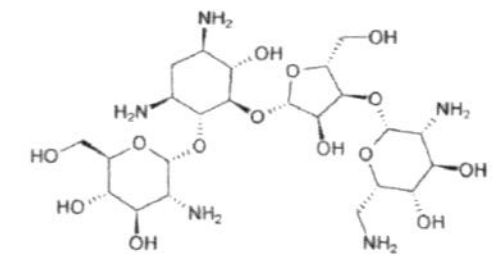	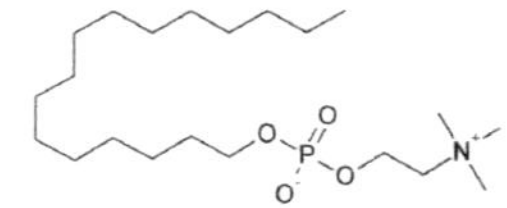
Physical properties	MW 525, AlogP[a] – 4.2	MW 680, AlogP – 3.8	MW 924, AlogP – 2.6	MW 616, AlogP – 8.7	MW 407, AlogP + 3.7
Route of delivery	Slow IV[b] or IM infusion	Slow IV or IM infusion	2 h IV infusion	IM	PO
Dose regimen	20 mg kg^{-1} day^{-1} 30 days	20 mg kg^{-1} day^{-1} 30 days	10–30 mg kg^{-1} total dose over 1–10 days	15–20 mg kg^{-1} day^{-1} 21 days	1.5–2.0 mg kg^{-1} day^{-1} 28 days
Clinical efficacy	35–95% depending on area, resistance in India	35–95% depending on area, resistance in India	>95% efficacy in India, variable response in Africa	93–95% India, 64–85% Africa	94–97%
Reported side effects	• painful injections • nephrotoxicity and cardiotoxicity • pancreatitis	• painful injections • cardiotoxicity • pancreatitis	• rigors and chills • nephrotoxicity • hypokalaemia • anaphylaxis	• painful injections • reversible nephrotoxicity, hepatotoxicity and ototoxicity	• teratogenic • gastrointestinal, nephrotoxicity and hepatotoxicity

[a]AlogP, calculated octanol–water partition coefficient.
[b]IV, intravenous; IM, intramuscular; PO, *per oral*.

treatment durations with the added benefit of perhaps reducing the risk of resistance. These efforts have met with some success, providing improved options for some patient populations, such as the combination of sodium stibogluconate and paromomycin as the recommended treatment for VL in East Africa. Further improvements for other patient populations may be possible through optimization of combination treatments, but providing a major step forward with efficacious, safe, well tolerated and convenient to use treatments which are genuinely adapted to the needs of patients will need a new generation of drugs designed for this purpose.

In this chapter, some of the limitations of the, largely injectable, current anti-leishmanial treatments are related to their chemical structures, and the potential of drug discovery efforts to deliver improved, orally acting medicines is discussed. The profiles of the leading candidates and drug classes are discussed to provide the reader with an overview of the current pipeline of possible new treatments and the goals for continued work in this field. The reader is directed to a detailed and comprehensive review by Nagle *et al.*[3] of the many other classes of compounds which have been explored as anti-leishmanial agents but have not yet yielded high levels of *in vivo* activity, and preclinical candidates.

2.2.1 Lessons Learned from the Properties of Current Drugs for Leishmaniasis

The medicinal chemistry literature provides ample guidance on the desirable, drug-like properties which are most likely to afford safe, effective drugs that can be readily formulated and are convenient for use by patients. The collected experience of many unsuccessful drug discovery and development projects provides further valuable guidance to drug discovery scientists, highlighting chemical elements, functional groups, and drug properties which are likely, at best, to hinder development or, in the worst case, result in failure.[4] The current drugs for leishmaniasis described in Table 2.1 are, with the exception of miltefosine, very polar with negative calculated octanol/water partition coefficients (AlogP) and have high (>500 Daltons) molecular weights (MWs). As a result of breaking two of the 'rule of five' guidelines proposed by Lipinski in his seminal reviews on the properties of drugs with oral bioavailability,[5] these drugs are predicted to be unlikely to be orally absorbed, which is consistent with their delivery by injection or infusion only.

2.3 Medicinal Chemistry Challenges and Strategies to Discover New Anti-leishmanial Drugs

In order to successfully design and develop new drugs to treat leishmaniasis, it is essential for medicinal chemists and drug discovery scientists to understand as much as possible about *Leishmania* parasites and the environments in which they exist. This will enable design of molecules with suitable drug-like properties to reach the relevant infected tissue compartments within patients and to kill the parasites. Although compounds can be tested for their ability to kill *Leishmania* parasites at different points in their zoonotic lifecycles, the most challenging and relevant one for treating human disease is the intracellular amastigote. *Leishmania* amastigotes concentrate inside acidic phagolysomes (with local pH as low as 5) within macrophages, presenting drugs with additional membranes and pH gradients to traverse and an environment rich in proteases and hydrolases which they must survive before reaching their target.[6]

The challenges in translating activity from *in vitro* and *in vivo* models to patients are described in Chapter 3, but a screening sequence in which compounds are first tested for anti-leishmanial activity and absorption, disposition, metabolism and elimination (ADME) characteristics *in vitro* before moving to *in vivo* pharmacokinetics (PK) and efficacy in preclinical species would seem sensible in order to identify the most promising candidates for future preclinical and clinical studies. The Drugs for Neglected Diseases *initiative* (DND*i*) has proposed target candidate profiles for new experimental compounds against visceral leishmaniasis, shown in Table 2.2, to help select the most promising optimized leads for preclinical development with a view to ultimately fulfilling the clinical target product profile.[7] The minimum acceptable profile at this discovery stage could ultimately provide a functional clinical cure for patients by bringing the infection under the control of the immune system without necessarily eliminating all the parasites. The more challenging ideal profile aims for a sterile cure resulting in complete clearance of infection, which would be highly desirable for patients with compromised immune status, such as those with HIV–VL co-infection, and to minimize the risk of patients developing post-kala-azar dermal leishmaniasis (PKDL). Similar target candidate profiles are being developed for cutaneous leishmaniasis (CL) to take into account the variety of parasite species which cause CL in both the New and Old World.

2.3.1 Phenotypic Drug Discovery Approaches

Adding *in vivo* anti-parasitic assessments to intracellular anti-leishmanial assays provides the basis of an effective phenotypic drug discovery platform. When these parasitology assays are coupled with medicinal chemistry, ADME–PK, drug safety evaluations and formulation development, it is possible to discover and optimize new compound classes to deliver preclinical candidates ready for further development. Some newly discovered candidates and series shown in Tables 2.3 and 2.4 are used to exemplify this approach in the following sections with chemical starting points arising from drug re-purposing, compound-class mining and high-throughput screening of diverse compound collections.

Table 2.2 Target candidate profile for new chemical entities for treatment of visceral leishmaniasis (VL).

	Acceptable (functional cure)	Ideal (sterile cure)
Efficacy		
In vivo:	>95% reduction in parasitaemia in liver and spleen in mouse or hamster model with *L. donovani*	100% reduction in parasitaemia in liver and spleen in mouse or hamster model–*L. donovani* and *L. infantum*
In vitro:	Consistent activity within tenfold *vs.* a panel of drug-sensitive and drug-resistant strains and isolates from India and East Africa	Consistent activity within tenfold *vs.* a panel of drug-sensitive and drug-resistant strains and isolates from India and East Africa
	In vitro: E_{max}>99%[a]	*In vitro:* E_{max} > 99%, Cidal mechanism of action
Safety		
In vitro:	No *in vitro* signals preventing development[b]	No *in vitro* signals preventing development
In vivo TI[c]:	(AUC at NOAEL[d])/(AUC at MED_{95}[e]) > 3[f]	(AUC at NOAEL)/(AUC at MED_{100}[g]) > 3
CMC[h]	Synthesis and formulation acceptable to enable PO or IV dosing for 1–10 days in human	Synthesis and formulation acceptable to enable PO or IV dosing for 1–10 days in human
DMPK[i]	Human dose prediction < 30 $mg\,kg^{-1}\,day^{-1}$ given daily or twice daily	Human dose prediction < 30 $mg\,kg^{-1}\,day^{-1}$ given daily or twice daily

[a]Compound able to give *in vitro* >99% reduction of intracellular amastigotes relative to untreated control.
[b]Includes: mammalian cytotoxicity, hERG (human ether-a-go-go-related gene), Ames, micronucleus, broad profiling.
[c]Therapeutic index (TI).
[d]No Observed Adverse Effect Level (NOAEL) determined in rat repeat dose toxicology for duration ≥ length of treatment in efficacy model.
[e]Minimum dose required to achieve >95% reduction in parasitaemia *in vivo*.
[f]Applies equally to *both* total AUC and free AUC comparisons.
[g]Minimum dose required to achieve 100% reduction in parasitaemia *in vivo*.
[h]Chemistry, manufacturing and controls (CMC).
[i]Drug metabolism and pharmacokinetics (DMPK).

2.3.1.1 Drug Re-purposing

The potential of drug re-purposing has been recently reviewed[8] and has already provided amphotericin B, miltefosine and paromomycin for treatment of leishmaniasis. The anti-parasitic drug fexinidazole was originally described in 1978 but was not developed beyond the early discovery stage.[9,10] This compound was re-discovered early in the next century by the DND*i* and further developed for the treatment of sleeping sickness.[11] The reported anti-leishmanial activity of fexinidazole,[12] combined with the encouraging progress in clinical development for sleeping sickness, indicated that fexinidazole might be re-purposed for treatment of leishmaniasis. Some of the key attributes of fexinidazole and its principal sulfone active metabolite are presented in Table 2.3 beside miltefosine for reference. Subsequently fexinidazole was studied in a small Phase II trial in VL patients in Sudan, but it failed to show conclusive efficacy in the majority of patients and no further development is planned for this indication.[13] The re-purposing of other nitroimidazole drugs including DNDI-VL-2098,[14] and two anti-tuberculosis drugs delamanid[15] (already on the market) and (R)-PA-824[16] (in late clinical development) have been proposed, but it may be difficult to achieve adequate plasma concentrations in humans due to limits of achievable exposure, safety considerations and sometimes complex pharmacokinetic–pharmacodynamic (PK–PD) relationships.[17]

2.3.1.2 Compound Class Mining

Nitroimidazoles. Some classes of compounds, such as nitroimidazoles, have well-known broad spectrums of application as anti-infective drugs. Nitroheterocylic drugs are used for the treatment of parasitic infections and bacterial infections, including tuberculosis. Examples of these drugs are nifurtimox (sleeping sickness and Chagas disease), benznidazole (Chagas disease), metronidazole (bacterial infections) and delamanid (tuberculosis). Building on the discovery of the potential of the nitroimidazole drug fexinidazole described in the preceding section, further mining and optimization yielded DNDI-VL-2098 as an optimized preclinical candidate with outstanding *in vitro* and *in vivo* activity against *Leishmania* parasites, and showed for the first time, to our knowledge, that it was possible to eliminate all the parasites from infected mice and hamsters, providing a sterile cure. The development of DNDI-VL-2098 was halted due to adverse effects observed during toxicology studies, but a similarly efficacious back-up compound DNDI-0690 which does not cause the same toxicity in rats was identified and selected as a preclinical candidate in September 2015, and is currently undergoing preclinical development.[20,21] The chemical structures and selected properties of DNDI-VL-2098 and DNDI-0690 are shown in Table 2.3. The mechanism of action of DNDI-0690 is believed to be mediated through bio-activation by parasitic nitroreductase NTR2.[17]

Table 2.3 Chemical structures and key properties of miltefosine, fexinidazole, DNDI-VL-2098 and DNDI-0690.[a]

Compound/property	Miltefosine	Fexinidazole (and principal sulfone active metabolite)	DNDI-VL-2098	DNDI-0690
Chemical structure	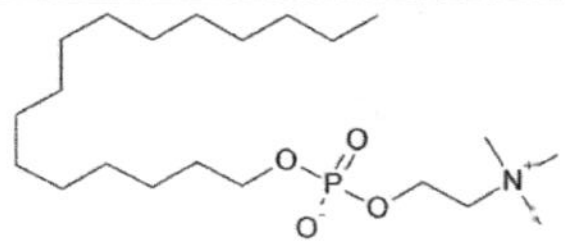	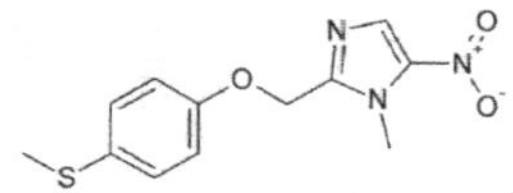 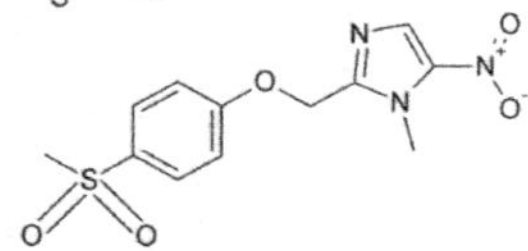	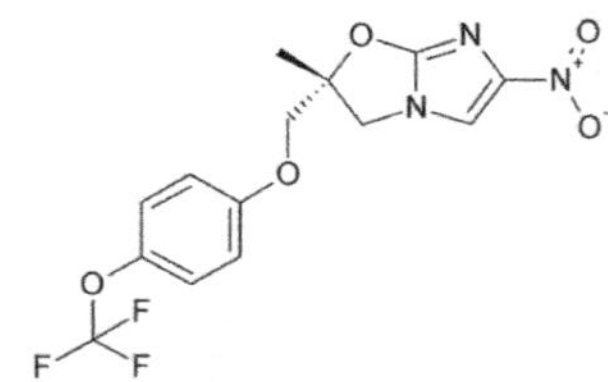	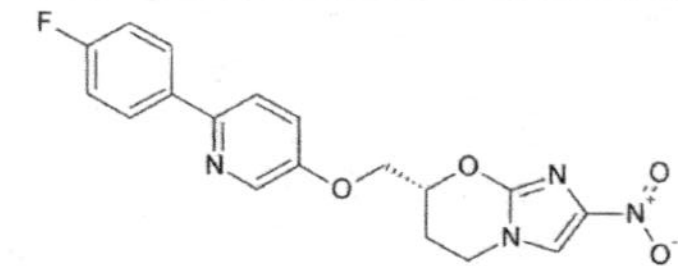
Physical properties	MW 407, ALogP + 3.7	Fexinidazole: MW 279, ALogP + 2.2 Sulfone: MW 311, ALogP + 1.2	MW 359, ALogP + 4.7	MW 370, ALogP + 3.4
In vitro IC_{50} (μM) *vs.* *L. donovani* HU3 strain amastigotes in macrophages	5.84 μM[18]	Fexinidazole: >50 μM Sulfone active metabolite: 5.3 μM[12]	0.09 μM[19]	0.03 μM[20]
In vivo MED[b] for mice infected with *L. donovani* HU3 strain	20 mg kg^{-1} for 5 days (DD8 strain)	Fexinidazole: 200 mg kg^{-1} for 5 days	6.25 mg kg^{-1} for 5 days	6.25 mg kg^{-1} for 3 days
Predicted or actual human dose	Recommended dose: 1.5–2.0 mg kg^{-1} day for 28 days	Fexinidazole dose tested in a Phase II trial: 1800 mg kg^{-1} day for 4 days then 1200 mg kg^{-1} day^{-1} for 6 days	Development halted at preclinical stage	Predicted dose range: 20–360 mg bid for 5–10 days

[a]50% inhibitory concentration.
[b]MED, minimum effective dose.

Benzoxaboroles. The application of cyclic, boron-containing drugs, termed benzoxaboroles, has been explored by the biotech company Anacor Pharmaceuticals which has developed a portfolio of compounds with activity against a wide range of pathogens that cause infectious diseases.[22] A screen of a library of prototypical benzoxaboroles, conducted at the Sandler Centre, identified hits with promising *in vitro* activity against kinetoplastid parasites. Subsequent lead optimization projects carried out by the DND*i* in collaboration with Anacor identified SCYX-7158,[23] which entered Phase II studies for sleeping sickness in November 2016,[11] and also DNDI-6148, selected as a preclinical candidate for visceral and cutaneous leishmaniasis in January 2016.[24] Although DNDI-6148 belongs to the substituted benzoxaborole class, its specific structure has not yet been disclosed. DNDI-6148 is orally active in mice infected with *L. donovani* HU3 strain with a minimum effective dose (MED) of 25 mg kg^{-1} twice daily (bid) for 10 days; further preclinical development of this drug candidate is underway.

2.3.1.3 *High-Throughput Screening of Chemical Diversity Collections*

Proteasome Inhibitors. In 2016, a team of researchers working at Novartis reported a high-throughput screen of three million compounds against *L. donovani* axenic amastigotes to identify a new class of anti-leishmanial compounds. This was followed by a phenotypic optimization programme making a further 3000 analogues, guided by an anti-leishmanial intra-macrophage assay and mouse *in vivo* model. The resulting optimized lead GNF-6702 shown in Table 2.4 displays outstanding *in vitro* and *in vivo* anti-parasitic activity against *L. donovani*, *Trypanosoma cruzi* and *Trypanosoma brucei* and is reported to be in preclinical development.[25] This class of compounds appears to show great promise for the treatment of both VL and CL, and it is encouraging to see further related compounds appearing in the patent literature.[26] Furthermore, the Novartis team conducted an elegant series of experiments, beginning with selection of drug-resistant parasites followed by whole-genome sequencing to demonstrate that this class of drugs kills kinetoplastid parasites through inhibition of the parasitic proteasome. This knowledge should result in further target-based drug discovery efforts focussed on the proteasome, exploiting the power of modern structure-assisted drug design.

Pyrazolopyrimidines. A high-throughput *in vitro* screen of a compound library against *L. donovani* in a partnership between the University of Dundee and DND*i* identified a novel series of pyrazolopyrimidines, which were subsequently optimized in a collaboration between the University of Dundee Drug Discovery Unit and the Diseases of the Developing World unit at GSK, Tres Cantos. The optimized lead GSK-3186899/DDD853651, shown in Table 2.4, has demonstrated excellent *in vitro* activity and *in vivo*

Table 2.4 Chemical structures and key properties of DNDI-6148/AN10705, GNF-6702 and GSK-3186899/DDD853651 and Compound **1**.

Compound/property	DNDI-6148	GNF-6702	GSK-3186899/ DDD853651	Compound **1**
Chemical structure	General benzoxaborole structure: 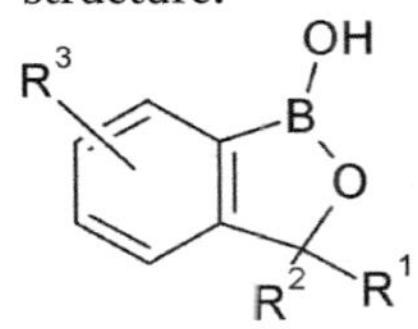	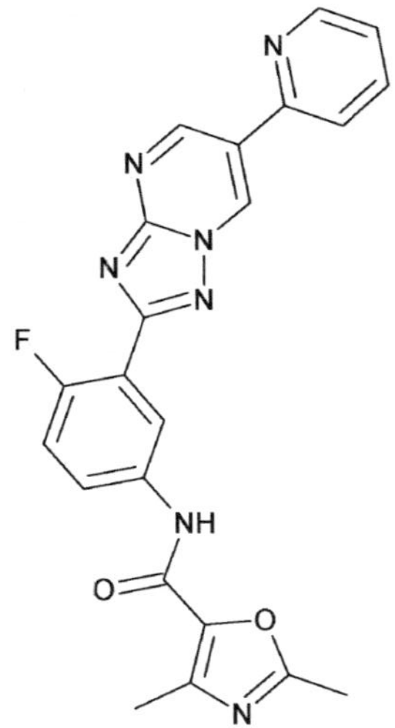	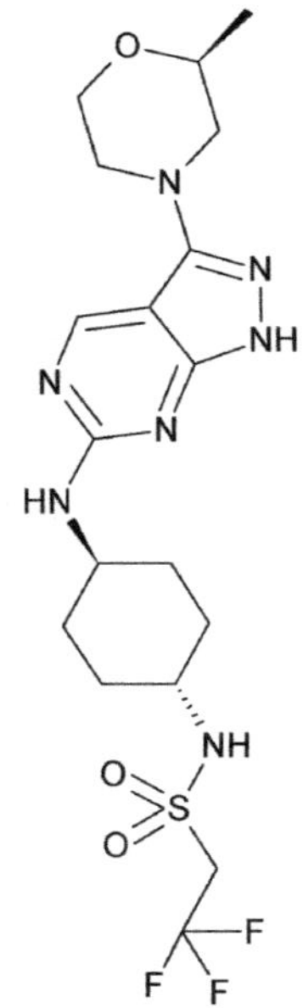	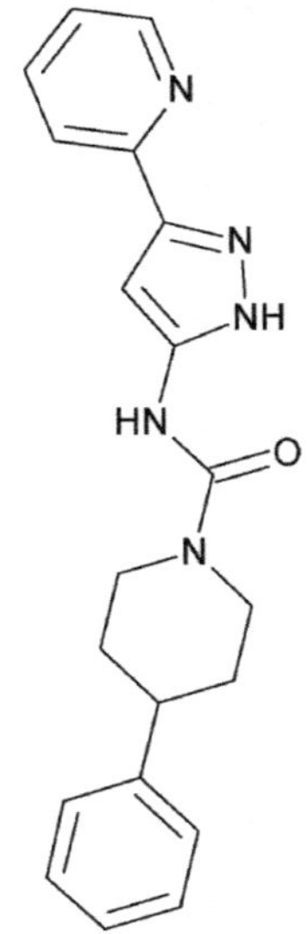
Physical properties	—	MW 429, ALogP + 2.4	MW 478, ALogP + 2.1	MW 347, ALogP + 3.1
In vitro IC_{50} (μM) *vs.* *L. donovani* HU3 strain amastigotes in macrophages	1.62	0.013	1.0 (LdBOB strain in THP-1 cells)	1.31 (L82 strain)
In vivo MED for mice infected with *L. donovani* HU3 strain	25 mg kg^{-1} twice daily for 10 days	10 mg kg^{-1} twice daily for 8 days	25 mg kg^{-1} twice daily for 10 days	50 mg kg^{-1} twice daily for 5 days in Hamsters infected with *L. infantum*

activity in a mouse model of VL and is now undergoing preclinical development.[27,28]

Aminopyrazoles. A novel aminopyrazole hit was identified from a high-throughput screen of a diverse collection of 95 000 compounds representing the broad chemical space of Pfizer's small-molecule collection. Optimization of this initial hit has yielded a series of ureas, such as compound **1** shown in Table 2.4, with excellent *in vivo* activity in a hamster model of VL after oral administration.[29] Further optimization of this series by DND*i* with support from Takeda Pharmaceutical Company Ltd and the Global Health Innovative Technology (GHIT) Fund aims to produce a preclinical candidate in 2017.[30]

2.4 Future Perspectives

In late 2015 and during 2016, five new lead series have been reported, which are orally bioavailable and highly effective at reducing parasitaemia in animal infection models, and at least four new candidates from these series have already advanced into preclinical development. It can be expected that further candidates and back-ups will follow from these efforts guided by phenotypic assays. Furthermore, as demonstrated by the elegant work of Supek and colleagues,[25] by careful deconvolution of the mechanism of action of these new compounds the validated biochemical targets and pathways can be revealed. This knowledge should enable a wave of target-based and structure-assisted drug discovery projects to find further optimized and differentiated drug candidates. All of these efforts remain essential if the research community wishes to build a high level of confidence in successfully delivering new treatments to patients. The chance of any new preclinical candidate surviving the challenges of preclinical toxicology studies, demonstrating acceptable human pharmacokinetics and safety characteristics and going on to successfully treat leishmaniasis patients before finally passing stringent review by the relevant regulatory authority is probably only in the range of 1 in 10–20. Furthermore, an optimal treatment may well require a combination of two such new agents to maximize efficacy and safety, minimize treatment duration and reduce the risk of emergence of drug resistance. So whilst these new candidates provide genuine hope of delivering new treatments, it is critical that these efforts continue to maintain a strong pipeline of development projects to maximize the chance of properly meeting the needs of leishmaniasis patients.

Acknowledgements

The DND*i* gratefully acknowledges financial support from the Bill & Melinda Gates Foundation, USA; the Department for International Development (DFID), UK; the Dutch Ministry of Foreign Affairs (DGIS), the Netherlands; European Union—Specific International Scientific Cooperation Activities

(FP7); the Federal Ministry of Education and Research (BMBF through KfW), Germany; the Global Health Innovative Technology (GHIT) Fund, Japan; Médecins Sans Frontières/Doctors without Borders, International; Medicor Foundation, Liechtenstein; the Spanish Agency for International Development Cooperation (AECID), Spain; the Swiss Agency for Development and Cooperation (SDC), Switzerland; and other private foundations and individuals.

References

1. WHO | Leishmaniasis. Available at: http://www.who.int/mediacentre/factsheets/fs375/en/(Accessed 11th January 2017).
2. About Leishmaniasis – DNDi. Available at: http://www.dndi.org/diseases-projects/leishmaniasis/ (Accessed: 11th January 2017).
3. A. S. Nagle, S. Khare, A. B. Kumar, F. Supek, A. Buchynskyy, C. J. N. Mathison, N. K. Chennamaneni, N. Pendem, F. S. Buckner, M. H. Gelb and V. Molteni, *Chem. Rev.*, 2014, **114**, 11305.
4. See for example: *The Handbook of Medicinal Chemistry: Principles and Practice*, ed. A. Davis and S. E. Ward, 2014, ISBN: 978-1-84973-625-1.
5. C. A. Lipinski, F. Lombardo, B. W. Dominy and P. J. Feeney, *Adv. Drug Delivery Rev.*, 2001, **46**, 3.
6. R. J. Burchmore and M. P. Barrett, *Int. J. Parasitol.*, 2001, **31**, 1311.
7. DNDi Target Product Profile for Visceral Leishmaniasis. Available at: http://www.dndi.org/diseases-projects/leishmaniasis/tpp-vl/(Accessed: 11th January 2017).
8. D. M. Klug, M. H. Gelb and M. P. Pollastri, *Bioorg. Med. Chem. Lett.*, 2016, **26**, 2569.
9. E. Winkelmann and W. Raether, *Arzneim. Forsch.*, 1978, **28**, 739.
10. E. Winkelmann and W. Raether, *Curr. Chemother. Infect. Dis., Proc. Int. Congr. Chemother.*, 11th, 1980, **2**, 969, .
11. A. Tarral, S. Blesson, O. V. Mordt, E. Torreele, D. Sassella, M. A. Bray, L. Hovsepian, E. Evène, V. Gualano, M. Felices and N. Strub-Wourgaft, *Clin. Pharmacokinet.*, 2014, **53**, 565.
12. S. Wyllie, S. Patterson, L. Stojanovski, F. R. C. Simeons, S. Norval, R. Kime, K. D. Read and A. H. Fairlamb, *Sci. Transl. Med.*, 2012, **4**, 119re1.
13. DNDi Fexindizaole. Available at http://www.dndi.org/diseases-projects/portfolio/fexinidazole-vl/(Accessed: 11th January 2017).
14. A. M. Thompson, P. D. O'Connor, A. Blaser, V. Yardley, L. Maes, S. Gupta, D. Launay, D. Martin, S. G. Franzblau, B. Wan, Y. Wang, Z. Ma and W. A. Denny, *J. Med. Chem.*, 2016, **59**, 2530.
15. S. Patterson, S. Wyllie, S. Norval, L. Stojanovski, F. R. Simeons, J. L. Auer, M. Osuna-Cabello, K. D. Read and A. H. Fairlamb, *Elife*, 2016, **5**, 334.
16. S. Patterson, S. Wyllie, L. Stojanovski, M. R. Perry, F. R. C. Simeons, S. Norval, M. Osuna-Cabello, M. De Rycker, K. D. Read and A. H. Fairlamb, *Antimicrob. Agents Chemother.*, 2013, **57**, 4699.

17. S. Wyllie, A. J. Roberts, S. Norval, S. Patterson, B. J. Foth, M. Berriman, K. D. Read and A. H. Fairlamb, *PLoS Pathog.*, 2016, **12**, e1005971.
18. R. García-Hernández, J. I. Manzano, S. Castanys and F. Gamarro, *PLoS Neglected Trop. Dis.*, 2012, **6**, e1974.
19. S. Gupta, V. Yardley, P. Vishwakarma, R. Shivahare, B. Sharma, D. Launay, D. Martin and S. K. Puri, *J. Antimicrob. Chemother*, 2015, **70**, 518.
20. S. Braillard and E. Chatelain, Poster, *Keystone Symposium on Drug Discovery for Parasitic Diseases*, Tahoe City, USA, 24–28 January, 2016.
21. DNDI-0690 Nitroimidazole. Available at: http://www.dndi.org/diseases-projects/portfolio/nitroimidazole/(Accessed: 11th January 2017).
22. Anacor: Neglected Diseases. Available at: https://www.anacor.com/R-and-D/neglected-diseases/ (Accessed: 11th January 2017).
23. R. T. Jacobs, B. Nare, S. A. Wring, M. D. Orr, D. Chen, J. M. Sligar, M. X. Jenks, R. A. Noe, T. S. Bowling, L. T. Mercer, C. Rewerts, E. Gaukel, J. Owens, R. Parham, R. Randolph, B. Beaudet, C. J. Bacchi, N. Yarlett, J. J. Plattner, Y. Freund, C. Ding, T. Akama, Y.-K. Zhang, R. Brun, M. Kaiser, I. Scandale and R. Don, *PLoS Neglected Trop. Dis*, 2011, **5**, e1151.
24. DNDI-6148 Oxaborole. http://www.dndi.org/diseases-projects/portfolio/oxaborole-dndi-6148/ (Accessed: 11th January 2017).
25. S. Khare, A. S. Nagle, A. Biggart, Y. H. Lai, F. Liang, L. C. Davis, S. W. Barnes, C. J. N. Mathison, E. Myburgh, M.-Y. Gao, J. R. Gillespie, X. Liu, J. L. Tan, M. Stinson, I. C. Rivera, J. Ballard, V. Yeh, T. Groessl, G. Federe, H. X. Y. Koh, J. D. Venable, B. Bursulaya, M. Shapiro, P. K. Mishra, G. Spraggon, A. Brock, J. C. Mottram, F. S. Buckner, S. P. S. Rao, B. G. Wen, J. R. Walker, T. Tuntland, V. Molteni, R. J. Glynne and F. Supek, *Nature*, 2016, **537**, 229.
26. S. Brand, E. V. Gaza, I. Gilbert, E. J. Ko, M. G. Thomas, M. Marco Martin, T. J. Miles and L. H. Sandberg, *PCT Int. Appl.*, WO 2016193111 A1, 2016.
27. The Development of DDD853651; a Potential Candidate for the Treatment of Visceral Leishmaniasis, M. Thomas, *EFMC-ISMC 2016; XXIV EFMC International Symposium on Medicinal Chemistry*; Manchester, UK; August 28 – September 1, 2016.
28. T. J. Miles and M. G. Thomas, *PCT Int. Appl*, WO 2016116563 A1, 2016.
29. C. E. Mowbray, S. Braillard, W. Speed, P. A. Glossop, G. A. Whitlock, K. R. Gibson, J. E. J. Mills, A. D. Brown, J. M. F. Gardner, Y. Cao, W. Hua, G. L. Morgans, P.-B. Feijens, A. Matheeussen and L. J. Maes, *J. Med. Chem.*, 2015, **58**, 9615.
30. DNDi Aminopyrazoles. Available at http://www.dndi.org/diseases-projects/portfolio/aminopyrazoles/(Accessed: 11th January 2017).

CHAPTER 3

From Bench to Bedside: Development and Optimization of Clinical Therapies for Visceral Leishmaniasis

FABIANA ALVES,*[a] JEAN-YVES GILLON,[a] BYRON ARANA[a] AND THOMAS P. C. DORLO[b]

[a] Drugs for Neglected Diseases *initiative*, 15 Chemin Louis-Dunant, Geneva, Switzerland; [b] Department of Pharmacy & Pharmacology, Antoni van Leeuwenhoek Hospital/The Netherlands Cancer Institute, Amsterdam, The Netherlands

*Email: falves@dndi.org

3.1 Introduction

Despite a considerable effort and investment in leishmaniasis drug discovery in the past several years, treatments for visceral leishmaniasis (VL) still rely on a few drugs that have limitations such as parenteral administration, poor tolerability and toxicity, long treatment duration and high cost. Given the scarcity of new compounds in the clinical pipeline in the last decade, efforts have been focussed on optimizing current available treatments according to the context of VL in different regions. In Southeast Asia, where pentavalent antimonials can no longer be used due to drug resistance, a single-dose regimen of liposomal amphotericin B and a combination regimen such as miltefosine–paromomycin for 10 days have proven to be

Drug Discovery Series No. 60
Drug Discovery for Leishmaniasis
Edited by Luis Rivas and Carmen Gil

Published by the Royal Society of Chemistry, www.rsc.org

highly efficacious and safe and are currently used on a routine basis. In Eastern Africa, the combination of sodium stibogluconate (SSG) with paromomycin for 17 days is recommended as first-line treatment; whilst in Brazil, where nearly 90% of VL cases from the Americas are reported, meglumine antimoniate given for 20 to 30 days is still used as first-line therapy, followed by treatment with liposomal amphotericin B over 7 days.

Despite improvements in current regimens through development of safer formulations or combinations, new oral treatments that are efficacious, safe, of short duration and easily implementable in remote areas where VL occurs are still badly needed. Unfortunately, the path that a new compound has to follow from early discovery to registration is costly and takes several years, and attrition rates due to safety issues or poor efficacy are high. In the era of new classes of anti-leishmanial compounds, there is a need to consider more systematic use of pharmacokinetic–pharmacodynamic (PK–PD) modelling and simulation for translation from preclinical discovery to early clinical development, and subsequently to further fine-tune and optimize the dosing regimen in later clinical development.

Hereafter, a summary of the *in vitro* and *in vivo* strategies usually used for VL in the early preclinical development stage to provide an indication of the susceptibility, target exposure and therapeutic index to support the transition from preclinical to clinical development and first-in-human testing is presented; followed by the example of miltefosine, the only oral drug currently available: its historical path through clinical development and post-registration, and the efforts to optimize its use for the treatment of VL using novel translational tools.

3.2 The Role of Preclinical Models in Preclinical to Clinical Translation

3.2.1 Strategies for *in vitro* Screening

Most current strategies for preclinical development assess "cidal" activity against *Leishmania. In vitro* assays can be categorized, depending on the stage (promastigotes, axenic amastigotes or intracellular amastigotes) of the parasite. However, whole-cell-based assays are generally considered the gold standard, as the intramacrophagic amastigote is the clinically relevant form of the parasite. A range of protocols has been developed by several groups, using slightly different conditions to measure parasite viability remaining after a defined exposure period to the drugs. Classical methods to detect parasites, such as Giemsa staining and microscopic counting or the use of the tetrazolium dye (3-(4,5-dimethylthiazol-2-yl)-2,5-diphenyltetrazolium bromide)[1] to assess viability are still in use, but development of phenotypic screening assays based on image analysis—known as high-content screening—has been a breakthrough in the recent years.[2,3] Laboratory strains of *Leishmania donovani* and *Leishmania infantum* are typically used here and, while promastigotes or axenic amastigotes are grown in culture medium, activity against intracellular amastigotes is determined in macrophages, the natural host cells targeted by *Leishmania.* Macrophages are either derived from mouse peritoneum[4,5] or bone-marrow, or are differentiated by chemical agents such as phorbol 12-myristate 13-acetate from mouse (*e.g.* J-774[6]) or human (THP-1) cell lines. New compounds are selected based on the 50% and 90% inhibitory concentrations (IC_{50} and IC_{90}), *i.e.* the concentrations of drugs that decrease the cell growth of a fixed inoculum of parasites by 50% and 90%, respectively, as compared with the negative control.

In vitro promastigote or axenic amastigote assays allow a fast and straightforward screening of large compound collections, as they are compatible with a relatively large-scale format and require minimal amounts of compounds. However, several laboratories have shown that they result in a relatively high rate of false positives and may not identify intracellular-stage-specific compounds.[7–9] This could tentatively be explained by the physiological differences between the two types of assays, including pH, media composition and incubation times, but also by differences in protein expression in relation to[7] parasite growth and duplication. Promastigotes are fast growing while intramacrophageal amastigotes are only slowly replicating. Also, in intracellular assays, drugs have to penetrate the parasitophorous vacuole, a macrophage sub-cellular compartment in which parasites survive and that is acidic and rich in microbicidal peptides and hydrolytic enzymes.[10] Lastly, in a panel of macrophages from different origins, the anti-leishmanial activity of various drugs, including amphotericin B deoxycholate, paromomycin and miltefosine was demonstrated to be host-cell-dependent.[11]

3.2.2 Preclinical Animal Models for Visceral Leishmaniasis

Animal models of VL have largely contributed to a better understanding of the pathophysiological processes underlying infection and host response to the disease. They also have played a paramount role in the selection of preclinical candidates, based on reduction of macrophageal parasite burden in the liver and/or spleen.

3.2.2.1 Murine Models

Experimental challenge murine models have been extensively used for the study of immunopathology[12,13] and genetic regulation[14] and for assessing the potential efficacy of new chemical entities (NCEs) against VL.[6,15] These models are produced by inoculation with either *L. donovani* or *L. infantum.* Murine models are highly susceptible to a variety of factors, including mouse strain, age and immunity status, virulence, infectivity and parasite stage, number of inoculated parasites and route of inoculation.[16] An important issue with murine models is that they do not directly translate to the human situation, because of the capacity of the murine immune system to reduce parasitic load.

3.2.2.2 Syrian Hamster Model

The hamster model is also used to assess activity of compounds against leishmaniasis; these animals exhibit severe clinical symptoms that are similar to those observed in naturally infected dogs and humans.[17] In this species, clinical symptoms of VL can range from mild to progressive fatal visceral disease.[18] Hamsters are usually infected by intracardiac or intraperitoneal inoculation of *L. donovani* or *L. infantum*, which progressively induces hepatosplenomegaly, anaemia, cachexia and immunodepression.[19] The hamster model is most frequently favoured, not only because disease evolution is close to the clinical situation in humans, but also because assessing drug effects on the liver, spleen and bone marrow may provide precious information for predicting efficacy in the clinic. However, the benefits of the model need to be balanced by the unusual pharmacokinetics of many compounds in hamsters.

3.2.3 Issues in Interpretation of Preclinical Models

Determination of *in vitro* cidal effects is essential for selection of the most promising new agents. However, parasite expression, host response and clinical manifestations vary considerably by endemic region and species, ranging from asymptomatic to life-threatening VL. While the use of laboratory *Leishmania* strains has the merit of helping standardize the assays and, as a result, ranking cidal activities of new agents, one should bear in mind that these strains may not adequately represent the diversity of parasite, virulence, infectivity and pathogenicity observed in clinical conditions. Therefore, the cidal activity may not translate to efficacy in clinical trials. Implementing a panel of strains in the process of drug screening may be a useful approach to better characterizing variability in drug activity.

Characterization of efficacy of the most interesting candidates in murine or hamster models is also a key step for the translation of potency in *in vitro* assays to potential efficacy in the host. In this simple, linear progression strategy, the BALB/c mouse and hamster models are considered to be acute and chronic models of VL respectively.[20] However, no preclinical model fully captures the complexity of disease states in human, and the models described above all have drawbacks. In addition, it is also sometimes challenging to make cross-study comparisons because of differences in experimental conditions. It is therefore important that models are well characterized, established and harmonized, resulting in a stable tissue infection and confirm that positive benchmark controls show consistent and reproducible efficacy.

Data should in any case be interpreted carefully, and absorption, disposition, metabolism and elimination (ADME) properties of the compounds, which may vary considerably between species, should be taken into account. Although not systematically performed, blood or plasma PK assessments should be included or performed in parallel under the same conditions as for *in vivo* efficacy, bearing in mind that inflammation and a variety of clinical manifestations may affect drug tissue penetration and distribution locally, as evidenced in brain tissue and cerebrospinal fluid in the presence of meningeal inflammation.[21]

3.2.4 How can Pre-clinical Data be Translated to Facilitate Clinical Development?

There is increasing attention being paid to pre-clinical to clinical translational science within the drug discovery and development paradigm for infectious diseases, and investments are being made to improve the understanding of PK–PD relationships in animal models to facilitate clinical development and to better understand the gaps between animal models and clinical application in humans.

For antibiotics, potency is generally defined by the minimum inhibitory concentration (MIC), which is the lowest concentration that completely inhibits the *in vitro* growth of a microorganism. Despite the fact that MIC is a good descriptor of a drug's antibiotic potential, it provides no information about the time course of anti-microbial activity. Integrating potency with PK characteristics, it is possible to define PK–PD parameters that describe antimicrobial effects in a dynamic way. Based on PK–PD profile, antibiotics can be roughly grouped into three categories. For drugs such as aminoglycosides the maximum plasma concentration ($C_{\max}$) in relation to the MIC appears to be important for its effect (concentration-dependence, defined by the peak concentration to MIC ratio). The higher the drug concentration, the greater the bactericidal effect. Others, like penicillins or cephalosporins, require plasma levels to be above the MIC for a certain period of time during the dosing interval (time-dependence). While yet others, like tetracyclines or vancomycin, exhibit persistent effects for efficacy both related to time and concentration, which means total drug exposure is important. Defining the ideal dosing regimen for appropriate efficacy therefore tends to maximize drug concentrations in plasma in the first case, the duration of exposure in the second and the amount of drug in the latter case. PK–PD tools and strategies have been developed[22–24] and are routinely used in antibiotic development to assess preclinical and clinical PK targets, to predict clinical efficacy based on preclinical data, and to optimize dosing regimens in clinical trials.

In the case of drug development for VL, there is now a lot of interest and investment in research to develop PK–PD tools to facilitate the translation of new compounds to clinical development. Nevertheless, most drugs currently in use were developed several decades ago, when these tools were clearly not available. The two latest drugs registered for VL, miltefosine and paromomycin, in 2002 and 2006 respectively, had been initially developed for other indications and pragmatically adapted for leishmaniasis. The use of PK–PD tools in clinical trials using the currently available drugs provides an opportunity to better characterize exposure–response relationships in different patient populations, and allows the use of state-of-the-art modelling and simulation to further optimize therapies. In order to illustrate this concept, we have chosen the case of the development of miltefosine for VL.

3.3 Challenges and Opportunities to Optimize Therapies for Leishmaniasis: The Case of Miltefosine

3.3.1 Discovery and Development History of Miltefosine for Leishmaniasis

The discovery of the anti-parasitic activity of alkylphosphocholine compounds, including miltefosine, was to a large extent serendipitous. A range of platelet-aggregating-factor analogues was initially synthesized in an effort to screen for anti-inflammatory properties, while at the same time their anti-tumour activity was screened in Germany as inhibitors of the membrane signalling pathway PI3K–Akt. In the 1980s it was discovered that miltefosine, one of these novel phospholipid compounds, was highly active against trypanosomatid parasites, and in 1987 Croft and colleagues reported its activity against *L. donovani* in intracellular amastigotes and *in vivo* after subcutaneous injection in a mouse infection model.[25]

In the pharmaceutical and clinical development of miltefosine, priority was given to its application in the treatment of cancer, particularly solid tumours. The various phase I and II studies that followed for oral miltefosine in severely ill cancer patients revealed good oral bioavailability of the compound in humans with a simple oral capsulated formulation. However, the high dose (150 mg day^{-1} and higher) and prolonged treatment courses needed to reach the systemic exposure expected to result in clinical efficacy in oncological patients were associated with dose-limiting gastrointestinal toxicity, such as nausea, vomiting and loss of appetite.[26–29] This led in the mid-1990s to the abandonment and discontinuation of development of oral miltefosine's anti-cancer application.[30]

The activity against VL was confirmed more or less concurrently in an oral miltefosine study in a BALB/c mouse model by Kuhlencord *et al.*, which showed excellent activity of 20 mg kg^{-1} day^{-1} oral miltefosine for 5 days and superiority over the standard drug sodium stibogluconate.[31] Based on the high oral bioavailability that was demonstrated in the phase I and II studies in cancer patients and the urgent need to have the first oral treatment for VL, a clinical development programme was pursued at Asta Medica. Considering the high number of VL patients reported in the Indian subcontinent (India, Bangladesh and Nepal), an agreement was reached in 1995 between Asta Medica (later the spin-off company Zentaris), the Special Programme for Research and Training in Tropical Diseases of the World Health Organization (TDR–WHO), the Indian government and various Indian academic collaborators. Given the available safety database from the Phase I and II studies in cancer patients, which contained tolerability data for over 200 oncological patients, healthy volunteer studies were therefore not performed within the clinical VL development programme. Clinical phase II studies started in adult VL patients,[32,33] including a dose-finding study, with initial dose regimens largely based on the 'flat' 150 mg day^{-1} dosages that were deemed maximally tolerable in terms of gastrointestinal toxicity directly taken from the initial oncological studies.

3.3.2 Changing Patterns in Clinical Efficacy

3.3.2.1 Clinical Development (Phase II–III Studies)

The clinical development of miltefosine started in India in the late 1990s and already in 2002 a large amount of data had been generated in five clinical trials in adult patients (over 12 years of age).[32–36] Gastrointestinal toxicity appeared to be much less severe in Indian VL patients than initially described in cancer patients, which, for instance, made lead-in dosing unnecessary.[30,33] Among the phase II trials in adults, different treatment regimens were tested (ranging from 50 to 200 mg day^{-1} for 14–42 days), and, in general, the daily dose of 100 mg day^{-1} for 21–28 days gave satisfactory efficacy (above 90%) and was well tolerated. In the randomized open-label phase III clinical trial the efficacy and safety of miltefosine (100 mg day^{-1} orally for 28 days) was compared with amphotericin B deoxycholate (1 mg kg^{-1} day^{-1} for 14 days) in adult VL patients.[35] The efficacy of miltefosine in this phase III trial was highly satisfactory: 98% cure rate at the end of therapy (2% of patients did not have parasitological assessment done, but were clinically well) and 94% cure rate (282 out of 299, 95% confidence interval (CI) 91–97%) at the end of the six-months follow-up by intention-to-treat (ITT) analysis. The actual relapse rate at six-months was 3% (9 out of 299 patients), whereas another 3% of the patients were lost to follow-up and defined as failures in the ITT analysis.

Miltefosine was registered in India in 2002 for VL indication for patients aged two years or older on the basis of published data for adults and further studies in children which were published later.[35] The dose was defined as a linear regimen of 2.5 mg kg^{-1} day^{-1} for 28 days. In practice, considering the capsule strength of 50 mg, adult patients with weights of over 25 kg received a 100 mg daily dose, and adults with weights of 25 kg or less received a 50 mg daily dose for 28 days, whereas children were dosed on the basis of 2.5 mg kg^{-1} day^{-1} that could be adjusted using the 10 mg capsule, when available. Due to this practical limitation, the actual dose administered ranged from 2 to 4 mg kg^{-1} day^{-1}.

Having an oral drug available for the treatment of anthroponotic VL, together with the ability to identify VL patients following an easy and reliable diagnostic algorithm based on clinical findings (fever for more than two weeks and presence of splenomegaly), and a user-friendly, field-adapted rapid diagnostic test (rK39), prompted the launch in 2005 of the Kala-azar

Elimination Initiative by India, Bangladesh and Nepal. Apart from case management, other pillars of the elimination programme were vector control, surveillance, social mobilization and operational research, aiming to reduce VL incidence to 1 case per 10 000 population at district or sub-district level by 2015 (subsequently changed to 2020).[37]

3.3.2.2 Post-registration Studies

Further studies were performed in India, Bangladesh and Nepal over the following years. A phase IV study performed in India (2007) aimed to investigate the feasibility of miltefosine treatment in real-life settings.[38] A total of 1132 VL patients (38% under 12 years of age, and 62% aged 12 years or over) were enrolled in the study, which was implemented as out-patient care in 13 centres in Bihar. Treatment was monitored through weekly visits to the clinic and compliance was assessed. Patients were counselled and clearly instructed about the nature of expected side effects, especially gastrointestinal reactions. The efficacy observed at six-months follow-up was 81.9% by ITT analysis (927 out of 1132), with 14.2% (161 patients) of failures which were lost to follow-up at the six-months visit. Among the patients who completed the study, the final cure rate was 95.5% (927 out of 971), with a significantly higher relapse rate observed in children (23 out of 358, 6.4%) as compared to adults (21 out of 613, 3.4%; $p = 0.03$).[38]

In 2012, a new study reported the evolution of miltefosine efficacy in India after a decade of use.[39] A total of 567 VL patients [135 (24%) under 12 years of age, and 432 (76%) aged 12 years and over] were treated in a hospital setting, with an efficacy at the end of six-months follow-up of 90.3% (512 out of 567). The reasons for failure in the ITT analysis were nine patients who had the treatment discontinued due to adverse events, four who withdrew consent, 39 cases of relapse (6.8%) and five deaths (including one patient with treatment discontinued due to adverse events and one case of relapse). This was a significant decline in efficacy from 94% in 2002 to 90% in 2012 ($p = 0.04$), with a twofold increase in relapse rate from 3% to 6.8% ($p = 0.02$). The reasons for failure were unlikely to be related to lack of compliance, as treatment was directly observed in this trial. However, in the field, as miltefosine was available in the market at an unaffordable price for the local impoverished population (US$ 145 for a 28-days treatment), practices of incomplete treatment, availability over the counter without proper prescription and non-compliance to the full treatment duration together with the long half-life of miltefosine raised important concerns about the risk of resistance development.[40] Resistance development had already compromised the use of antimonials in the region, now miltefosine, after less than a decade of use, was at risk if no strict measures to control access through a public distribution system and proper use were implemented.[40]

In Nepal, miltefosine treatment has been introduced as part of the regional elimination initiative, replacing antimonials, to which resistance had been clearly documented. From 2009 to 2011, a prospective cohort of 120 VL patients treated with miltefosine (as per recommendations in the guidelines) was followed up for 12 months after the end of treatment. Efficacy was assessed at the end of treatment, after 6- and 12-months follow-up, with cure rates of 95.8% (115 out of 120), 82.5% (99 out of 120) and 73.3% (88 out of 120), respectively. Failure attributed to relapses accounted for 10.8% at the six-months follow-up and 20% at the 12-months follow-up; these were confirmed to be due to the same parasite through fingerprinting (no re-infection), and no significant change in parasite susceptibility was identified pre *vs.* post treatment.[41] Relapse was most common among children less than 12 years old [incidence risk ratio (IRR) = 2.43, 95% CI 1.09–5.42].

Subsequently, data from the cohort of patients from Nepal and India (a total of 1016 VL patients) were analysed to identify clinical and epidemiological risk factors for failure. Paediatric patients were confirmed to be at higher risk of failure, with a 2.5-fold higher risk for the age group 10–14 years (95% CI 1.37–4.5) and 3.2-fold higher risk for the age group 2–9 years old (95% CI 1.77–5.77); male patients had a twofold higher risk of failure (95% CI 1.27–3.61). Possible factors considered to be associated with this difference in outcome in children were differences in immune response and drug pharmacokinetics.[42]

One of the approaches to maintaining a longer life for this drug was combining it with other anti-leishmanial treatments. The rationale for combining drugs was to reduce treatment duration, improve the safety profile and increase compliance, while maintaining high efficacy and avoiding the risk of resistance development. Different regimens were tested, including combining oral miltefosine (100 mg for adults of 25 kg or heavier, 50 mg for adults of less than 25 kg, and 2.5 $mg\,kg^{-1}\,day^{-1}$ for children less than 12 years of age) with paromomycin IM injections (11 $mg\,kg^{-1}\,day^{-1}$ base) over 10 days, or combining liposomal amphotericin B (one single injection IV, 5 $mg\,kg^{-1}$) with miltefosine treatment (as previously described) for 7 days. Both miltefosine combination regimens have proven to be highly efficacious, with ITT cure rates at six-months follow-up of 97.5% (156 out of 160, 95% CI 93.3–99.2) for liposomal amphotericin B and miltefosine; and 98.7% (157 out of 159, 95% CI 95.1–99.8%) for miltefosine and paromomycin, which were non-inferior to the amphotericin B deoxycholate comparator arm with a cure rate of 93% (146 out of 157, 95% CI 87.5–96.3%).

As described in Chapter 1, the epidemiology of VL is different in the three continents where the disease is endemic. Therefore, data from South Asia cannot be extrapolated to Eastern Africa or Latin America.

In Eastern Africa, miltefosine treatment was first described in Ethiopia by Médecins Sans Frontières (MSF), in the context of routine practice in a population with a high prevalence of HIV co-infection (29%). Adult male VL patients were randomized to receive miltefosine (100 mg for 28 days) or the standard-of-care treatment of sodium stibogluconate (SSG, 20 $mg\,kg^{-1}\,day^{-1}$ for 30 days). The cure rate at six-months follow-up among non-HIV-co-infected patients was 75.6% for miltefosine (95% CI 67.3–82.7%) and 77.4% for the SSG treatment (95% CI 69.4–84.1%).

More recently, in a phase II randomized multicentre clinical trial conducted in Kenya and Sudan, two study arms contained miltefosine, either as monotherapy (2.5 $mg\,kg^{-1}\,day^{-1}$) for 28 days, or in a combination of liposomal amphotericin B (10 $mg\,kg^{-1}$ single injection) with miltefosine (2.5 $mg\,kg^{-1}\,day^{-1}$) for 10 days. Efficacy at six-months follow-up was 72% (95% CI 60–85%) for miltefosine alone and 77% (95% CI 64–90%) for the liposomal amphotericin B and miltefosine combination. Although the study was not powered for sub-group analysis, there was significantly lower efficacy in children (less than 12 years of age) compared with adults in the monotherapy arm (59% *vs.* 86%, $p = 0.05$), and the same trend was observed in the liposomal amphotericin B and miltefosine combination arm (74% *vs.* 90%, $p = 0.159$).[43]

The differences in cure rate between Asia and Eastern Africa could be related to differences in parasite, host and/or drug exposure. The *L. donovani* population in Eastern Africa is genetically different from that in India.[44] While resistance mechanisms have been described *in vitro* for miltefosine,[45] there is nevertheless, to date, no evidence of widespread resistant strains that can be isolated from relapsed patients in the Indian subcontinent or Eastern Africa.[41,46] In previous studies, there was clearly a higher risk of relapse in children under 12 years of age in the two regions. Considering the differences in metabolism and drug clearance between age groups, it was imperative to examine the pharmacokinetics of miltefosine in order to provide insights into how miltefosine therapy might be optimized, especially for the paediatric population.

3.3.3 Dose Optimization of Miltefosine for Future Combination Strategies

During the initial phase of the clinical development of oral miltefosine for VL in India, only limited attention was given to the clinical pharmacokinetics of this drug: the various phase II dose-finding studies that were performed lacked extensive pharmacokinetic assessments, and only very sparse descriptive data were reported in the registration documents that were initially filed in India (2002) and Germany (2004). Overall, miltefosine pharmacokinetics are mainly characterized by a very slow initial, and even slower terminal, elimination; the conventional 2.5 $mg\,kg^{-1}\,day^{-1}$ for 28 days dosing regimen thus leads to a continued accumulation of the drug during the treatment period (Figure 3.1).[47] This means that individual end-of-treatment concentrations generally correlate to the overall exposure during treatment, but at the same time plasma concentrations in the first part of the treatment remain low and possibly suboptimal.

Miltefosine concentrations were analysed for the first time in VL patients enrolled in an early phase II dose-finding clinical trial in adults[33] and in a paediatric clinical trial in India,[48] however these data were not published at the time. Median miltefosine concentration around the end of treatment for

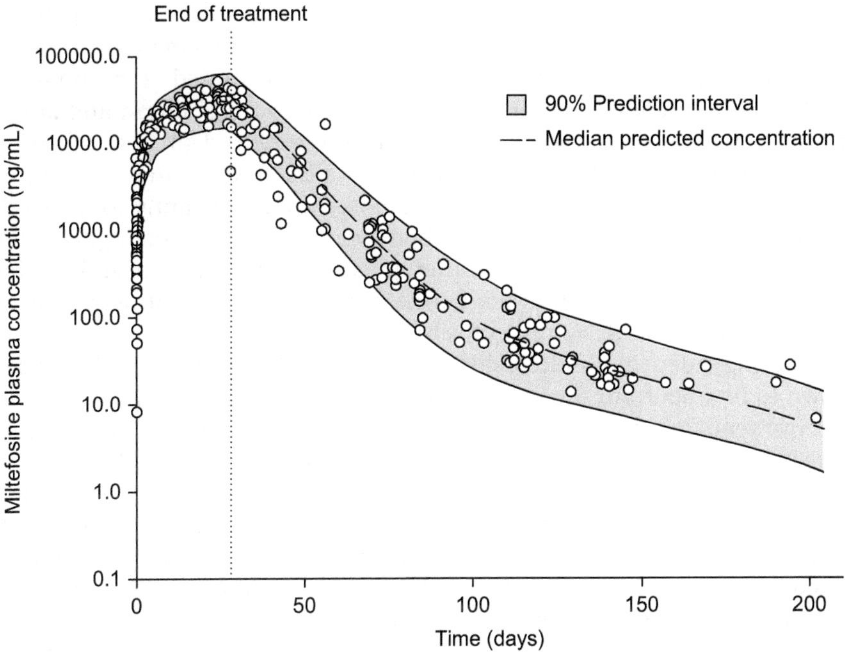

Figure 3.1 Visual predictive check of population pharmacokinetic model for miltefosine. Open circles represent observed data ($n = 382$) from 31 cutaneous leishmaniasis (*Leishmania major*) patients. All patients were treated with 150 mg day^{-1} miltefosine for 28 days. The grey area shows the 90% interval of the model predictions; the broken line indicates the median predicted concentrations.
Reproduced from Dorlo *et al.*, *Antimicrob. Agents Chemother.*, 2008, 52(8), 2855–2860 with permission.[47] Copyright © American Society for Microbiology.

adult patients treated with 100 mg day^{-1} (median 2.9 mg kg^{-1} day^{-1}) for 28 days was 70 μg ml^{-1} (day 23), whereas for a group of children treated with a median daily dose of 2.6 mg kg^{-1} day^{-1} for 28 days this was 24 μg ml^{-1} (between day 26 and 28). After 14 days of treatment, the adult group of patients had already accumulated miltefosine plasma concentrations of between 24 and 69 μg ml^{-1}. This is consistent with clinical findings from another phase II trial, where treatments for 14, 21 and 28 days with 100 mg kg^{-1} day^{-1} resulted in an efficacy at six-months follow-up of 89% (95% CI 71–100%), 100%, and 100% (95% CI 85–100), indicating substantial efficacy with 14 days of treatment.[49] No pharmacokinetic evaluation has, to our knowledge, been performed to date in healthy individuals to investigate, in a controlled manner, the effect of body weight on the distribution and clearance of the drug, or absorption-related issues, such as the effect of food on the relative oral bioavailability of the compound.

Other studies confirmed the apparent differences in drug exposure between adults and children using the conventional 2.5 mg kg^{-1} day^{-1} dose, with children generally defined as 12 years of age or less. A subset of patients of the Nepalese cohort study described above was included in a sparse pharmacokinetic assessment, which again demonstrated a pronounced difference in end-of-treatment concentrations between children and adults of approximately 30%.[50] Similar findings were observed in Eastern Africa, where patients with a lower body weight (less than 30 kg, mainly children between 7 and 12 years of age) were exposed to significantly lower levels of miltefosine than patients with a body weight of 30 kg or higher, both after 28 days of miltefosine monotherapy and after 10 days of miltefosine combination therapy, with average differences in end-of-treatment concentrations of 36% and 32%, respectively.[51]

Achieving adequate miltefosine drug exposure has nevertheless been shown to be important for VL, and various attempts have been made in the past five years to establish a PK–PD relationship, given the increased failure rates for this drug over time (see above). Treatment failure of miltefosine, in terms of relapse of disease within a 12-month follow-up period, was shown to be associated with lower drug exposure. The Nepalese cohort study mentioned earlier revealed that the probability of treatment failure was affected by the period of time that miltefosine concentrations where above ten times the *in vitro* IC_{50} susceptibility value of 17.9 $\mu g\,ml^{-1}$; the odds ratio for treatment failure decreased with increasing time-related drug exposure. As mentioned earlier, in previous studies Indian children barely reached this threshold value at day 28 of treatment, while Indian adults generally reached this value after just 14 days with 100 $mg\,kg^{-1}\,day^{-1}$.[52] Another exposure–response relationship was established for miltefosine in Eastern African VL patients, where miltefosine exposure time at levels greater than the IC_{90} inversely affected the relapse hazard and thus the time to relapse of infection.[53]

Many observations from various controlled and observational studies from both the Indian subcontinent and Eastern Africa indicate that children accumulate the drug to a lesser degree than adults under conventional 2.5 $mg\,kg^{-1}\,day^{-1}$ dosing, which is consistent with the clinical findings that children are more at risk of failing miltefosine treatment.[38,50,54] This, in combination with the established exposure–response relationship for miltefosine in VL, underlines the need to revise the miltefosine dosing guidelines for children.

From a mechanistic point of view, the lower miltefosine exposure in children can be explained by a standard physiological mechanism. It has long been recognized, over a wide range of animal species, that metabolic processes in the body are not linearly related to body mass or body size, but are related through a non-linear, so-called, allometric relationship between size and function.[55] From a PK perspective, this implies that clearance of drugs is not related to body weight linearly, but exponentially, with an allometric power-exponent of 3/4. This was also demonstrated for miltefosine when PK data from Indian and European patients with a wide distribution of body weights was combined; here, miltefosine clearance from the central compartment of distribution was most accurately estimated when scaled allometrically, based on fat-free mass.[52] Miltefosine clearance is thus higher per kg of body weight in children compared with adults, which means that giving the same mg kg^{-1} day^{-1} dose will lead to lower accumulation of the drug in children during the treatment period. Given the importance of adequate miltefosine exposure in VL patients, as illustrated by the aforementioned exposure–response relationship for treatment failure, allometric dosing was therefore proposed to overcome the disparity in miltefosine drug exposure between paediatric and adult patients.[52] Although the reasons for treatment failure in VL are probably multifactorial, it is expected that allometric dosing, and thus increased exposure in children, will result in fewer paediatric VL relapses. The safety and clinical pharmacokinetics of the proposed allometric dosing regimen are currently being evaluated in paediatric VL patients in Kenya and Uganda (Clinicaltrials.gov NCT02431143) and in post-kala-azar dermal patients in Bangladesh (Clinicaltrials.gov NCT02193022).

The differences in PK profiles between adult VL patients in Asia and Africa appear to be significant. While not much has been published on the pharmacokinetics of miltefosine in Indian VL patients, the reported median concentration on day 23 of miltefosine treatment at 100 mg day^{-1} (median 2.9 mg kg^{-1} day^{-1}) over 28 days was 70 μg ml^{-1} in India. In Africa, adults treated with 2.5 mg kg^{-1} day^{-1} had a mean end of treatment concentration (day 28) of 31 μg ml^{-1}.[43] Initial results from a population PK analysis of the Eastern African data indicated that bioavailability is significantly reduced in the first week of treatment, which may lead to the much lower levels of exposure to miltefosine in African patients compared with Indian patients.[53] The mechanism behind the reduced bioavailability is not yet well-understood, but this unexpected observation of non-linearity highlights the difficulty of externally validating a finding when extrapolating clinical and PK observations between patient populations in distinct geographical regions.

3.3.4 Translational Approach to Manage Miltefosine's Main Safety Concern: Teratogenicity

Miltefosine treatment is generally well tolerated. The most common adverse events are related to gastrointestinal effects, with vomiting occurring in at least 20–30% of patients treated. The events are in general mild, occur mainly in the first week(s) of treatment, and less than 1% of the patients are expected to discontinue treatment due to intolerance.[38] Other common adverse events include mild increases in liver enzymes and creatinine.

The major safety concern for miltefosine is its teratogenic potential, which hampers its application in the clinic, particularly in rural areas. Women of reproductive age have been shown to make up 6.2–26.3% of the total VL population, depending on geographic area.[56] While there is no human data available that confirms teratogenicity, animal models have indicated embryo–foetal toxicity, including death and teratogenicity. Preclinical reproductive toxicity studies in animals showed both embryotoxicity and fetotoxicity in rabbits and rats, while teratogenic effects were only demonstrated in rats at a lowest observed adverse effect level, which was lower than the recommended human dose.[57] Miltefosine is therefore contraindicated in pregnant women and contraceptive protection is strictly required in female patients of reproductive age. However, due to the extremely long elimination half-life of miltefosine (see above), making it detectable in the blood plasma of patients for as long as six months post-treatment, it remained unclear what would be an appropriate duration of contraception to avoid the risk of teratogenicity. To assess this, a translational animal-to-human PK modelling and simulation framework was designed to study and characterize the teratogenic risk, by translating the animal doses corresponding to the lowest observed adverse effect level in animals to human-equivalent doses, for which exposure was simulated in human female VL patients and compared with exposure after regular treatment.[58] This framework constituted a more rational teratogenic risk-management strategy and recommended increasing the current contraceptive cover periods to four months after the end of treatment for the standard 28-day miltefosine regimen, while for all shorter regimens (five, seven or ten days) two months may be considered adequate.

3.4 Final Remarks

This chapter has provided an overview of the different stages of progressing a new compound from early screening, through *in vivo* testing and potential translation, to clinical development, and the value of new PK–PD tools for supporting this process.

The case of miltefosine shows how PK–PD modelling and simulation can provide new insights into how to optimize currently available drugs, such as the use of allometric dosing of miltefosine to overcome the lower exposure in children and the practical example of the clinically recommended minimal duration of contraception for female patients of child-bearing age. The development of a gastro-resistant formulation may be explored for future optimization as a way to overcome the gastrointestinal effects of miltefosine. The new capsule may not only improve the tolerability of the treatment, but also potentially allow for improved regimens with loading dosing and shorter treatment duration.

These continuous improvements are necessary in the context of this single oral treatment that is available for VL, given the prospect that miltefosine may be the only option that can be considered in the near future for combination treatments with oral new chemical entities.

Acknowledgements

We are grateful to Charles Mowbray and Jean Robert Ioset (DND*i*, Geneva) for thoroughly reviewing the draft version of this chapter. We are also grateful to Louise Burrows for proof-reading the document.

References

1. P. Mukherjee, S. B. Majee, S. Ghosh and B. Hazra, *Int. J. Antimicrob. Agents*, 2009, **34**, 596.
2. J. L. Siqueira-Neto, S. Moon, J. Jang, G. Yang, C. Lee, H. K. Moon, E. Chatelain, A. Genovesio, J. Cechetto and L. H. Freitas-Junior, *PLoS Neglected Trop. Dis.*, 2012, **6**, e1671.
3. M. De Rycker, I. Hallyburton, J. Thomas, L. Campbell, S. Wyllie, D. Joshi, S. Cameron, I. H. Gilbert, P. G. Wyatt, J. A. Frearson, A. H. Fairlamb and D. W. Gray, *Antimicrob. Agents Chemother.*, 2013, **57**, 2913.
4. V. Yardley, F. Gamarro and S. L. Croft, *Antimicrob. Agents Chemother.*, 2010, **54**, 5356.
5. S. Shafi, F. Afrin, M. Islamuddin, G. Chouhan, I. Ali, F. Naaz, K. Sharma and M. S. Zaman, *Front. Microbiol.*, 2016, **7**, 1379.
6. S. Gupta, V. Yardley, P. Vishwakarma, R. Shivahare, B. Sharma, D. Launay, D. Martin and S. K. Puri, *J. Antimicrob. Chemother.*, 2015, **70**, 518.
7. G. De Muylder, K. K. H. Ang, S. Chen, M. R. Arkin, J. C. Engel and J. H. McKerrow, *PLoS Neglected Trop. Dis.*, 2011, **5**, e1253.
8. A. S. Nagle, S. Khare, A. B. Kumar, F. Supek, A. Buchynskyy, C. J. N. Mathison, N. K. Chennamaneni, N. Pendem, F. S. Buckner, M. H. Gelb and V. Molteni, *Chem. Rev.*, 2014, **114**, 11305.
9. A. Nühs, M. De Rycker, S. Manthri, E. Comer, C. A. Scherer, S. L. Schreiber, J. R. Ioset and D. W. Gray, *PLoS Neglected Trop. Dis.*, 2015, **9**, e0004094.
10. J. C. Antoine, E. Prina, T. Lang and N. Courret, *Trends Microbiol.*, 1998, **6**, 392.
11. K. Seifert, P. Escobar and S. L. Croft, *J. Antimicrob. Chemother.*, 2010, **65**, 508.
12. P. M. Kaye, M. Svensson, M. Ato, A. Maroof, R. Polley, S. Stager, S. Zubairi and C. R. Engwerda, *Immunol. Rev.*, 2004, **201**, 239.
13. C. R. Engwerda, M. Ato and P. M. Kaye, *Trends Parasitol.*, 2004, **20**, 524.
14. J. M. Blackwell, M. Fakiola, M. E. Ibrahim, S. E. Jamieson, S. B. Jeronimo, E. N. Miller, A. Mishra, H. S. Mohamed, C. S. Peacock, M. Raju, S. Sundar and M. E. Wilson, *Parasite Immunol.*, 2009, **31**, 254.
15. S. Patterson, S. Wyllie, L. Stojanovski, M. R. Perry, F. R. C. Simeons, S. Norval, M. Osuna-Cabello, M. De Rycker, K. D. Read and A. H. Fairlamb, *Antimicrob. Agents Chemother.*, 2013, **57**, 4699.
16. C. Loeuillet, A.-L. Bañuls and M. Hide, *Parasites Vectors*, 2016, **9**, 144.
17. M. Hommel, C. L. Jaffe, B. Travi and G. Milon, *Ann. Trop. Med. Parasitol.*, 1995, **89**(Suppl. 1), 55.
18. P. C. Melby, B. Chandrasekar, W. Zhao and J. E. Coe, *J. Immunol.*, 2001, **166**, 1912.
19. A. Nieto, G. Domínguez-Bernal, J. A. Orden, R. De La Fuente, N. Madrid-Elena and J. Carrión, *Vet. Res.*, 2011, **42**, 39.
20. S. Croft and V. Yadley, in *Handbook of Animal Models of Infection*, ed. O. Zak and M. A. Sande, Academic Press, 1999, pp. 783–787.

21. R. Nau, F. Sörgel and H. Eiffert, *Clin. Microbiol. Rev.*, 2010, **23**, 858.
22. T. Velkov, P. J. Bergen, J. Lora-Tamayo, C. B. Landersdorfer and J. Li, *Curr. Opin. Microbiol.*, 2013, **16**, 573.
23. G. L. Drusano, *Clin. Infect. Dis.*, 2007, **45**(Suppl. 1), S89.
24. E. Asin-Prieto, A. Rodriguez-Gascona and A. Isla, *J. Infect. Chemother.*, 2015, **21**, 319.
25. S. L. Croft, R. A. Neal, W. Pendergast and J. H. Chan, *Biochem. Pharmacol.*, 1987, **36**, 2633.
26. J. Verweij, A. Planting, M. van der Burg and G. Stoter, *J. Cancer Res. Clin. Oncol.*, 1992, **118**, 606.
27. J. Verweij, D. Gandia, A. S. Planting, G. Stoter and J. P. Armand, *Eur. J. Cancer*, 1993, **29A**, 778.
28. J. Verweij, K. Krzemieniecki, T. Kok, A. Poveda, C. van Pottelsberghe, M. van Glabbeke and H. Mouridsen, *Eur. J. Cancer*, 1993, **29A**, 208.
29. A. S. Planting, G. Stoter and J. Verweij, *Eur. J. Cancer*, 1993, **29A**, 518.
30. S. L. Croft and J. Engel, *Trans. R. Soc. Trop. Med. Hyg.*, 2006, **100**(Suppl. 1), S4.
31. A. Kuhlencord, T. Maniera, H. Eibl and C. Unger, *Antimicrob. Agents Chemother.*, 1992, **36**, 1630.
32. S. Sundar, F. Rosenkaimer, M. K. Makharia, A. K. Goyal, A. K. Mandal, A. Voss, P. Hilgard and H. W. Murray, *Lancet*, 1998, **352**, 1821.
33. T. K. Jha, S. Sundar, C. P. Thakur, P. Bachmann, J. Karbwang, C. Fischer, A. Voss and J. Berman, *N. Engl. J. Med.*, 1999, **341**, 1795.
34. S. Sundar, L. B. Gupta, M. K. Makharia, M. K. Singh, A. Voss, F. Rosenkaimer, J. Engel and H. W. Murray, *Ann. Trop. Med. Parasitol.*, 1999, **93**, 589.
35. S. Sundar, T. K. Jha, C. P. Thakur, J. Engel, H. Sindermann, C. Fischer, K. Junge, A. Bryceson and J. Berman, *N. Engl. J. Med.*, 2002, **347**, 1739.
36. S. Sundar, A. Makharia, D. K. More, G. Agrawal, A. Voss, C. Fischer, P. Bachmann and H. W. Murray, *Clin. Infect. Dis.*, 2000, **31**, 1110.
37. WHO, Regional Strategic Framework for Elimination of Kala-Azar From the South-East Asia Region (2005–2015). Regional Office for South-East Asia SEA-VBC-85 (Rev-1), New Delhi. World Health Organ, 2005.
38. S. K. Bhattacharya, P. K. Sinha, S. Sundar, C. P. Thakur, T. K. Jha, K. Pandey, V. R. Das, N. Kumar, C. Lal, N. Verma, V. P. Singh, A. Ranjan, R. B. Verma, G. Anders, H. Sindermann and N. K. Ganguly, *J. Infect. Dis.*, 2007, **196**, 591.
39. S. Sundar, A. Singh, M. Rai, V. K. Prajapati, A. K. Singh, B. Ostyn, M. Boelaert, J. C. Dujardin and J. Chakravarty, *Clin. Infect. Dis.*, 2012, **55**, 543.
40. S. Sundar and H. W. Murray, *Bull. W. H. O.*, 2005, **83**, 394.

41. S. Rijal, B. Ostyn, S. Uranw, K. Rai, N. R. Bhattarai, T. P. C. Dorlo, J. H. Beijnen, M. Vanaerschot, S. Decuypere, S. S. Dhakal, M. L. Das, P. Karki, R. Singh, M. Boelaert and J. C. Dujardin, *Clin. Infect. Dis.*, 2013, **56**, 1530.
42. B. Ostyn, E. Hasker, T. P. C. Dorlo, S. Rijal, S. Sundar, J. C. Dujardin and M. Boelaert, *PLoS One*, 2014, **9**, e100220.
43. M. Wasunna, S. Njenga, M. Balasegaram, N. Alexander, R. Omollo, T. Edwards, T. P. C. Dorlo, B. Musa, M. Hassan Sharaf Ali, M. Yasein Elamin, G. Kirigi, R. Juma, A. E. Kip, G. J. Schoone, A. Hailu, J. Olobo, S. Ellis, R. Kimutai, S. Wells, E. Awad Gasim Khalil, N. Strub Wourgaft, F. Alves and A. Musa, *PLoS Neglected Trop. Dis.*, 2016, **10**, e04880.
44. T. Gelanew, K. Kuhls, Z. Hurissa, T. Weldegebreal, W. Hailu, A. Kassahun, T. Abebe, A. Hailu and G. Schonian, *PLoS Neglected Trop. Dis.*, 2010, **4**, e889.
45. A. Mondelaers, M. P. Sanchez-ca, S. Hendrickx, E. Eberhardt, R. Garcia-Hernandez, L. Lachaud, J. Cotton, M. Sanders, B. Cuypers, H. Imamura, J. Dujardin, P. Delputte, P. Cos, G. Caljon, F. Gamarro, S. Castanys and L. Maes, *PLoS One*, 2016, **11**, e0154101.
46. V. K. Prajapati, S. Sharma, M. Rai, B. Ostyn, P. Salotra, M. Vanaerschot, J. C. Dujardin and S. Sundar, *Am. J. Trop. Med. Hyg.*, 2013, **89**, 750.
47. T. P. C. Dorlo, P. P. A. M. Van Thiel, A. D. R. Huitema, R. J. Keizer, H. J. C. De Vries, J. H. Beijnen and P. J. De Vries, *Antimicrob. Agents Chemother.*, 2008, **52**, 2855.
48. S. Sundar, T. K. Jha, H. Sindermann, K. Junge, P. Bachmann and J. Berman, *Pediatr. Infect. Dis. J.*, 2003, **22**, 434.
49. S. Sundar, A. Makharia, D. K. More, G. Agrawal, A. Voss, C. Fischer, P. Bachmann and H. W. Murray, *Clin. Infect. Dis.*, 2000, **31**, 1110.
50. T. P. C. Dorlo, S. Rijal, B. Ostyn, P. J. de Vries, R. Singh, N. Bhattarai, S. Uranw, J.-C. Dujardin, M. Boelaert, J. H. Beijnen and A. D. R. Huitema, *J. Infect. Dis.*, 2014, **210**, 146.
51. M. Wasunna, S. Njenga, M. Balasegaram, N. Alexander, R. Omollo, T. Edwards, T. P. C. Dorlo, B. Musa, M. H. S. Ali, M. Y. Elamin, G. Kirigi, R. Juma, A. E. Kip, G. J. Schoone, A. Hailu, J. Olobo, S. Ellis, R. Kimutai, S. Wells, E. A. G. Khalil, N. Strub Wourgaft, F. Alves and A. Musa, *PLoS Neglected Trop. Dis.*, 2016, **10**, e0004880.
52. T. P. C. Dorlo, A. D. R. Huitema, J. H. Beijnen and P. J. De Vries, *Antimicrob. Agents Chemother.*, 2012, **56**, 3864.
53. T. P. C. Dorlo, A. E. Kip, F. Alves, J. Alvar, A. Musa, E. Khalil, M. Wasunna and M. O. Karlsson, 2015, p. 24, Abstr 3567 [www.page-meeting.org/?abstract=3567].
54. B. Ostyn, E. Hasker, T. P. C. Dorlo, S. Rijal, S. Sundar, J.-C. Dujardin and M. Boelaert, *PLoS One*, 2014, **9**, e100220.
55. B. J. Anderson and N. H. G. Holford, *Drug Metab. Pharmacokinet.*, 2009, **24**, 25.

56. M. O. Harhay, P. L. Olliaro, M. Vaillant, F. Chappuis, M. A. Lima, K. Ritmeijer, C. H. Costa, D. L. Costa, S. Rijal, S. Sundar and M. Balasegaram, *Am. J. Trop. Med. Hyg.*, 2011, **84**, 543.
57. H. Sindermann and J. Engel, *Trans. R. Soc. Trop. Med. Hyg.*, 2006, **100**(Suppl. 1), S17.
58. T. P. C. Dorlo, M. Balasegaram, M. A. Lima, P. J. de Vries, J. H. Beijnen and A. D. R. Huitema, *J. Antimicrob. Chemother.*, 2012, **67**, 1996.

II. Methodologies and Medicinal Chemistry Strategies to Discover and Develop New Treatments

CHAPTER 4

Drug Assay Methodology in Leishmaniasis: From the Microplate to Image Analysis

VANESSA YARDLEY* AND MARKELLA KONIORDOU

Faculty of Infectious and Tropical Diseases, London School of Hygiene and Tropical Medicine, Keppel Street, London WC1E 7HT, UK
*Email: vanessa.yardley@lshtm.ac.uk

4.1 Introduction

Currently used anti-leishmanial drugs present a range of challenges to achieving satisfactory therapeutic outcomes, from being ineffective and allowing for relapses, to being toxic, expensive and inaccessible to those most in need. The strategies used so far for the development of currently used anti-leishmanials have also highlighted the need to increase drug discovery efforts and to adopt different strategies for drug discovery. With the exception of pentavalent antimonials, the three currently used anti-leishmanial drugs, Miltefosine, Paramomycin and Amphotericin B, have been outcomes of repositioning strategies rather than *de novo* discovery.[1] However, whilst repositioning has proved to be hugely time- and cost-effective, particularly because the safety profiles of approved drugs have been already established, it has also been deemed suitable mostly as a short-term strategy, as opposed to novel "hit and/or lead" drug discovery whose application can stretch further into the future.[2] Anti-leishmanial resistance, non-specificity and co-infections pose additional challenges that require new drug-discovery

Drug Discovery Series No. 60
Drug Discovery for Leishmaniasis
Edited by Luis Rivas and Carmen Gil

Published by the Royal Society of Chemistry, www.rsc.org

approaches and repositioning not only relies on the small number of approved drugs available, but can also can limit the potential for further understanding of new mechanisms of drug action.[2] Similarly, the longevity of impact of promising new therapeutic treatments under current clinical investigation (among which are re-formulations and combination therapies of existing anti-leishmanials), remains uncertain; again emphasizing the importance and urgency to search for novel drugs.

In order to respond to the growing need for new anti-leishmanial drugs, the field of anti-leishmanial drug discovery has made significant steps in recent years in two major ways. Firstly, by streamlining the drug discovery decision-making process through building target-profiles of drug candidates with optimal characteristics and setting clearly defined criteria for "hit" and "lead" compounds that are specific to *Leishmania*.[3–7] Secondly, by incorporating technological advancements in the methodologies used during pre-clinical stages of anti-leishmanial drug development. Amongst those methodologies are drug assays that provide invaluable data on the biological activity of potential new drug candidates and inform decision-making, in a progressively refined step-by-step process (Figure 4.1).

Drug assays play a central role in "hit" identification, the initial identification of a compound active against a receptor, enzyme or the whole parasite

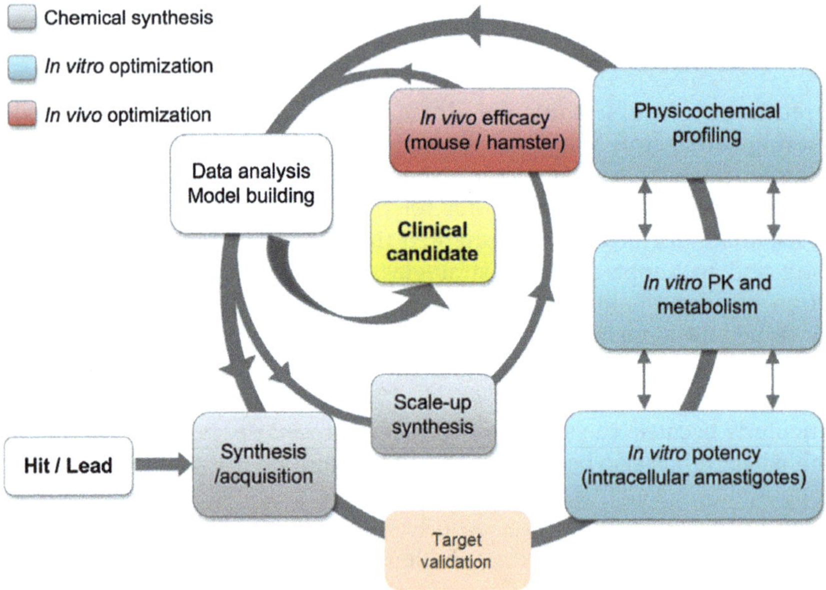

Figure 4.1 Example of a screening cascade.
Reprinted from *International Journal for Parasitology: Drugs and Drug Resistance*, Volume 2, L. H. Freitas-Junior, E. Chatelain, H. A. Kim and J. L. Siqueira-Neto, Visceral leishmaniasis treatment: What do we have, what do we need and how to deliver it?, 11–19, Copyright 2012, with permission from Elsevier.

and in "hit to lead" progression and "lead to drug candidate" optimization and selection, when further collection of data on efficacy, cytotoxicity and safety of novel compounds are required.[4] Specifically, the potency of a number of compounds is first tested in low-, medium- or high-throughput screening (HTS) assays against a non-animal, "*in vitro*" model of the infection. Once the activity of a compound is confirmed *in vitro*, it enters the next phase of evaluation *in vivo*, in an infected rodent model, for collection of efficacy data. Cytotoxicity assays for parasite-specific activity and mammalian cell safety are also performed in parallel. For the transition from hit to lead and lead to candidate, compounds undergo optimisation and chemical modification, entering successive rounds of *in vitro* and *in vivo* drug assays and data collection.

It is important to note that during hit discovery, drug assays inform both target-based and target-free approaches, yet the type and scale of assay, and the information each provides to the drug discovery process can be different and dependent on the approach.[8] A target-based approach involves the identification of a target small-molecule or a target pathway that is vital to the parasite, followed by development of compounds against such targets, whereas a target-free or phenotypic approach focuses on compounds that simply eliminate the parasite, without affecting the host. As such, target-based approaches may involve additional biochemical and enzymatic assays, whereas phenotypic approaches that may only assess the viability of the parasite in high-content screens (HCS) in a binary way (alive or dead), do not provide an understanding of the mechanism of action.[8] Despite the unprecedented access to genomic information and to dynamic computation and modelling that can uncover targets, persistent gaps in understanding of *Leishmania* biology still limit the number of possible targets.[9] Moreover, from identifying a target to validating it as a hit of anti-leishmanial activity involves sophisticated and expensive chemical modifications to enhance the putative agent's characteristics. It is important that the candidate compound has good permeability to cross membrane barriers and reach parasites inside the phagolysosomes of their host cells and that it is stable enough to withstand the low-pH environment.[8] Therefore—and in line with discovery of drugs against other infectious diseases[3]—most efforts in anti-leishmanial drug discovery have been based on phenotypic screen approaches.

This chapter will give an overview of phenotypic assay methodologies used for drug evaluation in *in vitro*, *ex vivo* and *in vivo* models of leishmaniasis. We will consider assay selection and detection mechanisms and touch upon how newer technologies are beginning to be incorporated into these protocols. Finally, we will conclude by summarizing constraints for conducting assays and the ideal assay requirements for optimal anti-leishmanial drug discovery understood to date. This is not a review of individual published studies as these are both numerous and readily available, but we will attempt to highlight the key considerations when embarking on a drug discovery programme for a new anti-leishmanial drug.

4.2 Drug Assay Methods

In vitro assays are able to measure the potency of a compound directly on the parasite [50% inhibitory concentration (IC_{50}) value] and *in vivo* experiments measure this by evaluating the parasite burden. Concurrent cytotoxicity assays, which will only briefly be mentioned in this chapter (Section 4.4.1), allow some selection to be made between the activity against the parasite and the toxicity to mammalian cells. This selectivity index (SI) can be used as a guide when choosing candidates for further evaluation.

New molecular, computational and image-based techniques have had extended applications in anti-leishmanial drug discovery, as will be seen in this section. However, to achieve a robust *in vitro* screening assay that overcomes the high attrition rates faced in drug discovery, several methodological features are optimally required. Such features include high throughput, high sensitivity and predictive ability of drug activity against the parasite, specificity to a clinically relevant form of the disease and consistency with older-used methods. A number of phenotypic, high-throughput screens (HTS) have been developed in order to assess large chemical libraries (see Chapter 5). These screens are formatted to a microtiter plate (96- or 384-well) to take automated systematic measurements of more than the single parameter of parasite viability, *i.e.* in multi-parametric high-content or image-based screens (HCS) that use robotic input. Anti-leishmanial drug screening originated from low-throughput settings and manual microscopic evaluation, which as a method remains especially useful for cross-validation of newer methods and strain sensitivity information. This method is also used in secondary screens, for confirmation of "hits" selected *via* the HTS platforms, forming part of the screening cascade[10] (Figure 4.1). Importantly, the selection of assay is dependent upon the availability of the model of infection.

This section will cover the different *in vitro*, *ex vivo* and *in vivo* models of infection for use in drug screening assays. It will then focus on the methods used for detecting viability or other parameters of anti-leishmanial drug activity, breaking down the technical basis of detection mechanisms and discussing the advantages and limitations of conventional and newer methods. Finally, it will look at innovative technological developments across the field of infectious disease research that may contribute to advancing anti-leishmanial drug assay methodology too, in the near future.

4.3 Models of Infection

4.3.1 *In Vitro*

In vitro phenotypic screening of anti-leishmanial drug activity has used axenic promastigotes, axenic amastigotes and intracellular amastigotes. The method selected to screen compounds often depended upon ease of use rather than relevance to a model of infection, resulting in many published studies using *Leishmania* promastigotes. There have been attempts to identify potent chemotypes through large-scale screening using *Leishmania* promastigotes and axenic amastigotes.[11–13] Unfortunately, however, these have not yet yielded a *bona fide* drug candidate. One of the most crucial considerations about promastigote and axenic amastigote screens is that translation of results from such screens to secondary screens using intracellular *Leishmania* infection screens (or later *in vivo* models of infection) are very poor.[14] Drug screening against axenic amastigotes, has been shown to have as low as a 4% confirmation rate in intracellular *Leishmania* screens, when, in contrast, of the compounds active in intracellular *Leishmania* screens, 50% were also active in parasite screens.[15] In order to test compounds in predictive and clinically relevant models of infection, it is of increased preference to use intracellular infections. Criteria for "hit to lead" progression for visceral leishmaniasis (VL) in fact require data on activity to be described in *in vitro* intracellular assays. In addition to observed differences against the stages of parasite used, drug susceptibilities can also significantly differ depending on the sub-species of *Leishmania* used.[16] Consequently, hit compounds identified in a screen should ideally be tested against other strains and/or species as part of a lead optimization plan. For example, if a primary screen uses the reference strain *Leishmania donovani HU3*, other *L. donovani* strains from different geographical areas, with different drug sensitivities or a recent clinical isolate could be used to evaluate the compound further.[5] In the case of cutaneous leishmaniasis (CL) species, activity against New World and Old World species should be assessed, as species differences in activity are even more pronounced for dermotropic *Leishmania.*[17,18] Finally, different host cells infected *in vitro* with the same pool of parasite suspension have been observed to respond differently when exposed to the same drugs.[19] Although reasons behind this are yet to be elucidated, discrete differences in the ways different host cells sustain or harbour the infection, differentiate into terminal macrophages and/or metabolise the drugs could be at play. Briefly, the appropriate lifecycle stage (amastigote), *Leishmania* species (VL or CL) and host cell type need to be considered carefully for their advantages and disadvantages when setting up a screening protocol.

For infections of a host cell macrophage, promastigotes and amastigotes from a *Leishmania*-infected mammalian host (rodent) can been used. Late metacyclic promastigotes cultured to a stationary phase have been identified as most suitable for mimicking the natural infection as caused by a sand fly bite and of highest infectivity.[20] In order to maintain virulence of the parasite (*i.e.* infection ability) the maintenance of promastigote cultures optimally involves "passaging" through a rodent host and harvesting amastigotes, which are then allowed to transform into stationary promastigotes in culture—primary promastigotes. Ideally, freshly harvested amastigotes are used to infect the host cells. This ensures a degree of predictability over intracellular numbers after 24 h of infection, and removes the confounding issue of extracellular promastigotes. Using amastigotes, however, is not always feasible, so the next best thing is to use primary promastigotes. It is important to keep the promastigote passage number low, under P8 for example, before discarding them and using a primary culture. It has been demonstrated that continual serial promastigote passage affects the virulence and probably influences other factors.[21] For CL species, amastigotes are harvested from infected skin (cutaneous lesion on ear, footpad or the base of tail), whereas for VL species, given the pathology of splenomegaly, amastigotes are usually sourced from the spleens of the infected animal.

Infection protocols follow incubation of the parasites with host-cell macrophages for a period of time that allows internalization of parasites as amastigotes within the macrophages, before exposure to compounds. This is done in order to achieve a level of intracellular infection that is neither so high as to cause macrophages to erupt and release more infective amastigotes (hence causing re-infection), nor so low that it is undetectable. Like most assays, *in vitro* infection assays require optimization steps in order to achieve a balance between infection ratios of parasite number to host cell number and the length of infection, depending on the outcome detected by the drug assay methodology used. Before optimising the infection ratios, ideally per species and per assay, it is also crucial to optimize host cell density per vessel selected. These steps can ensure the establishment of a standardised protocol that reliably detects the changes to the infection induced by drug exposure.

In terms of selecting which type of host cell macrophage to use, both monocytic cancer cell lines and primary macrophages from mammalian hosts have been used. Cell lines include the human leukaemia THP-1 cells induced to macrophages, whereas primary macrophages include mouse peritoneal exudate cells, murine bone marrow macrophages or human peripheral blood monocytes—induced to macrophages. The advantage of primary macrophages is that they are in their terminally differentiated, non-divisive form and do not need the addition of chemicals to prompt differentiation (*e.g.* PMA, phorbol 12-myristate 13-acetate, for THP-1 cells). The way compounds are used to differentiate monocytes to macrophages has been shown to interfere with the level of differentiation of cells and, theoretically by extension, the action of a compound under screening.[19] When it comes to cytotoxicity assays, cell types used usually involve cultures of cancer cell lines.[22]

Ideally, one would use primary macrophages and amastigotes harvested from a mammalian infected host in *in vitro* intracellular assays for *Leishmania* drug screening. This can be a challenge when scaling up a drug assay to a HTS format, as it would require high numbers of cells and parasites. Since methodological concessions of this kind might be necessary for HTS, they can be carefully accommodated in primary rounds of screening, leaving the more sophisticated models for secondary screens. Overall, advance optimisation of cell density and infection level per host cell and species, at the selected time points (24 h, 72 h or other) of drug exposure is vital for establishing a reproducible assay. What is also essential is to validate each assay against previous known methods and to include drug controls with well-known potency against *Leishmania* (Amphotericin B, Miltefosine), alongside the new compounds being screened in each assay.

4.3.2 *Ex Vivo*

Ex vivo models have a significant advantage over *in vitro* systems because they inherently reflect the true pathophysiology of the *Leishmania* infection. For CL species, explants are taken from the lymph nodes of infected rodents, whilst for VL species, they are extracted from the spleens.[8] In practice, less than a handful of studies have employed *Leishmania ex vivo* models of infection for drug screens,[23,24] mainly because involving animals is time-consuming and has high costs. The numbers of explants per animal can also vary, depending on the infection profile within the animal. Yet because a high yield of material extracted from a single animal can be used (*e.g.* one spleen yielding four 96 well plates[23]), *ex vivo* models can be particularly valuable for secondary screens.[8]

The chief advantage of explants is that they contain the entire set of cells, particularly different types of immune cells that are involved in the physiological host–parasite setting. The fundamental role of immune cell dynamics, beyond just the macrophage, is particularly prominent in recent investigations of the leishmanial infection.[25,26] This potential of higher clinical relevance and representative predictability of drug responses of *ex vivo* models, is also present *in vivo* models (Section 4.3.3) and has prompted several attempts to develop new *in vitro* models, further described in Section 4.5.

4.3.3 *In Vivo*

In vivo models of *Leishmania* infection have remained essentially unchanged for many years.[27,28] Permissive rodents are infected with the parasite and drug efficacy is measured against untreated controls. What has changed is the availability of some tools, which can interrogate the *in vivo* models in more detail. As well as yielding more information these advances also improves the "3R"s (replacement, reduction, refinement) principles[29] by maximising the output from a single experiment using fewer animals. For the purposes of compound evaluation, mouse models are typically used in the first instance, with the hamster being used at a later stage as a model for chronic, visceral infection.

4.3.3.1 Cutaneous Leishmaniasis

A number of permutations of mouse and *Leishmania* species have been studied, but not all pairings are suitable for drug evaluation purposes. For CL the development of lesions on the shaved rump of a mouse after infection should be a pre-requisite for a model. This will allow the lesion to be easily and closely monitored during the treatment regimen whilst causing minimal discomfort to the mouse and restricting access to the lesion of the host animal—important if dosing with a topical formulation. The inoculum size and time to lesion formation after subcutaneous injection should be standardized before commencing an experiment. For example, female BALB/c mice infected with 2×10^7 *Leishmania major* promastigotes will form nodules of 4–5 mm diameter after 7 days.[30] These parameters may not be identical in another location and should always be established and reproduced.

Low-passage promastigotes are used to infect mice. It is possible to use amastigotes but the numbers that can be harvested from donor mouse lesions are relatively low and do not appear to have an infectious advantage over low-passage promastigotes. For some *Leishmania* species there is an argument for selecting metacyclic, infective, promastigotes for injection but this is not an option for all CL species. For species and strains that are hard to infect the use of sand fly saliva, promastigote secretory gel (PSG), in the infective inoculum could be advantageous.[31] A better mimic of a sand fly bite would be to use PSG and a low-promastigote inoculum to generate lesions,[31] but this has not been validated in a drug evaluation context to date.

When evaluating anti-leishmanial activity of an experimental intervention options have been limited. Measurement of lesion size (and volume) is still used but is acknowledged to be highly subjective and the volume to be difficult to determine practically. Sampling or excising a lesion and extracting parasites into culture to gauge parasite load is also inexact.[32–34] Inflammation plays a major role in lesion development and can mask the anti-leishmanial potency of a drug.[28] A more rigorous way to evaluate drug efficacy would be to determine parasite load and lesion size progression. This would require tissue sampling and more animals having to be used, particularly if a number of time points are to be included. Quantitative PCR (qPCR) techniques, alongside lesion-size measurement have been shown to allow relatively quick, accurate and sensitive quantification of the parasite load in a CL lesion.[35] The development of recombinant bioluminescent *Leishmania* to follow *in vivo* infections can now be used to investigate drug efficiency by seeing, in real time, changes to parasite load and/or distribution in a single mouse.[36]

4.3.3.2 *Visceral Leishmaniasis*

The standard models for investigating anti-VL drug activity are *L. donovani* or *Leishmania infantum* in BALB/c mice. Amastigote infection via a lateral tail vein is the most dependable way to obtain a predictable level of infection. For example, inoculation of 2×10^7 *L. donovani* amastigotes into female BALB/c mice on day 0 will lead to a rising infection in the liver on days 7–18, the usual dosing period. Other mouse strains may also be used to mimic other host factors that may be present clinically, for example severe combined immunodeficiency (SCID) mice as a model for VL–HIV co-infection.[27] The *Leishmania*-hamster model is acknowledged to be more akin to human visceral leishmaniasis. A chronic liver and spleen infection becomes established, which will result in death after approximately 12 weeks if untreated. This model is usually employed when a candidate has been selected, as it is not widely available, more difficult to use and is more expensive than mouse models. More tools are being developed to interrogate this model further, such as intra-vital imaging, but the "gold standard" read out where this model is used remains microscopy.[37]

4.3.3.3 A Word on Pharmacokinetics and Pharmacodynamics

Information on the pharmacokinetics (PK) and the pharmacodynamics (PD) should be gathered alongside the anti-leishmanial data, where possible. Such information can be informative with regard to the target product profile of a potential drug and help to select the most "drug-like" candidate. Often the constraint for PK or PD experiments is cost, as these tests are carried out as part of a safety profile. It is possible to build in a "snapshot PK"[38] to an *in vivo Leishmania* model by including blood sampling at selected time points. If the compound detection methods are available, this can provide early information on drug distribution, half-life *etc.*, which can guide further medicinal chemistry around the compound.

4.4 Detection

Detection of anti-leishmanial activity in biological drug screens relies on methods to detect and quantify *Leishmania* parasite presence and/or the level of infection. All detection of drug activity operates under the paradigm of exposure and response, with the key detection parameter being that of parasite viability (*i.e.* presence). Assays measuring the responses of a single concentration per compound are sometimes used in primary screens, but preferably, more than three concentrations are examined per drug, so that an IC_{50} value can be established. In *in vitro* and *in vivo* assays, the IC_{50} or 50% effective dose (ED_{50}) value is derived from a dose–response curve plotted by concentrations of each compound against the marker detecting parasite viability. Cell viability is also the foundation for detection of drug toxicity in respective cytotoxicity assays. The dominant metric used to derive the IC_{50} value of a drug response is the percentage of inhibition of infection induced by exposure to the drug. This in turn can be derived from the percentage of infection, the amastigote burden (the number of amastigotes per host cell) or the increase or decrease of a detectable signal from a *Leishmania*-specific enzymatic, colorimetric, fluorimetric or luminometric reaction.

Current methods of detection either use techniques that directly quantify a signal of parasite presence, *i.e.* direct labelling of parasites, host cells *etc.*, or techniques that involve parasites genetically engineered to carry reporter genes that express a detectable signal. Quantification methods for drug activity have been developed against all infection models for *Leishmania* drug assays). Techniques used to detect *Leishmania in vitro*, *ex vivo* and *in vivo* have been based on cell-culture and imaging (from simple benchtop microscopy to advanced, confocal microscopy or other visualization tools), on immunology [spectroscopy, flow cytometry, enzyme-linked immunosorbent assay (ELISA), histology] or DNA and RNA polymerase chain reaction (PCR) and reverse transcription [(RT)–PCR]; or have involved a combination of these.

There are several decisions of assay design that are essential to consider in advance of establishing a detection system. Stringent quality control[39] can minimise variability between runs and can help prevent false-positive and false-negative results during drug screening. First, it is important to consider the use of controls at each step—negative controls (blank), positive signal controls and known drug controls. A second consideration regards the need for a strong signal that can be clearly distinguished from background noise. It is good practice to decide upon stringent thresholds of signal to noise ratio (for example, a tenfold difference between negative and positive). Thirdly, chemicals that can interfere with the infection or drug action and activity and multiple manipulation steps that can introduce bias or lead to loss of sample should be limited as much as possible. Fourthly, it remains essential that the detection tool selected is cross-validated against established methods and optimisation per model of infection takes place to standardise the method. Finally, not deviating from the established will help build a reproducible screening system. Ideally, a detection system will be cost-effective, timesaving and technically accessible. It will use a predictive infection model, have reproducible read out and limited toxic or fixative steps. The signal will be sensitive, quantifiable and distinguishable between parasite and host.

4.4.1 Conventional Methods

Considered to be the "gold standard" of techniques, staining fixed samples with the Giemsa dye is the most established method of assessing drug efficacy in intracellular drug assays.[40] The key metric of quantification is the percentage of infection, acquired through microscopic counting of infected versus uninfected host cells on fixed and Giemsa-stained 16-well slides (or alternative microplate surfaces). Another metric available from counting is the overall amastigote burden [*e.g.* the number of amastigotes for every 100 or 500 macrophages (The higher number is preferable, giving a larger sample size.) per well on a 16-well slide], which is a more robust predictor of drug activity than the percentage of infection and is therefore placed at the core of new, more advanced detection methodologies. The Giemsa dye is also used in *in vivo* models to quantify parasite loads on smear samples taken from infected (and drug-exposed) organs of animals. Altogether, despite the fact that microscopic counting with Giemsa staining is labour-intensive, time-consuming and prone to subjective interpretation, its distinct advantage is visual inspection. In addition to the interchange between percentage infection and amastigote burden, close examination of multiple parameters of the infected cells and tissue becomes possible, *i.e.* detecting changes of host cells or tissue under toxic drug concentrations or distinguishing between the levels of extracellular and intracellular parasites in the same sample. It is this capacity to distinguish between parasite and host characteristics that multi-parametric high-content screens aim to capture via imaging.

In order to quantify *Leishmania* in *in vitro* drug assays, a variety of methodological concepts have been explored. Some assays have been based on measuring the signal from a *Leishmania*-specific enzymatic reaction, such as on the ornithine decarboxylase enzyme activity (Callahan, 1997). Others have been based on light reactions (colorimetric, fluorimetric or luminometric reactions) including the use of resazurin dye (*e.g.* AlamarBlue® by Molecular Probes™) or 3-(4,5-dimethylthiazol-2-yl)-2,5-diphenyltetrazolium bromide/phenazine methosulfate (MTT/PMS), 2,3-bis-(2-methoxy-4-nitro-5-sulfophenyl)-2*H*-tetrazolium-5-carboxanilide (XTT) and 3-(4,5-dimethylthiazol-2-yl)-5-(3-carboxymethoxyphenyl)-2-(4-sulfophenyl)-2H-tetrazolium, inner salt (MTS) assays that provide a measure of whole-cell health viability (promastigotes or host cell toxicity).[41] Many assays have been based on the application of immunostaining or fluorescent labelling, and some have attempted to distinguish between the amastigote and host cells, by using differential labels. There is a breadth of labels available, such as fluorescent 4′,6-diamino-2-fenilindol (DAPI) or Hoechst 33342 that stain nuclei, carboxyfluorescein succinimidyl ester (CFSE) for cytoplasm or LysoTracker dyes that identify the parasitophorous vacuoles carrying *Leishmania* amastigotes.[42,43]

For the quantification of amastigotes *in vivo*, older cell-culture-based methods, such as excision of amastigotes from CL lesions by transformation into promastigotes on blood agar plates[33] have been gradually phased out. Similarly, the use of labelling techniques to detect *Leishmania* has been exploited in *in vivo* studies. Examples include the detection of amastigote antigens from extracted tissue by anti-*Leishmania* IgG labelled with peroxidase in an ELISA,[44] detection of kinetoplast DNA (kDNA) with the SYBR green dye on an RT–PCR protocol;[45] or employing combinations of immunological techniques, such as the development of a highly sensitive PCR–ELISA for *in vivo* amastigote quantification.[46]

Conventional methods have been applied usefully for the understanding of *Leishmania* infections and to anti-leishmanial drug discovery. Recently, high throughput, image-based processing, high-content screens against an intracellular *in vitro* model have been developed using labelling tools.[14,15] In one such screen, host and parasite parameters of infected cells on 384-well-plates were measured using a confocal microscope and data was captured using image-mining software. Using conventional methods for screening assays, however, has its limitations. Infected cells, in this image-based HTS needed to be fixed. Direct labelling most often requires fixing and permeabilising infected cells or tissue, whilst DNA and RNA techniques demand homogenisation. This not only means that intracellular structures may be disturbed but also that number of time points for drug activity detection is limited. Time-point restriction is also caused by terminal saturation in metabolic or immuno-binding reactions that are subject to the addition of substrates. Although many labels have been used in biological drug screens, not all are suitable for intracellular *Leishmania* studies. Labels can be cytotoxic, interfere with parasite replication and may be metabolised. All of these parameters can be controlled for, to some extent, but they can also lead to misinterpretation of results. Cost of reagents is not a trivial matter when using expensive stains in a high-throughput context and multiple steps always add time and reduce efficiency. The development of transgenic parasites that constitutively express a detectable signal through the course of an infection and drug treatment has provided detection alternatives of greater sensitivity, flexibility, speed and, ultimately, ease.

4.4.2 Transgenic *Leishmania* and Related Methods

Transgenic *Leishmania* parasites are genetically engineered to carry reporter genes that express proteins with a detectable catalytic, bioluminescent or fluorescent signal. The original series of reporter proteins includes catalytic proteins like chloramphenicol acetyl transferase (CAT) and β-galactosidase, bioluminescent luciferases that emit light upon addition of a substrate and a range of fluorescent proteins (*e.g.* green GFP, red DsRed, near-infrared IFP), that operate on different wavelengths of the visible spectrum and which emit a light signal following excitation from an external light source.[47]

There are numerous ways to detect the signal expressed from reporter genes, roughly split into imaging-based and non-imaging-based methods. A variety of non-imaging tools that use transgenic *Leishmania* in different models of infection have been applied to drug screening assays. For promastigote drug screening for example, the bioluminescence of promastigotes engineered to express an ATP-dependent β-galactosidase following drug exposure can be simply detected in a colorimetric assay using spectrophotometry.[48] Reporter-gene technology has been used in a multitude of studies using more relevant models of *Leishmania* infection *i.e. in vitro* intracellular, *ex vivo* and *in vivo* models.[47,49–56] Drug responses across these studies can be quantified using the same tools as one would with wild-type parasites, namely flow cytometry–often combined with markers for the host, qPCR against the known volume of tissue or combinations. The principle difference is that the readout is based upon the live parasite expressing the fluorescent or bioluminescent signal. Importantly, there have also been developments of *in vitro* high-throughput screens using such methods too.[50,57,58] Though they offer increased capacity for screening a larger number of compounds, they are still of limited flexibility when it comes to studying more interactions of parasites and host cells over time or in real-time, as is provided by an imaging HCS.[41]

The use of reporter gene technologies in combination with bio-imaging techniques has been revolutionary for the field of kinetoplastid disease, propelling the understanding of infection mechanisms and enabling more efficient compound screening. Imaging allows the detection of phenotypes [meeting the very definition of phenotypic screening (pheno-meaning, "appearing" in Greek)]. High-throughput, high-content screens for anti-leishmanials have been made possible through access to *in vitro* imaging tools. In one example host cells infected with transgenic *Leishmania* parasites that constitutively expressed the red fluorescent DsRed protein were multiply stained to generate a high-content image-based screen.[59]

The use of transgenic *Leishmania* finds its most crucial application in *in vivo* imaging models, where parasites can be visualized inside live animals in real-time. Animals are anesthetised, the bioluminescent substrate is injected, if appropriate, and the animals are placed into an imaging chamber system. The chamber system is linked to a charge-coupled device (CCD) camera able to sense the bioluminescent or fluorescent signal. Similarly to other forms of microscopic detection, the signal detected is representative of light intensity (photon number) and data retrieved are displayed on specialized computer software.[60] Many studies have focused on biophotonic whole-body imaging (BPI), enabled exclusively because of infection with *Leishmania* expressing a reporter gene. When it comes to deep tissue imaging, however, BPI remains limited. In contrast, intra-vital microscopy (IVVIS), which provides higher resolution, has been used to build stable models of infection and drug exposure in trypanosomatid disease.[61] New studies are exploring the use of intravital microscopy to gain a more complete understanding of anti-leishmanial activity. One example is through the development of a hamster model infected with luciferase-expressing *L. infantum chagasi* in which Miltefosine was accessed.[62] Another aimed at translating between *in vitro* and *in vivo* studies by deep-tissue imaging of fluorescence-expressing infected animals in conjunction with an image-based *in vitro* HCS/HTS.[41] Capturing ever higher resolution as required for individual cell visualization is the final aim and challenge of IV-VIS.[63]

There have been significant breakthroughs in applications of reporter gene technologies to anti-leishmanial drug screens despite the technical challenges of generating and maintaining stably expressing transgenic cultures and infected models. The design of the vector that carries the reporter gene needs careful advance planning. Beyond the selection of key reporter genes, the selection of the promoter sequence [episomal (kDNA) or nuclear 18S rRNA integration], flanking regions and antibiotic pressure genes also need to be considered. Following plasmid design, moreover, transformation and transfection protocols need to be optimized and the correct integration of reporter gene and reliable gene expression validated. Importantly, the features of the reporter-gene-expressing *Leishmania* line also need to correspond to the detection method that is eventually used. For use in drug assays, features that would ideally be incorporated in the design include: the expression of the reporter under the control of a nuclear promoter to give stable expression, the expression of a strong signal that can be detected throughout the parasite's lifecycle, in *in vitro* intracellular and *in vivo* models of infection, the ability to detect the reporter through both conventional and new tools (interchangeable), and, lastly, the absence of a need for antibiotic pressure, given that antibiotics could interfere with the new compounds under test.

The field is evolving quickly, developing new tools responsive to the needs of HTS and HCS and bio-imaging. Concerns about manipulation of the genome affecting parasite behaviour, causing loss of virulence and potential toxicity, or antibiotic pressure[49] are being addressed *e.g.* by the creation of new expression vectors and integration techniques, and considering the role of passage, both *in vitro* and *in vivo*, in the maintenance of parasite virulence. The variety of reporter proteins available also means that generating dual-reporter parasite lines that co-express more than one detectable signal is possible. This is particularly advantageous for measuring multiple variables or stages of the infection—co-labelling with fluorescence and bioluminescence will give the advantages of both systems of detection. The parameters of intra-vital imaging must always be standardized and validated, in each laboratory setting, in order to abrogate confounding factors such as background "noise", host auto-fluorescence, light penetration into target tissue, longevity post substrate (bioluminescence) or quenching time (fluorescence) of the signal, among others.

4.5 Technological Innovations and Developments

Newly investigated models of infection are key to the advancement of anti-leishmanial drug assay methodologies and come with exciting prospects. Attempts to develop *in vitro* models of *Leishmania* infection that can better simulate the true host–parasite interactions than current intracellular *in vitro* models look promising. These models also aim to limit the need for use of animals and set-up of *in vivo* assays, in conjunction with the 3R's principles.[29] The use of 2D and 3D *in vitro* cell culture to create a more complex *Leishmania* infection are currently being explored. In contrast to growing cells as a simple monolayer on the bottom surface of a vessel (16-well slide, 96-well plate *etc.*), cells here are cultured within a matrix with the intention to mimic tissue and provide more predictive information for subsequent *in vivo* tests.[64] At the same time, techniques to culture different cell types in this way are also being investigated. Co-cultures aim to mimic host tissue at the cellular level. For example co-culturing fibroblasts together with macrophages to mimic the skin environment or co-culturing a range of immune cells that are involved in the host's immune response against *Leishmania* could be used as an additional screening step for selected compounds.[65] The use of a 3D hepatic model[66] could provide useful information if VL-targeted compounds are being sought, as well as providing valuable PK and toxicity data.

Another refinement of *in vitro* models is the use of human stem cells differentiated to relevant host cells to better understand PD relationships upon drug exposure.[67] Such cells, infected with *Leishmania* could also be employed in the compound selection process. *In vitro* dynamic models have been used to examine the behaviour of immune cells[68] and of drugs in an infectious disease context.[69]

The use of humanized disease models is increasing.[70] Humanized murine models typically involve the re-population of organs or tissues with human cells in order to mimic human disease.[71] One such model has been used in CL to look at immune responses.[72] Therefore, as further humanized liver and spleen models to study disease dynamics and drug effects are being developed, this approach is also conceivable for VL.

Important new developments in the field also owe much to the rise of computational power and access to genomic and/or transcriptomic information-bioinformatics (see Chapter 7). Next-generation sequencing is being used for the development of screening.[73,74] In addition, a huge amount of previously tested compound data can now be used to build dynamic computational models that can predict efficacy of novel compounds in highly cost-effective virtual screens. Building such models relies on access to the data, which often remain locked under intellectual property ownership laws. However, open data access and sharing platforms are being established through public–private-partnerships or other forms of collaborations and the increased participation of disease-endemic countries.[4]

New tools for genetic manipulation can improve infection models in addition to finding drug targets (see Chapter 6). The clustered regularly interspaced short palindromic repeat (CRISPR)–CRISPR associated protein 9 (Cas9) editing system has been revolutionary for targeted mutagenesis studies and groups are inevitably exploring how to generate more stable transgenic *Leishmania* lines for use in infection models. Importantly, such tools can reach even wider and further into the field of leishmaniasis drug research. CRISPR–Cas9, for instance, is currently being recruited for target-based drug discovery approaches in the field of neglected-tropical disease research.[75] Another new means for identifying new drug targets against *Leishmania* specifically is the development of genomic screens such as Cros-Seq which exploit the particular genomic plasticity of *Leishmania* and the huge variation in gene copy number they display.[76] Resource availability is always a major consideration. The innovation and application of these tools will hopefully lead to more efficient and relevant anti-leishmanial screens with the added bonus of being cost effective. There is an observable shift to include more targeted strategies. A recent example is the use of new information on VL exosomes to identify a suitable drug target (casein kinase),[77] while a new class of anti-kinetoplastid compounds has been identified by multiple screening techniques.[9] Importantly, the unprecedented access to information about *Leishmania* parasites together with manipulation technologies are now making the identification of drug mechanism of action (MOA) by target-free approaches possible too. This is of considerable benefit, given the differing selectivity and sensitivity between *Leishmania* species, strains and clinical isolates.

4.6 Summary

Considering that sodium stibogluconate (SSG) has been in use since 1930, new drugs are still clearly needed. For anti-leishmanial drug discovery phenotypic screening assays remain at the core of the process, providing invaluable data about the potency of new compounds on the parasite. What is needed and what is achievable in each laboratory setting will influence the decisions over selecting a model and detection method, as a balance of cost, time and ease of access to reagents, animals and technology. The overarching principles for building a robust (reproducible and relevant) system are the same however, regardless of settings. It is important to try to use the most representative models of human infection and to standardize detection protocols, ensuring valid readouts that will accurately and reliably measure drug efficacy. Not all features can be accommodated due to setting-specific constraints, but if the limitations of a screen are clearly defined, results from the simplest primary screen can only add to the pool of information.

4.6.1 Optimism for the Future

New compounds, such as oxoboroles, aminopyrazoles and nitroimidazoles, are being evaluated (see Chapters 8–10). Drugs, such as fexinidazole, are further along the *Leishmania* drug development pipeline.[78,79] Besides those, building target product profiles for screening new compounds is proving to be hugely beneficial; helping to focus efforts to design assays that correspond to optimal therapeutic requirements at the clinic (dose, formulation, mode of administration, length of treatment *etc.*). The leads have been the result of collaborative networks of researchers, from Pharma, academia and governmental and non-governmental organisations and it has become apparent that this is an almost necessary approach for successful drug development in the neglected tropical diseases arena.

Finally, new techniques with more sophisticated tools are helping build powerful predictive systems. The range of new infection models, aligned with an emphasis on concurrent PK and PD data acquisition from the early stages of anti-leishmanial testing can increase predictability and relevance. The combination of genomic information, next-generation sequencing, new genetic manipulation tools and computer-based modelling capabilities[80] may bring about not only new forms of screening but, significantly, a change in the pattern of separating target-free and target-based approaches. Quick screening with high clinical relevance, *e.g.* clinical isolates and/or resistant strains, could be made possible.[74] It is no longer hard to envisage that such advances, already applied against other pathogens, will potentially translate into the field of anti-leishmanial drug discovery and lead to potent novel anti-leishmanial candidates.

Acknowledgements

We would like to kindly thank the authors and Elsevier for licensing the reuse of Figure 4.1.

References

1. M. Kaiser, P. Maser, L. P. Tadoori, J. R. Ioset and R. Brun, *PLoS One*, 2015, **10**, e0135556.
2. L. H. Freitas-Junior, E. Chatelain, H. A. Kim and J. L. Siqueira-Neto, *Int. J. Parasitol.: Drugs Drug Resist.*, 2012, **2**, 11.
3. K. Katsuno, J. N. Burrows, K. Duncan, R. Hooft van Huijsduijnen, T. Kaneko, K. Kita, C. E. Mowbray, D. Schmatz, P. Warner and B. T. Slingsby, *Nat. Rev. Drug Discovery*, 2015, **14**, 751.
4. S. Nwaka, B. Ramirez, R. Brun, L. Maes, F. Douglas and R. Ridley, *PLoS Neglected Trop. Dis.*, 2009, **3**, e440.
5. DNDi, Drugs for Neglected Diseases initiative: Target Product Profile for Visceral Leishmaniasis, 2016b. [29/11/2016]. Available from: http://www.dndi.org/diseases-projects/leishmaniasis/tpp-vl/.
6. DNDi, Drugs for Neglected Diseases initiative: Target Product Profile for Cutaneous Leishmaniasis, 2016a. [29/11/2016]. Available from: http://www.dndi.org/diseases-projects/leishmaniasis/tpp-cl/.
7. J. H. No, *Acta Trop.*, 2016, **155**, 113.
8. R. M. Reguera, E. Calvo-álvarez, R. Álvarez-Velilla and R. Balaña-Fouce, *Int. J. Parasitol.: Drugs Drug Resist.*, 2014, **4**, 355.
9. S. Khare, A. S. Nagle, A. Biggart, Y. H. Lai, F. Liang, L. C. Davis, S. W. Barnes, C. J. Mathison, E. Myburgh, M. Y. Gao, J. R. Gillespie, X. Liu, J. L. Tan, M. Stinson, I. C. Rivera, J. Ballard, V. Yeh, T. Groessl, G. Federe, H. X. Koh, J. D. Venable, B. Bursulaya, M. Shapiro, P. K. Mishra, G. Spraggon, A. Brock, J. C. Mottram, F. S. Buckner, S. P. Rao, B. G. Wen, J. R. Walker, T. Tuntland, V. Molteni, R. J. Glynne and F. Supek, *Nature*, 2016, **537**, 229.
10. A. Nuhs, M. De Rycker, S. Manthri, E. Comer, C. A. Scherer, S. L. Schreiber, J. R. Ioset and D. W. Gray, *PLoS Neglected Trop. Dis.*, 2015, **9**, e0004094.
11. E. R. Sharlow, D. Close, T. Shun, S. Leimgruber, R. Reed, G. Mustata, P. Wipf, J. Johnson, M. O'Neil, M. Grogl, A. J. Magill and J. S. Lazo, *PLoS Neglected Trop. Dis.*, 2009, **3**, e540.
12. A. S. Nagle, S. Khare, A. B. Kumar, F. Supek, A. Buchynskyy, C. J. Mathison, N. K. Chennamaneni, N. Pendem, F. S. Buckner, M. H. Gelb and V. Molteni, *Chem. Rev.*, 2014, **114**, 11305.
13. I. Peña, M. Pilar Manzano, J. Cantizani, A. Kessler, J. Alonso-Padilla, A. I. Bardera, E. Alvarez, G. Colmenarejo, I. Cotillo, I. Roquero, F. de Dios-Anton, V. Barroso, A. Rodriguez, D. W. Gray, M. Navarro, V. Kumar, A. Sherstnev, D. H. Drewry, J. R. Brown, J. M. Fiandor and J. Julio Martin, *Sci. Rep.*, 2015, **5**, 8771.
14. G. De Muylder, K. K. H. Ang, S. Chen, M. R. Arkin, J. C. Engel and J. H. McKerrow, *PLoS Neglected Trop. Dis.*, 2011, **5**, e1253.
15. J. L. Siqueira-Neto, S. Moon, J. Jang, G. Yang, C. Lee, H. K. Moon, E. Chatelain, A. Genovesio, J. Cechetto and L. H. Freitas, *PLoS Neglected Trop. Dis.*, 2012, **6**, e1671.

16. S. L. Croft, V. Yardley and H. Kendrick, *Trans. R. Soc. Trop. Med. Hyg.*, 2002, **96**, S127.
17. V. Yardley, S. L. Croft, S. D. Doncker, J.-C. Dujardin, S. Koirala, S. Rijal, C. Miranda, A. Llanos-Cuentas and F. Chappuis, *Am. J. Trop. Med. Hyg.*, 2005, **73**, 272.
18. J. Soto, B. A. Arana, J. Toledo, N. Rizzo, J. C. Vega, A. Diaz, M. Luz, P. Gutierrez, M. Arboleda, J. D. Berman, K. Junge, J. Engel and H. Sindermann, *Clin. Infect. Dis.*, 2004, **38**, 1266.
19. M. Koniordou, S. Patterson, S. Wyllie and K. Seifert, *Antimicrob. Agents Chemother.*, 2017, **61**, e01228.
20. D. L. Sacks and P. V. Perkins, *Science*, 1984, **223**, 1417.
21. L. Piel, P. Pescher, G. Bussotti, L. Ma, C. Bouchier and G. F. Spaeth, in *BSP Trypanosomiasis and Leishmaniasis Seminar*, The Biology Centre of the Czech Academy of Science, Ceske Buejovice, Czech Republic, 2016.
22. J. M. McKim Jr., *Comb. Chem. High Throughput Screening*, 2010, **13**, 188.
23. Y. Osorio, B. L. Travi, A. R. Renslo, A. G. Peniche and P. C. Melby, *PLoS Neglected Trop. Dis.*, 2011, **5**, e962.
24. A. G. Peniche, Y. Osorio, A. R. Renslo, D. E. Frantz, P. C. Melby and L. Travi, *Antimicrob. Agents Chemother.*, 2014, **58**, 78.
25. P. Kaye and P. Scott, *Nat. Rev. Microbiol.*, 2011, **9**, 604.
26. L. Beattie, A. Peltan, A. Maroof, A. Kirby, N. Brown, M. Coles, D. F. Smith and P. M. Kaye, *PLoS Pathog.*, 2010, **6**, e1000805.
27. S. L. Croft and V. Yardley, in *Handbook of Animal Models of Infection*, ed. O. Zak, Academic Press, New York, London, 1999, pp. 783–787.
28. V. Yardley and S. L. Croft, in *Handbook of Animal Models of Infection*, ed. O. Zak, Academic Press, London, 1999, pp. 775–781.
29. NC3Rs, National Centre for the Replacement Refinement & Reduction of Animals in Research, 2016. Available from: https://www.nc3rs.org.uk/.
30. K. Van Bocxlaer, V. Yardley, S. Murdan and S. L. Croft, *J. Pharm. Pharmacol.*, 2016, **68**, 862.
31. M. E. Rogers, T. Ilg, A. V. Nikolaev, M. A. Ferguson and P. A. Bates, *Nature*, 2004, **430**, 463.
32. F. Aguilar Torrentera, M. A. Lambot, J. D. Laman, M. Van Meurs, R. Kiss, J. C. Noel and Y. Carlier, *Am. J. Trop. Med. Hyg.*, 2002, **66**, 273.
33. J. O. Hill, R. J. North and F. M. Collins, *Infect. Immun.*, 1983, **39**, 1087.
34. R. G. Titus, M. Marchand, T. Boon and J. A. Louis, *Parasite Immunol.*, 1985, **7**, 545.
35. K. Van Bocxlaer, Cutaneous Leishmaniasis: Skin Barrier Properies and Drug Delivery Strategies. University College London, School of Pharmacy, 2015.
36. E. de La Llave, H. Lecoeur, A. Besse, G. Milon, E. Prina and T. Lang, *Cell Microbiol.*, 2011, **13**, 81.
37. A. Fortin, T. P. Dorlo, S. Hendrickx and L. Maes, *J. Antimicrob. Chemother.*, 2016, **71**, 1892.
38. B. Liu, J. Chang, W. P. Gordon, J. Isbell, Y. Zhou and T. Tuntland, *Drug Discovery Today*, 2008, **13**, 360.

39. Z. Zhang, N. Guan, T. Li, D. E. Mais and M. Wang, *Acta Pharm. Sin. B*, 2012, **2**, 429.
40. J. D. Berman, *J. Parasitol.*, 1984, **70**, 561.
41. C. L. Forestier, G. F. Spath, E. Prina and S. Dasari, *Microb. Pathog.*, 2015, **88**, 103.
42. J. L. Siqueira-Neto, O. R. Song, H. Oh, J. H. Sohn, G. Yang, J. Nam, J. Jang, J. Cechetto, C. B. Lee, S. Moon, A. Genovesio, E. Chatelain, T. Christophe and L. H. Freitas-Junior, *PLoS Neglected Trop. Dis.*, 2010, **4**, e675.
43. C. L. Forestier, G. F. Spath, E. Prina and S. Dasari, *Microb. Pathog.*, 2015, **88**, 103.
44. B. Ferrua, Y. Le Fichoux, I. Suffia, D. Rousseau, C. Roptin and J. Kubar, *J. Immunoassay Immunochem.*, 2001, **22**, 165.
45. L. Nicolas, E. Prina, T. Lang and G. Milon, *J. Clin. Microbiol.*, 2002, **40**, 1666.
46. T. Kobets, J. Badalova, I. Grekov, H. Havelkova, M. Svobodova and M. Lipoldova, *Nat. Protoc.*, 2010, **5**, 1074.
47. E. Calvo-Alvarez, R. Alvarez-Velilla, C. Fernandez-Prada, R. Balaña-Fouce and R. M. Reguera, *Drug Discovery Today*, 2015, **20**, 114.
48. A. Monte-Alegre, A. Ouaissi and D. Sereno, *Kinetoplastid Biol. Dis.*, 2006, **5**, 6.
49. A. Dube, R. Gupta and N. Singh, *Trends Parasitol.*, 2009, **25**, 432.
50. M. N. Rocha, C. M. Correa, M. N. Melo, S. M. Beverley, O. A. Martins, A. P. Madureira and R. P. Soares, *Diagn. Microbiol. Infect. Dis.*, 2013, **75**, 282.
51. E. Calvo-Alvarez, N. A. Guerrero, R. Alvarez-Velilla, C. F. Prada, J. M. Requena, C. Punzon, M. A. Llamas, F. J. Arevalo, L. Rivas, M. Fresno, Y. Perez-Pertejo, R. Balana-Fouce and R. M. Reguera, *PLoS Neglected Trop. Dis.*, 2012, **6**, e1927.
52. S. W. Kamau, F. Grimm and A. B. Hehl, *Antimicrob. Agents Chemother.*, 2001, **45**, 3654.
53. T. Lang, S. Goyard, M. Lebastard and G. Milon, *Cell Microbiol.*, 2005, **7**, 383.
54. G. Michel, B. Ferrua, T. Lang, M. P. Maddugoda, P. Munro, C. Pomares, E. Lemichez and P. Marty, *PLoS Neglected Trop. Dis.*, 2011, **5**, e1323.
55. P. Vacchina and M. A. Morales, *Antimicrob. Agents Chemother.*, 2014, **58**, 1825.
56. A. Bolhassani, T. Taheri, Y. Taslimi, S. Zamanilui, F. Zahedifard, N. Seyed, F. Torkashvand, B. Vaziri and S. Rafati, *Exp. Parasitol.*, 2011, **127**, 637.
57. R. Singh, M. Pittas, I. Heskia, F. Y. Xu, J. McKerrow, C. R. Caffrey and IEEE, in *2009 22nd IEEE International Symposium on Computer-Based Medical Systems*, IEEE, New York, 2009, pp. 198–205.
58. S. A. Pulido, D. L. Muñoz, A. M. Restrepo, C. V. Mesa, J. F. Alzate, I. D. Velez and S. M. Robledo, *Acta Trop.*, 2012, **122**, 36.
59. N. Aulner, A. Danckaert, E. Rouault-hardoin, J. Desrivot, O. Helynck and E. Prina, *PLoS Neglected Trop. Dis.*, 2013, **7**, e2154.
60. C. J. Thalhofer, J. W. Graff, L. Love-Homan, S. M. Hickerson, N. Craft, S. M. Beverley and M. E. Wilson, *J. Visualized Exp., JoVE*, 2010, 3.

61. H. Burrell-Saward, J. Rodgers, B. Bradley, S. L. Croft and T. H. Ward, *J. Antimicrob. Chemother.*, 2015, **70**, 510.
62. J. Q. Reimão, J. C. Oliveira, C. T. Trinconi, P. C. Cotrim, A. C. Coelho and S. R. B. Uliana, *PLoS Neglected Trop. Dis.*, 2015, **9**, e0003556.
63. S. I. Ellenbroek and J. van Rheenen, *Nat. Rev. Cancer*, 2014, **14**, 406.
64. R. Edmondson, J. J. Broglie, A. F. Adcock and L. Yang, *Assay Drug Dev. Technol.*, 2014, **12**, 207.
65. B. L. Duell, A. W. Cripps, M. A. Schembri and G. C. Ulett, *J. Biomed. Biotechnol.*, 2011, **2011**, 852419.
66. S. Chitrangi, P. Nair and A. Khanna, *Toxicol. In Vitro*, 2017, **38**, 8.
67. Y. Shi, H. Inoue, J. C. Wu and S. Yamanaka, *Nat. Rev. Drug Discovery*, 2017, **16**, 115.
68. L. Cucullo, N. Marchi, M. Hossain and D. Janigro, *J. Cereb. Blood Flow Metab.*, 2011, **31**, 767.
69. H. Box, J. Livermore, A. Johnson, L. McEntee, T. W. Felton, S. Whalley, J. Goodwin and W. W. Hope, *Antimicrob. Agents Chemother.*, 2015, **60**, 278.
70. N. C. Walsh, L. L. Kenney, S. Jangalwe, K. E. Aryee, D. L. Greiner, M. A. Brehm and L. D. Shultz, *Annu. Rev. Pathol.*, 2017, **12**, 187.
71. V. Soulard, H. Bosson-Vanga, A. Lorthiois, C. Roucher, J. F. Franetich, G. Zanghi, M. Bordessoulles, M. Tefit, M. Thellier, S. Morosan, G. Le Naour, F. Capron, H. Suemizu, G. Snounou, A. Moreno-Sabater and D. Mazier, *Nat. Commun.*, 2015, **6**, 7690.
72. A. K. Wege, C. Florian, W. Ernst, N. Zimara, U. Schleicher, F. Hanses, M. Schmid and U. Ritter, *PloS Neglected Trop. Dis.*, 2012, **6**, e1741.
73. A. C. Berger and S. Olsen, in *Genome-based Therapeutics: Targeted Drug Discovery and Development*, ed. Institute of Medicine, National Academies Press, Washington DC, USA, 2012, vol. 4.
74. P. Leprohon, C. Fernandez-Prada, E. Gazanion, R. Monte-Neto and M. Ouellette, *Int. J. Parasitol. Drugs Drug Resist.*, 2015, **5**, 26.
75. Z. Steinhart, Z. Pavlovic, M. Chandrashekhar, T. Hart, X. Wang, X. Zhang, M. Robitaille, K. R. Brown, S. Jaksani, R. Overmeer, S. F. Boj, J. Adams, J. Pan, H. Clevers, S. Sidhu, J. Moffat and S. Angers, *Nat. Med.*, 2017, **23**, 60.
76. M.-C. N. Laffitte, P. Leprohon, B. Papadopoulou and M. Ouellette, *F1000Research*, 2016, **5**, 2350.
77. E. Durieu, E. Prina, O. Leclercq, N. Oumata, N. Gaboriaud-Kolar, K. Vougogiannopoulou, N. Aulner, A. Defontaine, J. H. No, S. Ruchaud, A.-L. Skaltsounis, H. Galons, G. F. Späth, L. Meijer and N. Rachidi, *Antimicrob. Agents Chemother.*, 2016, **60**, 2822.
78. DNDi, Leishmaniasis Portfolio, 2017. Available from: http://www.dndi.org/diseases-projects/leishmaniasis/leish-portfolio/. [updated February 2017].
79. S. Sundar and J. Chakravarty, *Expert Opin. Invest. Drugs*, 2015, **24**, 43.
80. J. H. Sung, M. B. Esch and M. L. Shuler, *Expert Opin. Drug Metab. Toxicol.*, 2010, **6**, 1063.

CHAPTER 5

The Pursuit of Novel Anti-leishmanial Agents by High-throughput Screening (HTS) of Chemical Libraries

JULIO MARTIN,* JUAN CANTIZANI AND IMANOL PEÑA

GSK, Diseases of the Developing World, Tres Cantos, Spain
*Email: Julio.J.Martin@gsk.com

5.1 New Chemical Entities (NCEs) are an Unmet Need for Therapeutics of Kinetoplastid Diseases

Neglected diseases are broadly understood as those diseases affecting populations in mainly low-income countries. Despite the neglect suffered in terms of funding, research and policy, they are a leading cause of mortality, chronic disability and poverty. Amongst them, neglected tropical diseases (NTDs) are a group of infectious diseases categorized by the particular oblivion they have suffered in terms of investment in control measures, when compared with malaria and tuberculosis, "the big two" within the neglected diseases. Of the 850 new therapeutic products registered in 2000–2011, only 5 were indicated for NTDs, none of them being a new chemical entity (NCE).[1] However, the global disease burden caused by NTDs is high. According to World Health Organization (WHO), the number of disability-adjusted

Drug Discovery Series No. 60
Drug Discovery for Leishmaniasis
Edited by Luis Rivas and Carmen Gil

Published by the Royal Society of Chemistry, www.rsc.org

life-years (DALYs) caused by NTDs in 2004 was 18 325, just half of the figure for malaria or tuberculosis, and similar to the one for diabetes mellitus. The lack of investment in drug discovery for NTDs has been partially due to the absence of economic incentives. The issue of poverty is central to the nature of NTDs, since these diseases not only are due to poverty but also contribute to it through their socioeconomic impact.[2] The NTDs encompass a broad range of viral, bacterial and parasitic infections. There are 17 core NTDs as defined by the WHO.[3] The control of these diseases is overseen by the Control of Neglected Tropical Diseases program at WHO, which has drawn up a roadmap of their vision and goals to eliminate five diseases (leprosy, sleeping sickness, blinding trachoma, guinea worm disease and lymphatic filariasis) and to control another five (schistosomiasis, soil-transmitted helminthiases, visceral leishmaniasis, onchocerciasis and Chagas) by 2020.[2] All NTDs have been categorized as "tool ready", given that there are tools (*e.g.* drugs) that can be implemented now, yet also "tool deficient" because many of these tools (*e.g.* drugs and diagnostics) and implementation strategies are inadequate to achieve the desired goals.[3] New effective, safe, and affordable drugs, preferably oral, are needed. Indeed, more than one new drug for each is desirable, so that combination therapy can be employed to improve efficacy and avoid drug resistance and to provide backup drugs when resistance emerges. In order to address this imbalance, pharmaceutical industry and public institutions have recently started to join efforts to address the tool deficiency in NTDs through public–private partnerships as a response to their commitment to global health and social responsibility.

In January 2012, the WHO published a roadmap to guide the implementation of policies and strategies relating to their Global Plan on NTDs. This roadmap was launched at a high-profile event, "Uniting to Combat NTDs: Ending the Neglect and Reaching the 2020 Goals," held in London on January 30, 2012. Nine CEOs from the pharmaceutical industry shared a stage with the director general of the WHO, together with Bill Gates and political representatives of several member states to pledge to (1) sustain, expand and extend the drug supply; (2) accelerate research and development for new treatments; and (3) increase funding to improve implementation in order to control or eliminate ten NTDs. This became known as the London Declaration on Neglected Tropical Diseases.

Kinetoplastids are a group of flagellated protozoans that include the species of the genera *Leishmania* and *Trypanosoma*, which are human pathogens with devastating health and economic effects. The most common human diseases caused by kinetoplastids are included within the list of 17 NTDs declared by the WHO. They are human African trypanosomiasis (HAT, a.k.a. sleeping sickness), which is caused by three subspecies of *Trypanosoma brucei*; Chagas disease (a.k.a. American trypanosomiasis), which is caused by infection with *Trypanosoma cruzi*; and various clinical manifestations of leishmaniasis, which are caused by more than 20 species of *Leishmania*, *Leishmania donovani* and *Leishmania infantum–Leishmania chagasi* respectively being the species responsible of the life-threatening visceral leishmaniasis in Africa and East Asia, and South America and Mediterranean countries. Half a billion people, primarily in tropical and subtropical areas of the world, are at risk of contracting these diseases, and it is estimated that more than 20 million individuals are infected with the causative pathogens, resulting in extensive suffering and more than 80 000 deaths per year. Leishmaniasis is threatening 350 million people from endemic tropical regions and, after malaria, is the parasitic disease causing the second greatest number of deaths.

Although the human diseases caused by these protozoans are vector-borne, transmitted by different insects, and are clinically very distinctive, much of the molecular and cellular biology of the different kinetoplastids is very similar.[4] They are distinguished by the presence of a DNA-containing region, known as a "kinetoplast", in their single large mitochondrion. They have a similar genomic organization, similar cellular structures (*e.g.* a single flagellum for motion, and peroxisomes that perform the first step of glycolysis, therefore known as "glycosomes") and all undergo morphological changes during their life cycles in the insect and the vertebrate host. The genome of each parasite exceeds 8000 genes, more than 6000 of them being orthologues in common. Analysis of the genomes of *Leishmania major*, *T. cruzi* and *T. brucei* (also known as TriTryp) has revealed many common core metabolic functions as well as other pathways that might represent specific adaptations to environments associated with life in their respective insect and vertebrate hosts. Furthermore, this analysis of genomes offers unparalleled opportunities for the identification of novel drug targets. Indeed, new targets have already emerged from TriTryp as well as phenotypic screens. Strikingly, approximately 50% of these genomes encode hypothetical proteins that do not resemble orthologues in the human genome, and *vice versa*. For instance, no orthologues have been found for tyrosine kinases in the kinetoplastid genomes, whereas phospho-tyrosine residues are present in the kinetoplastid phospho-proteome. It has been suggested that there may be essential proteins that can be exploited as selective targets for chemotherapy.

5.2 The Challenge of Choosing the Right Screening Tactics and Assays

Accelerating research and development (R&D) for new treatments in NTDs encounters several fundamental barriers. The complex biology of many of these parasites and their need for vectors for development and transmission challenges traditional industrial-scale drug discovery programs. The general neglect that these diseases have encountered by the pharmaceutical industry has meant that basic research findings have not found their way into the drug discovery pipeline. For the sake of unprecedented novelty, the pursuit of NCEs starts with the screening of molecules of unknown activity. High-throughput screening (HTS) has become the classical approach to interrogate large chemical libraries (*i.e.* from hundreds of thousands to a few millions compounds) in a time- and cost-effective manner. Despite the basic research conducted on the biology of these organisms, the paucity of validated molecular targets and the lack of screening platforms and translational processes have prevented drug discovery and development. Access to large chemical libraries with drug-like properties has been partially hampered by the encumbrance of intellectual property and the lack of assay formats amenable to HTS which resemble the relevant biology of the parasites.

5.2.1 The Screening Approach: Target *vs.* Phenotypic Screening

Although target-based chemotherapeutic screens have been a focus of drug discovery programs in the post-genomic era, there has been a movement toward returning to phenotypic-based whole-organism screening assays.[5] This is based on the general consensus that target-based screens have been less successful than phenotypic screens in the discovery of novel drugs.[6] Moreover, with the exception of *T. brucei*, there is a paucity of genetic tools to validate drug targets in kinetoplastid parasites. Ultimately, in infectious diseases, the target is the infective parasite, and any compound coming from target-based screening is demanded to demonstrate activity in an antimicrobial assay. Observations cumulated over years from numerous organisations and research groups indicate a high attrition in target-based screening for anti-infectives, partly due to the poor translation from target activity to cellular and *in vivo* assays. Thus, running a target-agnostic phenotypic assay upfront in primary screening is a sensible and valid approach to early drug discovery. Furthermore, compounds identified in phenotypic screens not only represent a valuable starting point for lead optimization, but a chemical-biology probe that can be exploited for the identification of new targets, whose tractability and validation is already supported by the existence of these compounds.

Table 5.1 compares relevant features of the three different types of screens that can be designed to approach hit discovery for antimicrobial agents in general, and *Leishmania* in particular. Targeted cellular assays share pros and cons of both a target-based biochemical assay and a target-agnostic antimicrobial assay, thus being a reasonable compromise to encompass cellular activity and specificity. Furthermore, they may become a powerful tool to prove correct target engagement for biochemical hits with antimicrobial activity. Yeast-based systems have been proposed as an alternative cost-effective platform for target-directed screening of anti-leishmanials within a eukaryotic context.[7,8] The absence of permeability barriers in cell-free assays for the compound to reach the target biophase makes this assay format more sensitive to pick up active compounds that can interact with the molecular target even weakly. Thus it is usually

Table 5.1 Comparison of screening approaches to *Leishmania* hit discovery.

Feature	Targeted biochemical	Targeted cellular	Agnostic cellular
Assay			
Assay response	Direct or indirect measure of target function on cell-free systems	Direct or indirect measure of target-biased function or phenotype on parasite cellular milieu	Parasite growth or death
Permeability barriers for the compounds to reach the parasite biophase	None	Host-cell plasma membrane, phagolysosome, parasite plasma membrane (the latter not applicable if the target is a receptor, or plus organelle membrane if the target is not cytosolic). Uptake transporters and efflux pumps present	Host-cell plasma membrane, phagolysosome, parasite plasma membrane (the latter not applicable if the target is a receptor, or plus organelle membrane if the target is not cytosolic). Uptake transporters and efflux pumps present
Hit compounds			
Sensitivity/Potency	Medium–High (Low if target occupancy has to be high in order to give an assay response)	Medium (High if target occupancy to express phenotype is low or there is a driving force for compound uptake, *e.g.* transporters, pH-gradient)	Low–Medium (High if target occupancy to express phenotype is low or there is a driving force for compound uptake, *e.g.* transporters, pH-gradient)
Specificity	High	Medium	Low
Novelty	High	Medium	Low
Chemical and biological space of hits	Narrow	Narrow	Broad
Tractability and likelihood of translation into efficacy	Low (analogue search and synthesis might be required)	Medium (analogue search and synthesis is desirable)	High (analogue search and synthesis is desirable)
Target-assisted SAR	Yes	Possible	No
Doability			
Supply of bioreagents	Laborious (requires purified target protein)	Simple or Laborious (may require strain engineering)	Simple (requires parasite culture)
Assay development	Laborious (requires functional assay for the target protein)	Simple or Laborious (may require a biochemical pathway assay)	Easy (requires quantifying parasite growth)
Cost	High	Moderate	Low

expected that potency of compounds is higher in biochemical than cellular assays. Nevertheless, it might be the opposite when the biochemical assay is configured in such a way that very high occupancy of the target is required for the compound to exhibit statistically significant response, whereas low target occupancy is enough to give a cellular phenotype or there is even a driving force which concentrates the compound at the parasite biophase, such as pH-gradients or active transport. In this regard, it is worth noting that, in an intracellular assay where *Leishmania* parasites infect macrophages, active compounds have to pass through three compartments with different pHs and membrane transporters of small molecules (Figure 5.1). Compounds being actively transported or only ionisable at acidic pH may reach concentrations at the phagolysome or the parasite which are higher than those initially existing in the extracellular medium. A drawback of cellular screens is the possible insensitivity of compounds due to efflux pumps, which would result in missing out chemical starting points potentially optimizable.

In terms of logistics and economics, agnostic cellular assays do not require purification of a target protein and assay development is usually generic. Thus, it is expected to be a faster and more cost-effective approach for anti-infectives, unless *in vitro* culturing of the parasite is cumbersome or even unfeasible. Challenging all targets at once in an unbiased cellular milieu means casting the net more broadly for fishing out hits of relevance. However, targeted biochemical screening facilitates structure–activity relationships (SAR) based on molecular recognition of the target. Hence,

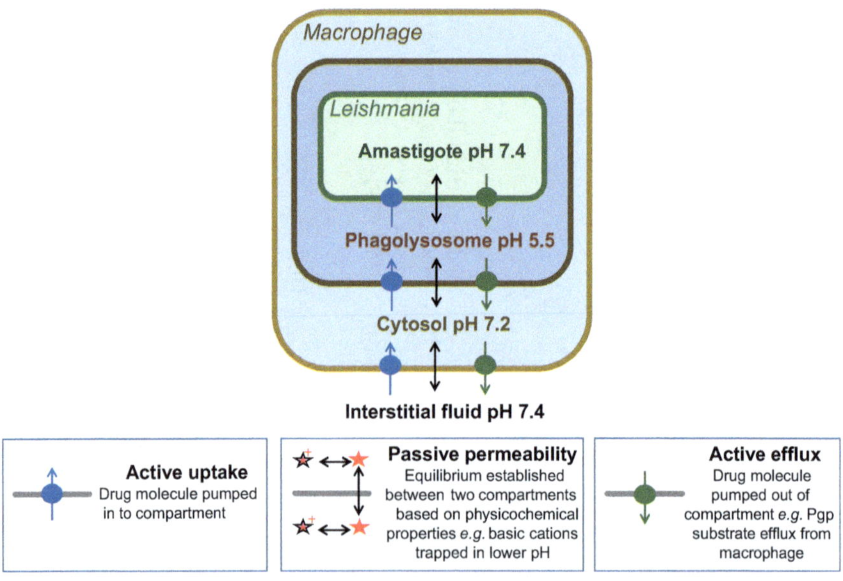

Figure 5.1 Permeability barriers and driving forces for compounds to reach the parasite biophase.

a posteriori target deconvolution for phenotypic hits would assist SAR in order to guide chemical hit optimization.

Since target-based screening methodologies are not particular to anti-infectives drug discovery, but characteristic of the target class itself, we will only focus on reviewing the specific aspects of agnostic cellular screening assays available for discovery of anti-leishmanials.

5.2.2 The Parasite

Leishmania is an intracellular parasite that undergoes adaptive transformations depending upon the host environment in which it lives. Significant differences may exist in drug sensitivity between different life-cycle stages, thus the choice of the most relevant *in vitro* assay in the disease milieu is key to succeeding in the translation of compound activity from *in vitro* to *in vivo* and eventually the clinic. Whereas the extracellular bloodstream trypomastigote form of *T. brucei* is the relevant stage for drug discovery in sleeping sickness, the intracellular amastigotes of *L. donovani* and *T. cruzi* in the host cells of infected tissues are the discovery focus for visceral leishmaniasis and Chagas disease, respectively. Over recent years, there have been significant advances in automated microscopy that have enabled compound testing against intracellular parasites in high-throughput mode.[9–16] Nevertheless, high-content screening methodologies have not reached enough throughput and cost-effectiveness yet in order to accomplish a primary blitz campaign against large chemical libraries comprising a few million compounds, such as the size of screening collections managed by big pharma companies. Alternatively, miniaturized assays have been developed with engineered or wild-type strains which allow us to monitor the growth of *Leishmania* and *Trypanosoma* in HTS.[17] Several researchers have appraised the suitability of free living *L. donovani* axenic amastigotes as a surrogate predictive for intracellular amastigote activity and compared with an intramacrophage high-content screening (HCS) assay format.[9,17] Although the presence of false-positives in a screening program is not ideal, most of these will probably be eliminated in subsequent screening models within the screening funnel. On the other hand, false-negatives, potentially interesting compounds, would be discarded at the first step. A large number of false-positives were observed with the screen using the free living parasites cultured axenically, but in this case, the resulting hits in both screens were the same, indicating that the problem of false negatives can be negated, whilst the size of the library for confirmatory assay was reduced by 95%.[9] Therefore there is potential for employing this axenic assay in a cost-effective manner as a pre-screen of a large library prior to using the more technically challenging intramacrophage assays. Nevertheless, De Muylder *et al.* have reported that naloxonazine is active against intracellular amastigotes, whereas it is inactive against promastigotes and axenic amastigotes, indicating that its activity is dependent on a macrophage function.[18]

Evaluation of *in vitro* drug susceptibility relies on *in vitro* assays that are able to recapitulate the environment of the parasite in the host along with the disease condition. Thus, choosing the most relevant stage pair of parasite and host cell is pivotal. Promastigote and amastigote life-cycle stages of *Leishmania* exhibit distinct features that affect susceptibility to drugs. *Leishmania* parasites have two major life cycle stages: the promastigote stage (inside the insect vector) and the amastigote stage (inside the parasitophorous vacuoles of the mammalian macrophage host). Since the intracellular form of the parasite causative of disease in the host is the amastigote, it is expected that assays testing *Leishmania* amastigotes are more disease-relevant and predictive. *L. donovani* clones that are able to be maintained and grown axenically in the lab have been isolated and they have constituted a powerful *in vitro* model to conciliate biological relevance and HTS amenability.[9]

5.2.3 The Host Cell

As for the host cell, *in vitro* drug testing models have been set up, which employ macrophages from primary or immortalized cells from different tissues and animal species. For instance, mouse peritoneal macrophages (PEMs), mouse bone marrow-derived macrophages (BMMΦ), human peripheral blood monocyte-derived macrophages (PBMΦ) and differentiated human THP-1 cells, human promonocytic U937 cells and murine J774.1 cells.[19] Although it has been shown that anti-leishmanial drug activity *in vitro* can be host-cell-dependent, it is worth saying that the magnitude of this dependence varies across compounds and is possibly related to the mode of action of each compound. While many different reasons could lie behind this effect, two key factors are the state of activation of the macrophage and the replicating stage of the parasite. For instance, it is noticeable that sodium stibogluconate, a compound whose action depends upon its oxidation state in the macrophage, seems to exhibit a high dependency, whereas amphotericin B, whose mechanism of action is more biophysical than biochemical, behaves quite consistently across the different macrophage sources. Despite the effort taken, the standardisation of assay conditions across primary macrophages from different sources is really cumbersome, and differences in parameters such as infectivity ratio, replication or metabolic state may account for apparent discrepancies otherwise attributed to the macrophage nature. In all, the differences seem to be more quantitative (*i.e.* potency values) than qualitative (*i.e.* activity *vs.* inactivity). In this regard, assuring reproducibility and standardisation may be more important than the macrophage source itself in order to support screening and compound profiling efforts for medicinal chemistry programs. Macrophages obtained from activation of immortalised cells, such as THP1-derived monocytes, are a reasonably well controlled source of host cells to warrant long-term consistency in hit and lead optimisation programs, despite apparently being less biologically relevant. Issues such as low replication rate and carry over of extracellular parasites can be overcome.[17,20] Seeking disease relevance of the model, *ex vivo* explants of macrophages accompanied by other immunological cells from infected animals (*i.e.* mice or hamsters) may constitute a better source than isolated primary macrophages from healthy animals. Such models have been recently reported for *L. donovani* and *L. infantum* in combination with engineered reporter systems (*e.g.* luciferase or infra-red fluorescent protein) for imaging quantification of parasite load which will allow monitoring of the translatability of compounds' activity using the same parasite strain through *in vitro*, *ex vivo* and *in vivo* models.[21–25]

5.2.4 The Screening Assay Format

In order to increase the throughput and HTS amenability of assays, numerous attempts have been made by using either different stages of wild type parasite or transgenic parasites, different readouts, different formats and different host cells. There is no "one-size fits all". Pros and cons have been summarised in Table 5.1. Likewise, herewith below, we will briefly describe different assay formats applied to actual HTS campaigns.

5.2.5 The Analysis of Screening Data

Experimentation is naturally and inevitably subjected to variability. When you are repeatedly testing or screening the same compound against a particular target or cells following the same assay procedure on a number of different occasions, you will not be able to get the same result for activity in all instances. If there is something certain for the screeners it is that certainty is false. In this context, we can see statistics as the science of decision making under uncertainty. Screening is about making decisions on how active or inactive one particular compound is against a biological system. Then, it seems clear that the usage of statistical tools in the analysis of screening experiments is the unavoidable way of doing a correct interpretation of the screening data, qualifying these with meaning, and hence converting them into valuable information that supports sound decision making.

Along the HTS workflow, there are three stages where key decisions have to be made based on experimental data:

a. Assay development: how good is the assay in terms of reliability, reproducibility and sensitivity to detect the kind of modulating compounds we are pursuing, which assay protocol is better? A review on statistical evaluation of assay quality is included in ref. 26. Since the publication of the *Z*-prime (Z') parameter,[27] this has become the corner stone of assay plate quality control and assay performance adopted by screening scientists:

$$Z' = 1 - 3 \times (\mathrm{SD_{signal}} + \mathrm{SD_{background}}) / |M_{\mathrm{signal}} - M_{\mathrm{background}}|$$

where SD is the standard deviation, *M* is the mean, signal is the response of an active assay (100% assay activity), and background is the response of an inactive assay (0% assay activity).

Nevertheless, by itself, *Z*-prime is not an indication of the robustness of the assay in the sense of reproducibility of the assay performance across days, scientists, equipment, laboratories, reagent batches, compounds, *etc.* For this purpose, evolution of *Z*-prime and other quality indicators for signal window and variability can be monitored. Even so, high *Z*-prime values do not guarantee reproducible percentages of response at single shot or XC50 values in dose–response experiments (XC50 is the compound concentration showing 50% of the total X response. When X corresponds to Inhibition, XC50 = IC50). Thus, more adequate statistical tools have been developed for the proper assessment of reproducibility and repeatability of single-shot (*e.g.* precision radius and accompanying statistical parameters, B-score and R-score), or correlation and agreement of dose–response screens (*e.g.* Minimum Significant Ratio, MSR).[28] Furthermore, we should bear in mind that the ultimate goal of HTS is the identification of novel compounds. The higher the sensitivity of the assay is for picking up modulators of weak potency, the better screen we have. The quality of the assay must be subordinated to this aim. In this regard, the optimisation of quality indicators of an assay based on making it insensitive to small fluctuations in the levels of the response that is measured might lead to a high *Z*-prime and a very robust but useless screen.

b. Prosecution of HTS campaign: is the screening process performing with the expected quality, are all quality indicators kept within the range of acceptance, are we maintaining the same quality and range of variability across all compounds tested, *i.e.* all wells within the plate, all plates within a daily run and all runs within the campaign? If the analysis is run retrospectively, we will try to identify patterns of experimental responses that are not randomly distributed in time (temporal patterns) or space (spatial patterns). Depending on the strength of the pattern and the knowledge of the rule quantifying the underlying order, we may think of correcting or rejecting the data. For a comprehensive review, see ref. 28.
c. Analysis of primary HTS data: which compounds are giving a positive response in the assay? Those deemed positive or hits will deserve additional investment and attention in further experiments. On the other hand, those qualified as negative will be abandoned, probably forever. It is not uncommon that selection of primary positives in HTS is simply made just on the basis of potency above a particular cut-off of activity that accommodates logistics. However, the assignment of the potency from a single-shot experiment is rather risky. Firstly, the reliability of the activity value highly depends upon the quality of the assay and the distribution of activity of the sample population tested. That is, the same threshold of activity does not have the same

meaning with respect to reliability in all assays. A hit selection process that minimised the rates of false positives and negatives, whatever their level of activity is, would eventually optimise the usage of the limited screening resources. False positives are annoying, but false negatives should be unacceptable, because they are usually abandoned forever. Highly potent positives that eventually turn out to be false are worthless, disappointing and can negatively bias the selection. On the other hand, true weak positives are valid for SAR and as starting points of hit-to-lead chemical programmes. Our preference is for giving the highest chances of picking up weak hits at the risk of realising that some of them are false in subsequent counter screens in the progression cascade. Therefore, we recommend the implementation of statistical cut-off as hit threshold. The most commonly adopted statistical method for hit selection in HTS experiments is the separation distance from screen noise (statistical significance), *i.e.* the mean $\pm k \cdot$ SD method and its variants, where SD is the standard deviation of the negative reference (*i.e.* background controls or inactive samples) and k is a multiplying scalar. It is commonly accepted that a distance of $3 \times$SD is the lowest statistical significance. The probability for the values greater than $3 \times$SD is 0.13% in a normal distribution, or in other words the probability for a compound falling further than $3 \times$SD from the mean of inactive samples of being a true positive is 99.87%.

Table 5.2 includes a summary of key quality and performance indicators to be considered during a HTS campaign.

5.2.6 Hit Triage: Progression Cascade and Selection Criteria

The successful identification of high-quality hit compounds that may become starting points for chemical optimization is the result of a smart combination of three factors: the chemical screening collection, the screening assays and the triage tactics. Hit rates of 1% in primary screening campaigns are not unusual. Therefore, chemical libraries of a few million compounds will generate tens of thousands of hits. Hence, as important as the quality of the primary screening, it is critical to implement an efficient and relevant hit progression cascade towards the selection of hits fulfilling the aspiration profile defined. Any compound that can be progressed in early drug discovery for antimicrobials should display three main attributes: efficacy (*i.e.* activity in the therapeutic model or assay), selectivity (*i.e.* no activity against mammalian cells and scarce activity on the homologous target) and tractability (*i.e.* reasonable properties for optimization by medicinal chemistry). Figure 5.2 displays a real example applied to the screening of a 1.8 million compounds collection.[17] In this case the aspirational hit profile was a compound with micromolar activity against *L. donovani* infecting macrophages, no overt cytotoxicity against human cells (*i.e.* HepG2

Table 5.2 Quality and performance indicators of small molecule screening campaigns.

Category	Parameter	Description
Assay	Type of assay	Targeted biochemical, targeted cellular, parasite growth
	Target	Molecular target or parasite species
	Primary measurement	Readout and assay principle
	Key reagents	Engineered strain, wild-type strain, target protein, developing reagent
	Assay protocol	As described in methods section
Library	Library size	Number of compounds
	Library composition	Diversity collection, focused sets, *etc.*
	Source	Supplier
Screen	Format	Plate type (density, description, supplier and reference catalogue)
	Concentration(s) tested	Compound concentration assayed
	Plate controls	DMSO, reference compound
	Reagent/compound dispensing system	Liquid dispensers used for compound and reagents addition
	Detection instrument and software	Plate reader and analysis software
	Assay validation/QC	Average Z' value
	Correction factors	Systematic error correction and pattern recognition tool
	Normalization	Data reduction calculations
Post-HTS analysis	Hit criteria	Hit cut-off or threshold, % (statistical or biological activity) (*e.g.* 30% Inhibition)
	Hit rate (%)	Number of hits per number of compounds tested
	Additional assay(s)	Counterscreen and secondary assays used for hit confirmation and qualification
	Confirmation of hit purity and structure	Criteria used
	Compound potency	$pIC_{50} = -\log(IC_{50})$, where IC_{50} is the molar compound concentration that exhibits 50% assay response
	Compound efficacy	Maximum asymptote of dose-response curve (E_{max}) and $pIC_{90} = -\log(IC_{90})$, where IC_{90} is the molar compound concentration that exhibits 90% assay response

hepatocytes and THP1-derived macrophages) and reasonable physicochemical properties (*i.e.* low molecular weight and lipophilicity). As aforementioned, primary screening with axenic amastigotes exhibits a low rate of false negatives relative to the intramacrophage assays (note: antimonials, which require oxidation, are one of the few examples). However one of its caveats is a high false-positive rate. For instance, only 25% of the primary hits in the axenic assay turned out to be active in the intramacrophage assay for the HTS campaign reported in ref. 17. Amongst the several reasons underpinning the attrition from a highly replicative axenic parasite to a slow-replicating intracellular parasite, a key factor is the differentiation between growth-slowing or static compounds and the more desirable cidal compounds. In order to sort out the former kind of compounds and prioritise the latter, a novel cytocidal-only axenic amastigote assay was developed by increasing the starting density of cells and lowering the limit of detection.[29]

Each tier of assays will have a capacity to process compounds in a timely and cost-effective manner. Thus, it is advisable to set quotas. Probably, the number of compounds passing through the thresholds of activity for each assay may exceed these quotas. For the sake of sampling the highest possible chemical diversity, compounds can be clustered by chemical similarities and just those exemplars of each group showing the best prospects selected. Likewise, it is worth getting rid of compounds containing odd moieties that show promiscuity throughout biological assays and targets of any nature.[30] Frequent hitters are not only potentially reactive compounds, but also aggregators, sequestering some biological reagents. They can be unveiled by several means: software tools (*e.g.* publically available lists), checking literature and assessing physico-chemical features (*e.g.* dynamic light scattering) or biological assays (*e.g.* β-lactamase and orthogonal readouts). Furthermore, dose–response curves should be inspected for aberrant fit or dubious shape (*i.e.* low slope or low maximum asymptote).

As aforementioned, the success of a screening campaign stems from an orchestrated combination of three components: the compounds, the assays and the triage tactics. A high-quality chemical library and set of assays may not deliver without a sensible critical path. Criteria of selection of hit compounds will dictate the assays and progression cascade to be implemented. There is not a single standard criteria set adopted by all the research community, and each screening campaign is customized according to the available resources and capabilities, as well as the sound judgement of the screening scientists.[9,17,29] In an attempt to seeking a common agreement on clear go/no-go criteria for the progression of compounds in drug discovery for infectious diseases of the developing world, the Japanese Global Health Innovative Technology (GHIT) Fund convened with experts from the Medicines for Malaria Venture, the Drugs for Neglected Diseases initiative (DNDi) and the TB Alliance, together with representatives from the Bill and Melinda Gates Foundation, have proposed disease-specific criteria for hits and leads for malaria, tuberculosis, visceral leishmaniasis and Chagas disease.[30] Progression cascade and selection criteria for hits (*i.e.* confirmed active compounds in dose–response out of a screening campaign) and tractable hits (*i.e.* hit compounds that deserve commitment to optimisation) proposed by the authors of this chapter are displayed in Figure 5.2 and Table 5.3.

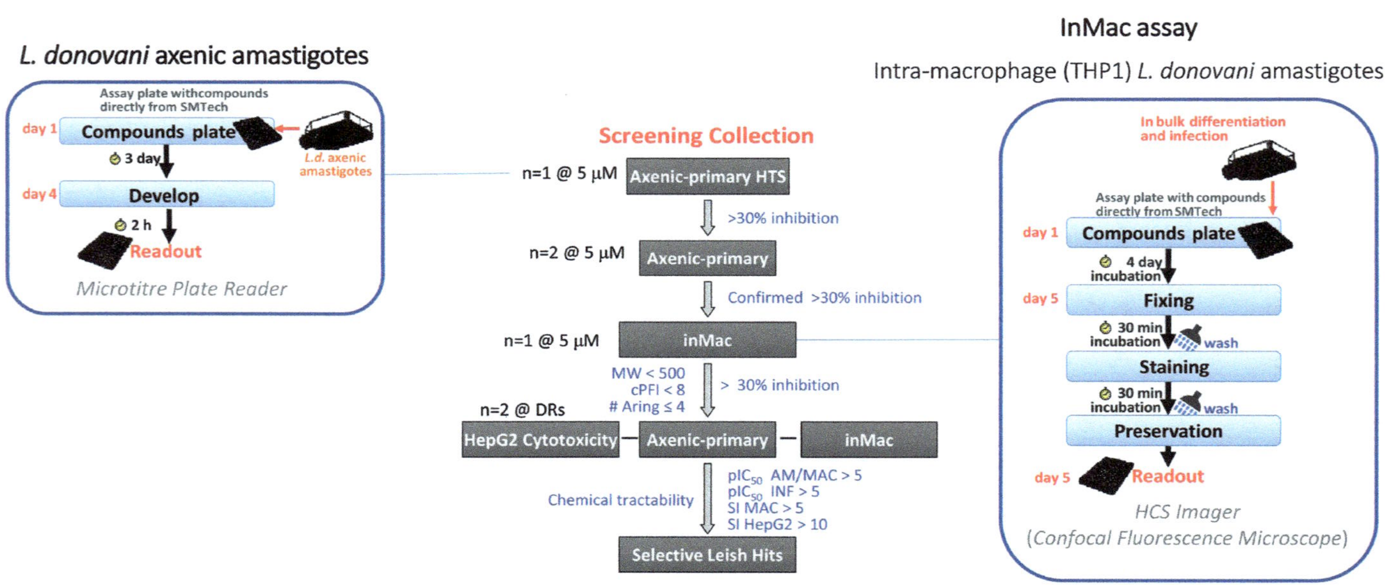

Figure 5.2 Progression cascade of HTS phenotypic hits towards new VL chemotherapeutic agents (*n.b.* Preservation step in InMac assay protocol is optional).

Table 5.3 Selection criteria for hit and tractable hit in screening of anti-leishmanial compounds. Note: Each criterion may be classified as a "Must" or "Nice-to-have" feature.

	Hit	Tractable hit
Efficacy – Biochemical[a]	**Assessment of genuine potency and selectivity**	
Activity against target protein	$pIC_{50} > 6$	$pIC_{50} > 6$ Correlation between biochemical and whole-cell activity or evidence of target engagement
Selectivity against mammalian (ideally human) target isoform	≥10 fold	≥10 fold
Efficacy – Whole Cell	**Assessment of potency and selectivity**	
Activity against axenic *Leishmania* amastigotes or promastigotes	$pIC_{50} > 5$ (preferable)	$pIC_{50} > 5$ (preferable)
Activity against intramacrophage *Leishmania* amastigotes	$pIC_{50} > 5$[b,c]	$pIC_{50} > 5$[b,c]
Selectivity against host cell (*e.g.* THP-1)	≥5-fold[d]	≥5-fold[d]
Selectivity against mammalian cells (*e.g.* HepG2)	≥10-fold	≥10-fold
Hit qualification	**Validation of *bona fide* activity**	
Orthogonal assay of biochemical or cellular assay	Active	Active
Interference and/or promiscuity assay	Inactive	Inactive
Frequent hitters and chemical reactivity	Preferably no alerts	Preferably no alerts
Developability	**Drug-likeness preferable but not essential**	
Structure of potential lead		Acceptable QC (>80%) and structure confirmation
SAR with first analogue iteration		Indication of SAR pattern + re-synthesis of the original hit

Table 5.3 (*Continued*)

	Hit	Tractable hit
Intellectual property	No conflict ideally	IP situation clarified with no hindrance to exploitation
Synthetic tractability		Amenable to synthetic modification (ideally less than eight steps)
Target engagement (MoA)		Preferred
Solubility (CLND)		>30 μM ideally
MW	<500	<500
Ligand efficiency (LE)[e]	>0.3 ideally	>0.3 ideally
Lipophilic ligand efficiency (LLE)[e]	Most drugs >5	Most drugs >5
Calculated PFI (ideally experimental)[f]	≤ 8	≤ 8 (ideally ≤ 7)
Number of aromatic rings	≤ 4	≤ 4
Safety profile	**Assay(s) to determine extent of the problem**	
Assessment of liabilities	No structural alerts for stability and reactivity or strategy to remove alerts	*In silico* prediction of alerts (*e.g.* DEREK) and *in vitro* evaluation on a selected panel of receptors, enzymes and ion channels

[a]Only applicable if target-based screening or target of phenotypic hits is known.
[b]Ideally in both readouts, *i.e.* number of amastigotes and percentage of infected host cells.
[c]Curves where maximum asymptote $<80\%$ and/or slope <0.8 discarded.
[d]If $pIC_{50}<5$, check curves and selectivity index between Axenic and HepG2.
[e]Note that for phenotypic assay, use just to track progression within a series.
[f]Property Forecast Index, PFI = Chrom log $D_{pH7.4}$ + number of aromatic rings.[40]

5.3 Case Studies of HTS Campaigns for Anti-Leishmanials

Quite a few screening campaigns seeking NCEs with anti-leishmanial activity have been run, ranging from a few thousands to millions of compounds, from natural products to synthetic organic chemicals, from academia to industry. We will review some of them as exemplars (Table 5.4). Testing compound concentrations in primary screening ranged from 4 to 10 μM.

Sharlow and colleagues at the University of Pittsburgh pioneered *Leishmania* HTS by prosecuting the screening of approximately 200 000 chemically diverse compounds against a *L. major* promastigote growth-inhibition assay based on fluorescent alamar blue detection in 384-well microtiter plate format.[31] Formerly, two smaller-scale screens with approximately 2100 and approximately 15 000 compounds had been performed also using *Leishmania* promastigotes.[18,25] The screen was purposely set-up for low-stringency, thus, not surprisingly, a high primary hit rate of 8.9% was obtained. Hits triage was assisted by computational chemical clustering and biological activity confirmed by flow cytometry (orthogonal readout) and against axenic amastigotes (alamar blue).

The largest HTS campaigns have been run by Novartis at Genomics Novartis Foundation (GNF)[32] and GSK at Diseases of the Developing World (DDW)[17] sites, which have screened their full chemical diversity decks of 2.9 and 1.8 million compounds respectively. Interestingly, both teams used a primary screening assay based on culturing axenic amastigotes of *L. donovani*. The two assays are replicative and detect the total amount of parasite cells by looking at metabolic markers, *i.e.* either total ATP content through coupling to luciferase and luminescent readout, or resazurin reduction with fluorescence readout. Hits inhibiting replication or killing the parasite are picked up, thus it is not possible to tell apart cidal and static anti-leishmanial compounds upfront. Secondary assays were implemented to ascertain these two types of mechanisms, as well as to confirm initial activity through non-replicative or hardly replicative assays with orthogonal readout which are not prone to the same kind of potential interferences. These campaigns used an intra-macrophage amastigote assay with primary murine peritoneal macrophages infected with *L. donovani* splenic amastigotes[32] or THP1-derived macrophages infected with GFP-expressing amastigotes (Figure 5.3).[17] Assays were performed in microtiter plates and after the incubation period the cells were fixed and stained for DNA (*e.g.* SYBR, DAPI or DRAQ5). Image collection and counting of macrophage cells and intracellular *L. donovani* amastigotes was performed using automated confocal microscope system and image analysis software. GFP-expressing parasite cells can be detected in a second fluorescence channel.[17] As a result of the HTS by GSK, a set of 192 anti-leishmanial compounds has been assembled and is provided as an open source for future lead discovery programs or chemical biology research. On the other hand, Novartis compounds are already providing

Table 5.4 Summary of HTS campaigns (>100 K compounds) prosecuted for the pursuit of new anti-leishmanial compounds.

		Primary assay						Confirmatory assays		
Institution	Chemical library size	Parasite form	Readout	Parasite species	Host cell	*Z*-prime	Confirmed hit rate (%)	Description	Host cell	Ref.
Novartis	2900 K	Axenic amastigote (replicative)	ATP (bioluminescence)	*L. donovani*	None	>0.4	0.9	Intramacrophage amastigotes (Automated imaging analysis of fluorescence confocal microscopy)	Primary peritoneal murine macrophages	32,33
GSK	1800 K	Axenic amastigote (replicative)	Resazurin (fluorescence)	*L. donovani*	None	0.7	1.8	Intramacrophage GFP-expressing amastigotes (Automated imaging analysis of fluorescence confocal microscopy)	THP1-derived macrophages	17
IPK (Institute Pasteur Korea)	26.5 K Extended up to 300 K	Promastigote	Resazurin (fluorescence)	*L. donovani*	None	0.5	0.5	Intramacrophage amastigotes (Automated imaging analysis of fluorescence confocal microscopy)	THP1-derived macrophages	8,34,41
		Intracellular amastigote	Automated imaging analysis of fluorescence confocal microscopy		THP1					
University of Pittsburgh	200 K	Promastigote (replicative)	Alamar blue (fluorescence)	*L. major*	None	0.9	8.9	Flow cytometry Axenic amastigotes	None	31
University of Dundee	26 K Extended up to +100 K	Axenic amastigote (replicative)	Resazurin (fluorescence)	*L. donovani*	None	>0.5	2.3	Intramacrophage GFP-expressing amastigotes (Automated imaging analysis of fluorescence confocal microscopy)	THP1-derived macrophages	7,39
		Axenic amastigote (cidal-only)	ATP (bioluminescence)	*L. donovani*	None	0.8	0.7			

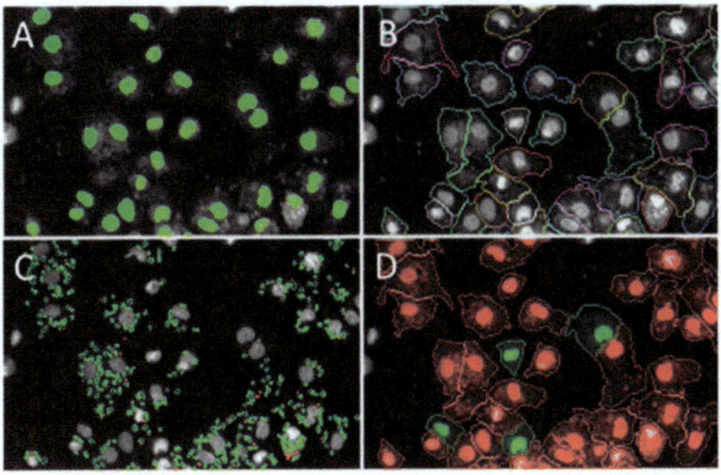

Figure 5.3 *Leishmania* intra-macrophage assay (inMac). Host cells: Human macrophages from differentiated THP1 cells; Parasite: *L. donovani* expressing eGFP from axenic culture of amastigotes. Conditions: Infection at multiplicity of infection (MOI) 10, 384 well plate format, 4 days incubation; Readout: confocal fluorescence imaging in Opera HCS reader, 20× air objective, two channels (DAPI-stained DNA and eGFP). Automatic imaging analysis for multiple results output: n Macrophages (MAC), n Amastigotes (AM), AM/MAC, percentage infected MAC: red/(red + green). Panel (A) Finding host cells nuclei, Panel (B) Finding host cells cytoplasm, Panel (C) Finding spots and selecting properties for parasite, Panel (D) Selecting populations and calculating results.

genetic and chemical validation of promising therapeutic targets for treatment of kinetoplastid infections.[32,33]

In order to circumvent the issues of high cost, complexity and low throughput of the intra-macrophage assays, the team at University of Dundee has developed a novel cytocidal-only axenic amastigote assay, developed by increasing the starting density of cells and lowering the limit of detection by using luminescent ATP-detection.[29] The assay shows higher predictivity of leishmanicidal intracellular activity than the sheer replicative assay. This new assay was validated as primary screen using a diversity-oriented synthesis library of approximately 10 000 compounds[20] and a diverse compound library of around 16 000 compounds previously screened in the classical replicative axenic format and in the intracellular assay.[7]

It is noteworthy that independent screening campaigns at Novartis, GSK and Dundee of different chemical libraries gave similar hit rates when the replicative axenic amastigote assay was used as the primary screen, *i.e.* 1–2%. Confirmation of actives in the intra-macrophage assay was around 25%. In comparison, the cytocidal axenic amastigote assay decreased the hit rate below 1%, whereas it increased the intra-macrophage confirmation above 50%.

Following up on former imaging assays set-up by Siqueira-Neto *et al.*[34] and De Muylder *et al.*,[18] scientists at Institut Pasteur Korea developed the first automated high-content, high-throughput image-based screening assay in 384-well microtiter plates targeting the intracellular amastigote stage of different species of *Leishmania* in infected human macrophages.[8] The assay was used to pilot a screen of a 26 500 compound library in order to compare it with one that used the promastigote insect-infective form. A hit rate of around 0.5% was obtained. Unlike other screening campaigns aforementioned, there was little overlap between the set of primary hits from the two assays. Of the hits selected against the intracellular amastigote, 50% were not selected in the promastigote screening, and only 4% of promastigote hits turned out to be active in the intracellular amastigote assay. Around 300 000 compounds have been screened under this format.[41]

De Muylder *et al.* at UCSF have prosecuted the screening of a privileged set of 909 bioactive compounds from Iconix Biosciences consisting largely of FDA-approved small molecules.[18] They compared two methods back-to-back, one targeting the free-living insect-infective promastigote stage of *L. donovani* (luminescent ATP detection) and the other targeting the intracellular amastigote stage (confocal fluorescence macroscopy with infected THP1 macrophages). Hit rates of the promastigote and intra-macrophage assays were around 6% and 3% respectively. Interestingly, the comparison showed that out of the 27 hits identified in the amastigote screen, 26 were also present in the promastigote screen. Only one compound, naloxonazine (a μ-opioid receptor antagonist), showed complete specificity for the intracellular amastigote stage, indicating a cell-pathogen target. Axenic amastigotes were used as a secondary assay. Compound activity against axenic amastigotes mostly correlated with that against promastigotes.

Natural products are an underexplored source of chemical diversity, which complements and hardly overlaps the synthetic and combinatorial chemical libraries (see Chapter 9). HTS of natural products requires a particular screening cascade which involves the chemical deconvolution of bioactive extracts and identification of the active principle by LC–MS de-replication and bioassay-guided extract fractionation. Since mixtures of several active compounds may be present and account for the activity of the mixture, the attrition of this process is usually high. Annang *et al.* have screened the natural products of Fundacion MEDINA against *L. donovani* by using a resazurin-based axenic amastigotes assay.[35] Primary hit rate was 2.2%, only slightly higher than the ones obtained for synthetic libraries and isolated compounds described. As in the cases described previously, an intra-macrophage assay was used as secondary screen, using THP1-derived macrophages and promastigotes expressing luciferase. The genetic engineering of reporter-expressing parasites allows intracellular amastigote screening to be run without imaging. An alternative method is based on back-transformation to promastigotes of amastigotes released from lysed macrophages.[36,37]

Aulner *et al.* at the Institut Pasteur Paris developed a high-content imaging assay based on the use of homogeneous populations of primary mouse macrophages hosting *Leishmania amazonensis* amastigotes.[12] The assay employs *Ds*Red2-expressing amastigotes and fluorescent staining with two vital cell-permeant dyes, *i.e.* Hoechst 33342 and LysoTracker. Image analysis is sequentially performed in three channels. Since the existence of giant *L. amazonensis*-containing vacuoles is a reliable indicator of the presence of intracellular proliferating amastigotes, overall the assay enumerates total and healthy macrophages, parasites and macrophage parasitophorous vacuoles. Screening validation was assessed with a pilot run comprising 60 compounds.

It is also worth noting that DNDi has established several academia and industry collaborative agreements for screening of large chemical libraries against visceral leishmaniasis, such as at Institut Pasteur Korea, the University of Dundee and the Genomics Novartis Foundation (GNF).[9]

Alternative assays to the ones described so far have been set up which are amenable to compounds screening. For instance, fluorescence microscopy image-based with pre-stained wild type *Leishmania* and THP-1 macrophages,[38] transgenic red fluorescent *L. major* and *L. donovani* promastigotes[39] or luciferase-expressing *L. infantum* to monitor amastigotes *in vitro*, *in vivo* and *ex vivo* by macroscopic luminometry.[22]

5.4 Conclusions

HTS has become a classical approach in early drug discovery for the pursuit of novel chemical starting points that can ultimately deliver innovative treatments. Current assay and cell culture methodologies have already enabled the screening of large chemical libraries comprising millions of compounds against live cells of *Leishmania.* A sophisticated panel of secondary assays are also available for qualification of hit compounds, which encompass susceptibility of intra-macrophage amastigotes monitored by automated microscopic imaging, amongst others. These screening efforts have started to deliver a pipeline of promising novel chemical series, which are deserving of further optimisation. The next challenge will be translating this *in vitro* anti-leishmanial activity into *in vivo* therapeutic efficacy in animal models and positive readout in phase II clinical trials. Until then, the real biological relevance of the *in vitro* and *in vivo* models used in drug discovery will not be properly assessed.

References

1. B. Pedrique, N. Strub-Wourgaft, C. Some, P. Olliaro, P. Trouiller, N. Ford, B. Pecoul and J. H. Bradol, *Lancet Global Health*, 2013, **1**, e371.
2. P. J. Hotez, A. Fenwick, L. Savioli and D. H. Molyneux, *Lancet*, 2009, **373**, 1570.

3. *World Health Organization. Global plan to combat neglected tropical diseases 2008–2015. 2007.* http://whqlibdoc.who.int/hq/2007/WHO_CDS_NTD_2007.2003_eng.pdf.
4. K. Stuart, R. Brun, S. Croft, A. Fairlamb, R. E. Gurtler, J. McKerrow, S. Reed and R. Tarleton, *J. Clin. Invest.*, 2008, **118**, 1301.
5. J. Kotz, *SciBX*, 2012, **5**, DOI: 10.1038/scibx.2012.380.
6. D. C. Swinney and J. Anthony, *Nat. Rev. Drug Discovery*, 2011, **10**, 507.
7. P. W. Denny and P. G. Steel, *J. Biomol. Screening*, 2015, **20**, 56.
8. J. L. Norcliffe, E. Alvarez-Ruiz, J. J. Martin-Plaza, P. G. Steel and P. W. Denny, *Parasitology*, 2014, **141**, 8.
9. M. De Rycker, I. Hallyburton, J. Thomas, L. Campbell, S. Wyllie, D. Joshi, S. Cameron, I. H. Gilbert, P. G. Wyatt, J. A. Frearson, A. H. Fairlamb and D. W. Gray, *Antimicrob. Agents Chemother.*, 2013, **57**, 2913.
10. J. L. Siqueira-Neto, S. Moon, J. Jang, G. Yang, C. Lee, H. K. Moon, E. Chatelain, A. Genovesio, J. Cechetto and L. H. Freitas-Junior, *PLoS Neg*lected *Trop. Dis.*, 2012, **6**, e1671.
11. R. Don and J. R. Ioset, *Parasitology*, 2014, **141**, 140.
12. N. Aulner, A. Danckaert, E. Rouault-Hardoin, J. Desrivot, O. Helynck, P. H. Commere, H. Munier-Lehmann, G. F. Spath, S. L. Shorte, G. Milon and E. Prina, *PLoS Neglected Trop. Dis.*, 2013, **7**, e2154.
13. J. Alonso-Padilla, I. Cotillo, J. L. Presa, J. Cantizani, I. Pena, A. I. Bardera, J. J. Martin and A. Rodriguez, *PLoS Neglected Trop. Dis.*, 2015, **9**, e0003493.
14. J. C. Engel, K. K. Ang, S. Chen, M. R. Arkin, J. H. McKerrow and P. S. Doyle, *Antimicrob. Agents Chemother.*, 2010, **54**, 3326.
15. S. Moon, J. L. Siqueira-Neto, C. B. Moraes, G. Yang, M. Kang, L. H. Freitas-Junior and M. A. Hansen, *PLoS One*, 2014, **9**, e87188.
16. L. L. Nohara, C. Lema, J. O. Bader, R. J. Aguilera and I. C. Almeida, *Parasitol. Int.*, 2010, **59**, 565.
17. I. Peña, M. Pilar Manzano, J. Cantizani, A. Kessler, J. Alonso-Padilla, A. I. Bardera, E. Alvarez, G. Colmenarejo, I. Cotillo, I. Roquero, F. de Dios-Anton, V. Barroso, A. Rodriguez, D. W. Gray, M. Navarro, V. Kumar, A. Sherstnev, D. H. Drewry, J. R. Brown, J. M. Fiandor and J. Julio Martin, *Sci. Rep.*, 2015, **5**, 8771.
18. G. De Muylder, K. K. Ang, S. Chen, M. R. Arkin, J. C. Engel and J. H. McKerrow, *PLoS Neglected Trop. Dis.*, 2011, **5**, e1253.
19. K. Seifert, P. Escobar and S. L. Croft, *J. Antimicrob. Chemother.*, 2010, **65**, 508.
20. D. Tegazzini, R. Diaz, F. Aguilar, I. Pena, J. L. Presa, V. Yardley, J. J. Martin, J. M. Coteron, S. L. Croft and J. Cantizani, *Antimicrob. Agents Chemother.*, 2016, **60**, 3524.
21. E. Calvo-Alvarez, K. Stamatakis, C. Punzon, R. Alvarez-Velilla, A. Tejeria, J. M. Escudero-Martinez, Y. Perez-Pertejo, M. Fresno, R. Balana-Fouce and R. M. Reguera, *PLoS Neglected Trop. Dis.*, 2015, **9**, e0003666.

22. G. Michel, B. Ferrua, T. Lang, M. P. Maddugoda, P. Munro, C. Pomares, E. Lemichez and P. Marty, *PLoS Neglected Trop. Dis.*, 2011, **5**, e1323.
23. Y. Osorio, B. L. Travi, A. R. Renslo, A. G. Peniche and P. C. Melby, *PLoS Neglected Trop. Dis.*, 2011, **5**, e962.
24. R. M. Reguera, E. Calvo-Alvarez, R. Alvarez-Velilla and R. Balana-Fouce, *Int. J. Parasitol.: Drugs Drug Resist.*, 2014, **4**, 355.
25. C. J. Thalhofer, J. W. Graff, L. Love-Homan, S. M. Hickerson, N. Craft, S. M. Beverley and M. E. Wilson, *J. Visualized Exp.*, 2010, 1980.
26. N. Malo, J. A. Hanley, S. Cerquozzi, J. Pelletier and R. Nadon, *Nat. Biotechnol.*, 2006, **24**, 167.
27. J. H. Zhang, T. D. Chung and K. R. Oldenburg, *J. Biomol. Screening*, 1999, **4**, 67.
28. I. Coma, J. Herranz and J. Martin, *Methods Mol. Biol.*, 2009, **565**, 69.
29. A. Nühs, M. De Rycker, S. Manthri, E. Comer, C. A. Scherer, S. L. Schreiber, J. R. Ioset and D. W. Gray, *PLoS Neglected Trop. Dis.*, 2015, **9**, e0004094.
30. K. Katsuno, J. N. Burrows, K. Duncan, R. Hooft van Huijsduijnen, T. Kaneko, K. Kita, C. E. Mowbray, D. Schmatz, P. Warner and B. T. Slingsby, *Nat. Rev. Drug Discovery*, 2015, **14**, 751.
31. E. R. Sharlow, D. Close, T. Shun, S. Leimgruber, R. Reed, G. Mustata, P. Wipf, J. Johnson, M. O'Neil, M. Grogl, A. J. Magill and J. S. Lazo, *PLoS Neglected Trop. Dis.*, 2009, **3**, e540.
32. S. Khare, A. S. Nagle, A. Biggart, Y. H. Lai, F. Liang, L. C. Davis, S. W. Barnes, C. J. Mathison, E. Myburgh, M. Y. Gao, J. R. Gillespie, X. Liu, J. L. Tan, M. Stinson, I. C. Rivera, J. Ballard, V. Yeh, T. Groessl, G. Federe, H. X. Koh, J. D. Venable, B. Bursulaya, M. Shapiro, P. K. Mishra, G. Spraggon, A. Brock, J. C. Mottram, F. S. Buckner, S. P. Rao, B. G. Wen, J. R. Walker, T. Tuntland, V. Molteni, R. J. Glynne and F. Supek, *Nature*, 2016, **537**, 229.
33. S. Khare, S. L. Roach, S. W. Barnes, D. Hoepfner, J. R. Walker, A. K. Chatterjee, R. J. Neitz, M. R. Arkin, C. W. McNamara, J. Ballard, Y. Lai, Y. Fu, V. Molteni, V. Yeh, J. H. McKerrow, R. J. Glynne and F. Supek, *PLoS Pathog.*, 2015, **11**, e1005058.
34. J. L. Siqueira-Neto, O. R. Song, H. Oh, J. H. Sohn, G. Yang, J. Nam, J. Jang, J. Cechetto, C. B. Lee, S. Moon, A. Genovesio, E. Chatelain, T. Christophe and L. H. Freitas-Junior, *PLoS Neglected Trop. Dis.*, 2010, **4**, e675.
35. F. Annang, G. Perez-Moreno, R. Garcia-Hernandez, C. Cordon-Obras, J. Martin, J. R. Tormo, L. Rodriguez, N. de Pedro, V. Gomez-Perez, M. Valente, F. Reyes, O. Genilloud, F. Vicente, S. Castanys, L. M. Ruiz-Perez, M. Navarro, F. Gamarro and D. Gonzalez-Pacanowska, *J. Biomol. Screening*, 2015, **20**, 82.
36. S. K. Jain, R. Sahu, L. A. Walker and B. L. Tekwani, *J. Visualized Exp.*, 2012, 4054.
37. D. Paape, A. S. Bell, W. P. Heal, J. A. Hutton, R. J. Leatherbarrow, E. W. Tate and D. F. Smith, *PLoS Neglected Trop. Dis.*, 2014, **8**, e3363.

38. M. J. Dagley, E. C. Saunders, K. J. Simpson and M. J. McConville, *Assay Drug Dev. Technol.*, 2015, **13**, 389.
39. P. Vacchina and M. A. Morales, *Antimicrob. Agents Chemother.*, 2014, **58**, 1825.
40. R. J. Young, D. V. Green, C. N. Luscombe and A. P. Hill, *Drug Discovery Today*, 2011, **16**, 822.
41. L. H. Freitas-Junior, E. Chatelain, H. A. Kim and J. L. Siqueira-Neto, *Int. J. Parasitol.: Drugs Drug Resist.*, 2012, **2**, 11.

CHAPTER 6

Omics and Their Impact on the Development of Chemotherapy Against Leishmania

CHRISTOPHER FERNÁNDEZ-PRADA,*[a] ISABEL M. VINCENT,[b] ÉLODIE GAZANION[c] AND RUBENS L. MONTE-NETO[d]

[a] Université de Montréal, Faculté de Médecine Vétérinaire – Département de Pathologie et Microbiologie, 3190 rue Sicotte, Saint-Hyacinthe J2S 2M2, Canada; [b] University of Glasgow, Institute of Infection, Immunity and Inflammation, Glasgow G12 8QQ, Scotland, UK; [c] Université de Montpellier – Maladies Infectieuses et Vecteurs: Ecologie, Génétique, Evolution et Contrôle -MiVEGEC (IRD224 – CNRS5290), Institut de Recherche pour le Développement (IRD), 911 Avenue Agropolis, Montpellier BP 64501, France; [d] Instituto René Rachou, Fundação Oswaldo Cruz, 1715, Av. Augusto de Lima, Belo Horizonte 30190-002, Brazil
*Email: christopher.fernandez.prada@umontreal.ca

6.1 Introduction

Omics-based technologies have emerged as invaluable tools in the quest to discover and develop innovative chemotherapeutic approaches against parasitic diseases. Omics have proven to be a major force by increasing the success rates in terms of drug discovery and drug repurposing. Despite the power of these tools, there are a number of limitations, outlined in this

Drug Discovery Series No. 60
Drug Discovery for Leishmaniasis
Edited by Luis Rivas and Carmen Gil

Published by the Royal Society of Chemistry, www.rsc.org

chapter, that should be considered when designing and performing omics studies involving *Leishmania* parasites.

The genus *Leishmania* encompasses several species of trypanosomatid parasites that cycle between the flagellar promastigote forms in the gut of the sandfly vector and the amastigote stage within the host's macrophages. While the host immune response plays a major role in *Leishmania* infection and the course of the disease,[1–3] the genetic background of each species also influences clinical outcomes.[4,5] Clinical disease ranges from self-healing cutaneous manifestations to life-threatening visceral leishmaniasis.[5–7] The emergence of resistance against the small number of drugs available, together with the notorious lack of funding devoted to drug research against *Leishmania*, makes the disease an ideal subject for large-scale drug repurposing studies.[8,9] Since 2005, when the 32.8-megabase haploid genome of the *Leishmania major* Friedlin strain was sequenced,[10] next-generation sequencing (NGS) has provided a massive amount of information about *Leishmania* genomic structure[11] and plasticity.[12,13] Similarly, novel high-throughput transcriptomic technologies have given insights into how molecular interactions occur between *Leishmania* and the mammalian host, as well as between these parasites and their sand fly vector.[14,15] The results of these studies have indicated that messenger RNA (mRNA) regulation is a key factor in *Leishmania* stage differentiation and adaptation to environmental stress. Moreover, as a direct consequence of polycistronic transcription of protein-coding genes in trypanosomatid parasites, steady-state levels of most mRNAs depend on posttranscriptional modifications (PTMs) for regulation. This, in fact, explained to a certain extent the poor correlation observed between the transcripts and the proteins expressed by these parasites.[16] Thereby, proteomics has proved a powerful tool in order to fulfil genomics' and transcriptomics' gaps and to lead to the discovery of novel targets and mechanisms of drug resistance.[17] These gaps include the inability to determine the subcellular localization of gene products and the failure to provide a complete picture of the relevant changes driving leishmanial immune bypass and parasite differentiation within macrophages. Transcriptomic and proteomic approaches have allowed researchers to partially reconstruct *Leishmania* metabolic-pathways and link them to stage-specific gene regulation mechanisms.[18] However, these studies do not take into consideration either the metabolic content of the parasite or the effect of environmental metabolites that may modulate the metabolic status of *Leishmania*. Recent advances in metabolomics technologies have tackled this problem by allowing the direct measurement of the cellular metabolome of *Leishmania* parasites under defined stress conditions, such as drug-perturbation, drug-resistance and stage-differentiation studies. Throughout this chapter we will discuss each of the above-mentioned omics tools (a brief summary of the main findings and recent techniques for the different omics is depicted in Figure 6.1) as well as their more significant strengths and weaknesses.

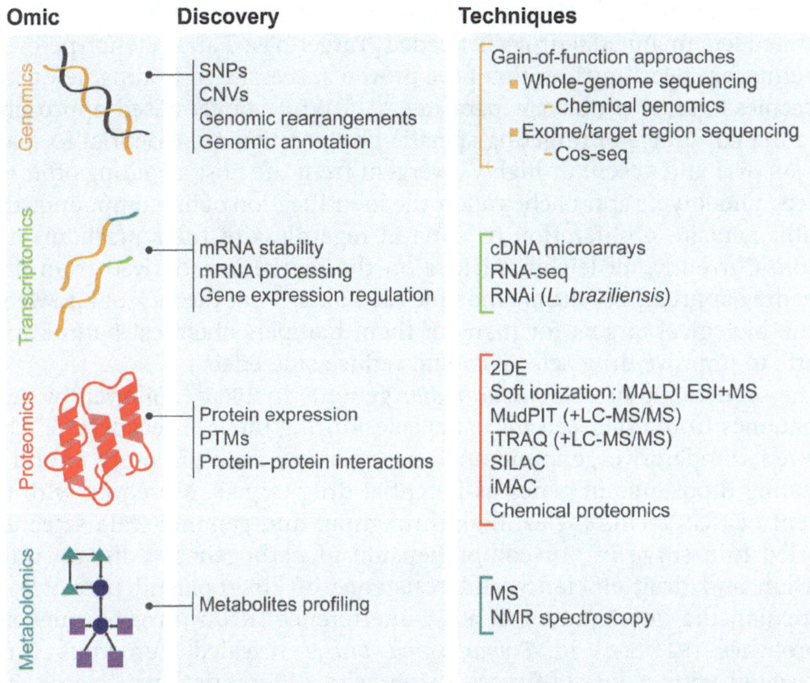

Figure 6.1 **Summary of the different omics following the flow of genetic information from the genome level to the metabolome level.** Each omic (left panel) allows for capturing different but essential information (middle panel). In order to maximize the quantity and quality of data derived from these omics, several associated techniques have been developed in the recent years (right panel). Each of these techniques is discussed in this chapter, taking particular account of their strengths and limitations.

6.2 Genomics

Through the drug-discovery process, the identification of drug targets is a major advantage for subsequent structure–activity relationship studies that greatly contribute to drug optimization and drug development.[19] The knowledge of biological targets allows mechanistic studies by modification of the chemical structure of compounds, in order to improve their potency and selectivity while reducing drug-related side effects. Furthermore, the use of known validated targets for drug screening is not only important for rational drug design but allows a better understanding of the biological mechanisms of disease.

6.2.1 Classical Genomic Strategies in *Leishmania*

In *Leishmania*, the development of drug resistance and/or side effects for some treatments due to off-target effects severely limits the efficacy of treatments available.[20] Hence, the development of novel selective and well-tolerated leishmanicidal drugs is needed. Target-based and phenotype-based screening are two strategies that have proven successful in finding new active molecules against protozoan parasites.[21–28] While target-based approaches have the advantage of targeting specific proteins that are essential for parasite survival and absent or highly divergent from the host, avoiding off-target effects, phenotypic approaches allow the identification of hit compounds that inhibit parasite proliferation or survival regardless of their mechanism of action. Currently, all leishmanicides on the market are derived from drug screening approaches with phenotypic read-outs,[26] but the lack of knowledge of the biological targets for many of them hampers chemical optimization efforts to improve drug selectivity and reduce side effects.

The sequencing of *Leishmania major* genome in 2005[10] followed by additional ones from other *Leishmania* species during the last decade[11,29–31] has allowed comparative genomic analysis, spurring drug discovery efforts by revealing thousands of genes as potential drug targets. Moreover, with the advent of NGS technologies, high-throughput and genome-scale screening has led to a surge in our comprehension of pathogenesis, disease transmission and drug efficiency and resistance of kinetoplastid parasites. In particular, the development of RNA interference (RNAi) target-sequencing approaches (RIT-Seq) in *Trypanosoma brucei* revealed numerous genes associated with a loss-of-fitness in parasite differentiation[32,33] and drug action.[34] In *Leishmania*, changes in gene expression, gene and chromosome copy number variations (CNVs) and single-nucleotide polymorphisms (SNPs) were detected and have brought new insights into transcription regulation,[35] population genetics[29,30] and mechanisms of drug resistance.[13]

6.2.2 Novel Gain-of-function Genomic Approaches

6.2.2.1 Cosmid Sequencing

While genome-wide RNAi-based screening approaches are limited to the *Leishmania* (*Viannia*) subgenus, since most species lack functional RNAi machinery,[36] gain-of-function screens based on the introduction of an episomal cosmid genomic library remain some of the most suitable methods to study the genetic basis of drug resistance and mode of action (MOA) in *Leishmania.*[37–39] An approach termed Cosmid Sequencing (Cos-Seq), that uses cosmid-based functional screening coupled to NGS, is a sensitive method for studying the dynamics of cosmid enrichment under selective drug pressure (Figure 6.2). This led to the identification of many previously unidentified genes as putative drug targets and resistance determinants.[40] After the introduction of the genomic library into drug-susceptible parasites, the pooled population of transfectants was subjected to incremental concentrations of anti-leishmanial drugs. At each selection step, cosmid-derived libraries were sequenced and analysed, allowing the genome-wide profiling of drug-enriched loci for each Cos-Seq selection. A first screen with the model drug methotrexate revealed an enrichment of cosmids containing a dihydrofolate reductase–thymidylate synthase (DHFR–TS),[40] a well-established target of the drug.[41,42] Subsequent, Cos-Seq screens with the five major anti-leishmanial drugs (organic antimonials, miltefosine, amphotericin B, paromomycin and pentamidine) highlighted many candidate targets to be investigated to determine whether some of these genes encode genuine drug targets,[40] thus supporting the hypothesis that current leishmanicides act on multiple minor targets. Some of the genes identified by Cos-Seq are clearly drug-resistance genes, such as the ATP-binding cassette (ABC) transporters MRPA (ABC-thiol transporter) and PRP1 (pentamidine resistance protein 1), whose implications in antimonial and pentamidine resistance, respectively, are well established.[43,44] The genes isolated by Cos-Seq in promastigote stages were also able to confer resistance in intracellular amastigotes.[40] Cos-Seq was developed for promastigotes, as they are much easier to handle, but is compatible with axenic and intracellular amastigotes, and may even be applied in *in vivo* animal models. This highly sensitive method may be also used with new active lead compounds with unknown MOA[24] to pinpoint their putative targets.

6.2.2.2 Chemical Genomics

A second promising approach for the identification of drug targets is to combine chemical screening with genome sequencing, a technique termed chemical genomics. This method usually involves the *in vitro* evolution of drug-resistant strains, followed by subsequent whole-genome analysis to find the targets of some of the many compounds discovered initially in whole-cell phenotypic and high-throughput screens. As a proof-of-mechanism, this method was recently illustrated in *Trypanosoma cruzi* leading to the discovery of cytochrome *b* as a novel drug target.[45] The compound GNF7686 was identified through a high-throughput screening on *Leishmania donovani* axenic amastigotes and further characterized for selectivity and efficiency against other kinetoplastid parasites. To investigate the mechanism of action, GNF7686-resistant *T. cruzi* parasites were generated using stepwise selection and sequenced by NGS. Whole-genome sequencing revealed a homozygous mutation in the cytochrome *b* gene, indicating a feasible inhibition of the mitochondrial respiratory chain by GNF7686. Further additional chemogenomic profiling experiments supported this hypothesis.[45] More recently, the same workflow was applied and led to the discovery of a selective inhibitor of the *T. cruzi* proteasome with *in vivo* efficacy against kinetoplastid parasites.[25] After whole-cell phenotypic and high-throughput screening of chemical libraries using proliferation assays and subsequent chemical optimization, the authors selected the lead compound GNF6702 which was validated *in vivo* against *L. donovani*, *L. major*, *T. cruzi* and *T. brucei*.[25] Following *in vitro* selection of *T. cruzi*-resistant strains against GNF6702 and subsequent whole-genome analysis by NGS, the authors identified two independent mutations in the β_4 subunit

■ Cos-seq

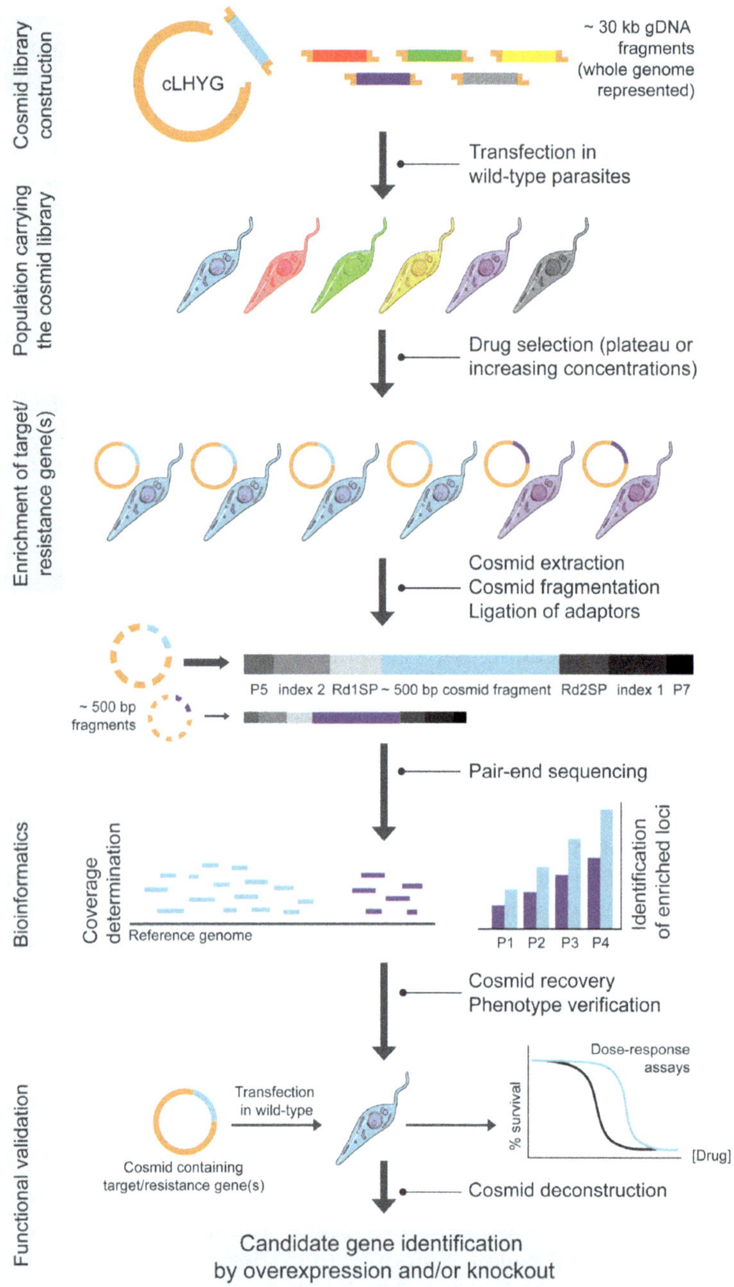
Cosmid library construction
cLHYG
~ 30 kb gDNA fragments (whole genome represented)
Transfection in wild-type parasites
Population carrying the cosmid library
Drug selection (plateau or increasing concentrations)
Enrichment of target/ resistance gene(s)
Cosmid extraction
Cosmid fragmentation
Ligation of adaptors
~ 500 bp fragments
P5 index 2 Rd1SP ~ 500 bp cosmid fragment Rd2SP index 1 P7
Pair-end sequencing
Bioinformatics
Coverage determination
Reference genome
P1 P2 P3 P4
Identification of enriched loci
Cosmid recovery
Phenotype verification
Functional validation
Transfection in wild-type
Cosmid containing target/resistance gene(s)
Dose-response assays
% survival
[Drug]
Cosmid deconstruction
Candidate gene identification by overexpression and/or knockout

of the proteasome, further validated by biochemical and cell proliferation assays.[25]

6.2.3 Current Limitations and Future Development of Genomic Approaches

Once a drug target has been identified, target validation can be performed using classical genetic methods, such as gene or transcript knockdown. However, in *Leishmania*, most species lack functional RNAi machinery.[36] Consequently, loss-of-function studies rely mainly on gene inactivation by targeted homologous-recombination-mediated gene replacement using a PCR fusion-based strategy.[46] The novel clustered regularly interspaced short palindromic repeats (CRISPR)–CRISPR-associated endonuclease (Cas) prokaryotic system for genome editing has been recently incorporated into the molecular toolbox for *Leishmania* parasites.[47,48] Despite its potentiality for a genome-wide targeted mutagenesis approach, extremely helpful for aiding drug discovery, it is not exempt from limitations. These include the absence of a non-homologous end-joining (NHEJ) pathway in *Leishmania*, as DNA is mostly repaired by homology-directed repair (HDR), and the difficult expression of a functional guide RNA (gRNA). Nevertheless, some of these issues could be solved soon thanks to the recently annotated Cpf1 nuclease, with potential advantages over Cas9 for its implementation in trypanosomatid parasites: (i) both Cpf1 nuclease and its associated gRNA are smaller than those present in the CRISPR–Cas9 system, and (ii) Cpf1 may improve the frequency of HDR over NHEJ by cleaving several times at the same site, as the sequence recognized by the gRNA was not deleted during the editing.[49]

Innovative combinations of phenotype-based screening with genome sequencing for the identification of drug MOA in *Leishmania*, highlight the genomics contribution to drug development for parasitic infections resulting in new active hits molecules and candidate drug targets for future research. However, efforts are still required in the development of loss-of-function genome-wide screening approaches, to complement gain-of-function studies, hence expanding the *Leishmania* genomics toolbox.

Figure 6.2 **The Cos-seq genomic approach.** A cosmid genomic DNA (gDNA) library covering the whole genome of wild-type *Leishmania* parasite is engineered in the cLHYG cosmid vector. This library is subsequently transfected into wild-type *Leishmania* parasites. The resultant heterogeneous population is then submitted to drug pressure (either increasing drug concentrations at each passage or a plateau protocol). Cosmids are isolated after each selection step and processed (fragmentation and ligation of adaptors) for subsequent paired-end Illumina sequencing. Bioinformatics pipelines allow the mapping of the NGS reads to the reference genome and calculation of gene abundance as well as differentially enriched loci. The final steps of this procedure involve the isolation of cosmids containing target or resistance candidate gene(s) for their phenotypic confirmation and functional validation.

6.3 Transcriptomics

Since trypanosomatid genomes are highly syntenic due to their polycistronic transcription,[50] transcriptomic approaches are useful to provide clues about mRNA stability, processing and gene expression regulation (see Chapter 12). Recent advances in drug target prospecting, identification and validation in *Leishmania* parasites are supported by basic research using molecular transcriptomics. Useful transcriptomic information to better understand parasite's behaviour under several microenvironmental conditions can be obtained not only by the tracking of gene expression modulation but also by the characterization of non-coding RNA panels. This strategy can be applied to study the mechanisms of action or resistance (MOAR) of anti-leishmanials in a systems biology manner, an expanding field where bioinformatics and data integration tools are essential. Together with genomic, proteomic and metabolomic data, transcriptome profiling offer hints to better understanding of parasite biology, detecting essential biochemical pathways for drug discovery and development purposes.

6.3.1 Transcriptomic Toolkit for Target Elucidation in *Leishmania*

The post-genomic era allowed the development of high-throughput hybridization techniques based on annotated genome sequences to monitor differential gene expression through complete cDNA microarrays in kinetoplastids.[51] This opened new avenues for obtaining the transcriptome profile to identify and implicate gene candidates in stage-specific features; as markers for vaccine studies; for drug target elucidation for drug development purposes and to better understand the mechanisms behind drug resistance in *Leishmania* parasites.[52–55] Since microarray technology relies on the use of probes, cross-hybridization, limited detection range and non-specific hybridization are some of the issues that limit its use.[56] Despite these technical limitations, together with the popularization of NGS technologies for RNA profiling, complete or customized cDNA microarrays are still in use and can be useful to detect a range of differentially expressed genes. Based on a previous study using cDNA microarray performed in antimony-unresponsive derived *L. donovani* field isolates,[57] Kaur and colleagues revealed the fatty acyl-CoA synthetase—an enzyme responsible for acyl-CoA formation through the hydrolysis of pyrophosphate—as a target candidate to be considered for drug development against leishmaniasis.[58] Known as a multifactorial phenomenon, drug resistance is a suitable condition to be explored by comparative transcript expression profiling. Indeed, miltefosine-resistant *L. donovani* from Indian patients, presented differentially expressed genes related to drug efflux, altered DNA replication and increased antioxidant defence, as revealed by customized transcript microarrays.[59] Yadav and colleagues recently applied the cDNA microarray in order to better understand the cell cycle in histone acetyltransferase HAT4 null

L. donovani mutants (HAT4-null). They demonstrated that HAT4-null exhibited a prolonged G2/M phase associated with the downregulation of *LdCdc20*, and thus a new regulatory mechanism to be investigated.[60]

Although RNA profiling by NGS technologies (RNA-Seq) does not require pre-designed complement probes, there is a high correlation of differential gene expression between microarray and RNA-Seq. The latter can be further applied to detect low-abundance transcripts, differentiate biologically critical isoforms and identify genetic variations, including polyadenylation and alternative splicing sites, on *Leishmania* mRNA sequences. Taking advantage of this high-throughput, massive and deep sequencing technology, RNA-Seq has been applied to study mRNA abundance in *Leishmania* parasites.[61,62] This allowed the massive identification of transcripts in *L. major* from which almost 20% did not match previously annotated genes; the characterization of processing spliced leader (SL) and polyadenylation addition sites, as well as the monitoring of mRNA stability based on untranslated region sequences as a condition for RNA translation.[62] Although *Leishmania* ribosomal RNA (rRNA) can be unusually polyadenylated during post-transcriptional processing,[63] libraries for RNA-Seq are constructed using PolyA$^+$ mRNA selection (Figure 6.3).[62]

Since rRNA are highly abundant, the rRNA-derived reads force an increase in read depth in order to obtain enough coverage to detect low differentially expressed genes. To circumvent this problem, the team headed by Dr Peter J. Myler (University of Washington, Seattle, USA) developed a protocol based on an SL-enriched mRNA library for mRNA-Seq profiling, where the 39-nucleotide mini-exon found at the 5′ end of all *Leishmania* mRNA specifically selects them for library preparation.[61] Several attempts were made to purify non-rRNAs using different commercially available approaches such as: RiboZero (Illumina), Terminator 5′-Phosphate Dependent Exonuclease (Epicentre), RiboMinus (ThermoFischer) or using non-commercial magnetic-based biotinylated rRNA removal probes (Rubens Monte-Neto, Philippe Leprohon and Marc Ouellette, unpublished data), however, "rRNA-free libraries" were often contaminated by rRNA. A non-rRNA library, including all other RNAs (*e.g.* mRNA, non-coding RNAs and micro RNAs) is yet to be described for *Leishmania* parasites. Thus, non-rRNA or non-mRNA library preparation must include purification by an affinity- or size selection step. In this regard, Eliaz and colleagues analysed the complete repertoire of small nucleolar RNA (snoRNAs) in *L. major* and revealed the role of snoRNA in rRNA processing by structural modifications.[64] Despite the aforementioned applications, RNA-Seq-derived transcriptomic profiles are also useful for resequencing, improving *Leishmania* reference genome annotation;[62,65] elucidating transcriptome remodelling across procyclic to infective metacyclic promastigotes[66] and monitoring metabolic biochemical reprograming under starvation to mimic microenvironmental changes.[67]

Although active RNAi machinery has been evolutionarily retained only in species of *Leishmania* belonging to the subgenus *Viannia*, it is a powerful loss-of-function tool to functionally manipulate gene expression.[36] The knockdown of amastins—claudin-related surface glycoproteins highly

■ RNA-seq

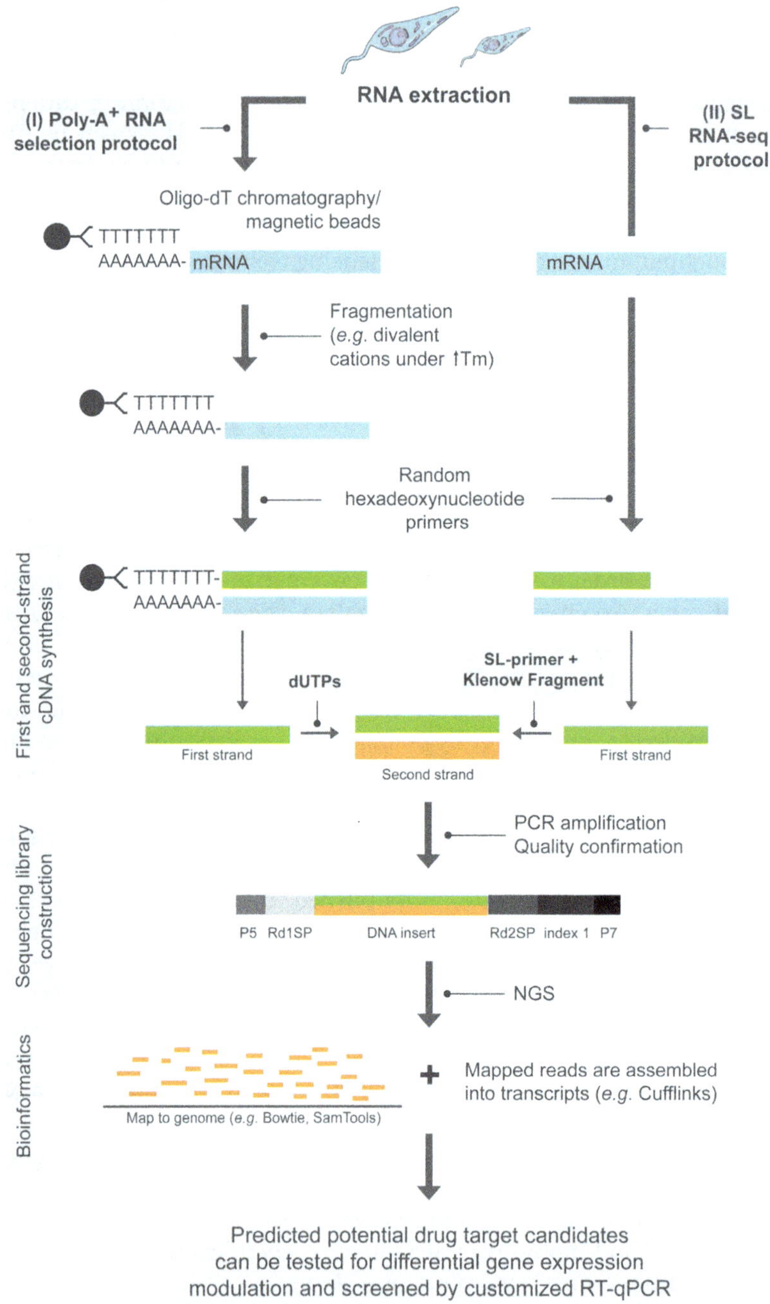
RNA extraction
(I) Poly-A+ RNA selection protocol
(II) SL RNA-seq protocol
Oligo-dT chromatography/ magnetic beads
TTTTTTT
AAAAAAA- mRNA
mRNA
Fragmentation (e.g. divalent cations under ↑Tm)
TTTTTTT
AAAAAAA-
Random hexadeoxynucleotide primers
TTTTTTT-
AAAAAAA-
First and second-strand cDNA synthesis
dUTPs
SL-primer + Klenow Fragment
First strand
Second strand
First strand
PCR amplification
Quality confirmation
Sequencing library construction
P5 Rd1SP DNA insert Rd2SP index 1 P7
NGS
Bioinformatics
Map to genome (e.g. Bowtie, SamTools)
Mapped reads are assembled into transcripts (e.g. Cufflinks)
Predicted potential drug target candidates can be tested for differential gene expression modulation and screened by customized RT-qPCR

expressed during the intracellular amastigote stage and potentially involved in tight junction complexes—by RNAi in *Leishmania braziliensis* led to an attenuated phenotype *in vitro* and *in vivo*, revealing a potential target for drug discovery studies.[68] It is important to note, however, that when choosing cDNA microarray, RNA-Seq or RNAi approaches, gene expression modulation must be validated by northern-blot or real-time quantitative PCR (RT-qPCR). Indeed, predicted potential drug target candidates can be tested for differential gene expression modulation and directly screened by customized RT-qPCR.

6.3.2 Recent Advances in and Perspectives on Transcriptomics for Studying Chemotherapy Against *Leishmania*

Since leishmaniasis is caused by dixenous parasites and clinical manifestations vary according to the etiological agent and to the host immunological conditions, microenvironmental changes would influence disease and/or treatment outcome. This will alter data interpretation of transcriptome profiling derived from field or cultured parasites.[69] Concomitant human and *L. braziliensis* transcriptome profiling from primary cutaneous lesions revealed a correlation between differential immune capacities of some patients and the development of metastatic mucosal leishmaniasis.[15] In a recent work, Christensen and colleagues simultaneously analysed human and *L. braziliensis* meta-transcriptomes from skin biopsies of early- and late-infected patients. They found that biopsies with detectable parasites presented increased B lymphocyte-specific and immunoglobulin transcripts, while amastin-like proteins were one of the *L. braziliensis*-derived highly expressed transcripts. Curiously, the most abundant parasite transcripts *in vivo* did not correlate with those from experimentally infected macrophages *in vitro*.[70]

Another interesting point concerns the role of non-coding RNAs (ncRNAs), widely involved in regulatory processes in several organisms. However, very little is known about their association with the outcome of leishmaniasis treatment. In order to facilitate the search for potential ncRNA targets, a

Figure 6.3 Overview of a classic RNA-seq experiment in *Leishmania* using PolyA$^+$ or SL mRNA enrichment. After total RNA extraction, mRNA is (I) enriched by oligo-dT based chromatography or magnetic beads or (II) selected by splicing leader (5′-SL) sequence. Captured RNA (I) or total unpurified RNA (II) is used for synthesizing the first cDNA strand by means of random primers. Next, the second cDNA strand is synthesized and both strands ends are then repaired and adenylated. The library is subsequently enriched in ligated fragments by PCR amplification, and its size and quality are confirmed prior to sequencing. After sequencing, reads are mapped to the *Leishmania* genome and assembled into transcripts. Finally, predicted potential drug target candidates can be tested for differential gene expression modulation and directly screened by customized RT-qPCR.

team headed by Dr Vinicius Maracaja-Coutinho (Universidad Mayor, Santiago, Chile) created a freely available database including structural ncRNA predictions mapped on *L. braziliensis* genome. This is part of their efforts on a *Leishmania* genome reannotation public database named LeishDB (www.leishdb.com).[71] Because of the growing access to NGS technologies, bioinformatic analysis is still the bottleneck for the advances in transcriptomic profiling for drug discovery purposes that need database resources integration.

To expand the toolbox for transcriptomic manipulation in *Leishmania* parasites, a revolutionary recent discovery by Dr Jennifer A. Doudna and colleagues (University of California, Berkeley, USA) may allow RNA interference mediated by CRISPR. They demonstrated that bacterial adaptive immune type VI CRISPR system uses C2c2 as RNA-guided Cas to target single-stranded RNA, and not DNA as originally described for Cas9.[72] This represents a new appealing tool for the study of functional transcriptomics, even in organisms lacking RNAi machinery, such as *Leishmania* species.

6.4 Proteomics

Modern proteomics allows the investigation of cellular components as well as secreted peptides and proteins, protein–protein interactions and post-translational modifications (PTMs). PTMs are of particular interest in *Leishmania* as the Trypanosomatidae regulate gene expression at the post-transcriptional and post-translational levels. Individual genes derived from polycistronic-precursor mRNAs have their expression regulated by mechanisms that stabilize the respective mRNA, trigger translation and, ultimately, stabilize the protein after translation.[16] Post-translational regulation is also associated with stage- or species-specific gene expression.[66,73,74] Genomics and transcriptomics techniques fail to reconstruct accurate signalling and metabolic pathways due to the poor correlation observed between mRNA content and the levels of translated proteins. Consequently, proteomics represents a preferable approach for the discovery of novel immunogenic proteins,[17,75] novel drug targets and drug-resistance mechanisms.[76,77]

6.4.1 Proteomics Methods

Current state-of-the-art proteomic technologies allow for a consistent, in-depth profiling of the *Leishmania* proteome. The first proteome maps of *Leishmania* were published more than 30 years ago using two-dimensional gel electrophoresis (2DE) techniques.[78,79] However, the major proteomic breakthrough came with the development of two soft ionization techniques: matrix assisted laser desorption/ionisation (MALDI)[80] and electrospray ionization (ESI),[81] which allowed for the identification of proteins and peptides by mass-spectrometry (MS) analysis. The second milestone in modern proteomics was achieved by the inclusion of fluorescent tags into the method pipelines. This turned 2DE into a highly sensitive and quantitative method.[82] Since then, several groups have tried to improve the methods used in *Leishmania* 2DE experiments. Brotherton and colleagues[83] used free-flow electrophoresis in isoelectric focusing mode (IEF–FFE) followed by protein separation using 2DE to increase proteome coverage of basic proteins of both promastigotes and axenic amastigotes of *Leishmania infantum*. This was an important improvement in the visualization of basic proteins using 2D gels. However, 2DE remains fairly insensitive to low-copy proteins and does not allow complete proteome coverage.[84] These limitations have driven the development and introduction of gel-free proteomics techniques, which are advantageous for the detection of low abundance proteins. Gel-free methods do, however, lack the ability to pre-select protein species or visualize differences in protein levels due to PTMs, which is possible with 2DE. Recently, Washburn and colleagues developed the multidimensional protein identification technology (MudPIT),[85] which uses two-dimensional liquid chromatographs directly coupled to tandem mass spectrometers (LC–MS/MS). The non-quantitative nature of MudPIT requires the introduction of several peptide-labelling complementary methods if quantification is desired. To this end, Rosenzweig and colleagues used isobaric tagging for relative and absolute quantifications (iTRAQ/LC–MS/MS) proteomic methodology in order to study PTMs of cellular proteins associated with stage differentiation in *L. donovani*.[86,87] Another tagging option used in *Leishmania* studies is the so-called stable isotope labelling amino acids in cell culture (SILAC). SILAC has been used to study protein secretion in *L. donovani*, finding non-classical secretion pathways, including the release of exosome-like microvesicles.[88] SILAC has also been applied to the study of drug-resistance mechanisms in *Leishmania*.[89,90] Finally, immobilized metal affinity chromatography (iMAC) has been used to reconstruct the whole *Leishmania* phosphoproteome and to identify phosphoproteins implicated in stress- and heat shock response, RNA and protein turnover, metabolism[82] and signalling;[91] as well as to identify and characterize novel virulence factors in these parasites.[92]

6.4.2 Proteomics in the Discovery of Modes-of-action and Drug-resistance Mechanisms in *Leishmania*

Resistance to the current available anti-leishmanial drugs is an important health issue. An understanding of drugs' MOAs and MOARs is essential in order to design improved public-health policies to prevent such resistance occurring. Proteomics has proved to be a powerful omic tool to unveil unknown mechanisms of action and/or drug resistance factors in *Leishmania* parasites. The first proteomics screen of drug-resistant *Leishmania* strains involved a 2DE comparative analysis of a genetically related pair of antimonial-sensitive and -resistant *L. donovani* field strains isolated from kala-azar patients. An important outcome was that programmed cell death (PCD) is altered in the antimonial-resistant parasites.[93] Two proteins, HSP83 and SKCRP14.1, were found to be involved in the MOA of several antileishmanial drugs, which would be further confirmed and expanded in several subsequent studies.[94,95] While upregulation of HSP83 is able to interfere with the mitochondrial membrane potential, resulting in a negative modulation of drug-mediated PCD; downregulation of SKCRP14.1 seems to protect resistant parasites against antimonials.[93] El Fadili and colleagues performed 2DE studies in axenic amastigotes of antimonial-resistant *L. infantum* strains in order to clarify the antimonial-resistance phenomena in the intracellular stage of the parasite.[96] The study revealed a major downregulation of the level of the kinetoplastid membrane protein KMP-11 in antimonial-resistant mutants, including one mutant where resistance was induced in the amastigote stage. Further proteomics studies have pinpointed a variety of important pathways involved in antimonial resistance. An iTRAQ comparative analysis found a marked upregulation of proteins involved in glycolysis in antimonial-resistant isolates, indicating that this pathway is a possible mechanism of defence, protecting resistant parasites from unusually higher oxidative stresses.[97] In addition to proteins involved in general stress responses, Walker and colleagues[98] found a novel mechanism involved in antimony resistance through the generation of several sub-proteomes of antimony-resistant *Leishmania panamensis* promastigotes using 2DE analyses. These upregulated proteins included the enzymes *S*-adenosylmethionine synthetase and *S*-adenosylhomocysteine hydrolase, both key enzymes in trypanothione synthesis.[98]

Certain proteomics studies have the limitation of under-representing membrane proteins. One of the key proteins in antimony resistance is the ABC transporter MRPA (ABCC3), commonly amplified on an extrachromosomal circular amplicon in both *in vitro* mutants[99,100] and field isolates.[101,102] However, this finding could not be confirmed at the protein level until Brotherton and colleagues used a specific proteomic analysis of membrane proteins through a SILAC approach combined with whole-genome sequencing to characterize a *L. infantum* resistant mutant.[89] *Leishmania* SILAC-based studies have also been essential in the elucidation of paromomycin and amphotericin B modes of action and resistance mechanisms.[90,103] Miltefosine, another leishmanicidal drug, was analysed through 2DE analyses followed by MS, revealing the potential involvement of eukaryotic initiation factor 4A in drug resistance in *L. donovani* promastigotes.[104] Further studies by Vincent and colleagues included the first 2DE–MS analysis of the *Leishmania* mitochondrial proteome, showing similar changes for some proteins pertaining to oxidative processes when the proteomic profiles from the antimonial and miltefosine-resistant mutants were compared. This work indicated ROS control to be a common mechanism of resistance against these two unrelated drugs.[105] Additionally, some preliminary proteomic studies involving the drugs sitamaquine and α-difluoromethylornithine (DFMO) in *L. donovani* have been published.[106,107] Interestingly, DFMO-resistant parasites revealed a dynamic modulation of critical enzymes involved in polyamine biosynthesis, protecting these parasites against drug-induced oxidative stress.[107]

6.4.3 The Future of Proteomics in Drug Discovery

In recent years chemical proteomics has arisen as a powerful complementary strategy for drug-target discovery.[108] This approach uses small drug-like molecules that can be bound to a ligand immobilized on a solid support or exposed to protein chips. Subsequently, those proteins bound to the ligand are identified as potential drug targets.[109] Wright and colleagues were the first group to adapt this technology to the study of a potential drug target in *Leishmania* parasites.[110] This study targeted *L. donovani's* global myristoylation and the suitability of the enzyme myristoyl-CoA:protein *N*-myristoyltransferase as a drug target in *Leishmania*, successfully validated in other protozoan parasites.[111,112]

Proteomic analyses tend to highlight the most abundant proteins in a sample, at the expense of less abundant, but no less important, proteins that may also contribute to a specific phenotype.[76,113] This issue has been partially resolved by studying independent sub-proteomes enriched with specific cellular components. Whole-cell proteomic studies encompass an important bottleneck associated with the difficulty of finding meaningful biological changes among the massive amounts of data generated. Consequently, proteomic-specific bioinformatics tools and databases have become essential for a more accurate interpretation of the *Leishmania* proteome. Databases holding vast amounts of protein–drug information (*e.g.*, PubChem:[114] https://pubchem.ncbi.nlm.nih.gov; DrugBank:[115] http://www.drugbank.ca; ChEMBL: https://www.ebi.ac.uk/chembl/) are being screened in an automated manner to identify potential opportunities for drug repurposing.[116] This represents a novel and inexpensive approach to pre-screen entire leishmanial proteomes against known bioactive compounds. Finally, TDR targets (http://tdrtargets.org), an innovative chemogenomics resource for neglected diseases, integrates many of these databases as a single online resource for the rapid identification of molecular targets for drug development.[117] In this way, the leishmanial DNA topoisomerase IB, a rare heterodimer essential to *Leishmania*,[118] has been recently validated as a promising target for drug intervention against visceral leishmaniasis.[119,120]

6.5 Metabolomics

Metabolomics is the newest of the omics technologies and has been fundamental in the discovery and validation of pathways in the parasitic protozoa.[121] The technology is at the forefront of cellular metabolism and gives a real-time readout of metabolites involved in cellular processes in a much more dynamic way than genomics, transcriptomics or proteomics do, as cells move through different life cycle stages or different nutritional environments. As a new technology, however, metabolomics has been the most difficult to standardise. As a result, there are numerous competing technologies, methods and software packages in use, even within the relatively small field of *Leishmania* metabolomics.[121]

Metabolomics technologies are gaining in popularity in protozoan parasite research, as large compound libraries become available to academia as part of the London Declaration on Neglected Tropical Diseases. Metabolomics is an important tool to shed light on the phenotypic screens for those drugs whose targets are often unknown. Analysis of the drug-dependent perturbations to the metabolome can provide clues to the MOA of a compound highlighted from a phenotypic screen.

6.5.1 Metabolomics Methods

The metabolite content of a cell is generally considered to comprise small molecules of between 50 and 1400 Da. These metabolites are the substrates and products of enzymatic processes as well as cofactors, toxins and other molecules. To analyse the metabolome of a cell population, cultures are generally quenched to immediately cease metabolism before cells are concentrated and lysed and proteins and membranes are removed. Cell extracts are fractionated using chromatography and individual metabolite levels are measured using either MS or nuclear magnetic resonance (NMR). Data deconvolution is very diverse and most research groups use their own preferred tools. For a review of metabolomics techniques used in parasitic protozoa see Vincent and Barrett.[121]

Data analysis is the most complex step of any metabolomics experiment and most labs have their own preferred software packages to pick out mass features and identify metabolites. Metabolite identification relies on the collection of various orthogonal data to increase identification confidence. The metabolomics standards initiative provides a set of recommendations on how these data should be,[122] but they generally include:

1. Feature mass either matched to an authentic standard or a database.
2. Retention time or retention index of feature compared with standard retention times.
3. Natural isotope distribution either alone or compared with an authentic standard.
4. Fragmentation of the parent ion into its constituent parts compared with a database (*e.g.* mzCloud – https://www.mzcloud.org) or an authentic standard.
5. Fragmentation to produce ion trees for *de novo* identification.

Isotopically labelled carbon or nitrogen atoms are tracked though pathways in real time or allowed to reach a state of equilibrium.[123] Fragmentation of labelled metabolites downstream of a precursor can allow the tracking of every atom from a precursor metabolite.

Controls in metabolomics experiments are extremely important. The biochemistry of a parasite will change over the course of its life cycle[124] as the population becomes denser, so controls for cell-cycle and cell-density effects must be included in each experiment. Test compounds must also be analysed for purity, for in-source fragmentation, and for common adducts, in order to exclude drug artefacts from the changes seen in the drug-treated metabolome.

6.5.2 Metabolomics and Anti-leishmanial Drugs

To date, all metabolomics experiments on anti-leishmanial drug-treated parasites have been performed on the insect promastigote stage. The biochemistry of promastigotes is likely to be very different from that of the vertebrate-infective amastigote stage, so a caveat must be raised regarding the use of promastigotes to analyse the effects of drugs.[125] It is certainly more difficult to culture axenic amastigotes and their biochemistry is not necessarily closer to that of intracellular amastigotes.[126,127]

6.5.2.1 Organic Antimonials

The MOA of antimonials was analysed by Canuto and colleagues.[128] In this study capillary electrophoresis (CE) MS was used to analyse promastigotes treated with 120 μM Sb^{III} for 12 h.[128] Following treatment, there was a large increase in thiol-containing metabolites including glutathione and cystathionine and *S*-adenosyl-homocysteine of the reverse *trans*-sulphuration pathway. There were also increases in some amino acids.[128] Glutathione is not the main thiol in *Leishmania*, instead two glutathione molecules are joined by a polyamine linker (spermidine) to form trypanothione, a molecule not detected on this platform. Antimony(III) is known to be pumped out of *Leishmania* through conjugation to trypanothione, which also helps to reduce the initial Sb^{V} into Sb^{III}, its active form.[129–132] The glutathione that was detected in this study was in its disulphide form, which would indicate a reduced capacity to deal with oxidative stress after drug treatment.

6.5.2.2 Miltefosine (Hexadecylphosphocholine)

Miltefosine is a phospholipid analogue, so analyses of its MOA should include a detailed lipidomic analysis of the plasma membrane as well as internal lipid metabolism. Two metabolite extraction methods (methanol–chloroform–water and methanol–water) and three chromatography methods: gas chromatography (GC), liquid chromatography (LC) and capillary electrophoresis (CE) were used by Canuto and colleagues to analyse the MOA of miltefosine in *L. donovani* promastigotes.[133] Unfortunately, a reversed-phase analysis was missing from their repertoire, so lipid species were not included in their dataset. They did detect, however, a decrease in levels of arginine, ornithine and citrulline, all nitrogenous amino acids of the polyamine pathway.[133] Trypanothione disulphide (the reduced end product of the polyamine pathway) was also decreased in treated parasites.[133]

Another study of miltefosine action used LC–MS to analyse *L. infantum* promastigotes treated with miltefosine over five hours at its concentration giving 90% inhibition (IC_{90}; 20.6 μM).[134] Broad changes to the metabolome were detected, including increases in alkane fragments and sugars, DNA damage and changes to thiol and polyamine levels. A lipid analysis showed no changes to phospholipids after treatment, indicating that changes were related to internal lipid metabolism accompanied by oxidative stresses.[134] An analysis of membrane lipids by Imbert and colleagues, however, did show changes to several lipid classes after 48 hours of 10 μM miltefosine treatment of *L. donovani* promastigotes,[135] which may indicate differences in drug action dependent on the drug concentration used or treatment exposure time.

6.5.2.3 *Amphotericin B*

Polyene amphotericin B is the first-line anti-leishmanial agent (either alone or in combination with miltefosine) for treating visceral leishmaniasis in highly endemic regions of India where organic antimonial derivatives have become obsolete due to drug resistance.[136] The MOA of amphotericin B encompasses the preferential binding to ergosterol and formation of channel-like pores through the plasmatic membrane, causing a leakage of metabolites that damages the cell and triggers parasite killing.[137] The MOAR of amphotericin B involves an altered ability of the parasite to synthesize ergosterol, which blocks the binding of the drug and its pharmacological effect.[138] A recent metabolomics study targeting lipids in amphotericin-B-resistant parasites revealed by electrospray ionisation-mass spectrometry (ES-MS) the protective effect derived from the increase of several inositol-phosphoceramide species (IPCs).[139] In addition to IPCs, Fernandez-Prada and colleagues found a marked increase in fatty acid C19Δ, which seems to provide a substantial survival gain by protecting against environmental stress in other organisms,[140] after derivatisation of the total fatty acid content into fatty acid methyl esters and subsequent quantification by GC–MS.[139] More interestingly, Fernandez-Prada and colleagues found that lipid remodelling was not only present in amphotericin-B-resistant parasites but in miltefosine-resistant ones too. These changes seem to be driven by different mutations in a common P-type ATPase transporter, thus evidencing the risk of cross-resistance emergence against current sequential co-treatments against visceral leishmaniasis.[139]

6.5.3 Resistance and Synergy

Studies on resistance can be problematic as lab mutants are generally selected for resistance on promastigote forms of the parasite, which may not be indicative of real-life resistance mechanisms. Clinical resistant isolates also pose problems as they frequently lack a clean genetic background for comparing metabolic responses.

Despite these issues, there have been several attempts to study the metabolome of both laboratory-selected resistant promastigotes and field isolates. Studies on antimonial resistance by Berg and colleagues found an increase in the potential for cells to cope with oxidative stresses.[124] This seems to be a common resistance mechanism as *Leishmania* resistant to a single leishmanicide is often tolerant to other leishmanicides, increasing the potential to develop a multidrug resistance phenotype.[141] It is therefore pertinent to study drug combinations in *Leishmania*. Another study by Berg and colleagues attempted to deconvolute the mechanisms behind multidrug resistance.[142] Again, protection against oxidative stresses was highlighted in the metabolomics data, as well as changes in membrane fluidity.[142]

Reduced drug uptake is another common resistance mechanism. Deficient miltefosine uptake has been shown in a metabolomics analysis of miltefosine-resistant cells by Vincent and colleagues.[134] Mutations in the miltefosine transporter had already been detected in relation to the loss of drug import.[143] More studies of drug uptake and metabolism will be useful as new lead compounds enter the development pipeline.

Drug synergy has not been studied in *Leishmania*, but it has in a single study in the related parasite *T. brucei*.[144] Here, the nifurtimox–eflornithine combination, used in the field, was tested *in vitro*. This combination was not synergic either for inhibitory concentration, or for metabolomics assay.[144] Similar studies conducted in *Leishmania* will allow drug combinations to be assessed *in vitro* before testing in experimentally infected animal models.

6.5.4 The Future of Metabolomics in *Leishmania* Drug Development

There is a pressing need for *Leishmania* drug research to move into macrophage-dwelling cells. To study drug action, infected macrophages can be treated as a single parasitological unit, as metabolic changes upon drug pressure may occur at the parasite–macrophage interface or within both cellular entities separately. There has been substantial progress in the malaria field in the deconvolution of parasite and host cell biochemistry,[121] but *Leishmania* research has yet to catch up. Resistance to compounds should also be selected in infected macrophages or in animals[145] to better reflect the field situation while obtaining a clean, tractable genetic background.

Given the worrisome levels of drug resistance in the field and the need for new combination therapies, drug synergy studies should be developed in *Leishmania*. Researchers developing these should be mindful of potential resistance mechanisms[142] and design rational drug combinations to prevent multidrug resistance from developing.

There is a large heterogeneity in the response of patients and *Leishmania* strains to chemotherapy. The detection of likely responders to chemotherapy is a new field of pharmacogenomics in parasitology. Biomarkers in blood or urine are being sought to detect people that will respond well to a certain treatment and those that will not, avoiding prescribing toxic treatments to people that will not benefit.

Another potential use of metabolomics in anti-leishmanial drug development is predicting drug toxicity in human cell lines. The identification of the mechanisms of toxicity may allow some drugs, otherwise side-lined, to be modified to reduce their toxic effects.

6.6 Concluding Remarks

The application of omics approaches to *Leishmania* research has provided invaluable insights into the biological process that drive the replication, differentiation and transmission of these pathogens. Furthermore, these techniques constitute a cornerstone of the discovery and development of novel molecularly targeted therapeutic approaches, as well as a fundamental toolbox for the study and appraisal of drug-resistance mechanisms. Advances in sample preparation, labelling and instrumentation have made it possible, *inter alia*, to identify previously unknown aspects of *Leishmania* biology, define and re-visit complex metabolic pathways with more sensitivity than ever before and discover novel targets for drug intervention. Thanks to the panoply of omics methods currently adapted to *Leishmania* (together with invaluable online resources compiling all the omics-derived data), it is possible to adapt these technologies to address issues that were unreachable to us until now.

Moreover, we have widened the frame of study, being able, for example to span from analyses focused on single specific metabolites to analyses covering the whole metabolome of the parasite. While many technical and biological challenges remain, the implementation of sophisticated bioinformatics tools for shifting these data, along with consortium initiatives and the growing participation of the pharmaceutical industry, will allow omics toolboxes to be used in the discovery and development of novel leishmanicides. As discussed above, these toolboxes allow the prediction of toxic effects of novel therapeutic agents in human cell lines, making if feasible to rationally modify these compounds to reduce their toxic effects and accelerate the development of more effective treatments. Finally, as the emergence of resistance restricts the durability of treatment response, the use of the omics for the appraisal of field strains isolated from treated patients could be a valuable tool in order to identify specific resistance factors within a *Leishmania* population and redirect both treatment and drug discovery towards alternatively activated metabolic pathways.

Acknowledgements

The authors would like to thank Dr Mínguez-Menéndez for the conceptualization and graphic design of all the figures included in this chapter.

C.F.P is supported by a Natural Sciences and Engineering Research Council of Canada (NSERC) Discovery Grant (RGPIN-2017-04480).

References

1. M. A. Hartley, S. Drexler, C. Ronet, S. M. Beverley and N. Fasel, *Trends Parasitol.*, 2014, **30**, 412.
2. D. Liu and J. E. Uzonna, *Front. Cell Infect. Microbiol.*, 2012, **2**, 83.
3. M. Podinovskaia and A. Descoteaux, *Future Microbiol.*, 2015, **10**, 111.
4. H. W. Murray, J. D. Berman, C. R. Davies and N. G. Saravia, *Lancet*, 2005, **366**, 1561.
5. R. Reithinger, J. C. Dujardin, H. Louzir, C. Pirmez, B. Alexander and S. Brooker, *Lancet Infect. Dis.*, 2007, **7**, 581.
6. L. I. McCall, W. W. Zhang and G. Matlashewski, *PLoS Pathog.*, 2013, **9**, e1003053.
7. I. Okwor and J. Uzonna, *Am. J. Trop. Med. Hyg.*, 2016, **94**, 489.
8. N. K. Copeland and N. E. Aronson, *Curr. Opin. Infect. Dis.*, 2015, **28**, 426.
9. S. Sundar and J. Chakravarty, *Expert Opin. Pharmacother.*, 2013, **14**, 53.
10. A. C. Ivens, C. S. Peacock, E. A. Worthey, L. Murphy, G. Aggarwal, M. Berriman, E. Sisk, M. A. Rajandream, E. Adlem, R. Aert, A. Anupama, Z. Apostolou, P. Attipoe, N. Bason, C. Bauser, A. Beck, S. M. Beverley, G. Bianchettin, K. Borzym, G. Bothe, C. V. Bruschi, M. Collins, E. Cadag, L. Ciarloni, C. Clayton, R. M. Coulson, A. Cronin, A. K. Cruz, R. M. Davies, J. De Gaudenzi, D. E. Dobson, A. Duesterhoeft, G. Fazelina, N. Fosker, A. C. Frasch, A. Fraser, M. Fuchs, C. Gabel, A. Goble, A. Goffeau, D. Harris, C. Hertz-Fowler, H. Hilbert, D. Horn, Y. Huang, S. Klages, A. Knights, M. Kube, N. Larke, L. Litvin, A. Lord, T. Louie, M. Marra, D. Masuy, K. Matthews, S. Michaeli, J. C. Mottram, S. Muller-Auer, H. Munden, S. Nelson, H. Norbertczak, K. Oliver, S. O'Neil, M. Pentony, T. M. Pohl, C. Price, B. Purnelle, M. A. Quail, E. Rabbinowitsch, R. Reinhardt, M. Rieger, J. Rinta, J. Robben, L. Robertson, J. C. Ruiz, S. Rutter, D. Saunders, M. Schafer, J. Schein, D. C. Schwartz, K. Seeger, A. Seyler, S. Sharp, H. Shin, D. Sivam, R. Squares, S. Squares, V. Tosato, C. Vogt, G. Volckaert, R. Wambutt, T. Warren, H. Wedler, J. Woodward, S. Zhou, W. Zimmermann, D. F. Smith, J. M. Blackwell, K. D. Stuart, B. Barrell and P. J. Myler, *Science*, 2005, **309**, 436.
11. C. S. Peacock, K. Seeger, D. Harris, L. Murphy, J. C. Ruiz, M. A. Quail, N. Peters, E. Adlem, A. Tivey, M. Aslett, A. Kerhornou, A. Ivens, A. Fraser, M. A. Rajandream, T. Carver, H. Norbertczak, T. Chillingworth, Z. Hance, K. Jagels, S. Moule, D. Ormond, S. Rutter, R. Squares, S. Whitehead, E. Rabbinowitsch, C. Arrowsmith, B. White, S. Thurston, F. Bringaud, S. L. Baldauf, A. Faulconbridge, D. Jeffares, D. P. Depledge, S. O. Oyola, J. D. Hilley, L. O. Brito, L. R. Tosi, B. Barrell, A. K. Cruz, J. C. Mottram, D. F. Smith and M. Berriman, *Nat. Genet.*, 2007, **39**, 839.

12. J. M. Ubeda, F. Raymond, A. Mukherjee, M. Plourde, H. Gingras, G. Roy, A. Lapointe, P. Leprohon, B. Papadopoulou, J. Corbeil and M. Ouellette, *PLoS Biol.*, 2014, **12**, e1001868.
13. P. Leprohon, C. Fernandez-Prada, E. Gazanion, R. Monte-Neto and M. Ouellette, *Int. J. Parasitol.: Drugs Drug Resist.*, 2015, **5**, 26.
14. L. Beattie, M. d'El-Rei Hermida, J. W. Moore, A. Maroof, N. Brown, D. Lagos and P. M. Kaye, *Cell Host Microbe*, 2013, **14**, 357.
15. A. C. Maretti-Mira, J. Bittner, M. P. Oliveira-Neto, M. Liu, D. Kang, H. Li, C. Pirmez and N. Craft, *PLoS Neglected Trop. Dis.*, 2012, **6**, e1816.
16. S. Haile and B. Papadopoulou, *Curr. Opin. Microbiol.*, 2007, **10**, 569–577.
17. D. Paape and T. Aebischer, *J. Proteomics*, 2011, **74**, 1614.
18. T. Lahav, D. Sivam, H. Volpin, M. Ronen, P. Tsigankov, A. Green, N. Holland, M. Kuzyk, C. Borchers, D. Zilberstein and P. J. Myler, *FASEB J.*, 2011, **25**, 515.
19. G. C. Terstappen, C. Schlupen, R. Raggiaschi and G. Gaviraghi, *Nat. Rev. Drug Discovery*, 2007, **6**, 891.
20. S. L. Croft, S. Sundar and A. H. Fairlamb, *Clin. Microbiol. Rev.*, 2006, **19**, 111.
21. R. J. Pierce, F. Dubois-Abdesselem, J. Lancelot, L. Andrade and G. Oliveira, *Curr. Pharm. Des.*, 2012, **18**, 3567.
22. E. Durieu, E. Prina, O. Leclercq, N. Oumata, N. Gaboriaud-Kolar, K. Vougogiannopoulou, N. Aulner, A. Defontaine, J. H. No, S. Ruchaud, A. L. Skaltsounis, H. Galons, G. F. Spath, L. Meijer and N. Rachidi, *Antimicrob. Agents Chemother.*, 2016, **60**, 2822.
23. F. Annang, G. Perez-Moreno, R. Garcia-Hernandez, C. Cordon-Obras, J. Martin, J. R. Tormo, L. Rodriguez, N. de Pedro, V. Gomez-Perez, M. Valente, F. Reyes, O. Genilloud, F. Vicente, S. Castanys, L. M. Ruiz-Perez, M. Navarro, F. Gamarro and D. Gonzalez-Pacanowska, *J. Biomol. Screening*, 2015, **20**, 82.
24. I. Pena, M. Pilar Manzano, J. Cantizani, A. Kessler, J. Alonso-Padilla, A. I. Bardera, E. Alvarez, G. Colmenarejo, I. Cotillo, I. Roquero, F. de Dios-Anton, V. Barroso, A. Rodriguez, D. W. Gray, M. Navarro, V. Kumar, A. Sherstnev, D. H. Drewry, J. R. Brown, J. M. Fiandor and J. Julio Martin, *Sci. Rep.*, 2015, **5**, 8771.
25. S. Khare, A. S. Nagle, A. Biggart, Y. H. Lai, F. Liang, L. C. Davis, S. W. Barnes, C. J. Mathison, E. Myburgh, M. Y. Gao, J. R. Gillespie, X. Liu, J. L. Tan, M. Stinson, I. C. Rivera, J. Ballard, V. Yeh, T. Groessl, G. Federe, H. X. Koh, J. D. Venable, B. Bursulaya, M. Shapiro, P. K. Mishra, G. Spraggon, A. Brock, J. C. Mottram, F. S. Buckner, S. P. Rao, B. G. Wen, J. R. Walker, T. Tuntland, V. Molteni, R. J. Glynne and F. Supek, *Nature*, 2016, **537**, 229.
26. J. Muller and A. Hemphill, *Expert Opin. Drug Discovery*, 2016, **11**, 815–824.
27. F. J. Gamo, L. M. Sanz, J. Vidal, C. de Cozar, E. Alvarez, J. L. Lavandera, D. E. Vanderwall, D. V. Green, V. Kumar, S. Hasan, J. R. Brown, C. E. Peishoff, L. R. Cardon and J. F. Garcia-Bustos, *Nature*, 2010, **465**, 305.

28. E. R. Sharlow, T. A. Lyda, H. C. Dodson, G. Mustata, M. T. Morris, S. S. Leimgruber, K. H. Lee, Y. Kashiwada, D. Close, J. S. Lazo and J. C. Morris, *PLoS Neglected Trop. Dis.*, 2010, **4**, e659.
29. T. Downing, H. Imamura, S. Decuypere, T. G. Clark, G. H. Coombs, J. A. Cotton, J. D. Hilley, S. de Doncker, I. Maes, J. C. Mottram, M. A. Quail, S. Rijal, M. Sanders, G. Schonian, O. Stark, S. Sundar, M. Vanaerschot, C. Hertz-Fowler, J. C. Dujardin and M. Berriman, *Genome Res.*, 2011, **21**, 2143.
30. M. B. Rogers, J. D. Hilley, N. J. Dickens, J. Wilkes, P. A. Bates, D. P. Depledge, D. Harris, Y. Her, P. Herzyk, H. Imamura, T. D. Otto, M. Sanders, K. Seeger, J. C. Dujardin, M. Berriman, D. F. Smith, C. Hertz-Fowler and J. C. Mottram, *Genome Res.*, 2011, **21**, 2129.
31. F. Real, R. O. Vidal, M. F. Carazzolle, J. M. Mondego, G. G. Costa, R. H. Herai, M. Wurtele, L. M. de Carvalho, R. Carmona e Ferreira, R. A. Mortara, C. L. Barbieri, P. Mieczkowski, J. F. da Silveira, M. R. Briones, G. A. Pereira and D. Bahia, *DNA Res.*, 2013, **20**, 567.
32. S. Alsford, D. J. Turner, S. O. Obado, A. Sanchez-Flores, L. Glover, M. Berriman, C. Hertz-Fowler and D. Horn, *Genome Res.*, 2011, **21**, 915.
33. B. M. Mony, P. MacGregor, A. Ivens, F. Rojas, A. Cowton, J. Young, D. Horn and K. Matthews, *Nature*, 2014, **505**, 681.
34. S. Alsford, S. Eckert, N. Baker, L. Glover, A. Sanchez-Flores, K. F. Leung, D. J. Turner, M. C. Field, M. Berriman and D. Horn, *Nature*, 2012, **482**, 232.
35. H. G. van Luenen, C. Farris, S. Jan, P. A. Genest, P. Tripathi, A. Velds, R. M. Kerkhoven, M. Nieuwland, A. Haydock, G. Ramasamy, S. Vainio, T. Heidebrecht, A. Perrakis, L. Pagie, B. van Steensel, P. J. Myler and P. Borst, *Cell*, 2012, **150**, 909.
36. L. F. Lye, K. Owens, H. Shi, S. M. Murta, A. C. Vieira, S. J. Turco, C. Tschudi, E. Ullu and S. M. Beverley, *PLoS Pathog.*, 2010, **6**, e1001161.
37. J. Clos and K. Choudhury, *Mini-Rev. Med. Chem.*, 2006, **6**, 123.
38. P. A. Genest, A. Haimeur, D. Legare, D. Sereno, G. Roy, N. Messier, B. Papadopoulou and M. Ouellette, *Mol. Biochem. Parasitol.*, 2008, **158**, 95.
39. N. Marquis, B. Gourbal, B. P. Rosen, R. Mukhopadhyay and M. Ouellette, *Mol. Microbiol.*, 2005, **57**, 1690.
40. E. Gazanion, C. Fernandez-Prada, B. Papadopoulou, P. Leprohon and M. Ouellette, *Proc. Natl. Acad. Sci. U. S. A.*, 2016, **113**, E3012.
41. J. A. Coderre, S. M. Beverley, R. T. Schimke and D. V. Santi, *Proc. Natl. Acad. Sci. U. S. A.*, 1983, **80**, 2132.
42. J. M. Ubeda, D. Legare, F. Raymond, A. A. Ouameur, S. Boisvert, P. Rigault, J. Corbeil, M. J. Tremblay, M. Olivier, B. Papadopoulou and M. Ouellette, *Genome Biol.*, 2008, **9**, R115.
43. A. C. Coelho, S. M. Beverley and P. C. Cotrim, *Mol. Biochem. Parasitol.*, 2003, **130**, 83.
44. D. Legare, D. Richard, R. Mukhopadhyay, Y. D. Stierhof, B. P. Rosen, A. Haimeur, B. Papadopoulou and M. Ouellette, *J. Biol. Chem.*, 2001, **276**, 26301.

45. S. Khare, S. L. Roach, S. W. Barnes, D. Hoepfner, J. R. Walker, A. K. Chatterjee, R. J. Neitz, M. R. Arkin, C. W. McNamara, J. Ballard, Y. Lai, Y. Fu, V. Molteni, V. Yeh, J. H. McKerrow, R. J. Glynne and F. Supek, *PLoS Pathog.*, 2015, **11**, e1005058.
46. A. Mukherjee, G. Roy, C. Guimond and M. Ouellette, *Mol. Microbiol.*, 2009, **74**, 914.
47. L. Sollelis, M. Ghorbal, C. R. MacPherson, R. M. Martins, N. Kuk, L. Crobu, P. Bastien, A. Scherf, J. J. Lopez-Rubio and Y. Sterkers, *Cell Microbiol.*, 2015, **17**, 1405.
48. W. W. Zhang and G. Matlashewski, *mBio*, 2015, **6**, e00861.
49. B. Zetsche, J. S. Gootenberg, O. O. Abudayyeh, I. M. Slaymaker, K. S. Makarova, P. Essletzbichler, S. E. Volz, J. Joung, J. van der Oost, A. Regev, E. V. Koonin and F. Zhang, *Cell*, 2015, **163**, 759.
50. N. M. El-Sayed, P. J. Myler, G. Blandin, M. Berriman, J. Crabtree, G. Aggarwal, E. Caler, H. Renauld, E. A. Worthey, C. Hertz-Fowler, E. Ghedin, C. Peacock, D. C. Bartholomeu, B. J. Haas, A. N. Tran, J. R. Wortman, U. C. Alsmark, S. Angiuoli, A. Anupama, J. Badger, F. Bringaud, E. Cadag, J. M. Carlton, G. C. Cerqueira, T. Creasy, A. L. Delcher, A. Djikeng, T. M. Embley, C. Hauser, A. C. Ivens, S. K. Kummerfeld, J. B. Pereira-Leal, D. Nilsson, J. Peterson, S. L. Salzberg, J. Shallom, J. C. Silva, J. Sundaram, S. Westenberger, O. White, S. E. Melville, J. E. Donelson, B. Andersson, K. D. Stuart and N. Hall, *Science*, 2005, **309**, 404.
51. R. C. Duncan, P. Salotra, N. Goyal, N. S. Akopyants, S. M. Beverley and H. L. Nakhasi, *Curr. Mol. Med.*, 2004, **4**, 611.
52. R. L. do Monte-Neto, A. C. Coelho, F. Raymond, D. Legare, J. Corbeil, M. N. Melo, F. Frezard and M. Ouellette, *PLoS Neglected Trop. Dis.*, 2011, **5**, e1167.
53. A. Rochette, F. Raymond, J. M. Ubeda, M. Smith, N. Messier, S. Boisvert, P. Rigault, J. Corbeil, M. Ouellette and B. Papadopoulou, *BMC Genomics*, 2008, **9**, 255.
54. P. Leprohon, D. Legare, I. Girard, B. Papadopoulou and M. Ouellette, *Eukaryotic Cell*, 2006, **5**, 1713.
55. P. Leprohon, D. Legare, F. Raymond, E. Madore, G. Hardiman, J. Corbeil and M. Ouellette, *Nucleic Acids Res.*, 2009, **37**, 1387.
56. S. Zhao, W. P. Fung-Leung, A. Bittner, K. Ngo and X. Liu, *PLoS One*, 2014, **9**, e78644.
57. N. Singh, R. Almeida, H. Kothari, P. Kumar, G. Mandal, M. Chatterjee, S. Venkatachalam, M. K. Govind, S. K. Mandal and S. Sundar, *Parasitology*, 2007, **134**, 777.
58. J. Kaur, R. Tiwari, A. Kumar and N. Singh, *Mol. Biol. Int.*, 2011, **2011**, 278051.
59. A. Kulshrestha, V. Sharma, R. Singh and P. Salotra, *Parasitol. Res.*, 2014, **113**, 1171.
60. A. Yadav, U. Chandra and S. Saha, *Sci. Rep.*, 2016, **6**, 27510.

61. A. Haydock, M. Terrao, A. Sekar, G. Ramasamy, L. Baugh and P. J. Myler, *Methods Mol. Biol.*, 2015, **1201**, 207.
62. A. Rastrojo, F. Carrasco-Ramiro, D. Martin, A. Crespillo, R. M. Reguera, B. Aguado and J. M. Requena, *BMC Genomics*, 2013, **14**, 223.
63. S. Decuypere, J. Vandesompele, V. Yardley, S. De Donckeri, T. Laurent, S. Rijal, A. Llanos-Cuentas, F. Chappuis, J. Arevalo and J. C. Dujardin, *Parasitology*, 2005, **131**, 321.
64. D. Eliaz, T. Doniger, I. D. Tkacz, V. K. Biswas, S. K. Gupta, N. G. Kolev, R. Unger, E. Ullu, C. Tschudi and S. Michaeli, *RNA Biol.*, 2015, **12**, 1222.
65. G. Alonso, A. Rastrojo, S. Lopez-Perez, J. M. Requena and B. Aguado, *Parasites Vectors*, 2016, **9**, 74.
66. L. A. Dillon, K. Okrah, V. K. Hughitt, R. Suresh, Y. Li, M. C. Fernandes, A. T. Belew, H. Corrada Bravo, D. M. Mosser and N. M. El-Sayed, *Nucleic Acids Res.*, 2015, **43**, 6799.
67. J. L. Martin, P. A. Yates, R. Soysa, J. F. Alfaro, F. Yang, K. E. Burnum-Johnson, V. A. Petyuk, K. K. Weitz, D. G. Camp, 2nd, R. D. Smith, P. A. Wilmarth, L. L. David, G. Ramasamy, P. J. Myler and N. S. Carter, *PLoS Pathog.*, 2014, **10**, e1003938.
68. R. M. de Paiva, V. Grazielle-Silva, M. S. Cardoso, B. N. Nakagaki, R. P. Mendonca-Neto, A. M. Canavaci, N. Souza Melo, P. M. Martinelli, A. P. Fernandes, W. D. daRocha and S. M. Teixeira, *PLoS Pathog.*, 2015, **11**, e1005296.
69. P. J. Alcolea, A. Alonso, M. Dominguez, V. Parro, M. Jimenez, R. Molina and V. Larraga, *PLoS Neglected Trop. Dis.*, 2016, **10**, e0004693.
70. S. M. Christensen, L. A. Dillon, L. P. Carvalho, S. Passos, F. O. Novais, V. K. Hughitt, D. P. Beiting, E. M. Carvalho, P. Scott, N. M. El-Sayed and D. M. Mosser, *PLoS Neglected Trop. Dis.*, 2016, **10**, e0004992.
71. F. Torres, R. Arias-Carrasco, A. Barral, V. Maracaja-Coutinho and A. T. L. Queiroz, *Database*, 2017, DOI: 10.1093/database/bax047.
72. A. East-Seletsky, M. R. O'Connell, S. C. Knight, D. Burstein, J. H. Cate, R. Tjian and J. A. Doudna, *Nature*, 2016, **538**, 270.
73. N. Boucher, Y. Wu, C. Dumas, M. Dube, D. Sereno, M. Breton and B. Papadopoulou, *J. Biol. Chem.*, 2002, **277**, 19511.
74. C. E. Clayton, *EMBO J.*, 2002, **21**, 1881.
75. J. Schroeder and T. Aebischer, *Hum. Vaccines*, 2011, **7**(Supp. l), 10–15.
76. J. B. de Jesus, C. Mesquita-Rodrigues and P. Cuervo, *Subcell. Biochem.*, 2014, **74**, 323.
77. P. Cuervo and J. B. de Jesus, in *Drug Resistance in Leishmania Parasites*, ed. A. E. A. Ponte-Sucre, Springer-Verlag, Wien, 2012, pp. 215–236.
78. E. Handman, G. F. Mitchell and J. W. Goding, *J. Immunol.*, 1981, **126**, 508.
79. N. G. Saravia, M. A. Gemmell, S. L. Nance and N. L. Anderson, *Clin. Chem.*, 1984, **30**, 2048.
80. M. Karas and F. Hillenkamp, *Anal. Chem.*, 1988, **60**, 2299.

81. K. Tanaka, H. Waki, Y. Ido, S. Akita, Y. Yoshida and T. Yoshida, *Rapid Commun. Mass Spectrom.*, 1988, **2**, 151.
82. M. A. Morales, R. Watanabe, C. Laurent, P. Lenormand, J. C. Rousselle, A. Namane and G. F. Spath, *Proteomics*, 2008, **8**, 350.
83. M. C. Brotherton, G. Racine and M. Ouellette, *Methods Mol. Biol.*, 2015, **1201**, 247.
84. G. Baggerman, E. Vierstraete, A. De Loof and L. Schoofs, *Comb. Chem. High Throughput Screening*, 2005, **8**, 669.
85. M. P. Washburn, D. Wolters and J. R. Yates, 3rd, *Nat. Biotechnol.*, 2001, **19**, 242.
86. D. Rosenzweig, D. Smith, P. J. Myler, R. W. Olafson and D. Zilberstein, *Proteomics*, 2008, **8**, 1843.
87. D. Zilberstein, *Methods Mol. Biol.*, 2015, **1201**, 261.
88. J. M. Silverman, S. K. Chan, D. P. Robinson, D. M. Dwyer, D. Nandan, L. J. Foster and N. E. Reiner, *Genome Biol.*, 2008, **9**, R35.
89. M. C. Brotherton, S. Bourassa, P. Leprohon, D. Legare, G. G. Poirier, A. Droit and M. Ouellette, *PLoS One*, 2013, **8**, e81899.
90. M. C. Brotherton, S. Bourassa, D. Legare, G. G. Poirier, A. Droit and M. Ouellette, *Int. J. Parasitol.: Drugs Drug Resist.*, 2014, **4**, 126.
91. A. L. Foucher, G. F. Spath and I. K. Pemberton, *Arch. Physiol. Biochem.*, 2010, **116**, 163.
92. M. Oppenheimer, A. L. Valenciano and P. Sobrado, *Biochem. Biophys. Res. Commun.*, 2011, **407**, 552.
93. B. Vergnes, B. Gourbal, I. Girard, S. Sundar, J. Drummelsmith and M. Ouellette, *Mol. Cell. Proteomics*, 2007, **6**, 88.
94. F. S. Matrangolo, D. B. Liarte, L. C. Andrade, M. F. de Melo, J. M. Andrade, R. F. Ferreira, A. S. Santiago, C. P. Pirovani, R. A. Silva-Pereira and S. M. Murta, *Mol. Biochem. Parasitol.*, 2013, **190**, 63.
95. S. Moreira Dde, P. Pescher, C. Laurent, P. Lenormand, G. F. Spath and S. M. Murta, *Proteomics*, 2015, **15**, 2999.
96. K. El Fadili, J. Drummelsmith, G. Roy, A. Jardim and M. Ouellette, *Exp. Parasitol.*, 2009, **123**, 51.
97. N. Biyani, A. K. Singh, S. Mandal, B. Chawla and R. Madhubala, *Mol. Biochem. Parasitol.*, 2011, **179**, 91.
98. J. Walker, R. Gongora, J. J. Vasquez, J. Drummelsmith, R. Burchmore, G. Roy, M. Ouellette, M. A. Gomez and N. G. Saravia, *Mol. Biochem. Parasitol.*, 2012, **183**, 166.
99. M. Ouellette and P. Borst, *Res. Microbiol.*, 1991, **142**, 737.
100. K. Grondin, B. Papadopoulou and M. Ouellette, *Nucleic Acids Res.*, 1993, **21**, 1895.
101. M. K. Mittal, S. Rai, Ashutosh, Ravinder, S. Gupta, S. Sundar and N. Goyal, *Am. J. Trop. Med. Hyg.*, 2007, **76**, 681.
102. A. Mukherjee, P. K. Padmanabhan, S. Singh, G. Roy, I. Girard, M. Chatterjee, M. Ouellette and R. Madhubala, *J. Antimicrob. Chemother.*, 2007, **59**, 204.

103. B. Chawla, A. Jhingran, A. Panigrahi, K. D. Stuart and R. Madhubala, *PLoS One*, 2011, **6**, e26660.
104. G. Singh, H. D. Chavan and C. S. Dey, *Int. J. Antimicrob. Agents*, 2008, **31**, 584.
105. I. M. Vincent, G. Racine, D. Légaré and M. Ouellette, *Proteomes*, 2015, **3**, 328.
106. L. Imbert, S. Cojean, D. Libong, P. Chaminade and P. M. Loiseau, *Biomed. Pharmacother.*, 2014, **68**, 893.
107. A. K. Singh, S. Roberts, B. Ullman and R. Madhubala, *J. Proteomics*, 2014, **102**, 44.
108. B. F. Cravatt, A. T. Wright and J. W. Kozarich, *Annu. Rev. Biochem.*, 2008, **77**, 383.
109. M. Ehrmann, F. Kaschani and M. Kaiser, *Chem. Biol.*, 2015, **22**, 309.
110. M. H. Wright, D. Paape, E. M. Storck, R. A. Serwa, D. F. Smith and E. W. Tate, *Chem. Biol.*, 2015, **22**, 342.
111. J. A. Frearson, S. Brand, S. P. McElroy, L. A. Cleghorn, O. Smid, L. Stojanovski, H. P. Price, M. L. Guther, L. S. Torrie, D. A. Robinson, I. Hallyburton, C. P. Mpamhanga, J. A. Brannigan, A. J. Wilkinson, M. Hodgkinson, R. Hui, W. Qiu, O. G. Raimi, D. M. van Aalten, R. Brenk, I. H. Gilbert, K. D. Read, A. H. Fairlamb, M. A. Ferguson, D. F. Smith and P. G. Wyatt, *Nature*, 2010, **464**, 728.
112. M. H. Wright, B. Clough, M. D. Rackham, K. Rangachari, J. A. Brannigan, M. Grainger, D. K. Moss, A. R. Bottrill, W. P. Heal, M. Broncel, R. A. Serwa, D. Brady, D. J. Mann, R. J. Leatherbarrow, R. Tewari, A. J. Wilkinson, A. A. Holder and E. W. Tate, *Nat. Chem.*, 2014, **6**, 112.
113. J. Petrak, R. Ivanek, O. Toman, R. Cmejla, J. Cmejlova, D. Vyoral, J. Zivny and C. D. Vulpe, *Proteomics*, 2008, **8**, 1744.
114. S. Kim, P. A. Thiessen, E. E. Bolton, J. Chen, G. Fu, A. Gindulyte, L. Han, J. He, S. He, B. A. Shoemaker, J. Wang, B. Yu, J. Zhang and S. H. Bryant, *Nucleic Acids Res.*, 2016, **44**, D1202.
115. D. S. Wishart, C. Knox, A. C. Guo, S. Shrivastava, M. Hassanali, P. Stothard, Z. Chang and J. Woolsey, *Nucleic Acids Res.*, 2006, **34**, D668.
116. A. Sateriale, K. Bessoff, I. N. Sarkar and C. D. Huston, *J. Am. Med. Inform. Assoc.*, 2014, **21**, 238.
117. M. P. Magarinos, S. J. Carmona, G. J. Crowther, S. A. Ralph, D. S. Roos, D. Shanmugam, W. C. Van Voorhis and F. Aguero, *Nucleic Acids Res.*, 2012, **40**, D1118.
118. C. F. Prada, R. Alvarez-Velilla, R. Diaz-Gonzalez, C. Prieto, Y. Perez-Pertejo, R. Balana-Fouce and R. M. Reguera, *Biochim. Biophys. Acta*, 2012, **1820**, 2062.
119. R. Balana-Fouce, C. F. Prada, J. M. Requena, M. Cushman, Y. Pommier, R. Alvarez-Velilla, J. M. Escudero-Martinez, E. Calvo-Alvarez, Y. Perez-Pertejo and R. M. Reguera, *Antimicrob. Agents Chemother.*, 2012, **56**, 5264.
120. C. F. Prada, R. Alvarez-Velilla, R. Balana-Fouce, C. Prieto, E. Calvo-Alvarez, J. M. Escudero-Martinez, J. M. Requena, C. Ordonez,

A. Desideri, Y. Perez-Pertejo and R. M. Reguera, *Biochem. Pharmacol.*, 2013, **85**, 1433.
121. I. M. Vincent and M. P. Barrett, *J. Biomol. Screening*, 2015, **20**, 44.
122. L. W. Sumner, A. Amberg, D. Barrett, M. H. Beale, R. Beger, C. A. Daykin, T. W. Fan, O. Fiehn, R. Goodacre, J. L. Griffin, T. Hankemeier, N. Hardy, J. Harnly, R. Higashi, J. Kopka, A. N. Lane, J. C. Lindon, P. Marriott, A. W. Nicholls, M. D. Reily, J. J. Thaden and M. R. Viant, *Metabolomics*, 2007, **3**, 211.
123. A. Chokkathukalam, D. H. Kim, M. P. Barrett, R. Breitling and D. J. Creek, *Bioanalysis*, 2014, **6**, 511.
124. M. Berg, M. Vanaerschot, A. Jankevics, B. Cuypers, I. Maes, S. Mukherjee, B. Khanal, S. Rijal, S. Roy, F. Opperdoes, R. Breitling and J. C. Dujardin, *Mol. Microbiol.*, 2013, **90**, 428.
125. G. De Muylder, K. K. Ang, S. Chen, M. R. Arkin, J. C. Engel and J. H. McKerrow, *PLoS Neglected Trop. Dis.*, 2011, **5**, e1253.
126. T. R. Holzer, W. R. McMaster and J. D. Forney, *Mol. Biochem. Parasitol.*, 2006, **146**, 198.
127. A. Rochette, F. Raymond, J. Corbeil, M. Ouellette and B. Papadopoulou, *Mol. Biochem. Parasitol.*, 2009, **165**, 32.
128. G. A. Canuto, E. A. Castilho-Martins, M. Tavares, A. Lopez-Gonzalvez, L. Rivas and C. Barbas, *Electrophoresis*, 2012, **33**, 1901.
129. S. Wyllie, M. L. Cunningham and A. H. Fairlamb, *J. Biol. Chem.*, 2004, **279**, 39925.
130. R. Mukhopadhyay, S. Dey, N. Xu, D. Gage, J. Lightbody, M. Ouellette and B. P. Rosen, *Proc. Natl. Acad. Sci. U. S. A.*, 1996, **93**, 10383.
131. S. Yan, F. Li, K. Ding and H. Sun, *J. Biol. Inorg. Chem.*, 2003, **8**, 689.
132. K. El Fadili, N. Messier, P. Leprohon, G. Roy, C. Guimond, N. Trudel, N. G. Saravia, B. Papadopoulou, D. Legare and M. Ouellette, *Antimicrob. Agents Chemother.*, 2005, **49**, 1988.
133. G. A. Canuto, E. A. Castilho-Martins, M. F. Tavares, L. Rivas, C. Barbas and A. Lopez-Gonzalvez, *Anal. Bioanal. Chem.*, 2014, **406**, 3459.
134. I. M. Vincent, S. Weidt, L. Rivas, K. Burgess, T. K. Smith and M. Ouellette, *Int. J. Parasitol.: Drugs Drug Resist.*, 2014, **4**, 20.
135. L. Imbert, R. G. Ramos, D. Libong, S. Abreu, P. M. Loiseau and P. Chaminade, *Anal. Bioanal. Chem.*, 2012, **402**, 1169.
136. P. L. Olliaro, *Curr. Opin. Infect. Dis.*, 2010, **23**, 595.
137. N. Mbongo, P. M. Loiseau, M. A. Billion and M. Robert-Gero, *Antimicrob. Agents Chemother.*, 1998, **42**, 352.
138. M. Ouellette, J. Drummelsmith and B. Papadopoulou, *Drug Resist. Updates*, 2004, **7**, 257.
139. C. Fernandez-Prada, I. M. Vincent, M. C. Brotherton, M. Roberts, G. Roy, L. Rivas, P. Leprohon, T. K. Smith and M. Ouellette, *PLoS Neglected Trop. Dis.*, 2016, **10**, e0005171.
140. L. Shabala and T. Ross, *Res. Microbiol.*, 2008, **159**, 458.
141. W. Moreira, P. Leprohon and M. Ouellette, *Cell Death Dis.*, 2011, **2**, e201.

142. M. Berg, R. Garcia-Hernandez, B. Cuypers, M. Vanaerschot, J. I. Manzano, J. A. Poveda, J. A. Ferragut, S. Castanys, J. C. Dujardin and F. Gamarro, *Antimicrob. Agents Chemother.*, 2015, **59**, 2242.
143. F. J. Perez-Victoria, M. P. Sanchez-Canete, K. Seifert, S. L. Croft, S. Sundar, S. Castanys and F. Gamarro, *Drug Resist. Updates*, 2006, **9**, 26.
144. I. M. Vincent, D. J. Creek, K. Burgess, D. J. Woods, R. J. Burchmore and M. P. Barrett, *PLoS Neglected Trop. Dis.*, 2012, **6**, e1618.
145. V. Gomez Perez, R. Garcia-Hernandez, V. Corpas-Lopez, A. M. Tomas, J. Martin-Sanchez, S. Castanys and F. Gamarro, *Int. J. Parasitol.: Drugs Drug Resist.*, 2016, **6**, 133.

CHAPTER 7

In silico *Tools for Target Identification and Drug Molecular Docking in* Leishmania

CARLOS ROCA,[†] VÍCTOR SEBASTIÁN-PÉREZ[†] AND NURIA E. CAMPILLO*

Centro de Investigaciones Biológicas (CSIC), Ramiro de Maeztu 9, 28040 Madrid, Spain
*Email: nuria.campillo@csic.es

7.1 Introduction

Leishmaniasis is one of the most common neglected parasitic diseases and is also one of the most neglected diseases in terms of drug discovery. This disease is caused by trypanosomatid protozoan parasites that belong to the genus *Leishmania*; the disease affects about 350 million people, with two million new cases annually.[1]

Leishmaniasis is a complex disease caused by several species of the *Leishmania* genus, ranging in severity from cutaneous and mucocutaneous lesions to the chronic visceral form that, if untreated, can cause death. One of the main problems in leishmaniasis is the limited number of drug options, together with the adverse effects that they can cause.

[†]These authors have contributed equally.

Drug Discovery Series No. 60
Drug Discovery for Leishmaniasis
Edited by Luis Rivas and Carmen Gil

Published by the Royal Society of Chemistry, www.rsc.org

A great variety of drug discovery strategies for neglected tropical diseases, specifically for leishmaniasis, have been employed during recent years to exploit two possible sets of targets: those that have been validated in other organisms and diseases, and those that have not. The strategies used to obtain a drug can be grouped mainly into three approaches: drug repurposing, *i.e.* extending the indications of existing drugs for other conditions to tropical diseases; piggy-back discovery, in which the discovery of new drugs is focused on one or a few classes of well-studied and validated targets; and *de novo* drug discovery.[2]

These strategies can be approached in a 'wet' (biological–chemical approaches) or 'dry' (*in silico* approach) way in order to determine whether a ligand is biologically active and, if it is a promising hit, to perform an optimisation study to achieve an effective drug. The classical approach of identifying an essential protein as a potential target involves experimental characterisation using gene knock-out and knock-down strategies in the target organism.[3]

Herein, we will focus on computational strategies to develop effective anti-leishmanial drugs, used in the discovery of essential targets and the identification and optimisation of drugs.

7.2 *Leishmania* Targets Identification

Target discovery is an initial step for effective drug development. In order to be considered as a good druggable target, it has to meet several criteria. Among these criteria are essentiality, druggability, assayability and specificity or selectivity. One of the first essential criteria for drug target selection is its ability to be disease-modifying. In the case of *Leishmania,* this means that the target must be essential to the viability of the parasite or its inhibition must significantly affect its virulence; thus, it is necessary to select a target relevant to human health. Another key criterion for drug target selection is its druggability, *i.e.* the likelihood of small molecules being able to modulate the target. In addition, it is important that the putative drug target should be either absent in the host or markedly different from the host homologue so it can be exploited as a drug target.

In the case of *Leishmania* spp. and for parasites in general, there are few validated molecular targets, in some cases due to the lack of translation from target-based activity to whole-cell assays or *in vivo* activities. Recently, the availability of the complete genome sequence of *Leishmania* species such as *Leishmania major*,[4] *Leishmania donovani*, *Leishmania infantum*, *Leishmania braziliensis*[5] and *Leishmania panamensis*[6] has led to a better understanding of the biology of the parasite. The genome sequence conjugated with comparative genomics, proteomics and *in silico* tools can provide molecules and pathways that lead to the discovery of novel drug targets.

In a general way, the approaches employed in target identification and validation can be grouped into two classes: molecular and whole-cell approaches.[7]

The whole-cell or system approach employs the whole organism to obtain useful data. The information obtained is derived from the clinics and *in vivo* animal studies in the areas of physiology, pathology and epidemiology. The molecular approach is geared towards the identification of druggable targets using the cellular mechanism underlying the disease phenotype of interest.[7]

The development of the computational field in recent decades has helped with the identification of new targets using *in silico* tools. One approach to the identification of targets is by means of a homology search; that is, to find orthologous in the organism of interest that could be essential by comparing its sequences against a list of essential genes in other known organisms (Figure 7.1).

Different resources and databases have been developed as very useful tools for an exploratory search of essentiality of a particular protein.[8] The database of essential genes (DEG, available at http://www.essentialgene.org), constructed in 2003 and continually updated with new information of essential gene records of bacteria, archaea and eukaryotes, can also be used to perform a BLAST search for the target of interest.[9,10] The DrugBank database (http://www.drugbank.ca) is another online database containing extensive biochemical and pharmacological information about drugs, their mechanisms of action and their targets.[11,12] Others interesting and useful tools to perform homology search are BLAST or the TDR Targets Database. It is necessary to take into account that proteins inside the cell perform specific functions depending on their interactions, and these interactions can vary between species. Therefore, a deep analysis of protein interactions is crucial in order to understand complex pathogens such as *Leishmania* spp. These interactions can be described as a network, known as the interactome, that describes all possible molecular interactions among proteins.[8] Proteins involved in metabolism constitute another important source of drug targets. In the case of *Leishmania* spp., many of these pathways are similar to those in humans and therefore the identification of suitable targets is a challenge.[3]

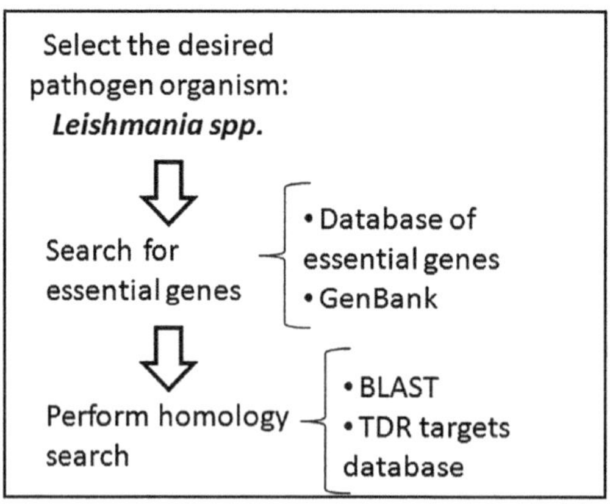

Figure 7.1 General flowchart for performing homology searching.

The approach used to model pathway metabolism is quite different from those that model the interactome. Therefore, the interactions between enzymes depend upon the rate of turnover of molecules or fluxes, not necessarily through physical interactions as described for the interactome.[8] In order to build a metabolic network, it is necessary to list all the substances with their concentrations and the reactions among them. An interesting example is the reconstruction of the metabolic network of *L. major* by Chavali *et al.*[13] This network reconstruction accounts for 560 genes, 1112 reactions, 1101 metabolites and 8 unique subcellular localisations using a systems-based approach.

Another powerful tool to target identification is reverse or inverse docking.[14,15] It involves docking a small-molecule drug or ligand in the potential binding cavity of the clinically relevant targets. Detailed analyses of the binding characteristics allow for ranking the targets according to the tightness of binding. This process can potentially identify novel molecular targets for the drug or ligand which may be relevant to its mechanism of action and/or side effect profile.[3,16] An interesting case of the application of a reverse docking methodology to identify targets was published by Britto da Mata *et al.*[17] A target-fishing procedure based on docking studies using 15 different enzymes from *L. braziliensis* and two from *Leishmania amazonensis* was performed to identify probable targets for the active compounds identified by *in vivo* studies. After comparison between the score for docking and the 50% inhibitory concentration (IC_{50}) values, the authors concluded that hexokinase was the most probable target in *L. braziliensis*.

Other interesting computational tools used to obtain data to identify targets are machine learning techniques (ML), such as artificial neural networks (ANNs). ANNs are able to identify patterns on protein targets such as domains or post-translational modifications that can be related to a specific kind of compound. Then, the artificial system will learn these patterns and, when challenged by proteins from the organism of interest, it will predict the potential association with a particular compound. Two studies have applied this strategy, employing different techniques, such as support vector machines (SVM) and Bayesian classifiers (BC).[18,19]

7.3 Anti-leishmanial Drug Development

Once the protein target has been identified, the next step in the drug discovery process is finding an inhibitor of the target. Drug discovery is an iterative cycle of identifying promising hits followed by lead optimisation. This is an expensive process, which finally results in a few hits that can advance into further drug development stages.

Based on the knowledge of the actors involved in the drug discovery process (target and ligand), the different computational tools can be grouped into two important approaches, *i.e.* the structure-based drug design (SBDD) and ligand-based drug design (LBDD) approaches.

7.3.1 Ligand-based Drug Design

LBDD approaches are powerful methods based on small-molecule information using a series of known active and inactive compounds against a target or species (when the target structure is unknown). This strategy can be used alone or in combination with SBDD. In anti-leishmanial drug development, it is quite important because there is not a great number of validated targets. Ligand-based methods are based on finding common significant structural and physicochemical characteristics, assuming that structurally similar compounds display similar biological activity. Similarity searching methods might be more readily available for similarity or substructure searching and can be used to find compounds that could share common features with other known active drugs.[20] Common LBDD techniques are quantitative structure–activity relationships (QSAR), pharmacophore-based methods, ANNs and similarity searching methods.

QSAR studies are based on the premise that changes in bioactivity are associated with structural and molecular variations in a set of compounds. A statistical model is generated from this correlation to develop and mathematically predict the desired biological properties of different compounds. QSAR techniques can be classified according to the different dimensions they represent, nD-QSAR ($n = 3$–6).[21] 3D and 4D-QSAR are the most commonly used techniques in *Leishmania* drug design. On the one hand, 3D-QSAR studies represent each ligand molecule by a single 3D entity, playing a key role in the identification of the bioactive conformation, orientation and the protonation state. On the other hand, in 4D-QSAR approaches, the ligands are supplied as an ensemble of conformations, orientations and protonation states, amplifying the number of different possible active conformations for each ligand, increasing the calculation complexity and required time, but at the same time improving the evaluation of the structures, producing a more complete result than in 3D-QSAR studies. Pharmacophore modelling is another useful *in silico* tool that is based on the concept that binding site information can be extracted from the analysis of the bioactive conformation of candidate compounds. Its main aim is to screen ligand sets in order to appreciate the ligand–target affinity (or its absence), or to analyse the binding mode for a set of compounds in order to comprehend their mechanism of action. Bioactive chemical structures with different scaffolds, but with a similar 3D arrangement of their functional groups, can be aligned, finding the common key interaction groups and their tridimensional disposition. Pharmacophore modelling can be performed in the molecular alignment stage of 3D-QSAR modelling, *i.e.* hybrid 3D-QSAR-based pharmacophore modelling studies, improving the results obtained with both techniques.[22]

One alternative approach to LBDD is machine-learning techniques, such as ANNs. These strategies can be applied in *Leishmania* drug development in two ways: on the one hand, ANNs can be used to screen chemical compounds in order to appreciate different physicochemical properties or to associate active compounds with their *Leishmania* targets, while on the other hand, it is possible to find patterns on protein targets that are related to a specific kind of compound in order to obtain new drug targets and the potential drug that binds to them. ANNs are mathematical methods that present learning ability and the capability of solving problems using mathematical models when simple mathematical functions do not solve them. By using the learning algorithm in a neural network, one can derive useful information from training data samples, and build up knowledge about the weights that connect different neurons together.[23]

Several researchers have performed QSAR studies in order to obtain predictive models for obtaining leishmanicidal compounds. The study of the structural features that correlate with high inhibition activity of farnesyl pyrophosphate synthase (FPP) for bisphosphonates derivatives[24] used two approximations based on 3D-QSAR techniques: comparative molecular field analysis (CoMFA)[25] and comparative molecular similarity index (CoMSIA).[26] The CoMSIA fields showed key features, such as positive charge in the bisphosphonate side chain and a hydrophobic feature that significantly contributed to activity. This study showed that the position of ring nitrogens is related to the activity or the likely effects of resonance stabilisation. Another example is the model of anti-leishmanial activity and cytotoxicity prediction that was generated using the topological molecular computational design-computer-aided-rational-drug-design (TOMOCOMD-CARDD) approach,[27] obtaining an overall 93.4% correct classification for the active and inactive training set and predicting the activity of 14-hydroxylunularin (**1**). The accuracy of the prediction was tested by assaying 14-hydroxylunularin *in vitro* against the amastigote stages of *Leishmania* spp., showing high activity *in vitro* against intracellular forms of *Leishmania* spp. (IC_{50} *L. amazonensis*, 1.5 μM), in the absence of cytotoxicity against mammalian cells. 4D-QSAR has been used to develop models for a series of synthetic dihydrobenzofuran lignans and related benzofurans as active compounds against *L. donovani*,[28] showing that the individual energy contributions identified the van der Waals-electrostatic and solvation energy terms as the most influential on the total predicted free binding energy, 82% of which was explained by these factors. This study led to the characterisation of an *n*-butyl-3,3′-deoxydihydrobenzofuran derivative (**2**) as a promising lead compound with anti-leishmanial activity (IC_{50}) of 0.12 $\mu g\,ml^{-1}$ against *L. donovani* axenic amastigotes and 0.19 $\mu g\,ml^{-1}$ against the intracellular ones (Figure 7.2).

An interesting work was published by Prado-Prado *et al.*[29] The author used a multi-target quantitative structure–activity relationships (mt-QSAR) study in order to predict the probability with which a compound is active against different pathogen parasite species with only a model. The authors concluded that the mt-QSAR methodology with a large data set improved the results significantly.

Delfin *et al.* developed a 3D pharmacophore model in order to describe the anti-leishmanial activity of dinitroaniline sulphonamides.[30] Later, this model was used to screen 55 000 molecules of the Maybridge compound

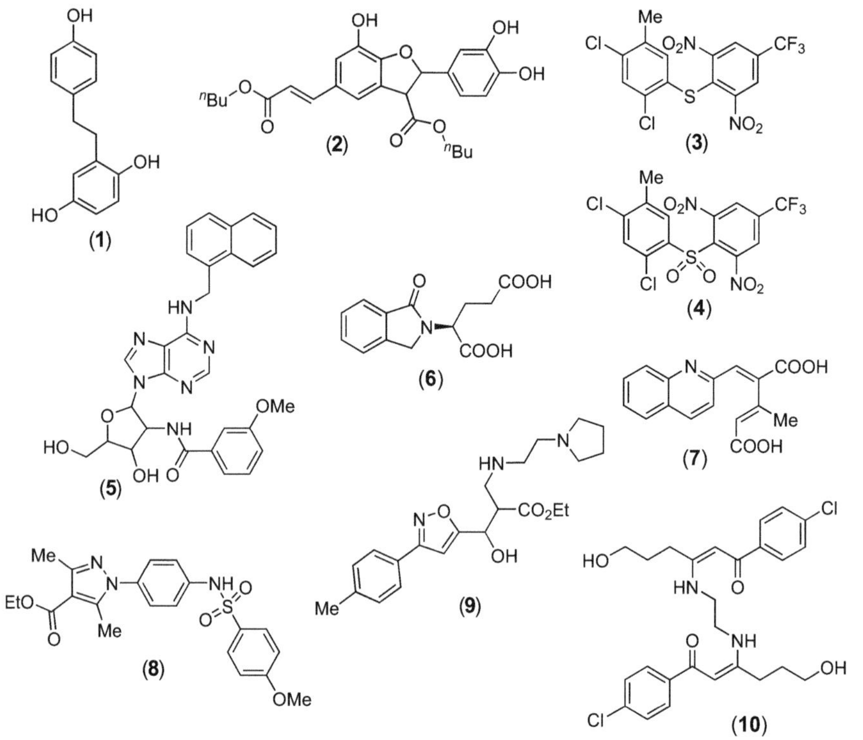

Figure 7.2 Selection of hits discovered using *in silico* tools.

database (http://www.maybridge.com/html/home.htm), and the 19 best compounds were tested against *L. donovani*, obtaining two highly active anti-leishmanial compounds and five moderately active compounds. The highly active compounds identified, BTB 06237 (2,4-dichloro-5-methylphenyl 2,6-dinitro-4-(trifluoromethyl)phenyl sulphide (**3**) and BTB 06256 (2-[(2,4-dichloro-5-methylphenyl)sulfonyl]-1,3-dinitro-5-(trifluoromethyl)benzene) (**4**), were active, with IC_{50} values against *L. donovani* amastigotes of 0.5 ± 0.2 and 2.3 ± 0.8 μM, respectively. BTB 06237 also reduced parasite burdens in *Leishmania mexicana*-infected J774 macrophages at low micromolar concentrations. In another study performed to test the activity of the compounds synthesised by Bharate *et al.*,[31] a pharmacophore model for 3-3′-diindolylmethanes was developed using PHASE,[32] with a correlation value (r^2) of 0.88 between the experimental and predicted values for anti-leishmanial activity.

The coupling of 3D-QSAR and pharmacophore modelling was performed to identify novel pteridine reductase inhibitors and used to perform virtual screening (VS) in order to predict new potent inhibitory scaffolds.[33]

Several works have developed ANN models in the *Leishmania* field. Jamal and Scaria[34] published a biological activity predictive model for novel anti-leishmanial compounds, using a dataset of pyruvate kinase inhibitors for *L. mexicana*, with different classifier algorithms (naïve Bayes, random forest, J48 and sequential minimum optimization (SMO) in order to find the best method; it was found that the random forest method was the most robust and accurate predictive model. Agarwal *et al.*[35] used ANNs to study the influence of steric, electronic and hydrophobic components of pentamidine congeners that possess a benzofuran motif in terms of the activity against *L. donovani*. Another study performed by Garkani-Nejad and Ahmadi-Roudi[36] demonstrated the role of geometric and steric features of nitro-hydrazine derivatives in anti-leishmanial activity.

7.3.2 Structure-based Drug Design

The structure-based or target approach includes computational tools that use structural knowledge of the target in order to identify or design specific hits. The philosophy of the target-based approach is to gain insight into the interaction of the target and possible hits using 3D structural knowledge. Therefore, one of the key points is to have an available 3D structure of the target.

X-ray crystallography and nuclear magnetic resonance (NMR) spectroscopy are two of the most important experimental techniques for elucidating the three-dimensional structural information of target drugs. Currently, 285 experimental protein structures of *Leishmania* are available in the Protein Data Bank (PDB), http://www.rcsb.org/pdb. Tables 7.1 and 7.2 show the results of a search in the PDB for the 3D-structures of potential *Leishmania* targets. Over 50% of all structures, found in the PDB, correspond to *L. major*. The complete genome of this species was the first published and has been the most widely studied among the trypanosomatid species. A large number of crystal structures from *L. mexicana* are also present, representing more than 20% of all protein crystals from *Leishmania*. The rest of the species represent less than 10% in terms of number of crystal structures.

Regarding the enzyme class, the most represented type of enzyme is transferases with 89 proteins, followed by oxidoreductases with almost 30% of all crystal structures. Ordered by number of structures, in this classification we can also find hydrolases, isomerases, lyases and ligases (Table 7.2).

Table 7.1 Classification of the PDB 3D structures by organism.

Leishmania spp.	PDB structures [number (%)]
L. major	159 (55.8%)
L. mexicana	59 (20.7%)
L. donovani	28 (9.8%)
L. infantum	17 (6.0%)
L. braziliensis	8 (2.8%)
L. tarentolae	6 (2.1%)
Other	8 (2.8%)

Table 7.2 Classification of the PDB 3D structures from *Leishmania* by enzyme function.

Enzyme type	PDB structures [number (%)]
Transferases	89 (40.9%)
Oxidoreductases	65 (29.8%)
Hydrolases	24 (11%)
Isomerases	24 (11%)
Lyases	9 (4.1%)
Ligases	7 (3.2%)

X-ray crystallography or NMR spectroscopy are time-consuming techniques and not successful with all proteins, especially with membrane proteins. A successful approach to contribute to the theoretical knowledge of the conformation of the proteins is the prediction of the 3D structure from its amino acid sequence. These structures can be achieved using different types of approaches; the first and most accurate approach is "comparative" or "homology" modelling.

Homology modelling is a very useful tool in SBDD, for those cases in which a 3D structure of the target of interest is not available experimentally. Therefore, the main aim of homology modelling is to predict a structure from its sequence with an accuracy that is similar to the results obtained experimentally. Homology modelling is a multistep process that can be summarised as follow (1) template selection, (2) single or multiple sequence target–template alignments, (3) model building, (4) model refinement (gap deletions and additions) and (5) model validation. There are several programs and web servers available for homology modelling able to build complete 3D structure models from query sequences, such as Modeller,[37,38] SwissModel,[39] PrISM[40] and CONGEN.[41]

Molecular docking is a useful tool to predict both structurally the most likely binding mode, and energetically the binding affinity of a small ligand onto the target. Molecular docking has a wide variety of uses and applications, including structure–activity studies, hit identification or lead optimisation, providing binding hypotheses to facilitate predictions for multiple studies. Mainly, it has been used to identify novel chemical probes and hits that can be optimised into lead molecules and candidate drugs. Some of the most widely used docking tools are AutoDock,[42] Glide,[43] DOCK,[44] FlexX[45] and GOLD,[46] which use different searching methods and scoring functions.

Virtual screening (VS) is an *in silico* technique used to improve the drug search by using computational chemistry. The goal of this technique is to identify novel molecules that could bind the target of interest and to predict a ranking in which active compounds are above inactive molecules. This process is performed in several steps by the use of different filters, such as physicochemical properties, drug-like properties, pharmacophore models or docking, and using commercially available or in-house chemical libraries over the structure of the target.

Molecular dynamics (MD) methods allow simulation of dynamic biological and chemical events at the atomic level. Molecular dynamics has replaced the early view of proteins as relatively rigid structures by a dynamic model in which internal conformational changes under physiological conditions provide further information on the complete system.

In recent years, interesting examples of the application of the structure-based approach to discover potent inhibitors of different *Leishmania* spp. validated targets have been published.

The polyamine metabolism of trypanosomatids is an interesting pathway for drug development, because the metabolites involved play important roles in vital processes of the parasite such as the biosynthesis of nucleic acids and proteins, proliferation and cell differentiation, and protection against oxidative damage.[47,48] Spermidine synthase (SS) is an important enzyme in the synthetic pathway of polyamines in *Leishmania,* and is an essential element for the survival of this protozoan. Studies have shown that spermidine synthase is required for the virulence of *L. donovani.*[49] Grover *et al.*[50] used a computational approach to identify inhibitors of SS. The first step was homology modelling studies and molecular dynamics simulations to validate the 3D models of SS. Despite high identity of the protein to the human homologue, active site analysis revealed that the binding pocket was larger in the human enzyme, which led to the discovery of selective inhibitors. In this study, high-throughput virtual screening was used with a described compound collection and docking studies and identified two promising compounds, a chromene and a triazol derivative, which, according to the results of computational and MD studies, significantly inhibited *Leishmania* spermidine synthase with high selectivity.

Trypanothione reductase (TR) is an enzyme critical to the maintenance of the thiol redox balance in trypanosomatids. This target has been of interest for the development of antiprotozoal pharmaceutical molecules, such as pentavalent antimonials that interfere with the trypanothione [$T(SH)_2$] metabolism *in vivo.*[51] Compounds that disrupt the maintenance of levels of $T(SH)_2$, either through conjugation with $T(SH)_2$, or inhibition of the synthesis of polyamines (mainly TR), have the potential to be effective and selective anti-trypanosomal agents.[52,53] Diverse docking and MD studies have been carried out for different families of compounds such as imidazoles,[54,55] tricyclic and quinone derivatives[56] or chromene-like compounds.[57] In the case of imidazoles, initially 984 nitroimidazole analogues were docked in the enzyme, selecting top docking hits for absorption, distribution, metabolism, excretion and toxicity (ADMETox) properties studies and molecular mechanics energies combined with the generalized Born and surface area continuum solvation (MM–GBSA) methods. MD simulations were carried out for the two analogues with the best MM–GBSA values to ensure protein–ligand interactions, concluding that one of the nitroimidazole compounds has great potential to inhibit the enzyme and could be developed into an antileishmanial drug.

For the quinone derivatives, all the derivatives showed a similar binding mode in the coenzyme FAD binding site of the enzyme. The acridine and phenothiazine compounds were tested *in vitro* using the *Trypanosoma cruzi* homologue enzyme because of the high similarity, showing a submicromolar inhibitory constant (K_i) of 0.78 μM for the most potent compound.

The chromene-2-thione derivatives were synthesised and docked into the active site of TR, showing moderate anti-leishmanial activity in the micromolar range, and the most active compounds were effective at between 17 and 25 μM in intracellular amastigotes. The most active compound was [6-bromo-3-(4-methoxybenzoyl)-2*H*-chromene-2-thione], identifying a chemical space for designing compounds with improved activity based on this scaffold.

The glycolytic pathway has been extensively studied as a target for SBDD. The enzymes involved in this pathway provide an excellent approach because this process is one of the most important sources of ATP.[58] Docking is the most used tool in drug discovery for assessing structure–activity relationships. Some examples in *Leishmania* spp. are pyruvate kinase (PyK),[59] glyceraldehyde-3-phosphate dehydrogenase (GAPDH)[60,61] and hexokinase.[17] In the case of PyK of *L. major*, furanose sugar amino amides were analysed by docking studies to explain the different structure–activity relationships of the derivatives, identifying an *N,N′*-substituted-1-amino-2,5-anhydro-1-deoxy-1-D-mannonamide derivative as an starting point inhibitor of the enzyme in the micromolar range ($IC_{50} = 26$ μM).

For GAPDH, the most potent compound of the adenosine analogues was *N*^{6}-(2-methylbenzyl)-2′-deoxy-2′-(3-methoxybenzamido)-adenosine with an activity of 4 μM for *L. mexicana* GAPDH and an overall greater than 10^4-fold affinity improvement over adenosine, the lead compound. Docking studies permitted structure-based drug design using adenosine as the starting point, to assess the binding mode of the compounds.[60] These analogues were improved in further studies,[61] achieving submicromolar activity with the compound *N*^{6}-(1-naphthalenemethyl)-2′-(3-methoxybenzamido) (**5**), with full inhibition of the enzyme at 0.28 μM (Figure 7.2).

In the case of hexokinase, a target-fishing procedure based on docking methodology using different enzymes from *L. braziliensis* and *L. amazonensis* was implemented to identify the biochemical target of the active compounds.[17] These molecules are a family of dialkylphosphorylhydrazones, with several compounds presenting anti-leishmanial activity in the submicromolar range for phenotypic screening. The most active compound monoglyceride lipase (MGL), was a phosphorohydrazidic acid, *N′*-[(1*E*)-(3,5-dichloro-4-pyridinyl)methylene]-di-*sec*-butyl ester, with an IC_{50} value in *L. braziliensis* of 0.03 μM.

Another interesting study case is the design of inhibitors of LdLip3, a lipase from *L. donovani. Leishmania* is a lipid-scavenging organism, so the lipid metabolism of the pathogen plays a vital role in energy storage, virulence, transmission, cell cycle and modulation of the host response. Additional roles of lipid metabolism have been implicated in cellular signalling as well as organisation, dynamics and trafficking of membrane components. It is believed that this enzyme, LdLip3, participates in key biological processes that include host lipid degradation, alterations of the phagolysosome membrane and structural remodelling of membrane lipids important in pathogen survival and infection. This enzyme might favour parasite growth and development in the host.[62]

Similarity-based virtual screening was used to identify potential inhibitors using an LdLip3 3D structure that had been previously modelled. The NCI (National Cancer Institute) diversity set of compounds and small molecules from the ZINC database (http://zinc.docking.org) were collected, using a total of 22 919 molecules. Molecules with higher binding affinity toward LdLip3 than human monoglyceride lipase (MGL), were considered the top hits, and after the first screening, a similarity analysis of the top ten hits of the initial database was carried out on ZINC to identify novel hits selectively targeting LdLip3. Finally, four compounds were selected for further anti-leishmanial *in vitro* assays, showing IC_{50} values of between 5.2 μM and 17.3 μM, values in relation with the free energy binding predicted by docking. The protocol and the utility of the *in silico* tools were validated in the screen of 22 919 compounds, leading to four active compounds tested *in vitro* in promastigote leishmanial cells.[63] The active compounds present different chemical structures, such as phosphonate [(*S*)-2-ammonio-2-benzyl-4-phosphonatobutanoate), isoquinoline (2-(3,3-dimethyl-3,4-dihydro-isoquinolin-1-ylamino)-propionic acid), isoindole (2-(1-oxo-1,3-dihydro-isoindol-2-yl)-pentanedioic acid) (**6**) and quinoline (*E*,4*E*)-3-methyl-4-(2-quinolylmethylene)pent-2-enedioic acid] (**7**) (Figure 7.2).

One of the most studied enzymes in protozoa is NMT (*N*-myristoyl-transferase), an enzyme related to post-translational modifications of proteins; this protein facilitates the association of substrate proteins with membranes.[64,65] This enzyme has been validated pre-clinically as a target for the treatment of fungal and parasitic infections; also, genetic evidence has been provided that NMT is essential for viability in the *Leishmania* life cycle.[66,67] A number of 3D structures have been deposited in the PDB, and studies regarding the ligand binding mode of thienopyrimidine, piperidinylindole, aminoacylpyrrolidine and a biphenyl derivative were compared according to their binding mode in the *L. major* NMT (LmNMT) crystal structure.[68] The existence of these crystal structures allowed structure-based studies.[69,70] In the first study, virtual screening calculations against LmNMT were performed to select compounds to be further tested *in vitro*; docking calculations were also used to probe the possible binding pattern of protein-ligand. Compounds were selected to be tested in the parasite. The most active ones were pyrazole (**8**), ethyl 1-(4{[(4-methoxyphenyl)sulphonyl]amino}phenyl)-3,5-dimethyl-1*H*-pyrazole-4-carboxylate (Figure 7.2), and one sulfonamide derivative with activities of 13.8 and 15.7 μM, respectively. In the second study, potent LdNMT and LmNMT were synthesised and tested in a structure-guided design. Hybridisation of the two independent binding modes of hits led to the discovery of highly potent inhibitors of LdNMT; the most potent compounds were pyrrolidine derivatives that presented nanomolar values for K_i and an IC_{50} value for the enzyme of up to 1.6 nM.

Another important metabolic pathway is the sterol biosynthesis (SB) pathway. This pathway produces a special class of sterols, including ergosterol and other sterols, that are required for growth and viability but absent in mammalian host cells. Sterol 14α-demethylase (CYP51) is a key enzyme in the synthesis of ergosterol, essential to the membranes of the parasite. This enzyme has been found to be indispensable in *L. donovani,* using both biological and pharmacological approaches.[71] Several studies have been performed using the sterol 24-C methyltransferase (SMT) enzyme. As no 3D structure was found in PDB, homology modelling is necessary to perform docking studies or molecular dynamics studies; physicochemical and functional characterisation has also been reported for this target.[72,73] The IC_{50} values have been reported for a series of lanosterol and cholesterol derivatives at the micromolar and submicromolar range. In the studies of Azam *et al.*,[72,73] two different docking tools (FRED and GOLD) were used to compare binding modes and energies of 57 mechanism-based inactivators and to study protein–ligand interactions. Also, MD simulations using Amber software were carried out with the most active compounds. In an analogous way, the second study performed docking studies using the Autodock tool and MD simulations. Both studies provided a detailed insight into the mechanism of binding and confirmed the experimental findings, thereby encouraging rational drug design against the SMT enzyme involved in leishmanial disease.

The protein kinases are important regulatory molecules that play roles in the signal transduction cascade of eukaryotic cells. Several kinases have been shown to be essential in the *Leishmania* parasite. Casein kinase has been defined as a valid drug target.[74] Cyclin-dependent kinase (CRK3) and glycogen synthase kinase (GSK-3) are also essential for cell-cycle progression as they are involved in the G2–M transition in promastigotes and in cell control pathways leading to apoptosis-like death, respectively.[75,76] Also, the dramatic effect of these kinases inhibitors on *Leishmania* further validates these drug targets.[77] Docking studies of active compounds have been performed for both enzymes.[76,78] The reported compounds that inhibited parasite GSK-3 were mainly 6-bromo substituted indirubin derivatives; the IC_{50} values for the most potent compound was 90 nM for LdGSK-3 from *L. donovani*. Also, some of these compounds were powerful inhibitors of both *L. donovani* promastigote and intracellular amastigote growth in the submicromolar range. Docking of this derivative and some representatives was studied to determine the binding mode and key interactions related to its activity. Other interesting examples can be found in the design of inhibitors of mitogen activated protein kinases (MAPKs). MAPKs play important roles in the parasite, as this family of proteins are often the final kinases in signal transduction cascades ultimately leading to changes in protein expression profiles. LmxMPK4 from *L. mexicana* is highly conserved throughout different *Leishmania* spp. and is essential for the parasite, as it plays a crucial role in the proliferation and survival of promastigotes and amastigotes; it is thus a potential drug target for the treatment of leishmaniasis.[79–81] Due to the absence of a crystal structure of this protein, several approaches to homology modelling have been carried out, with Modeller software as the most common tool.[82,83] To validate the model, common tools like ProSa and Procheck as well as MD simulations were carried out using the CHARMM module of the Discovery Studio package. Studies of the cavity depth, electrostatic and lipophilic potential surfaces of *Leishmania* MAPK and human extracellular signal-regulated kinase 2 (ERK2) suggested that there are important differences in the ATP binding site. These differences in the ATP binding pocket of the proteins can help with the design of selective inhibitors to MAPK. Also, a series of ATP competitive nanomolar inhibitors of the CRK3 cyclin-dependent kinase of *L. mexicana* with antiparasitic activity[77] were docked in the *L. mexicana* MAPK model due to the high similarity of the CRK3 and MAPK binding sites in the parasite. Taking into account the effective dose (ED_{50}) values assayed and docking energies obtained in the study, it was concluded that a reasonably good correlation exists between both values for indirubin inhibitors.

The folate metabolic pathway provides numerous opportunities for targeted chemotherapy, with a strong potential for 'repurposing' of compounds. The formation of reduced folate *via* dihydrofolate reductase–thymidylate synthase (DHFR–TS) or pteridine reductase 1 (PTR1), and 10-formyl-tetrahydrofolate, a key compound of mitochondrial protein synthesis, *via* methylene-H_4F dehydrogenase–methenyl-H_4F cyclohydrolase (DHCH) are essential processes suitable for chemotherapeutic attack.[84,85] However, the 3D DHFR–TS complex is not available in the PDB and molecular modelling studies are required to carry out virtual screening studies. Rajasekaran and Chen[86] identified a lead compound with a tricyclic chemical structure of a benzoimidazothiazol by employing an efficient virtual screening protocol. The first step was the modelling of a 3D model of *L. major* DHFR–TS using as a template the structure of DHFR–TS in *T. cruzi* with a sequence identity of 66%. In this protocol, flexible docking and ADME property filtering among other steps were developed, and allowed the investigators to screen 126 923 compounds in the Chemicals Available for Purchase (CAP) database. This was performed in order to obtain a lead molecule that was analysed using MD simulations with a good interaction profile and exhibiting a strong interaction with the receptor through hydrogen bonds. Thus, this molecule and its analogues could be a promising lead compound for the development of anti-leishmanial drugs.

Dipeptidylcarboxypeptidase (DCP), an angiotensin converting enzyme (ACE)-related metallopeptidase, has been characterised and established as a drug target for anti-leishmanial drug discovery. DCP has an important role in parasite stage differentiation, during which protease activity is increased, and plays an indirect role in nutrition and pathogenesis.[87–89] Captopril was able to inhibit DCP activity as well as parasite growth, which strongly indicates that this enzyme is a drug target in *Leishmania*. Some studies have focused on the identification of novel inhibitors of *Leishmania* DCP. One of them[90] started with a database of 15 000 compounds to finally assay 46 compounds. The protocol included drug-like filtering in the first step to avoid non-drug-like compounds, and pharmacophore filtering based on a model developed with four different features that are important for ligand–protein recognition. Finally, compounds were assayed both with the enzyme and in promastigotes and amastigotes. This study led to the discovery of four compounds that inhibit the enzyme in the nanomolar range and the growth of the parasite at the micromolar range, such as ethyl 3-hydroxy-2-(((2-(pyrrolidin-1-yl)ethyl)amino)methyl)-3-(3-(*p*-tolyl)isoxazol-5-yl)propanoate (**9**) and (2*Z*,2′*Z*)-3,3′-(ethane-1,2-diylbis(azanediyl))bis(1-(4-chlorophenyl)-6-hydroxyhex-2-en-1-one) (**10**).

Another important enzyme related to the protease family is the CAAX prenyl protease, of which two isoforms in *L. donovani* (CAAX I and CAAX II) are known, involved in the processing of various physiologically important proteins. One of the most extensive studies[91] performed homology modelling studies, molecular and induced-fit docking, virtual screening and MD analyses as well as binding-free energy calculations. This study provided new and more potent inhibitors with good ADME profiles for this protein.

7.3.3 Drug Repurposing

The repurposing of drugs is a strategy that implies the expansion of the applications of existing drugs against a new disease. This strategy facilitates rapid and cost effective drug development because preclinical and clinical development can use already existing data.

In the last decade, the number of potential therapies against *Leishmania* spp. has risen *via* drug repurposing. As a summary, approved and experimental drugs at the clinical stage for the treatment of leishmaniasis can be classified according to their original action: antifungals (amphotericin B, itraconazole and posaconazole), antibiotics (paromomycin and azithromycin), antitumorals (miltefosine, imiquimod, sunitinib, sorafenib and lapatinib), bisphosphonates used as concomitant therapies for cancer (pamidronate), antidepressants (mianserin) and antihypertensives (sitamaquine).[92]

Traditionally, studies on drug repositioning have focused on *in vitro* screening, such as the drugs mentioned above; however, *in silico* screening methods can be useful tools due to their rationality and cost saving capabilities. *In silico* methods, including SBDD and LBDD strategies, have been applied to drug repurposing in several studies, using different tools that have been mentioned before, such as molecular 2D similarity, docking studies, pharmacophores and machine-learning methods used in drug discovery for *Mycobacterium tuberculosis*[93] and *Plasmodium* spp.[93,94]

7.4 Target Databases

The creation and use of different databases has allowed the integration and weighting of all the information available from genomic, proteomic and structural data, as well as its prioritisation and applicability to the drug discovery process.

Databases such as GenBank,[95] the Database of Essential Genes and OrthoMCL-DB can provide the necessary resources to predict and identify new possible *Leishmania* targets. When novel targets are identified, it is important to cross-reference data between these drug targets and the biological data of known drugs, in order to obtain several drugs that could be active on the homologous targets identified. The STITCH 2 database,[79] DrugBank database[11] and Binding Data Base (BDD)[96] are useful databases that compile known and predicted drug–target relationships with biological target data and pharmacological data.

As mentioned before, another interesting tool for the detailed exploration of drug targets in neglected tropical diseases is the TDR Targets Database, a chemogenomics resource for neglected tropical diseases.[97] This database identifies proteins that might be considered good drug targets because of their accordance with different criteria such as dissimilarity to human proteins and structural similarity to other druggable proteins, *i.e.* functions that are easy to measure. TDR allows filtering of targets by taking into consideration several key features such as essentiality, phylogenetic distribution, validation data, druggability or assayability. As a result of one of the searches, in Table 7.3 we can observe enzyme prioritisation according to the existing information. Most of these important parasite proteins are part of metabolic pathways that have been studied extensively in the literature.

Another interesting database for target identification is the Potential Drug Target Database (PDTD),[98] which contains over 1100 protein entries presented in the PDB, with other data extracted from the literature and several databases such as the Therapeutic Target Database (TTD),[99] DrugBank and Thomson Pharma, containing 28 targets for parasitic infectious diseases. This database is integrated with TarFisDock,[100] a docking tool that can be used to identify binding proteins for small molecules in conjunction with PDTD, by performing reverse-docking techniques. Target identification is not the only issue in anti-leishmanial drug discovery. SBDD needs a three-dimensional structure of the target. The most important database of 3D structures experimentally obtained is the Protein Data Bank. As mentioned before, there are few *Leishmania* 3D structures deposited in the PDB

Table 7.3 Top ranking of the validated targets according to a search in the TDR Targets Database.

Ranking	Gene name	Gene product	Weight
1	LmjF29.0820	Cysteine peptidase C	416
2	LmjF05.0350	Trypanothione reductase	386
2	LmjF06.0860	Dihydrofolate reductase–thymidylate synthase	386
2	LmjF23.0050	Cyclophilin	386
2	LmjF25.0910	Cyclophilin	386
2	LmjF06.0120	Cyclophilin	386
2	LmjF18.0270	Protein kinase, glycogen synthase kinase, putative	386
8	LmjF36.1960	Phosphomannomutase	366
8	LmjF23.0270	Pteridine reductase 1	366

(approximately 250). Several databases, such as the Protein Model Portal (PMP)[101] or ModBase[102] may play an important role in drug discovery. ModBase is a database of annotated protein structure models that contains almost 30 million reliable models for domains in 4.7 million unique protein sequences, and also provides information on ligand binding sites, single-nucleotide polymorphism annotation and protein–protein interactions. The PMP provides structural information for proteins by combining experimental structures in the PDB and theoretical models from various modelling resources such as ModBase and the SWISS-MODEL Repository.[103] Additionally, PMP provides an interface to several protein structure prediction servers such as HHPredB,[104] I-TASSER,[105] ModWEB, M4T[106] and SWISS-MODEL.[39]

7.5 Concluding Remarks and Outlook

Through this chapter, the most relevant tools for finding drugs and targets in leishmaniasis employing computational approaches have been described. Leishmaniasis, along with other parasitic diseases, continues to be harmful to human health, particularly in tropical regions, the development of new drugs being of utmost importance.

In recent years, the efforts of the research community in terms of computational analysis have been focused on the development of computational approaches to reveal the targets responsible for phenotypic results and in the identification and validation of a large number of *Leishmania* targets. Thus, tools such as protein interaction maps together with metabolic flux balance analysis and inverse docking are becoming useful alternatives to understand protein function and identify drugs.

Regarding identification and design of inhibitors, this can be approached both from point of view of the target and the ligand. Ligand-based approaches, such as QSAR or machine learning, are a promising option to identify anti-leishmanial drugs, especially when a target is not available.

Moreover, it is known that phenotypic assays are the fastest way to get results for neglected tropical diseases. Thus, similarity searching methods might be more readily available for similarity or substructure searching and can be used to find compounds that could share common features to other known active drugs. Special considerations should be made with the use of mt–QSAR models. There are many of pathogen parasite species with different susceptibility profiles to anti-parasitic drugs. QSAR models predict the biological activity of drugs against only one parasite species. Therefore, to predict the probability with which a drug is active against different species with a single unify model, such as mt–QSAR, is a goal of major importance.

The target-based approach is a very useful strategy when the target is known. Virtual screening studies together with the development of commercial databases (Table 7.4) have allowed an important advance in drug development.

Table 7.4 Summary of databases presented in this chapter.

Databases	Database information
Database of Essential Genes GenBank	Genetic sequences
OrthoMCL-DB	Database for orthologous searching
TDR Targets Database	Database for orthologous searching with specific data for neglected tropical diseases
Potential Drug Target Database	Database for target identification implemented with a reverse docking software
Protein Data Bank	3D structures experimentally obtained by X-ray crystallography and NMR
Protein Model Portal ModBase	3D structures of different protein models
STITCH 2 DrugBank database Binding Data Base Therapeutic Target Database	Drug–target relationship, interactions, pharmacological data and biological data.

It is important to remark that any computational experiment is exploratory and experimental validation is necessary; however, the use of *in silico* tools is valuable in drug discovery in order to reduce cost and time in this complex process.

Acknowledgements

Financial support from EC 7th Framework Programme (FP7-HEALTH-2013-INNOVATION-1, PDE4NPD no. 602666), Redes de Investigación Cooperativa Instituto de Salud Carlos III RICET (RD16/0027/0010) and FEDER funds is acknowledged. We also thank MECD (FPU15/01465 to V. S.).

References

1. WHO Technical Report Series, 2010.
2. G. J. Crowther, D. Shanmugam, S. J. Carmona, M. A. Doyle, C. Hertz-Fowler, M. Berriman, S. Nwaka, S. A. Ralph, D. S. Roos, W. C. Van Voorhis and F. Aguero, *PLoS Neglected Trop. Dis.*, 2010, **4**, e804.
3. R. Rajasekaran and Y. P. Chen, *Drug Discovery Today*, 2015, **20**, 958.
4. A. C. Ivens, C. S. Peacock, E. A. Worthey, L. Murphy, G. Aggarwal, M. Berriman, E. Sisk, M. A. Rajandream, E. Adlem, R. Aert, A. Anupama, Z. Apostolou, P. Attipoe, N. Bason, C. Bauser, A. Beck, S. M. Beverley, G. Bianchettin, K. Borzym, G. Bothe, C. V. Bruschi, M. Collins, E. Cadag, L. Ciarloni, C. Clayton, R. M. Coulson, A. Cronin, A. K. Cruz, R. M. Davies, J. De Gaudenzi, D. E. Dobson, A. Duesterhoeft, G. Fazelina, N. Fosker, A. C. Frasch, A. Fraser, M. Fuchs, C. Gabel, A. Goble, A. Goffeau, D. Harris, C. Hertz-Fowler, H. Hilbert, D. Horn, Y. Huang, S. Klages, A. Knights, M. Kube, N. Larke, L. Litvin, A. Lord, T. Louie, M. Marra, D. Masuy, K. Matthews, S. Michaeli, J. C. Mottram,

S. Muller-Auer, H. Munden, S. Nelson, H. Norbertczak, K. Oliver, S. O'Neil, M. Pentony, T. M. Pohl, C. Price, B. Purnelle, M. A. Quail, E. Rabbinowitsch, R. Reinhardt, M. Rieger, J. Rinta, J. Robben, L. Robertson, J. C. Ruiz, S. Rutter, D. Saunders, M. Schafer, J. Schein, D. C. Schwartz, K. Seeger, A. Seyler, S. Sharp, H. Shin, D. Sivam, R. Squares, S. Squares, V. Tosato, C. Vogt, G. Volckaert, R. Wambutt, T. Warren, H. Wedler, J. Woodward, S. Zhou, W. Zimmermann, D. F. Smith, J. M. Blackwell, K. D. Stuart, B. Barrell and P. J. Myler, *Science*, 2005, **309**, 436.
5. C. S. Peacock, K. Seeger, D. Harris, L. Murphy, J. C. Ruiz, M. A. Quail, N. Peters, E. Adlem, A. Tivey, M. Aslett, A. Kerhornou, A. Ivens, A. Fraser, M. A. Rajandream, T. Carver, H. Norbertczak, T. Chillingworth, Z. Hance, K. Jagels, S. Moule, D. Ormond, S. Rutter, R. Squares, S. Whitehead, E. Rabbinowitsch, C. Arrowsmith, B. White, S. Thurston, F. Bringaud, S. L. Baldauf, A. Faulconbridge, D. Jeffares, D. P. Depledge, S. O. Oyola, J. D. Hilley, L. O. Brito, L. R. Tosi, B. Barrell, A. K. Cruz, J. C. Mottram, D. F. Smith and M. Berriman, *Nat. Genet.*, 2007, **39**, 839.
6. A. Llanes, C. M. Restrepo, G. Del Vecchio, F. J. Anguizola and R. Lleonart, *Sci. Rep.*, 2015, **5**, 8550.
7. M. A. Lindsay, *Nat. Rev. Drug Discovery*, 2003, **2**, 831.
8. A. F. Florez, S. Watowich and C. Muskus, in *Current Topics in Tropical Medicine*, ed. A. Rodriguez-Morales, 2012, ch. 16, pp. 253–272.
9. R. Zhang, H. Y. Ou and C. T. Zhang, *Nucleic Acids Res.*, 2004, **32**, D271.
10. F. Gao, H. Luo, C. T. Zhang and R. Zhang, *Methods Mol. Biol.*, 2015, **1279**, 219.
11. D. S. Wishart, C. Knox, A. C. Guo, S. Shrivastava, M. Hassanali, P. Stothard, Z. Chang and J. Woolsey, *Nucleic Acids Res.*, 2006, **34**, D668.
12. V. Law, C. Knox, Y. Djoumbou, T. Jewison, A. C. Guo, Y. Liu, A. Maciejewski, D. Arndt, M. Wilson, V. Neveu, A. Tang, G. Gabriel, C. Ly, S. Adamjee, Z. T. Dame, B. Han, Y. Zhou and D. S. Wishart, *Nucleic Acids Res.*, 2014, **42**, D1091.
13. A. K. Chavali, J. D. Whittemore, J. A. Eddy, K. T. Williams and J. A. Papin, *Mol. Syst. Biol.*, 2008, **4**, 177.
14. D. Rognan, *Mol. Inf.*, 2010, **29**, 176.
15. J. L. Jenkins and J. J. Tanner, *Acta Crystallogr., Sect. D: Biol. Crystallogr.*, 2006, **62**, 290.
16. P. S. Kharkar, S. Warrier and R. S. Gaud, *Future Med. Chem.*, 2014, **6**, 333.
17. C. B. da Matta, A. C. de Queiroz, M. S. Santos, M. S. Alexandre-Moreira, V. T. Goncalves, N. Del Cistia Cde, C. M. Sant'Anna and J. B. DaCosta, *Eur. J. Med. Chem.*, 2015, **101**, 1.
18. A. Bulashevska, M. Stein, D. Jackson and R. Eils, *Comput. Biol. Chem.*, 2009, **33**, 457.
19. R. R. Thangudu, M. Tyagi, B. A. Shoemaker, S. H. Bryant, A. R. Panchenko and T. Madej, *BMC Bioinf.*, 2010, **11**, 365.

20. P. Willett, *Drug Discovery Today*, 2006, **11**, 1046.
21. M. G. Damale, S. N. Harke, F. A. Kalam Khan, D. B. Shinde and J. N. Sangshetti, *Mini-Rev. Med. Chem.*, 2014, **14**, 35.
22. C. C. Melo-Filho, R. C. Braga and C. H. Andrade, *Curr. Comput.-Aided Drug Des.*, 2014, **10**, 148.
23. M. Wesolowski and B. Suchacz, *J. AOAC Int.*, 2012, **95**, 652.
24. J. M. Sanders, A. O. Gomez, J. Mao, G. A. Meints, E. M. Van Brussel, A. Burzynska, P. Kafarski, D. Gonzalez-Pacanowska and E. Oldfield, *J. Med. Chem.*, 2003, **46**, 5171.
25. R. D. Cramer, D. E. Patterson and J. D. Bunce, *J. Am. Chem. Soc.*, 1988, **110**, 5959.
26. G. Klebe, U. Abraham and T. Mietzner, *J. Med. Chem.*, 1994, **37**, 4130.
27. V. Roldos, H. Nakayama, M. Rolon, A. Montero-Torres, F. Trucco, S. Torres, C. Vega, Y. Marrero-Ponce, V. Heguaburu, G. Yaluff, A. Gomez-Barrio, L. Sanabria, M. E. Ferreira, A. Rojas de Arias and E. Pandolfi, *Eur. J. Med. Chem.*, 2008, **43**, 1797.
28. S. Van Miert, S. Van Dyck, T. J. Schmidt, R. Brun, A. Vlietinck, G. Lemiere and L. Pieters, *Bioorg. Med. Chem.*, 2005, **13**, 661.
29. F. J. Prado-Prado, X. Garcia-Mera and H. Gonzalez-Diaz, *Bioorg. Med. Chem.*, 2010, **18**, 2225.
30. D. A. Delfin, A. K. Bhattacharjee, A. J. Yakovich and K. A. Werbovetz, *J. Med. Chem.*, 2006, **49**, 4196.
31. S. B. Bharate, J. B. Bharate, S. I. Khan, B. L. Tekwani, M. R. Jacob, R. Mudududdla, R. R. Yadav, B. Singh, P. R. Sharma, S. Maity, B. Singh, I. A. Khan and R. A. Vishwakarma, *Eur. J. Med. Chem.*, 2013, **63**, 435.
32. S. L. Dixon, A. M. Smondyrev, E. H. Knoll, S. N. Rao, D. E. Shaw and R. A. Friesner, *J. Comput.-Aided Mol. Des.*, 2006, **20**, 647.
33. D. Dube, V. Periwal, M. Kumar, S. Sharma, T. P. Singh and P. Kaur, *J. Mol. Model.*, 2012, **18**, 1701.
34. S. Jamal and V. Scaria, *BMC Bioinf.*, 2013, **14**, 329.
35. A. Agarwal, R. Mishra and S. Paliwal, *Int. J. Univers. Pharm. Bio Sci.*, 2013, **2**, 628.
36. Z. Garkani-Nejad and B. Ahmadi-Roudi, *Eur. J. Med. Chem.*, 2010, **45**, 719.
37. A. Sali and T. L. Blundell, *J. Mol. Biol.*, 1993, **234**, 779.
38. R. Sanchez and A. Sali, *Proteins*, 1997, (Suppl 1), 50.
39. L. Bordoli, F. Kiefer, K. Arnold, P. Benkert, J. Battey and T. Schwede, *Nat. Protoc.*, 2009, **4**, 1.
40. A. S. Yang and B. Honig, *Proteins*, 1999, (Suppl 3), 66.
41. H. Li, R. Tejero, D. Monleon, D. Bassolino-Klimas, C. Abate-Shen, R. E. Bruccoleri and G. T. Montelione, *Protein Sci.*, 1997, **6**, 956.
42. D. S. Goodsell and A. J. Olson, *Proteins*, 1990, **8**, 195.
43. R. A. Friesner, J. L. Banks, R. B. Murphy, T. A. Halgren, J. J. Klicic, D. T. Mainz, M. P. Repasky, E. H. Knoll, M. Shelley, J. K. Perry, D. E. Shaw, P. Francis and P. S. Shenkin, *J. Med. Chem.*, 2004, **47**, 1739.

44. T. J. Ewing, S. Makino, A. G. Skillman and I. D. Kuntz, *J. Comput.-Aided Mol. Des.*, 2001, **15**, 411.
45. M. Rarey, B. Kramer and T. Lengauer, *J. Comput.-Aided Mol. Des.*, 1997, **11**, 369.
46. G. Jones, P. Willett, R. C. Glen, A. R. Leach and R. Taylor, *J. Mol. Biol.*, 1997, **267**, 727.
47. M. L. Cunningham and A. H. Fairlamb, *Eur. J. Biochem.*, 1995, **230**, 460.
48. F. Frezard, C. Demicheli, C. S. Ferreira and M. A. Costa, *Antimicrob. Agents Chemother.*, 2001, **45**, 913.
49. C. Gilroy, T. Olenyik, S. C. Roberts and B. Ullman, *Infect. Immun.*, 2011, **79**, 2764.
50. A. Grover, S. P. Katiyar, S. K. Singh, V. K. Dubey and D. Sundar, *Biochim. Biophys. Acta*, 2012, **1824**, 1476.
51. P. Baiocco, G. Colotti, S. Franceschini and A. Ilari, *J. Med. Chem.*, 2009, **52**, 2603.
52. R. L. Krauth-Siegel and O. Inhoff, *Parasitol. Res.*, 2003, **90**, S77.
53. G. A. Holloway, W. N. Charman, A. H. Fairlamb, R. Brun, M. Kaiser, E. Kostewicz, P. M. Novello, J. P. Parisot, J. Richardson, I. P. Street, K. G. Watson and J. B. Baell, *Antimicrob. Agents Chemother.*, 2009, **53**, 2824.
54. R. K. Pandey, D. Sharma, T. K. Bhatt, S. Sundar and V. K. Prajapati, *J. Biomol. Struct. Dyn.*, 2015, **33**, 2541.
55. M. C. O'Sullivan, T. B. Durham, H. E. Valdes, K. L. Dauer, N. J. Karney, A. C. Forrestel, C. J. Bacchi and J. F. Baker, *Bioorg. Med. Chem.*, 2015, **23**, 996.
56. S. K. Venkatesan, A. K. Shukla and V. K. Dubey, *J. Comput. Chem.*, 2010, **31**, 2463.
57. R. K. Verma, V. K. Prajapati, G. K. Verma, D. Chakraborty, S. Sundar, M. Rai, V. K. Dubey and M. S. Singh, *ACS Med. Chem. Lett.*, 2012, **3**, 243.
58. C. L. Verlinde, V. Hannaert, C. Blonski, M. Willson, J. J. Perie, L. A. Fothergill-Gilmore, F. R. Opperdoes, M. H. Gelb, W. G. Hol and P. A. Michels, *Drug Resist. Updates*, 2001, **4**, 50.
59. M. W. Nowicki, L. B. Tulloch, L. Worralll, I. W. McNae, V. Hannaert, P. A. Michels, L. A. Fothergill-Gilmore, M. D. Walkinshaw and N. J. Turner, *Bioorg. Med. Chem.*, 2008, **16**, 5050.
60. A. M. Aronov, C. L. Verlinde, W. G. Hol and M. H. Gelb, *J. Med. Chem.*, 1998, **41**, 4790.
61. A. M. Aronov, S. Suresh, F. S. Buckner, W. C. Van Voorhis, C. L. Verlinde, F. R. Opperdoes, W. G. Hol and M. H. Gelb, *Proc. Natl. Acad. Sci. U. S. A.*, 1999, **96**, 4273.
62. A. M. Shakarian, G. C. McGugan, M. B. Joshi, M. Stromberg, L. Bowers, C. Ganim, J. Barowski and D. M. Dwyer, *Mol. Cell. Biochem.*, 2010, **341**, 17.
63. S. Parameswaran, P. Saudagar, V. K. Dubey and S. Patra, *J. Mol. Graphics Modell.*, 2014, **49**, 68.

64. E. W. Tate, A. S. Bell, M. D. Rackham and M. H. Wright, *Parasitology*, 2014, **141**, 37.
65. M. H. Wright, D. Paape, E. M. Storck, R. A. Serwa, D. F. Smith and E. W. Tate, *Chem. Biol.*, 2015, **22**, 342.
66. J. A. Brannigan, B. A. Smith, Z. Yu, A. M. Brzozowski, M. R. Hodgkinson, A. Maroof, H. P. Price, F. Meier, R. J. Leatherbarrow, E. W. Tate, D. F. Smith and A. J. Wilkinson, *J. Mol. Biol.*, 2010, **396**, 985.
67. H. P. Price, M. R. Menon, C. Panethymitaki, D. Goulding, P. G. McKean and D. F. Smith, *J. Biol. Chem.*, 2003, **278**, 7206.
68. J. A. Brannigan, S. M. Roberts, A. S. Bell, J. A. Hutton, M. R. Hodgkinson, E. W. Tate, R. J. Leatherbarrow, D. F. Smith and A. J. Wilkinson, *IUCrJ*, 2014, **1**, 250.
69. N. Singh, P. Shah, H. Dwivedi, S. Mishra, R. Tripathi, A. A. Sahasrabuddhe and M. I. Siddiqi, *Mol. Biosyst.*, 2016, **12**, 3711.
70. J. A. Hutton, V. Goncalves, J. A. Brannigan, D. Paape, M. H. Wright, T. M. Waugh, S. M. Roberts, A. S. Bell, A. J. Wilkinson, D. F. Smith, R. J. Leatherbarrow and E. W. Tate, *J. Med. Chem.*, 2014, **57**, 8664.
71. L. I. McCall, A. El Aroussi, J. Y. Choi, D. F. Vieira, G. De Muylder, J. B. Johnston, S. Chen, D. Kellar, J. L. Siqueira-Neto, W. R. Roush, L. M. Podust and J. H. McKerrow, *PLoS Neglected Trop. Dis.*, 2015, **9**, e0003588.
72. S. S. Azam, A. Abro and S. Raza, *J. Biomol. Struct. Dyn.*, 2015, **33**, 2563.
73. S. S. Azam, A. Abro, S. Raza and A. Saroosh, *Mol. Biol. Rep.*, 2014, **41**, 4279.
74. N. Rachidi, J. F. Taly, E. Durieu, O. Leclercq, N. Aulner, E. Prina, P. Pescher, C. Notredame, L. Meijer and G. F. Spath, *Antimicrob. Agents Chemother.*, 2014, **58**, 1501.
75. P. Hassan, D. Fergusson, K. M. Grant and J. C. Mottram, *Mol. Biochem. Parasitol.*, 2001, **113**, 189.
76. E. Xingi, D. Smirlis, V. Myrianthopoulos, P. Magiatis, K. M. Grant, L. Meijer, E. Mikros, A. L. Skaltsounis and K. Soteriadou, *Int. J. Parasitol.*, 2009, **39**, 1289.
77. K. M. Grant, M. H. Dunion, V. Yardley, A. L. Skaltsounis, D. Marko, G. Eisenbrand, S. L. Croft, L. Meijer and J. C. Mottram, *Antimicrob. Agents Chemother.*, 2004, **48**, 3033.
78. E. Reznickova, A. Popa, T. Gucky, M. Zatloukal, L. Havlicek, V. Bazgier, K. Berka, R. Jorda, I. Popa, A. Nasereddin, C. L. Jaffe, V. Krystof and M. Strnad, *Bioorg. Med. Chem. Lett.*, 2015, **25**, 2298.
79. D. Kuhn and M. Wiese, *Mol. Microbiol.*, 2005, **56**, 1169.
80. M. Wiese, *Int. J. Parasitol.*, 2007, **37**, 1053.
81. Q. Wang, I. M. Melzer, M. Kruse, C. Sander-Juelch and M. Wiese, *Kinetoplastid Biol. Dis.*, 2005, **4**, 6.
82. M. Awale, V. Kumar, P. Saravanan and C. G. Mohan, *J. Mol. Model.*, 2010, **16**, 475.
83. C. L. Gupta, M. K. Khan, M. F. Khan and A. K. Tiwari, *Interdiscip. Sci.*, 2013, **5**, 136.

84. S. M. Beverley, T. E. Ellenberger and J. S. Cordingley, *Proc. Natl. Acad. Sci. U. S. A.*, 1986, **83**, 2584.
85. S. M. Murta, T. J. Vickers, D. A. Scott and S. M. Beverley, *Mol. Microbiol.*, 2009, **71**, 1386.
86. R. Rajasekaran and Y. P. Chen, *J. Mol. Model.*, 2012, **18**, 4089.
87. M. S. Baig, S. Gangwar and N. Goyal, *Cell. Mol. Biol.*, 2011, **57**, 56.
88. M. S. Baig, A. Kumar, M. I. Siddiqi and N. Goyal, *J. Comput.-Aided Mol. Des.*, 2010, **24**, 77.
89. N. Goyal, R. Duncan, A. Selvapandiyan, A. Debrabant, M. S. Baig and H. L. Nakhasi, *Mol. Biochem. Parasitol.*, 2006, **145**, 147.
90. S. Gangwar, M. S. Baig, P. Shah, S. Biswas, S. Batra, M. I. Siddiqi and N. Goyal, *Chem. Biol. Drug Des.*, 2012, **79**, 149.
91. S. Singh, S. Vijaya Prabhu, V. Suryanarayanan, R. Bhardwaj, S. K. Singh and V. K. Dubey, *J. Biomol. Struct. Dyn.*, 2016, **34**, 2367.
92. M. L. Sbaraglini, M. C. Vanrell, C. L. Bellera, G. Benaim, C. Carrillo, A. Talevi and P. S. Romano, *Curr. Top. Med. Chem.*, 2016, **16**, 2201.
93. S. Ekins, J. Bradford, K. Dole, A. Spektor, K. Gregory, D. Blondeau, M. Hohman and B. A. Bunin, *Mol. Biosyst.*, 2010, **6**, 840.
94. J. C. Williams, J. P. Zeelen, G. Neubauer, G. Vriend, J. Backmann, P. A. Michels, A. M. Lambeir and R. K. Wierenga, *Protein Eng.*, 1999, **12**, 243.
95. D. A. Benson, M. Cavanaugh, K. Clark, I. Karsch-Mizrachi, D. J. Lipman, J. Ostell and E. W. Sayers, *Nucleic Acids Res.*, 2013, **41**, D36.
96. M. K. Gilson, T. Liu, M. Baitaluk, G. Nicola, L. Hwang and J. Chong, *Nucleic Acids Res.*, 2016, **44**, D1045.
97. M. P. Magarinos, S. J. Carmona, G. J. Crowther, S. A. Ralph, D. S. Roos, D. Shanmugam, W. C. Van Voorhis and F. Aguero, *Nucleic Acids Res.*, 2012, **40**, D1118.
98. Z. Gao, H. Li, H. Zhang, X. Liu, L. Kang, X. Luo, W. Zhu, K. Chen, X. Wang and H. Jiang, *BMC Bioinf.*, 2008, **9**, 104.
99. C. Qin, C. Zhang, F. Zhu, F. Xu, S. Y. Chen, P. Zhang, Y. H. Li, S. Y. Yang, Y. Q. Wei, L. Tao and Y. Z. Chen, *Nucleic Acids Res.*, 2014, **42**, D1118.
100. H. Li, Z. Gao, L. Kang, H. Zhang, K. Yang, K. Yu, X. Luo, W. Zhu, K. Chen, J. Shen, X. Wang and H. Jiang, *Nucleic Acids Res.*, 2006, **34**, W219.
101. J. Haas, S. Roth, K. Arnold, F. Kiefer, T. Schmidt, L. Bordoli and T. Schwede, *Database*, 2013, **2013**, bat031.
102. U. Pieper, B. M. Webb, G. Q. Dong, D. Schneidman-Duhovny, H. Fan, S. J. Kim, N. Khuri, Y. G. Spill, P. Weinkam, M. Hammel, J. A. Tainer, M. Nilges and A. Sali, *Nucleic Acids Res.*, 2014, **42**, D336.
103. F. Kiefer, K. Arnold, M. Kunzli, L. Bordoli and T. Schwede, *Nucleic Acids Res.*, 2009, **37**, D387.
104. J. Soding, A. Biegert and A. N. Lupas, *Nucleic Acids Res.*, 2005, **33**, W244.
105. A. Roy, A. Kucukural and Y. Zhang, *Nat. Protoc.*, 2010, **5**, 725.
106. D. Rykunov, E. Steinberger, C. J. Madrid-Aliste and A. Fiser, *J. Struct. Funct. Genomics*, 2009, **10**, 95.

CHAPTER 8

Medicinal Chemistry Strategies to Discover New Leishmanicidal Drugs

ANA MARTINEZ AND CARMEN GIL*

Centro de Investigaciones Biológicas (CSIC), Ramiro de Maeztu 9, 28040 Madrid, Spain
*Email: carmen.gil@csic.es

8.1 Introduction

Leishmaniasis is a major health problem faced by many countries, caused by infection with *Leishmania* parasite and transmitted to humans by the bite of infected female sand flies. Although the majority of the disease burden is found in poverty-stricken countries within Southeast Asia, East Africa and Latin America, it is also endemic in several Mediterranean countries.[1] Recent epidemiological studies have shown an increased prevalence in Europe, largely caused by an increase in international travel, difficulty in eradicating leishmanial infection in acquired immune deficiency syndrome (AIDS) patients and the use of immunosuppressive medications.[2]

Despite ongoing attempts to produce a molecular vaccine, at present treatment and control of this neglected disease rely on chemotherapy. To date, there are few drugs available for treating leishmaniasis: pentavalent antimonials (sodium stibogluconate and meglumine antimoniate), liposomal amphotericin B (polyene antibiotic), pentamidine, paramomycin (aminoglycoside antibiotic) and miltefosine (hexadecylphosphocholine) (Figure 8.1).

Drug Discovery Series No. 60
Drug Discovery for Leishmaniasis
Edited by Luis Rivas and Carmen Gil

Published by the Royal Society of Chemistry, www.rsc.org

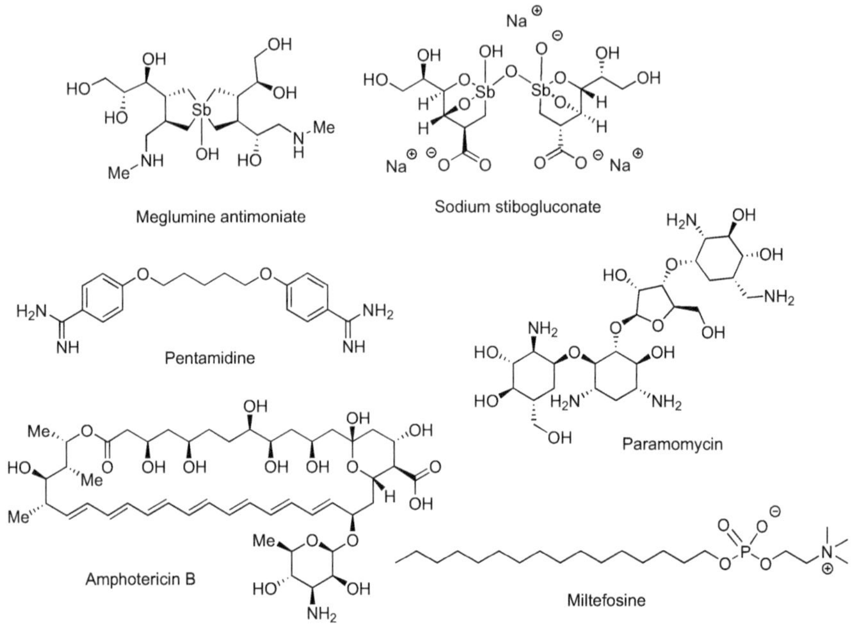

Figure 8.1 Chemical structures of leishmanicidal drugs.

These drugs present major limitations, such as non-oral routes of administration (apart from miltefosine), high toxicity of antimonials, non-encapsulated amphotericin B and pentamidine, and teratogenity for miltefosine. Moreover, the high cost of liposomal amphotericin B leads to patients in developing countries withdrawing from treatment. In addition, the number of drug-resistant strains has increased sharply.[3,4]

In the search for more effective drugs to treat leishmaniasis researchers have chosen to follow different approaches.[5,6] To date, phenotypic screening has been the most used methodology.[7] Different *Leishmania* strains are *in vitro* cultured and their viability after treatment with different compounds is the readout measured to find a new therapeutic agent. Among the different strategies to discover new drugs, natural products or privileged structures are without any doubt excellent starting points.[8,9] Throughout the history of medicinal chemistry natural products have played and continue to play an invaluable role in drug discovery.[10] Pharmacophores derived from natural products are excellent candidates for biologically relevant chemical libraries. Such is the case of quinones or nitrogen heterocycles that have been shown to be promising as new drugs for the treatment of leishmaniasis, as summarized in this chapter. Moreover, finding new uses for existing drugs, so called drug repurposing,[11] and different new target-based approaches[12] are interesting and fruitful methods for discovering and developing new drugs for neglected tropical diseases, including leishmaniasis.

8.2 Functional Scaffolds for Drug Discovery Against *Leishmania*

8.2.1 Quinones

Quinones can be considered as privileged structures and are secondary metabolites present in several plant families extensively used in folk medicine due to a broad range of biological activities. Particularly important are their effects as anticancer or antibacterial agents among others.[13] The molecular structure of these cyclic compounds is responsible for redox cycling due to the ability to accept one or two electrons, being involved in different biological oxidative processes.[14]

Their antiprotozoal activities have been reported, and several of them have been identified as lead candidates for drug development (Figure 8.2). Moreover, related quinone compounds occur with high frequency in natural products, such as lapachol (**1**)[15] or plumbagin (**2**),[16] have shown activity against trypanosomatid. Parasitic protozoa of the order Kinetoplastida are particularly sensitive to oxidative stress, and, based on this, reduction of naphthoquinones by parasitic flavoenzymes can be considered as a promising strategy for the development of new drugs.[17,18]

Plumbagin (**2**), a naphthoquinone extracted from *Pera benensis,* and its derivative 2-methoxy-1,4-naphthoquinone (**3**), inhibited the growth of *Leishmania donovani* promastigotes and axenic amastigotes. Moreover, these effects were found to be related to the disruption of the redox homeostasis of the parasite.[16] Other naphthoquinones such as 2,3-dichloro-5,8-dihydroxy-1,4-naphthoquinone (**4**) and 2,3-dibromo-1,4-naphthoquinone (**5**) were found to be able to induce killing of promastigotes and intracellular amastigotes of *L. donovani* with a similar potency to sodium stibogluconate but with higher therapeutic index.[19]

The natural product lapachol (**1**), originally isolated from the heartwood of trees of the Bignoniaceae family, was used as a source of inspiration to generate a small library of derivatives, which allowed the identification of a phenoxy derivative (**6**) as a hit candidate to be further developed.[20] Naphthoquinone **6** presented an IC_{50} (50% inhibitory concentration) value (1.26 μM) against *L. donovani* axenic amastigotes, four times higher than miltefosine (0.31 μM) although with a selective index of only 4.7. The optimization of this hit led to the synthesis of a library of quinones (naphthoquinones and anthraquinones) conjugated with polyamines. Among this new series of quinones emerged the potent analogue **7**, able to inhibit the growth of *Trypanosoma brucei* and *L. donovani*. Studies to decipher the mechanism of action of these conjugates showed that anthraquinone **7** was able to inhibit TbrTR (*T. brucei* trypanothione reductase) *in vitro* and, with respect to *Leishmania*, to cause mitochondrial dysfunction.[21]

Also inspired by lapachol (**1**), 1,4-naphthoquinones coupled to 1,2,3-triazoles were synthesized and as a result, triazole naphthoquinones (**8** and **9**) with selectivity indexes in the range of 10–15 and activity on different species of *Leishmania*, including resistant strains, were described.[22]

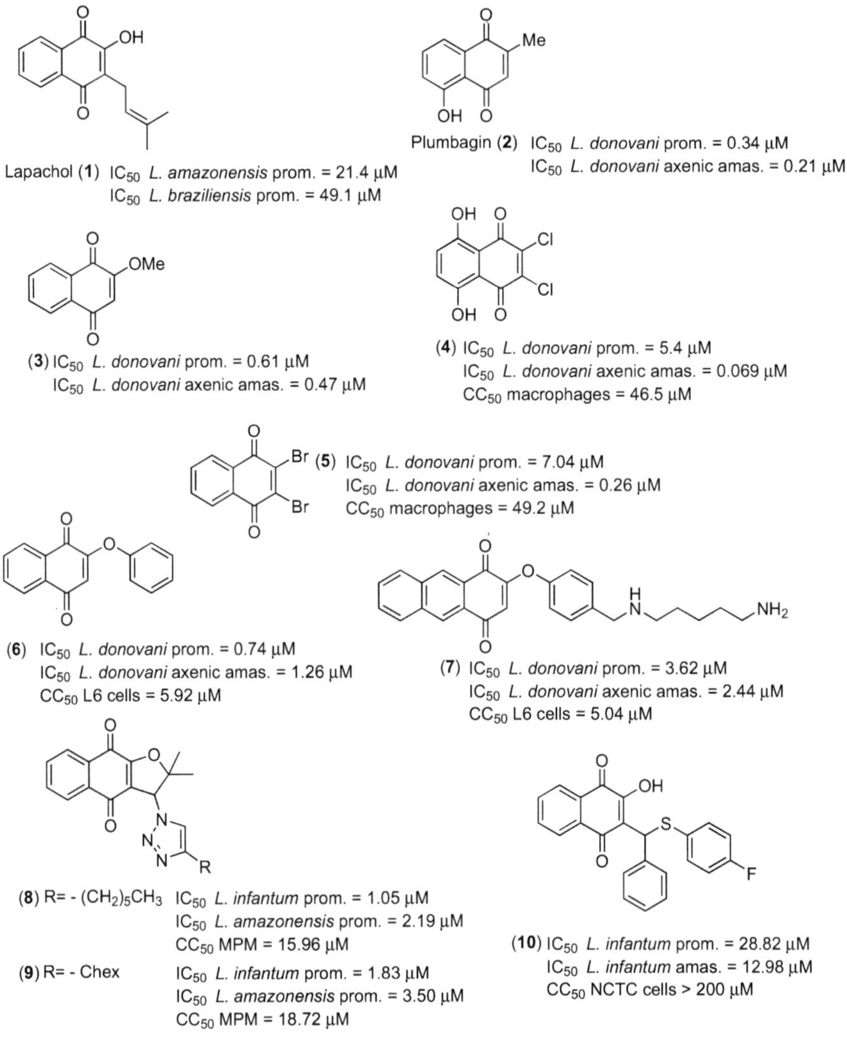

Figure 8.2 Chemical structures of quinones with leishmanicidal activity[15,16,19–23] (prom., promastigotes; amas., amastigotes; MPM, murine peritoneal macrophages).

A series of 36 compounds having naphthoquinone as a core system were synthesized and evaluated against the extracellular and intracellular form of the parasite. 2-Hydroxy-3-phenylsulfanylmethyl-[1,4]-naphthoquinone (**10**) was the most interesting compound because of its IC_{50} (28.82 and 12.98 μM against the promastigote and amastigote respectively) and selectivity index higher than 15. The analysis of the biological activity of the 36 derivatives was interesting from the point of view of the structure–activity relationships. The most promising compounds required two phenyl groups but the presence of nitro substituents led to mammalian cell toxicity. It is noteworthy that the fluorine atom from compound **10** is quite important to improve the selectivity index of the final compounds.[23]

The capacity of quinones to produce free oxygen radicals is dramatically influenced by the nature and position of substituents and contributes both to the therapeutic and toxic action of these substances.[24] In summary, the aim in this field is to obtain modified quinones carefully derivatised in order to avoid toxicity while maintaining biological activity.

8.2.2 Nitroaromatic Compounds

Despite the potential toxicity issues associated with compounds bearing a nitro group, nowadays we are experiencing a renaissance of this kind of compounds for the treatment of trypanosomatid diseases.[25]

Many of the nitroaromatics in development result from drug repurposing programs (Figure 8.3). This is the case for the use of nifurtimox (traditionally

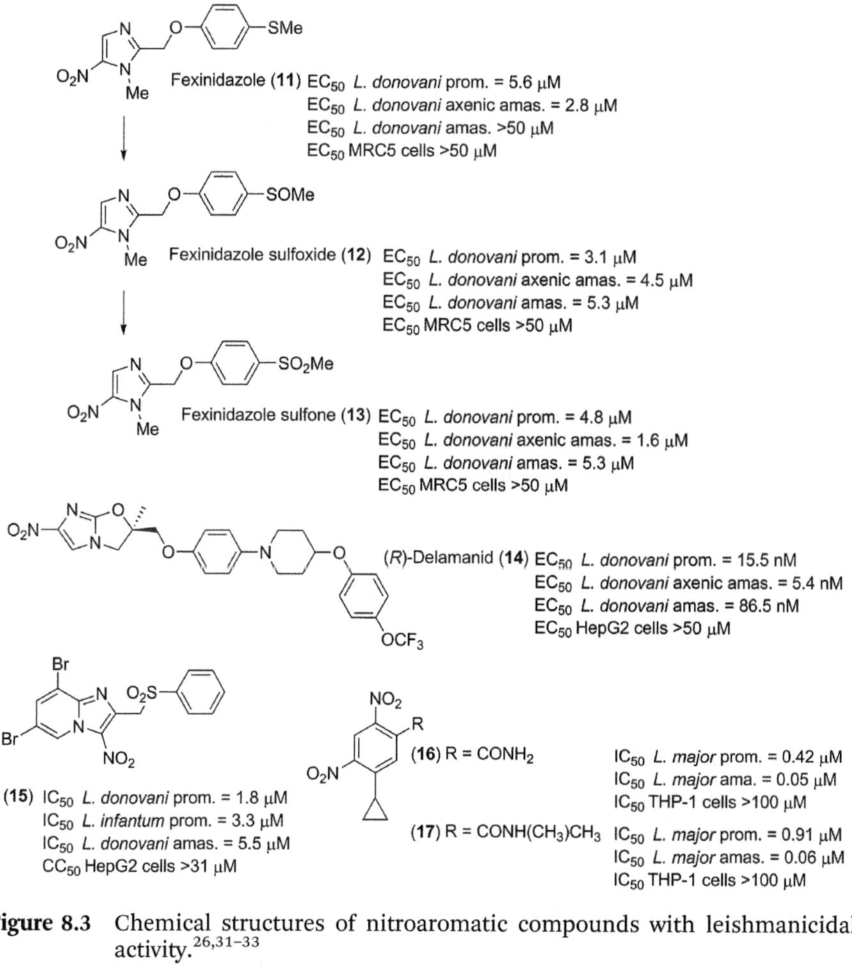

Figure 8.3 Chemical structures of nitroaromatic compounds with leishmanicidal activity.[26,31–33]

used for Chagas disease) to treat human African trypanosomiasis (HAT) or the development of fexinidazole (**11**) for visceral leishmaniasis.[26] The associated risk of a project based on nitroaromatic compounds is reduced significantly by using a nitroaromatic with safety proven previously, because not all nitro drugs are toxic. The unique properties of nitro groups are mainly that they are electron-withdrawing, small, polar and able to form hydrogen bonds, together with the possibility of being bioactivated by enzymatic reduction to give reactive species. These reactive species could be responsible for the biological effects, the nitro compounds acting as pro-drugs, or even could cause toxic effects. For this reason, it is quite important to define the enzymes responsible for activation of nitro drugs in order to reduce toxicity.

The metabolism of fexinidazole (**11**) has been extensively studied. After a rapid metabolism, the blood concentrations of its two main metabolites, sulfoxide (**12**) and/or sulfone (**13**), exceeded that of fexinidazole (**11**). Both metabolites were active against *L. donovani* amastigotes grown in macrophages, whereas the parent compound was inactive. Moreover, after daily treatment for five days a suppression of infection of 98.4% was observed in a mouse model of visceral leishmaniasis.[26] Regarding the mechanism of action, the pivotal role of a type I nitroreductase (NTR), absent in mammalian cells, in fexinidazole activation was confirmed.[27]

As the regulatory toxicology of fexinidazole showed an excellent safety profile,[28] and studies in healthy volunteers were also satisfactory,[29] in 2013 a phase II proof of concept clinical trial for visceral leishmaniasis was conducted in Sudan. However, it failed to show conclusive efficacy in the majority of patients.[30] Novel clinical trials to explore the use of fexinidazole in combination with miltefosine are currently in progress.

Another example of a clinically approved drug that could be repurposed as a visceral leishmaniasis therapy is delamanid (**14**). This nitro aromatic compound is an approved drug for multi-drug-resistant tuberculosis that has proven to be a potent inhibitor of *L. donovani* both *in vitro* and *in vivo*. Although the particular mechanism of action is not fully known, the importance of the nitro group is well established due to the lack of activity of des-nitro-delamanid.[31]

Among other nitroaromatic compounds with interesting scaffolds to be further studied due to their anti-*Leishmania* activities are imidazo[1,2-*a*]-pyridine derivative **15**,[32] aziridinyl nitrobenzamide compounds (**16** and **17**)[33] (Figure 8.3), or nitrotriazole-based agents.[34–36] In all these series, the nitro group was mandatory for the leishmanicidal properties and NTR appears to be the enzyme responsible of its mechanism of action. The fact that the activation of nitroheterocyclic prodrugs can be catalyzed by NTR, absent from most eukaryotes and with trypanosomatids as a major exception, have led to a renewed interest in such compounds as antiparasitic agents.[37]

8.2.3 Nitrogen Heterocycles

Experience has shown that compounds with biological activity are often derived from heterocyclic structures.[9] Nitrogen heterocycles in particular exhibit a number of biological and pharmacological activities due in part to similarities with many natural and synthetic molecules with a broad range of biological properties, including antiparasitic activities.

8.2.3.1 Bicyclic Nitrogen Heterocycles

Quinoline derivatives, both synthetic and of natural origin, exhibited a broad spectrum of biological activities, including appealing *in vitro* and *in vivo* activities against *Leishmania*. A huge number of analogs bearing a large diversity of substituents at several positions of the heterocyclic core have been synthesized and tested in several studies. In general, substitution at any position of the quinoline ring could lead to active compounds, although improved activity is usually associated with multiple substitutions.[38] It is noteworthy also that nitro groups were successfully introduced into the quinoline ring, leading to an 8-nitroquinoline series with anti-leishmanial potential.[39,40] However, the general mechanism of action of these derivatives remains unknown with the exception of some selected quinolines.

Sitamaquine (**18**) (Figure 8.4), is an orally active 8-aminoquinoline that was administered in a Phase II clinical trials conducted in India and Africa for treatment of visceral leishmaniasis.[41,42] Although the molecular target of this tri-substituted quinoline is not completely identified, it was envisaged that this was due to interaction with the mitochondrial respiration of the parasite. In particular, its ability to inhibit the respiratory chain complex II.[43] The future of sitamaquine (**18**) depends on the results of new clinical trials to ensure the safety of the new drug, mainly at the level of methemoglobinemia and nephrotoxicity.[44]

Another 8-aminoquinoline, tafenoquine (**19**) (Figure 8.4), also has potential for the oral treatment of leishmaniasis. Tafenoquine (**19**) has shown similar potency to sitamaquine (**18**), in both *in vitro* and *in vivo* models.[45] Moreover, this new quinoline targeted *Leishmania* respiratory complex III, leading to an apoptosis-like death process.[46]

Despite the fact that numerous compounds on quinoline scaffolds have been reported to be leishmanicidals, the targets for most of them remain elusive and this lack of knowledge is a serious difficulty for the development of a rational program of drug design and development.

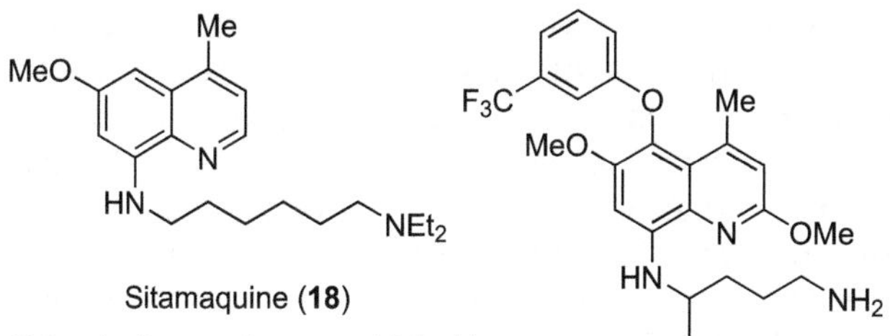

Sitamaquine (**18**)

EC_{50} *L. donovani* prom. = 19.8 μM

Tafenoquine (**19**)

Figure 8.4 Chemical structures of quinoline derivatives with leishmanicidal activity.[43]

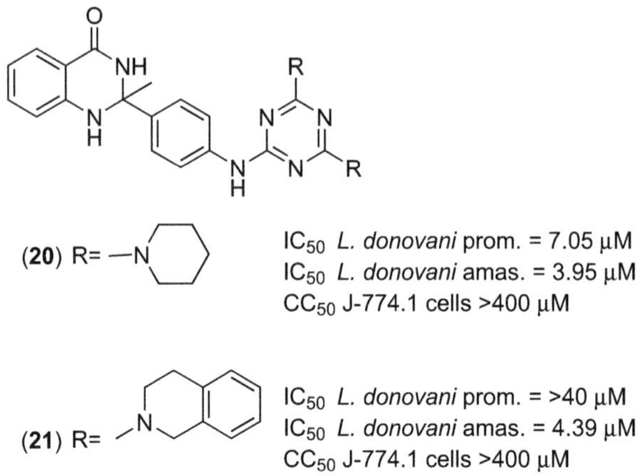

Figure 8.5 Chemical structures of quinazoline derivatives with leishmanicidal activity.[47]

The quinazoline ring is another benzoannelated heterocycle presents in a number of natural products and drug-like molecules of synthetic origin. Due to the high potential of this privileged scaffold it was chosen to be hybridized with pyrimidine, triazine or triazole rings. The hybridization of two bioactive pharmacophores led to the discovery of novel and potent compounds against intracellular *Leishmania* amastigotes with a high selectivity index. Quinazolinone–triazine hybrids **20** and **21** even showed *in vivo* efficacy in the golden hamster model for visceral leishmaniasis (Figure 8.5).[47]

8.2.3.2 Monocyclic Nitrogen Heterocycles

Azole antifungals have been used to treat cutaneous leishmaniasis with variable success. These drugs act through inhibition of the synthesis of ergosterol (the main fungal sterol), offering an interesting opportunity for drug repurposing for the treatment of leishmaniasis. This possibility is endowed by the fact that the parasites synthesize also 24-substituted sterols such as ergosterol, in contrast to mammals, which synthesize cholesterol.[48] Besides the use of azole antifungals, new azole-based anti-leishmanial compounds have also been designed and synthesized (Figure 8.6). Such is the case of the series of furanyl and thiophenyl azoles developed through rational drug design approaches. The most interesting compound was the furanyl azole **22** with an appealing selective amastigote activity (>10) better than those of the thiophenyl azoles. The presence of the imidazole ring was also crucial for the activity, as its replacement by triazole decreased the activity notably.[49] It is noteworthy that a recent report of the anti-leishmanial activity of azole drugs containing an imidazole ring, bifonazole (**23**), clotrimazole (**24**), and econazole (**25**), corroborates the potential anti-leishmanial activity of imidazoles and showed for the

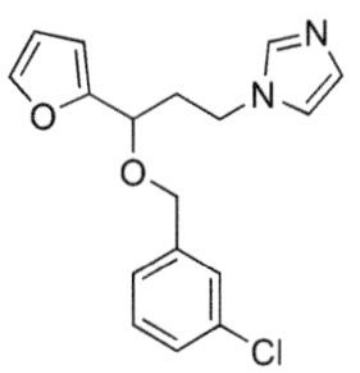

(**22**) IC_{50} *L. donovani* amas. = 3.04 μM
CC_{50} J-774.1 cells = 60.21 μM

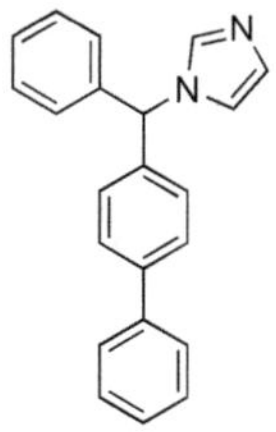

Bifonazole (**23**) IC_{50} *L. donovani* prom. = 8.97 μM
IC_{50} *L. donovani* amas. = not active
CC_{50} NCTC cells = 41.81 μM

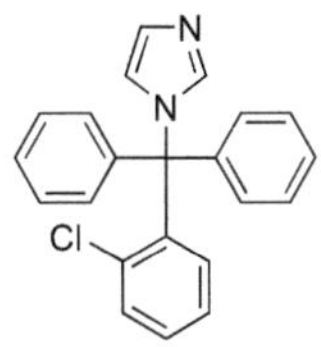

Clotrimazole (**24**) IC_{50} *L. donovani* prom. = 2.81 μM
IC_{50} *L. donovani* amas. = not active
CC_{50} NCTC cells = 39.24 μM

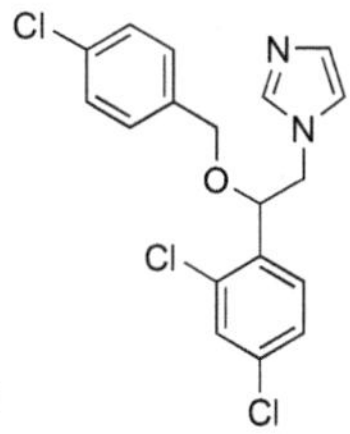

Econazole (**25**) IC_{50} *L. donovani* prom. = 5.36 μM
IC_{50} *L. donovani* amas. = 11.95 μM
CC_{50} NCTC cells = 42.69 μM

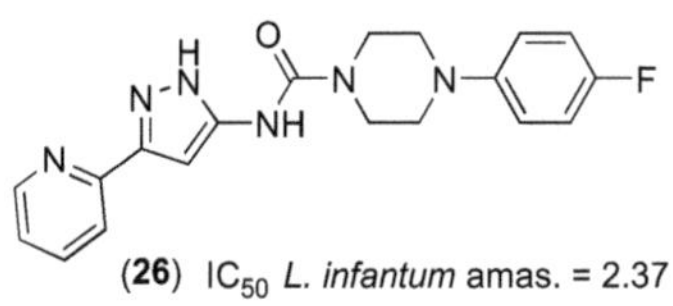

(**26**) IC_{50} *L. infantum* amas. = 2.37 μM
IC_{50} *L. donovani* amas. = 1.31 μM
CC_{50} MRC5 cells >64 μM

Figure 8.6 Chemical structures of imidazole and pyrazole derivatives with leishmanicidal activity.[49,50,54]

first time, to our knowledge, their potential action to affect the regulation of reactive oxygen species (ROS) of the parasites.[50]

The pyrazole ring is a versatile heterocyclic pharmacophore with broad range of biological properties, including anti-leishmanial activity.[51–53] Therefore, a novel series of amino-pyrazole amides was initially identified from an HTS campaign. The hit optimization process led to amino-pyrazole ureas with improved potency against intracellular *Leishmania infantum* amastigotes and metabolic stability respect to the amide derivatives. Moreover, the selected candidate, compound **26**, demonstrated high levels of *in vivo* efficacy in a hamster model of visceral leishmaniasis.[54]

8.3 Targets for Leishmaniasis Drug Discovery

Without any doubt, what significantly accelerates drug discovery in any field is the knowledge of validated targets. In this sense, there are increasing efforts to rationalize the discovery of new leishmanicidal agents based not only on well-established targets or routes, but also on new emerging targets that are currently at different validation stages. These targets are, for example, enzymes involved in *Leishmania* polyamine synthesis, thiol, sterol and folic acid metabolism, together with different parasite leishmanial-specific proteases, protein kinases or phosphodiesterases (some of these targets and others are specifically addressed in other chapters of this volume). New compounds specifically design to interact with them have started to appear in the arena of the future drugs for the treatment of this neglected disease.

8.3.1 Enzymes of Polyamine Biosynthesis

Polyamine metabolism is essential for parasite survival. Polyamines are essential metabolites in eukaryotes for cell growth and proliferation, and in trypanosomatid protozoa play an additional role in the synthesis of the critical thiol, trypanothione [bis(glutathionyl)spermidine].[55] In particular, spermidine, the final product of the polyamine biosynthesis pathway, serves as a precursor for trypanothione, the key molecule for parasite defense against oxidative stress. Hence, the enzymes responsible for the biosynthesis and conjugation of spermidine are promising targets for the development of new drugs against leishmaniasis.[56]

Arginase, the first enzyme of the polyamine pathway in *Leishmania*, is a binuclear Mn(II) metalloenzyme that initiates *de novo* polyamine biosynthesis by catalyzing the hydrolysis of L-arginine to L-ornithine and urea. Ornithine is used in the polyamine pathway, being essential for cell proliferation and ROS detoxification by trypanothione. It was recently established that arginine uptake and arginase activity are important in establishing and maintaining *Leishmania* infection.[57] The essential role of arginase in promastigotes proliferation, but not in the intracellular form, has been recently reported using null mutants.[58] Furthermore, deletion of the arginase gene in *Leishmania major* demonstrated the importance of this enzyme, as the parasite becomes auxotrophic for polyamines.[59]

Based on this background, arginase inhibitors have been reported as potent leishmanicidal agents (Figure 8.7).[60] Among them, different flavonoids isolated from different natural products,[61] specifically, fisetin (**27**), quercetin (**28**), luteolin (**29**) and 7,8-hydroxyflavone (**30**) are selective inhibitors of *L. amazonensis* arginase (LaARG) and have shown high leishmanicidal activity.[62,63] Due to the IC_{50} values of these flavonoids and their therapeutic window, they may be used as food supplements in leishmaniasis treatment.[64] Moreover, natural compounds present in green tea, such as epigallocatechin gallate (**31**), catechin (**32**) and epicatechin (**33**), have been also reported as LaARG inhibitors.[65]

The crystal structure of arginase from *Leishmania mexicana* (LmARG) and different inhibitor complexes, including 2(*S*)-amino-6-boronohexanoic acid (**34**), have been reported, deciphering some important clues not only for arginase inhibition but also for design of specific leishmanicidal agents.[66]

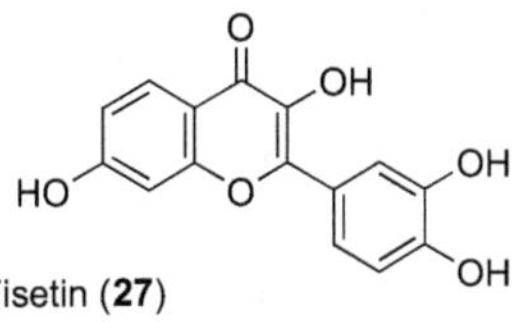

Fisetin (**27**)
IC_{50} *L. donovani* axenic amas. = 2.1 μM
CC_{50} L6 cells = 134.6 μM

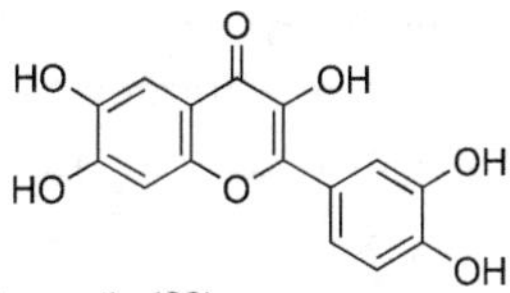

Quercetin (**28**)
IC_{50} *L. amazonensis* ARG = 4.0 μM
IC_{50} *L. donovani* axenic amas. = 3.3 μM
CC_{50} L6 cells = 122.8 μM

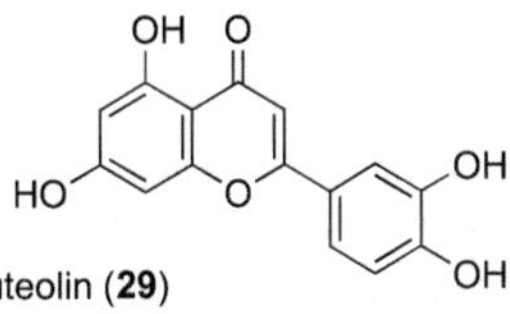

Luteolin (**29**)
IC_{50} *L. donovani* axenic amas. = 2.8 μM
CC_{50} L6 cells = 32.9 μM

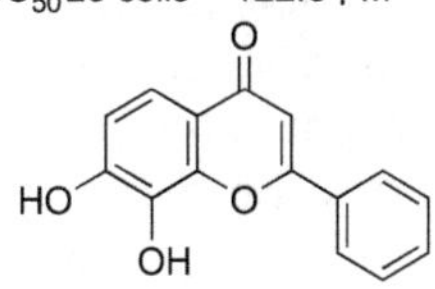

7,8-hydroxyflavone (**30**)
IC_{50} *L. donovani* axenic amas. = 6.7 μM
CC_{50} L6 cells = 31.1 μM

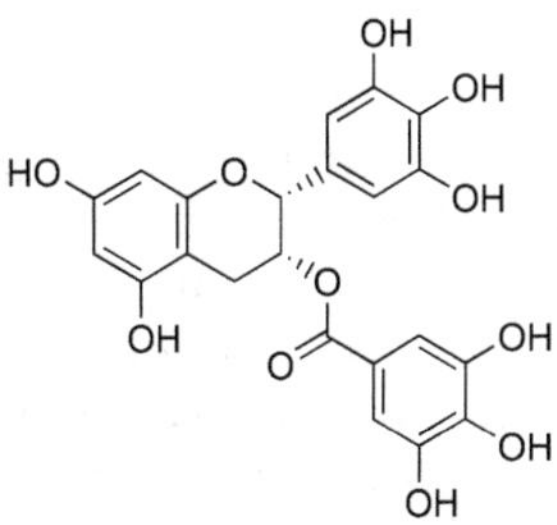

Epigallocatechin gallate (**31**)
IC_{50} *L. amazonensis* ARG = 3.8 μM
IC_{50} *L. donovani* axenic amas. = 41.7 μM
CC_{50} L6 cells = 32.1 μM

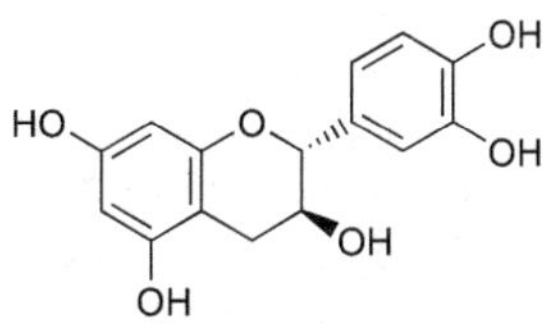

Catechin (**32**)
IC_{50} *L. amazonensis* ARG = 1.7 μM
IC_{50} *L. donovani* axenic amas. >100 μM
CC_{50} L6 cells >310 μM

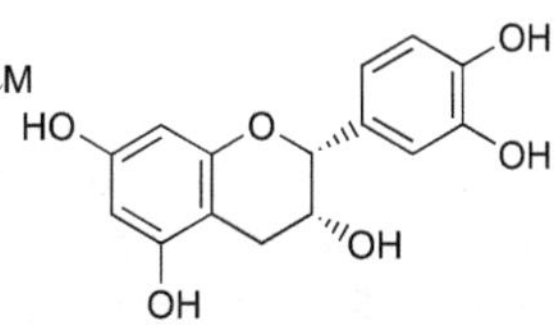

Epicatechin (**33**) IC_{50} *L. amazonensis* ARG = 1.6 μM
IC_{50} *L. donovani* axenic amas. >100 μM
CC_{50} L6 cells >310 μM

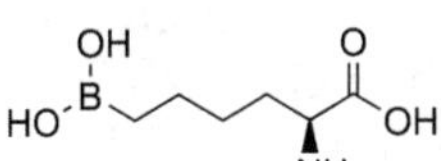

(**34**)

Figure 8.7 Arginase inhibitors.[61,63,65]

Subtle differences in the three-dimensional structures among arginases from different species indicate that it is possible to maintain affinity while fine-tuning intermolecular interactions. This strategy may provide, in the near future, isozyme-specific and species-specific arginase inhibitors.[67]

Other enzymes, involved also in the crucial polyamine biosynthesis such as ornithine decarboxylase (ODC), *S*-adenosylmethionine decarboxylase (SAM-DC), or spermidine synthase (SpdSyn), together with different polyamine transporters, are also promising targets that regulate the intracellular polyamine level. In fact, several specific leishmanial polyamine transporters such as LmPOT1, that transport putrescine and spermidine,[68] or the high affinity arginine-specific transporter LdAAP3,[69] have been cloned and are available for research (see Chapter 14). Hence, development of inhibitors to stop polyamine biosynthesis and transport may offer new effective drugs in the near future.

8.3.2 Enzymes of Thiol Metabolism

An approach to treat leishmaniasis patients is to explore the metabolic pathways of *Leishmania* which are either absent or differ in mammals. This family of protozoan parasites possesses a trypanothione-based redox metabolism that provides the reducing equivalents for several vital processes, such as the biosynthesis of DNA precursors and the detoxification of hydroperoxyde.[70] Most of the enzymes of the redox pathway proved to be essential and therefore fulfill this crucial prerequisite to be a potential drug target (see Chapter 16).[71]

The biosynthesis of trypanothione depends on L-cysteine bioavailability which is regulated by its biosynthesis pathway. The *de novo* cysteine biosynthesis pathway is comprised of serine *O*-acetyltransferase (SAT) and cysteine synthase (CS) enzymes which sequentially mediate two consecutive steps of cysteine biosynthesis, and are absent from the mammalian host. Recently, an increase in LdSAT and LdCS expression and/or activity has been reported in amastigotes, showing potential targets for disrupting the redox homeostasis of *Leishmania*.[72] Furthermore, in pathogenic trypanosomatids, a single enzyme called trypanothione synthetase (TryS) catalizes trypanothione biosynthesis, being indispensable for parasite survival. By other hand, trypanothione reductase (TryR) reduces trypanothione. Both enzymes are present in all trypanosomatids but absent from the mammalian host, which, in addition to their essentiality, renders them highly specific and selective targets.[73] Thus, the search of potent inhibitors of TryS and TryR is an active area in the fight against neglected diseases. Furthermore their crystal structures are available, allowing rational drug design.[74,75]

The screen of a chemical library against *L. Infantum* TryS (LiTryS) identified novel chemical scaffolds as substituted paullones such as the compounds FS-544 (**35**) and MOL2008 (**36**) (Figure 8.8).[76] These LiTryS inhibitors halt the proliferation of *L. infantum* promastigotes, although the selectivity indexes regarding murine macrophages are far from optimal and require further optimization.

Clerodanediterpenes from *Croton cajucara* Benth, a plant found in the Amazonian region used in folk medicine, showed promising *in vitro* anti-leishmanial effects against *L. amazonensis*, in particular, *trans*-dehydrocrotonin (DCTN, **37**) presented the best profile against promastigotes, axenic amastigotes and intracellular amastigotes with no macrophage toxicity (Figure 8.9). In addition, this compound is an inhibitor of *L. amazonensis* TryR.[77]

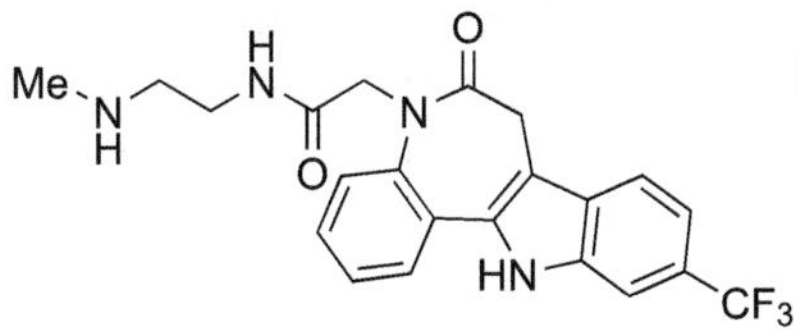

FS-554 (**35**)
IC_{50} *L. infantum* TryS = 0.35 μM
EC_{50} *L. infantum* prom. = 112.3 μM

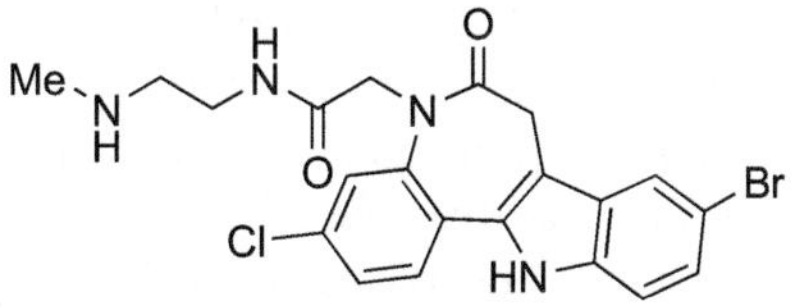

MOL2008 (**36**)
IC_{50} *L. infantum* TryS = 0.15 μM
EC_{50} *L. infantum* prom. = 12.6 μM

Figure 8.8 Trypanothione synthase inhibitors.[76]

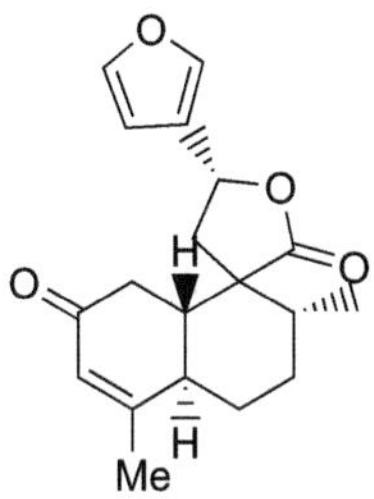

DCTN (**37**)
IC_{50} *L. amazonensis* prom. = 38.4 μM
IC_{50} *L. amazonensis* axenic amas. = 63.6 μM

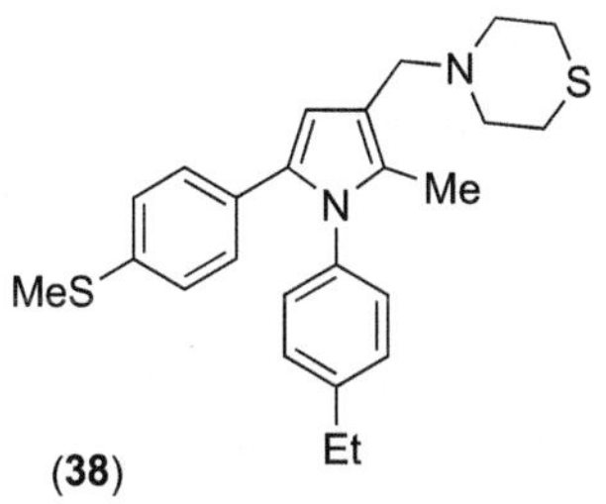

(**38**)
IC_{50} *L. donovani* amas. = 13.77 μM
CC_{50} KB cells = 12.94 μM

Figure 8.9 Trypanothione reductase inhibitors.[77,78]

A phenotypic assay on *L. donovani* intracellular amastigotes allowed the identification of a novel class of azole-based derivatives as promising compounds. Enzymatic assays on *L. infantum* trypanothione reductase (LiTR) showed that the diaryl-pirrole derivative **38** inhibited this enzyme (Figure 8.9).[78] Docking studies together with crystallization of LiTR–**38** complex identified the key molecular interactions within the protein that may boost further optimization. Also the naphthoquinone plumbagin (2) (Figure 8.2), is reported to be a LiTryR inhibitor.[16]

8.3.3 Enzymes of Folate Metabolism

Leishmania parasites are autotrophic for both folate and non-conjugated pteridines. *Leishmania* obtain these metabolites from their two hosts through different transporters. Within the parasite, folates are reduced by a bifunctional dihydrofolate reductase (DHFR)–thymidylate synthase (TS) and by a novel pteridin reductase (PTR1). Although not essential, the loss of several of these metabolic enzymes results in attenuation or loss of virulence in animal models. Thus, the folate metabolic pathway offers different target opportunities for drug discovery programs, with a strong potential for re-purposing compounds developed for several human cancer treatments or for other infectious diseases.[79]

Potent pteridine reductase inhibitors were discovered under a screening strategy using a folate-based chemical library.[80] After assays against folate-dependent *Leishmania* enzymes including PTR1, DHFR and TS, methotrexate was discovered to be a *L. major* TS (LmTS) inhibitor, while compounds **39** and **40** are potent PTR1 inhibitors (Figure 8.10). Further assays of optimized derivatives in combination with pyremethamine, a DHFR inhibitor, showed an essential synergic effect, because successful antifolate therapy in *Leishmania* required the inhibition of both enzymes simultaneously.[81] Moreover, the search for inhibitors of PTR1 led to the discovery of triazine mimetics by structural similarity between the reported PTR1 inhibitors and pentamidine,[82] and also a novel thiadiazole hit was discovered by virtual screening.[83] The triazine dimer **41** was highly active against intracellular amastigotes *in vitro* and also *in vivo,* while PTR1 binding was suggested by molecular docking analysis. A virtual screening approach searching for compounds with a feasible interaction with *L. major* PTR1, but not with

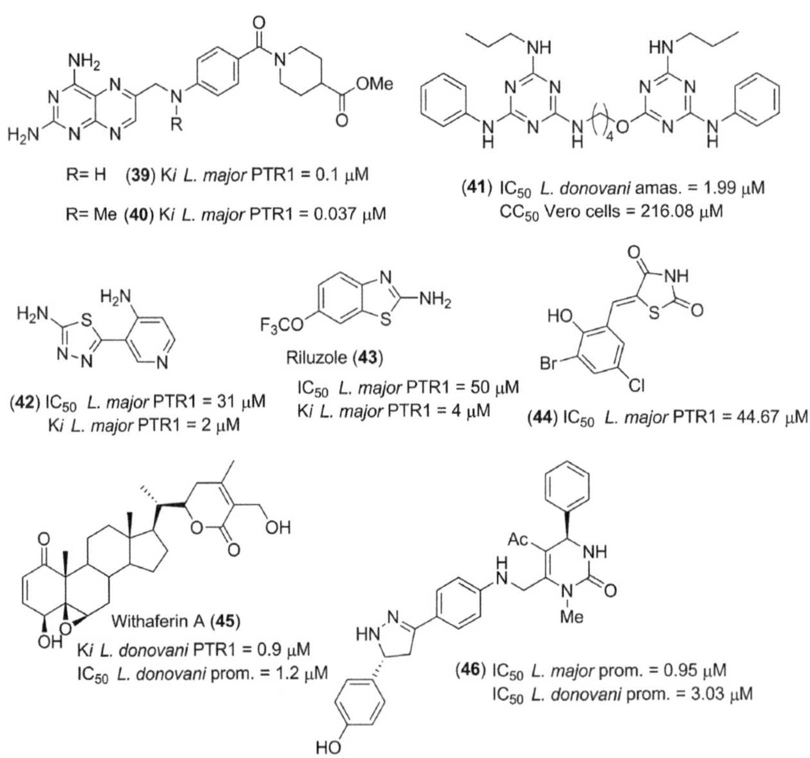

Figure 8.10 *Leishmania* PTR1 inhibitors.[80,82–86]

human DHFR, led to the identification of non-folate-like inhibitors like thiadiazole (**42**) and also the benzothiazole (**43**), known as riluzole (Figure 8.10). Both compounds were tested in promastigotes of *L. mexicana* and *L. major*, using single compounds and in combination with pyrimethamine. As riluzole (**43**), is a drug approved for amyotrophic lateral sclerosis, their combination with pyrimethamine is suitable for early preclinical evaluation under a drug-repurposing strategy.

More recently, in an effort to design 2,4-diamino purine derivatives that mimic the pteridine ring, the thiadiazolidinone ring was chosen for bioisosteric replacement of the pteridine or purine ring. Following this trend, a series of thiazolidine-2,4-dione derivatives were synthetized and screened against LmPTR1 through thermal shift assays, being the most promising molecule **44**. This is the first non-competitive inhibitor of LmPTR1 described (Figure 8.10).[84] The steroidal lactone Withaferin A (**45**), abundant in *Withania somnifera* leaves and in a lower abundance in roots, traditionally used in ayurvedic medicine, was recently reported to be leishmanicidal. It was identified as a PTR1 inhibitor by molecular modeling. Further experimental studies confirmed its non-competitive mode of inhibition.[85]

In an attempt to explore novel and more potent anti-leishmanial compounds to diversify the current inhibitors, and based on the reported X-ray structure of PTR1 from *L. major*, a number of dihydropyrimidine-based derivatives able to make specific interactions with the PTR1 active site have been designed and synthetized.[86] The lead compound **46** has shown *in vitro* anti-leishmanial activity against promastigotes of both *L. major* and *L. donovani*.

8.3.4 Sterol Biosynthesis Pathway

Sterols are essential constituents of eukaryotic cells, playing an important role in the normal structure and function of cell membranes, and also acting as precursors for the synthesis of several different molecules, such as steroid hormones. Cholesterol is the main sterol found in mammalian cells. However, ergosterol and some derivatives predominate in lower eukaryotes such as fungi or protozoa. For trypanosomatids, their essential requirement for ergosterol and other 24-alkyl sterols, absent in mammals, is well established.[48] The biosynthesis of ergosterol and cholesterol involve a number of enzymes, with some steps involving specific enzymes with different counterparts in mammalian cells and trypanosomatids. Squalene synthase (SQS) or sterol C14α-demethylase (CYP51) stand out among the enzymes in this pathway. Moreover, the repositioning of known clinical drugs that interfere with sterol biosynthesis, such as anti-hypercholesterolemia or anti-fungal drugs, may feed the leishmaniasis therapy pipeline.[87]

Furthermore, new approaches, such as the synergistic combination of drugs implicated in the same pathway may reduce resistance and minimize toxic effects. In this sense, the aryl-quinuclidine E5700 (**47**), which inhibits

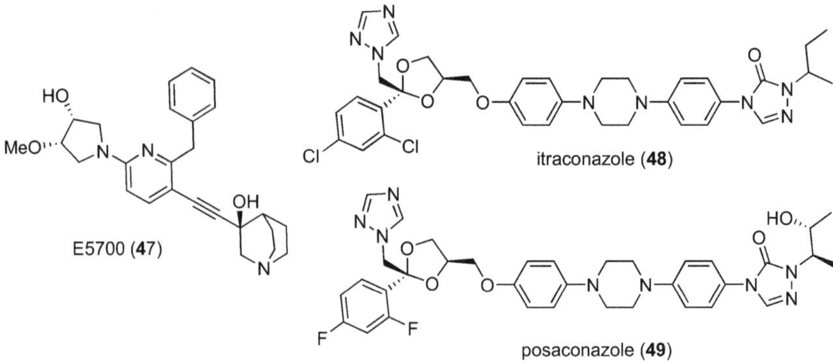

Figure 8.11 Ergosterol biosynthesis inhibitors.

SQS, in combination with two antifungal azoles inhibitors of CYP51, itraconazole (**48**), and posaconazole (**49**), improved the efficacy against *L. amazonensis* significantly (Figure 8.11).[88]

8.3.5 *Leishmania* Proteases

Leishmania genome sequences have enriched the knowledge of the biology of this parasite and simultaneously have provided ideal drug targets that exist distinctly in the parasite. Some proteases have been extensively investigated and found to be indispensable for parasite survival and disease pathogenesis.[89] Hence, inhibitors for these enzymes are appealing candidates for the development of anti-leishmanial drugs (Figure 8.12).[90]

Different approaches to discover these potential drugs are being followed. Different plant extracts and natural products have been assayed as leishmanicidal agents with protease inhibition properties. The natural bioflavonoid fukugetin (**50**), isolated from *Garcinia brasiliensis*, a tree native to Brazilian forest, was assayed *in vitro* as a *Leishmania* protease inhibitor ($IC_{50} = 3.02$ μM) but was inactive against living *L. amazonensis*.[91] However, other bioflavonoids and their semisynthetic modifications have proved to be potent leshmanicidal agents with protease inhibition profiles.[92] The accessibility to a recombinant *L. mexicana* cathepsin rCPB2.8 has facilitated the development of a screening program on natural compounds. Agathisflavone (**51**), tetrahydro-robustaflavone and the terpene 3-oxo-urs-12-en-28-oic acid (**52**) inhibited this protease with Ki values in the low micromolar range,[93] through a partial non-competitive inhibition. More recently, extracts of *Coccinia grandis*, a medicinal plant known in traditional folk medicine for its anti-leishmanial efficacy have been reported to show activity against *L. donovani* promastigotes by direct inhibition of serine proteases.[94]

As several protease inhibitors have been approved for the treatment of other infections such as human immunodeficiency virus (HIV) infection,

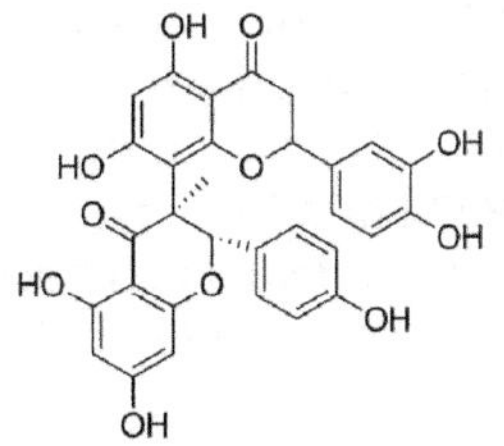

Fukugetin (**50**)
IC_{50} *Leishmania* proteases = 3.02 μM

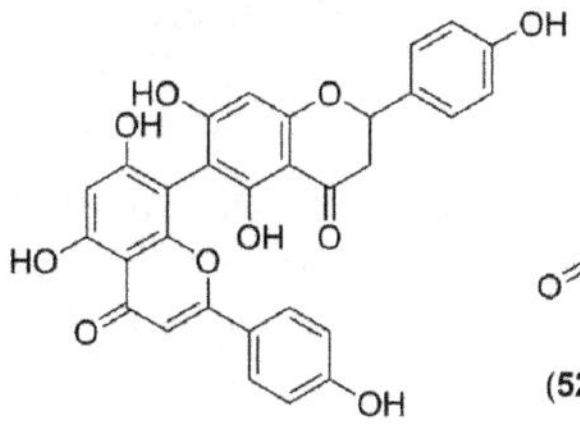

Agathisflavone (**51**)
IC_{50} *L. mexicana* rCPB2.8 = 0.43 μM
Ki L. mexicana rCPB2.8 = 0.14 μM

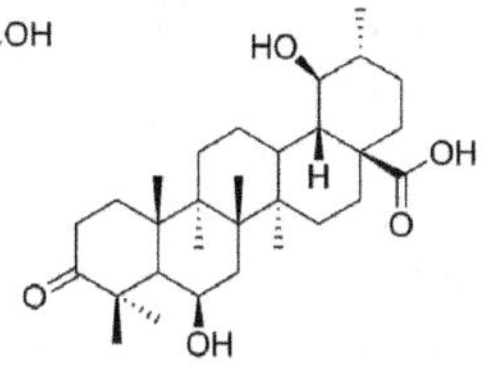

(**52**) IC_{50} *L. mexicana* rCPB2.8 = 3.78 μM *Ki L. mexicana* rCPB2.8 = 1.26 μM

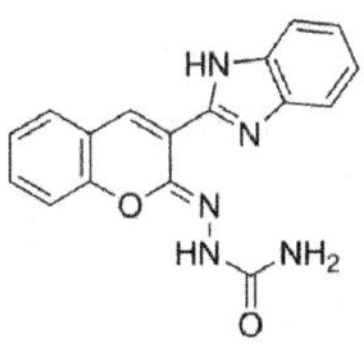

CP229988 (**53**)
IC_{50} *L. mexicana* CPB2.8ΔCTE = 0.01 μM
Ki L. mexicana CPB2.8ΔCTE = 0.005 μM

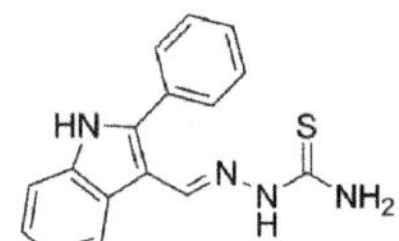

CP247128 (**54**)
IC_{50} *L. mexicana* CPB2.8ΔCTE = 0.06 μM
Ki L. mexicana CPB2.8ΔCTE = 0.04 μM

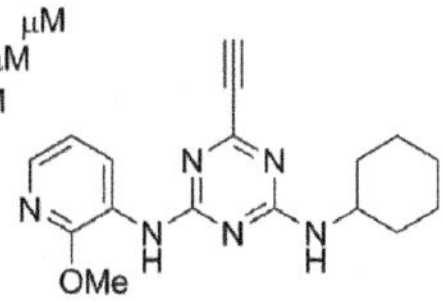

CP241026 (**55**)
IC_{50} *L. mexicana* CPB2.8ΔCTE = 1.1 μM
Ki L. mexicana CPB2.8ΔCTE = 0.57 μM

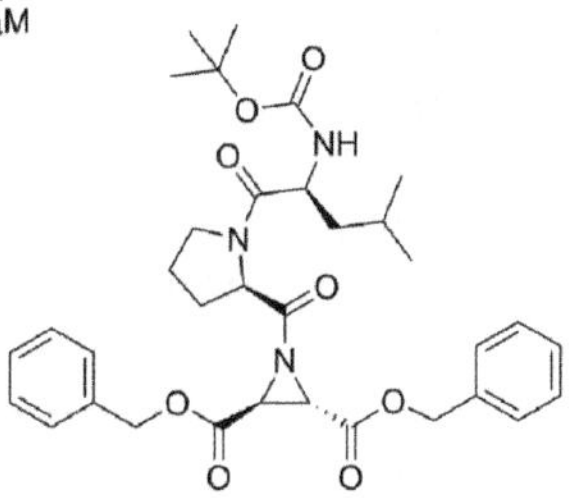

(**56**) *Ki L. mexicana* CPB2.8 = 3.8 μM
Ki L. major CatB = 18.2 μM
IC_{50} *L. major* prom. = 37.4 μM
CC_{50} J774.1 cells = 68 μM

Figure 8.12 Inhibitors of *Leishmania* proteases.[91,93,96,97]

repurposing strategies have been applied. Such is the case for the aspartic protease inhibitors nelfinavir and saquinavir that have been assayed in different species of *Leishmania*[95] Nelfinavir was capable of reducing the proliferation of many species at 94% being able to inhibit the aspartic peptidase activity of all species of *Leishmania* tested. These results add new possibilities in order to generate novel medications to treat *Leishmania*–HIV-coinfection.

Finally, specific medicinal chemistry programs have been designed to look for selective parasite proteases. Using chemoinformatic tools and the structures of specific proteases from humans and *Leishmania*, virtual screening methodology was applied to discover specific inhibitors. That is the case of semicarbozone CP229988 (**53**), thiosemicarbazone CP247128 (**54**) and nitrile CP241026 (**55**) found as inhibitors of *L. mexicana* cysteine protease CPB.[96] In other cases, some previous human proteases inhibitors were taken as starting scaffolds to be finely tuned as inhibitors for *Leishmania* proteases. In such a way, aziridine-2,3-dicarboxylates peptidomimetic compounds were optimized, resulting in the compound **56** that inhibited *L. mexicana* cathepsin B (LmaCatB) and *L. mexicana* CPB2 LmaCPB.2, but not mammalian cathepsins. This compound displayed leishmanicidal activities against *L. major* promastigotes and amastigotes, being a good candidate for future *in vivo* studies.[97]

8.3.6 *Leishmania* Kinases

The characterization of the kinome of several parasites, such as *Leishmania*, has uncovered deep divergences with those of their mammalian host, being a new source of important targets for discovering new leishmanicidal agents.[98] Among others, leishmanial mitogen activated protein kinases (MAPKs),[99] cdc2-related kinase 3 (CRK3),[100] glycogen synthase kinase-3 (GSK-3)[101] and casein kinase 1 (CK1)[102] have emerged as potential drug targets.

Starting from known human protein kinase inhibitors, such as the well-known induribins, a medicinal chemistry program has been developed to shift indirubin-specific inhibition towards leishmanial glycogen synthase kinase 3 (LGSK-3) relative to cyclin-dependent kinase 1 (LCRK3). Due to the fact that 6-bromo-indirubin-3′-oxime (6-BIO, **57**) was identified as a more potent inhibitor of LCRK3 than of LGSK-3,[103] an inhibitor-driven study was carried out by using an indirubin library composed of 35 derivatives against *L. donovani* promastigotes and intracellular amastigotes. Afterwards, indirubins with anti-leishmanial activity were tested against both enzymes, purified from homologous expression systems.[101] In that way, a 3′-bulky amino substituted 6-BIO derivative, compound **58**, with enhanced selectivity towards LGSK-3 resulted as a new scaffold for the design of leishmanicidal drugs (Figure 8.13). Moreover, the solved structure of LGSK-3 from *L. major* may open the way towards structure-based design of selective LGSK-3 inhibitors.[104]

Chemoinformatics has been applied to identify LCRK3 inhibitors due to the relevant role that CRK3 has in the regulation of proliferation and viability inside the macrophage. In this way, a fragment-based QSAR model used pyrazole-derived compounds exhibiting inhibitory activity against *Leishmania* CRK3.[105] The analysis carried out provides new clues for the design of new pyrazole-based anti-leishmanial drugs previously described as potent and selective CRK3 inhibitors like derivative **59**.[106]

An established human CK1 inhibitor, D4476 (**60**), able to inhibit also *L. major* CK1.2 (LmCK1.2), was used to study the effects of CK1 kinase inhibition on axenic and intracellular amastigotes.[102] Afterwards, following a target-based drug discovery program, a kinase-focused library with 5019 molecules was screened to identify novel leishmanicidal compounds, yielding two candidates (**61** and **62**) whose primary target seems to be LmCK1.2.[107]

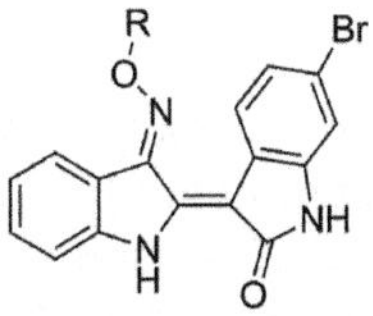

6-BIO (**57**) R=H IC$_{50}$ LGSK-3 = 0.15 μM
IC$_{50}$ LCRK3 = 0.02 μM
GI$_{50}$ *L. donovani* prom. = 0.85 μM
GI$_{50}$ *L. donovani* amas. = 0.70 μM
GI$_{50}$ J774.1 cells >25 μM

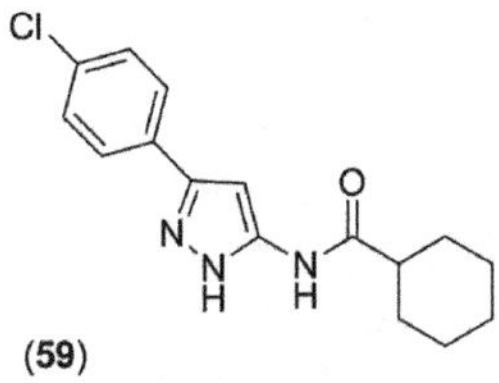

(**59**)
IC$_{50}$ LCRK3-CYC6 = 0.34 μM

(**58**) R= 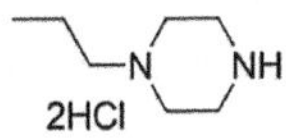
2HCl

IC$_{50}$ LGSK-3 = 0.10 μM
IC$_{50}$ LCRK3 >3.33 μM
GI$_{50}$ *L. donovani* prom. = 0.75 μM
GI$_{50}$ *L. donovani* amas. = 0.59 μM
GI$_{50}$ J774.1 cells = 6.25 μM

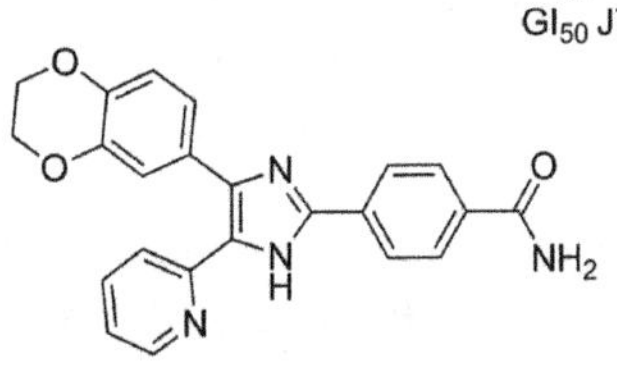

D4476 (**60**) IC$_{50}$ *L. major* CK1.2 = 0.09 μM
EC$_{50}$ *L. donovani* prom. = 30 μM
EC$_{50}$ *L. donovani* amas. = 42 μM
EC$_{50}$ macrophages >100 μM

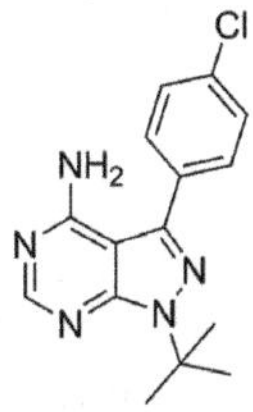

(**61**) IC$_{50}$ *L. major* CK1.2 = 1.60 μM
EC$_{50}$ *L. donovani* prom. >10 μM
EC$_{50}$ *L. donovani* axenic amas. >50 μM
EC$_{50}$ *L. donovani* amas. = 1 μM

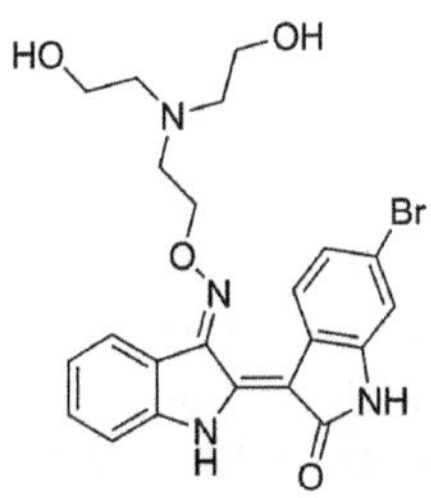

(**62**) IC$_{50}$ *L. major* CK1.2 = 0.93 μM
EC$_{50}$ *L. donovani* prom. = 2.0 μM
EC$_{50}$ *L. donovani* axenic amas. >10 μM
EC$_{50}$ *L. donovani* amas. = 0.06 μM

Figure 8.13 Inhibitors of *Leishmania* protein kinases.[101–103,106,107]

8.3.7 *Leishmania* Phosphodiesterases

The structural similarity between human and protozoan phosphodiesterases (PDEs), together with the vast expertise available regarding human PDE inhibitors paved the way for the development of disease-specific PDE inhibitors as new anti-protozoan drugs.[108] *Leishmania* parasites are currently known to express five PDEs that have been identified and cloned, although ultimate validation as drug targets is in progress.[109] Up to now, no specific leishmanial PDE inhibitor is known. However, the crystal structure of *L. major* PDEB1 has been solved, showing great similarity with the catalytic site of humans PDEs but with a unique pocket next to the active site that may permit the design of parasite-selective inhibitors for the treatment of leishmaniasis.[110]

8.4 Conclusions

Leishmaniasis is a major public health problem faced by many countries. The available treatments are less than optimal due to significant limitations, such as the toxicity of the drugs used or the emergence of resistant strains. Thus, discovery and development of novel drug candidates with better pharmacological profiles are still required.

Searching for leishmanicidal drugs is a challenging task for medicinal chemists. Ideally, drug screening procedures should use amastigotes residing in macrophages, although many laboratories have successfully used promastigotes or axenic amastigotes as primary tests to find a novel leishmanicidal agent. Concerning the search for new chemical entities, natural products or privileged structures, such as nitrogen heterocycles are pivotal in drug discovery programs because they exhibit diverse biological and pharmacological activities, offering a high degree of structural diversity. Undoubtedly, what significantly accelerates medicinal chemistry programs is better knowledge of validated targets. Regarding *Leishmania*, the emergence of new validated targets together with well-established ones, allows the rationalization of the discovery of new leishmanicidal agents by establishing clear structure–activity relationships.

Nowadays, selection of compounds from different sources on the basis of target-based and/or whole-cell assays, together with repurposing of approved drugs, provides future candidates for the chemotherapy of this neglected disease.

Acknowledgements

Funding from the EC 7th Framework Programme (FP7-HEALTH-2013-INNOVATION-1, PDE4NPD no. 602666), the Spanish Ministry of Economy and Competitiveness (MINECO, project no. SAF2015-65740-R), Redes de Investigación Cooperativa Instituto de Salud Carlos III RICET (RD16/0027/0010) and FEDER funds are acknowledged.

References

1. D. Savoia, *J. Infect. Dev. Ctries.*, 2015, **9**, 588.
2. D. Pace, *J. Infect.*, 2014, **69**(Suppl. 1), S10.
3. J. H. No, *Acta Trop.*, 2016, **155**, 113.
4. S. Hendrickx, P. J. Guerin, G. Caljon, S. L. Croft and L. Maes, *Parasitology*, 2016, DOI: 10.1017/S0031182016002031.
5. H. Hussain, A. Al-Harrasi, A. Al-Rawahi, I. R. Green and S. Gibbons, *Chem. Rev.*, 2014, **114**, 10369.
6. A. S. Nagle, S. Khare, A. B. Kumar, F. Supek, A. Buchynskyy, C. J. Mathison, N. K. Chennamaneni, N. Pendem, F. S. Buckner, M. H. Gelb and V. Molteni, *Chem. Rev.*, 2014, **114**, 11305.

7. R. M. Reguera, E. Calvo-Alvarez, R. Alvarez-Velilla and R. Balana-Fouce, *Int. J. Parasitol. Drugs Drug Resist.*, 2014, **4**, 355.
8. *Drug Discovery from Natural Products*, ed. O. Genilloud and F. Vicente, The Royal Society of Chemistry, Cambridge (UK), 2012.
9. *Privileged Scaffolds in Medicinal Chemistry*, ed. S. Bräse, The Royal Society of Chemistry, Cambridge (UK), 2016.
10. D. J. Newman and G. M. Cragg, *J. Nat. Prod.*, 2016, **79**, 629.
11. L. G. Ferreira and A. D. Andricopulo, *Drug Discovery Today*, 2016, **21**, 1699.
12. T. S. Skinner-Adams, S. D. Sumanadasa, G. M. Fisher, R. A. Davis, D. L. Doolan and K. T. Andrews, *Drug Discovery Today*, 2016, **21**, 725.
13. P. Babula, V. Adam, L. Havel and R. Kizek, *Curr. Pharm. Anal.*, 2009, **5**, 47.
14. A. V. Pinto and S. L. de Castro, *Molecules*, 2009, **14**, 4570.
15. N. M. Lima, C. S. Correia, L. L. Leon, G. M. Machado, F. Madeira Mde, A. E. Santana and M. O. Goulart, *Mem. Inst. Oswaldo Cruz*, 2004, **99**, 757.
16. N. Sharma, A. K. Shukla, M. Das and V. K. Dubey, *Parasitol. Res.*, 2012, **110**, 341.
17. A. Cavalli, F. Lizzi, S. Bongarzone, F. Belluti, L. Piazzi and M. L. Bolognesi, *FEMS Immunol. Med. Microbiol.*, 2010, **58**, 51.
18. L. Salmon-Chemin, E. Buisine, V. Yardley, S. Kohler, M. A. Debreu, V. Landry, C. Sergheraert, S. L. Croft, R. L. Krauth-Siegel and E. Davioud-Charvet, *J. Med. Chem.*, 2001, **44**, 548.
19. C. M. Lezama-Davila, A. P. Isaac-Marquez, G. Kapadia, K. Owens, S. Oghumu, S. Beverley and A. R. Satoskar, *Biol. Pharm. Bull.*, 2012, **35**, 1761.
20. M. L. Bolognesi, F. Lizzi, R. Perozzo, R. Brun and A. Cavalli, *Bioorg. Med. Chem. Lett.*, 2008, **18**, 2272.
21. F. Lizzi, G. Veronesi, F. Belluti, C. Bergamini, A. Lopez-Sanchez, M. Kaiser, R. Brun, R. L. Krauth-Siegel, D. G. Hall, L. Rivas and M. L. Bolognesi, *J. Med. Chem.*, 2012, **55**, 10490.
22. T. T. Guimaraes, C. Pinto Mdo, J. S. Lanza, M. N. Melo, R. L. do Monte-Neto, I. M. de Melo, E. B. Diogo, V. F. Ferreira, C. A. Camara, W. O. Valenca, R. N. de Oliveira, F. Frezard and E. N. da Silva, Jr., *Eur. J. Med. Chem.*, 2013, **63**, 523.
23. E. G. Pinto, I. O. Santos, T. J. Schmidt, S. E. Borborema, V. F. Ferreira, D. R. Rocha and A. G. Tempone, *PLoS One*, 2014, **9**, e105127.
24. S. L. de Castro, F. S. Emery and E. N. da Silva Junior, *Eur. J. Med. Chem.*, 2013, **69**, 678.
25. S. Patterson and S. Wyllie, *Trends Parasitol.*, 2014, **30**, 289.
26. S. Wyllie, S. Patterson, L. Stojanovski, F. R. Simeons, S. Norval, R. Kime, K. D. Read and A. H. Fairlamb, *Sci. Transl. Med.*, 2012, **4**, 119re1.
27. S. Wyllie, S. Patterson and A. H. Fairlamb, *Antimicrob. Agents Chemother.*, 2013, **57**, 901.

28. E. Torreele, B. Bourdin Trunz, D. Tweats, M. Kaiser, R. Brun, G. Mazue, M. A. Bray and B. Pecoul, *PLoS Neglected Trop. Dis.*, 2010, **4**, e923.
29. A. Tarral, S. Blesson, O. V. Mordt, E. Torreele, D. Sassella, M. A. Bray, L. Hovsepian, E. Evene, V. Gualano, M. Felices and N. Strub-Wourgaft, *Clin. Pharmacokinet.*, 2014, **53**, 565.
30. https://clinicaltrials.gov/ct2/show/NCT01980199.
31. S. Patterson, S. Wyllie, S. Norval, L. Stojanovski, F. R. Simeons, J. L. Auer, M. Osuna-Cabello, K. D. Read and A. H. Fairlamb, *eLife*, 2016, **5**, e09744.
32. C. Castera-Ducros, L. Paloque, P. Verhaeghe, M. Casanova, C. Cantelli, S. Hutter, F. Tanguy, M. Laget, V. Remusat, A. Cohen, M. D. Crozet, P. Rathelot, N. Azas and P. Vanelle, *Bioorg. Med. Chem.*, 2013, **21**, 7155.
33. A. A. Voak, K. Seifert, N. A. Helsby and S. R. Wilkinson, *Antimicrob. Agents Chemother.*, 2014, **58**, 370.
34. M. V. Papadopoulou, W. D. Bloomer, H. S. Rosenzweig, E. Chatelain, M. Kaiser, S. R. Wilkinson, C. McKenzie and J. R. Ioset, *J. Med. Chem.*, 2012, **55**, 5554.
35. M. V. Papadopoulou, W. D. Bloomer, H. S. Rosenzweig, I. P. O'Shea, S. R. Wilkinson, M. Kaiser, E. Chatelain and J. R. Ioset, *Bioorg. Med. Chem.*, 2015, **23**, 6467.
36. M. V. Papadopoulou, W. D. Bloomer, H. S. Rosenzweig, I. P. O'Shea, S. R. Wilkinson and M. Kaiser, *Eur. J. Med. Chem.*, 2015, **103**, 325.
37. B. S. Hall, X. Wu, L. Hu and S. R. Wilkinson, *Antimicrob. Agents Chemother.*, 2010, **54**, 1193.
38. K. A. Reynolds, W. A. Loughlin and D. J. Young, *Mini Rev. Med. Chem.*, 2013, **13**, 730.
39. L. Paloque, P. Verhaeghe, M. Casanova, C. Castera-Ducros, A. Dumetre, L. Mbatchi, S. Hutter, M. Kraiem-M'rabet, M. Laget, V. Remusat, S. Rault, P. Rathelot, N. Azas and P. Vanelle, *Eur. J. Med. Chem.*, 2012, **54**, 75.
40. C. Kieffer, A. Cohen, P. Verhaeghe, S. Hutter, C. Castera-Ducros, M. Laget, V. Remusat, M. K. M'Rabet, S. Rault, P. Rathelot, N. Azas and P. Vanelle, *Eur. J. Med. Chem.*, 2015, **92**, 282.
41. T. K. Jha, S. Sundar, C. P. Thakur, J. M. Felton, A. J. Sabin and J. Horton, *Am. J. Trop. Med. Hyg.*, 2005, **73**, 1005.
42. M. K. Wasunna, J. R. Rashid, J. Mbui, G. Kirigi, D. Kinoti, H. Lodenyo, J. M. Felton, A. J. Sabin, M. J. Albert and J. Horton, *Am. J. Trop. Med. Hyg.*, 2005, **73**, 871.
43. L. Carvalho, J. R. Luque-Ortega, C. Lopez-Martin, S. Castanys, L. Rivas and F. Gamarro, *Antimicrob. Agents Chemother.*, 2011, **55**, 4204.
44. P. M. Loiseau, S. Cojean and J. Schrevel, *Parasite*, 2011, **18**, 115.
45. V. Yardley, F. Gamarro and S. L. Croft, *Antimicrob. Agents Chemother.*, 2010, **54**, 5356.
46. L. Carvalho, J. R. Luque-Ortega, J. I. Manzano, S. Castanys, L. Rivas and F. Gamarro, *Antimicrob. Agents Chemother.*, 2010, **54**, 5344.

47. M. Sharma, K. Chauhan, R. Shivahare, P. Vishwakarma, M. K. Suthar, A. Sharma, S. Gupta, J. K. Saxena, J. Lal, P. Chandra, B. Kumar and P. M. Chauhan, *J. Med. Chem.*, 2013, **56**, 4374.
48. W. de Souza and J. C. Rodrigues, *Interdiscip. Perspect. Infect. Dis.*, 2009, **2009**, 642502.
49. V. K. Marrapu, M. Mittal, R. Shivahare, S. Gupta and K. Bhandari, *Eur. J. Med. Chem.*, 2011, **46**, 1694.
50. J. T. Mesquita, T. A. da Costa-Silva, S. E. Borborema and A. G. Tempone, *Mol. Cell Biochem.*, 2014, **389**, 293.
51. M. S. dos Santos, M. L. Oliveira, A. M. Bernardino, R. M. de Leo, V. F. Amaral, F. T. de Carvalho, L. L. Leon and M. M. Canto-Cavalheiro, *Bioorg. Med. Chem. Lett.*, 2011, **21**, 7451.
52. R. K. Marra, A. M. Bernardino, T. A. Proux, K. S. Charret, M. L. Lira, H. C. Castro, A. M. Souza, C. D. Oliveira, J. C. Borges, C. R. Rodrigues, M. M. Canto-Cavalheiro, L. L. Leon and V. F. Amaral, *Molecules*, 2012, **17**, 12961.
53. Z. Dardari, M. Lemrani, A. Sebban, A. Bahloul, M. Hassar, S. Kitane, M. Berrada and M. Boudouma, *Arch. Pharm. Chem. Life. Sci.*, 2006, **339**, 291.
54. C. E. Mowbray, S. Braillard, W. Speed, P. A. Glossop, G. A. Whitlock, K. R. Gibson, J. E. Mills, A. D. Brown, J. M. Gardner, Y. Cao, W. Hua, G. L. Morgans, P. B. Feijens, A. Matheeussen and L. J. Maes, *J. Med. Chem.*, 2015, **58**, 9615.
55. G. Colotti and A. Ilari, *Amino Acids*, 2011, **40**, 269.
56. A. Ilari, A. Fiorillo, P. Baiocco, E. Poser, G. Angiulli and G. Colotti, *Mini Rev. Med. Chem.*, 2015, **15**, 243.
57. M. F. da Silva and L. M. Floeter-Winter, *Subcell. Biochem.*, 2014, **74**, 103.
58. J. M. Boitz, C. A. Gilroy, T. D. Olenyik, D. Paradis, J. Perdeh, K. Dearman, M. J. Davis, P. A. Yates, Y. Li, M. K. Riscoe, B. Ullman and S. C. Roberts, *Infect. Immun.*, 2016, **85**, e00554.
59. R. M. Reguera, R. Balana-Fouce, M. Showalter, S. Hickerson and S. M. Beverley, *Mol. Biochem. Parasitol.*, 2009, **165**, 48.
60. Y. A. Ivanenkov and N. V. Chufarova, *Pharm. Pat. Anal.*, 2014, **3**, 65.
61. L. R. de Sousa, S. D. Ramalho, M. C. Burger, L. Nebo, J. B. Fernandes, M. F. da Silva, M. R. Iemma, C. J. Correa, D. H. de Souza, M. I. Lima and P. C. Vieira, *J. Nat. Prod.*, 2014, **77**, 392.
62. E. R. da Silva, C. Maquiaveli Cdo and P. P. Magalhaes, *Exp. Parasitol.*, 2012, **130**, 183.
63. D. Tasdemir, M. Kaiser, R. Brun, V. Yardley, T. J. Schmidt, F. Tosun and P. Ruedi, *Antimicrob. Agents Chemother.*, 2006, **50**, 1352.
64. L. C. Manjolin, M. B. dos Reis, C. Maquiaveli Cdo, O. A. Santos-Filho and E. R. da Silva, *Food Chem.*, 2013, **141**, 2253.
65. M. B. dos Reis, L. C. Manjolin, C. Maquiaveli Cdo, O. A. Santos-Filho and E. R. da Silva, *PLoS One*, 2013, **8**, e78387.
66. E. L. D'Antonio, B. Ullman, S. C. Roberts, U. G. Dixit, M. E. Wilson, Y. Hai and D. W. Christianson, *Arch. Biochem. Biophys.*, 2013, **535**, 163.

67. Y. Hai and D. W. Christianson, *Acta Crystallogr., Sect. F: Struct. Biol. Commun.*, 2016, **72**, 300.
68. M. P. Hasne and B. Ullman, *J. Biol. Chem.*, 2005, **280**, 15188.
69. I. Darlyuk, A. Goldman, S. C. Roberts, B. Ullman, D. Rentsch and D. Zilberstein, *J. Biol. Chem.*, 2009, **284**, 19800.
70. A. E. Leroux and R. L. Krauth-Siegel, *Mol. Biochem. Parasitol.*, 2016, **206**, 67.
71. K. Singh, G. Garg and V. Ali, *Curr. Drug Metab.*, 2016, **17**, 897.
72. K. Singh, K. P. Singh, A. Equbal, S. S. Suman, A. Zaidi, G. Garg, K. Pandey, P. Das and V. Ali, *Biochimie*, 2016, **131**, 29.
73. L. S. Bernardes, C. L. Zani and I. Carvalho, *Curr. Med. Chem.*, 2013, **20**, 2673.
74. P. Baiocco, S. Franceschini, A. Ilari and G. Colotti, *Protein Pept. Lett.*, 2009, **16**, 196.
75. P. Saudagar and V. K. Dubey, *Biol. Chem.*, 2011, **392**, 1113.
76. D. Benitez, A. Medeiros, L. Fiestas, E. A. Panozzo-Zenere, F. Maiwald, K. C. Prousis, M. Roussaki, T. Calogeropoulou, A. Detsi, T. Jaeger, J. Sarlauskas, L. Peterlin Masic, C. Kunick, G. R. Labadie, L. Flohe and M. A. Comini, *PLoS Negl. Trop. Dis.*, 2016, **10**, e0004617.
77. G. S. Lima, D. B. Castro-Pinto, G. C. Machado, M. A. Maciel and A. Echevarria, *Phytomedicine*, 2015, **22**, 1133.
78. P. Baiocco, G. Poce, S. Alfonso, M. Cocozza, G. C. Porretta, G. Colotti, M. Biava, F. Moraca, M. Botta, V. Yardley, A. Fiorillo, A. Lantella, F. Malatesta and A. Ilari, *ChemMedChem*, 2013, **8**, 1175.
79. T. J. Vickers and S. M. Beverley, *Essays Biochem.*, 2011, **51**, 63.
80. A. Cavazzuti, G. Paglietti, W. N. Hunter, F. Gamarro, S. Piras, M. Loriga, S. Allecca, P. Corona, K. McLuskey, L. Tulloch, F. Gibellini, S. Ferrari and M. P. Costi, *Proc. Natl. Acad. Sci. U. S. A.*, 2008, **105**, 1448.
81. P. Corona, F. Gibellini, A. Cavalli, P. Saxena, A. Carta, M. Loriga, R. Luciani, G. Paglietti, D. Guerrieri, E. Nerini, S. Gupta, V. Hannaert, P. A. Michels, S. Ferrari and P. M. Costi, *J. Med. Chem.*, 2012, **55**, 8318.
82. K. Chauhan, M. Sharma, R. Shivahare, U. Debnath, S. Gupta, Y. S. Prabhakar and P. M. Chauhan, *ACS Med. Chem. Lett.*, 2013, **4**, 1108.
83. S. Ferrari, F. Morandi, D. Motiejunas, E. Nerini, S. Henrich, R. Luciani, A. Venturelli, S. Lazzari, S. Calo, S. Gupta, V. Hannaert, P. A. Michels, R. C. Wade and M. P. Costi, *J. Med. Chem.*, 2011, **54**, 211.
84. F. H. Leite, P. B. Santiago, T. Q. Froes, J. da Silva Filho, S. G. da Silva, R. M. Ximenes, A. R. de Faria, D. J. Brondani, J. F. de Albuquerque and M. S. Castilho, *Eur. J. Med. Chem.*, 2016, **123**, 639.
85. S. Chandrasekaran, J. Veronica, R. K. Gundampati, S. Sundar and R. Maurya, *J. Enzyme. Inhib. Med. Chem.*, 2016, **31**, 1029.
86. U. Rashid, R. Sultana, N. Shaheen, S. F. Hassan, F. Yaqoob, M. J. Ahmad, F. Iftikhar, N. Sultana, S. Asghar, M. Yasinzai, F. L. Ansari and N. A. Qureshi, *Eur. J. Med. Chem.*, 2016, **115**, 230.
87. S. T. de Macedo-Silva, W. de Souza and J. C. Rodrigues, *Curr. Med. Chem.*, 2015, **22**, 2186.

88. S. T. de Macedo-Silva, G. Visbal, J. A. Urbina, W. de Souza and J. C. Rodrigues, *Antimicrob. Agents Chemother.*, 2015, **59**, 6402.
89. P. Das, M. N. Alam, D. Paik, K. Karmakar, T. De and T. Chakraborti, *Indian J. Biochem. Biophys.*, 2013, **50**, 363.
90. A. P. Lima, F. C. Reis and T. F. Costa, *Curr. Med. Chem.*, 2013, **20**, 3152.
91. I. O. Pereira, D. M. Assis, M. A. Juliano, R. L. Cunha, C. L. Barbieri, L. V. do Sacramento, M. J. Marques and M. H. dos Santos, *J. Med. Food*, 2011, **14**, 557.
92. V. S. Gontijo, W. A. Judice, B. Codonho, I. O. Pereira, D. M. Assis, J. P. Januario, E. E. Caroselli, M. A. Juliano, A. de Carvalho Dosatti, M. J. Marques, C. Viegas Junior and M. Henrique dos Santos, *Eur. J. Med. Chem.*, 2012, **58**, 613.
93. L. R. de Sousa, H. Wu, L. Nebo, J. B. Fernandes, M. F. da Silva, W. Kiefer, T. Schirmeister and P. C. Vieira, *Exp. Parasitol.*, 2015, **156**, 42.
94. P. Das, D. Paik, A. Pramanik, T. De and T. Chakraborti, *Indian J. Exp. Biol.*, 2015, **53**, 740.
95. L. O. Santos, B. S. Vitorio, M. H. Branquinha, C. M. Pedroso e Silva, A. L. Santos and C. M. d'Avila-Levy, *J. Antimicrob. Chemother.*, 2013, **68**, 348.
96. J. Schröder, S. Noack, R. J. Marhofer, J. C. Mottram, G. H. Coombs and P. M. Selzer, *PLoS One*, 2013, **8**, e77460.
97. C. Schad, U. Baum, B. Frank, U. Dietzel, F. Mattern, C. Gomes, A. Ponte-Sucre, H. Moll, U. Schurigt and T. Schirmeister, *Antimicrob. Agents Chemother.*, 2016, **60**, 797.
98. C. Merritt, L. E. Silva, A. L. Tanner, K. Stuart and M. P. Pollastri, *Chem. Rev.*, 2014, **114**, 11280.
99. M. Dacher, M. A. Morales, P. Pescher, O. Leclercq, N. Rachidi, E. Prina, M. Cayla, A. Descoteaux and G. F. Spath, *Mol. Microbiol.*, 2014, **93**, 146.
100. P. Hassan, D. Fergusson, K. M. Grant and J. C. Mottram, *Mol. Biochem. Parasitol.*, 2001, **113**, 189.
101. A. Efstathiou, N. Gaboriaud-Kolar, D. Smirlis, V. Myrianthopoulos, K. Vougogiannopoulou, A. Alexandratos, M. Kritsanida, E. Mikros, K. Soteriadou and A. L. Skaltsounis, *Parasit. Vectors*, 2014, **7**, 234.
102. N. Rachidi, J. F. Taly, E. Durieu, O. Leclercq, N. Aulner, E. Prina, P. Pescher, C. Notredame, L. Meijer and G. F. Spath, *Antimicrob. Agents Chemother.*, 2014, **58**, 1501.
103. E. Xingi, D. Smirlis, V. Myrianthopoulos, P. Magiatis, K. M. Grant, L. Meijer, E. Mikros, A. L. Skaltsounis and K. Soteriadou, *Int. J. Parasitol.*, 2009, **39**, 1289.
104. K. K. Ojo, T. L. Arakaki, A. J. Napuli, K. K. Inampudi, K. R. Keyloun, L. Zhang, W. G. Hol, C. L. Verlinde, E. A. Merritt and W. C. Van Voorhis, *Mol. Biochem. Parasitol.*, 2011, **176**, 98.
105. S. Goyal, J. K. Dhanjal, C. Tyagi, M. Goyal and A. Grover, *Chem. Biol. Drug Des.*, 2014, **84**, 54.
106. L. A. Cleghorn, A. Woodland, I. T. Collie, L. S. Torrie, N. Norcross, T. Luksch, C. Mpamhanga, R. G. Walker, J. C. Mottram, R. Brenk,

J. A. Frearson, I. H. Gilbert and P. G. Wyatt, *ChemMedChem*, 2011, **6**, 2214.

107. E. Durieu, E. Prina, O. Leclercq, N. Oumata, N. Gaboriaud-Kolar, K. Vougogiannopoulou, N. Aulner, A. Defontaine, J. H. No, S. Ruchaud, A. L. Skaltsounis, H. Galons, G. F. Spath, L. Meijer and N. Rachidi, *Antimicrob. Agents Chemother.*, 2016, **60**, 2822.
108. T. Seebeck, G. J. Sterk and H. Ke, *Future Med. Chem.*, 2011, **3**, 1289.
109. A. Johner, S. Kunz, M. Linder, Y. Shakur and T. Seebeck, *BMC Microbiol.*, 2006, **6**, 25.
110. H. Wang, Z. Yan, J. Geng, S. Kunz, T. Seebeck and H. Ke, *Mol. Microbiol.*, 2007, **66**, 1029.

CHAPTER 9

Natural Products as a Source of New Drugs Against Leishmania

JOÃO HENRIQUE G. LAGO*[a] AND ANDRE G. TEMPONE[b]

[a] Centro de Ciências Naturais e Humanas, Universidade Federal do ABC, Santo André, SP, 09210-180, Brazil; [b] Centre of Parasitology and Mycology, Instituto Adolfo Lutz, São Paulo, SP, 01246-000, Brazil
*Email: joao.lago@ufabc.edu.br

9.1 *Leishmania*—A Challenging Parasite

The management of leishmaniasis has been a challenge for Public Health systems especially for developing countries. Although over the last few years some new therapeutic approaches have been proposed, including the first oral drug miltefosine, new drug combinations and different drug regimens, the current clinical alternatives still illustrate a worrisome scene. This includes a limited number of available drugs with important limitations due to elevated toxicity, costs of treatment and long periods of administration.[1] However, making new anti-parasitic drugs will not be easy because the disease, the parasite and the host are not simple and isolated entities.

Important issues still need to be addressed when considering new approaches against leishmaniasis, such as: (i) differences between the cutaneous and viscerotropic species, (ii) local differences among strains and naturally resistant strains, (iii) *Leishmania* strains naturally infected with viruses (Totiviridae family), which exacerbate and disseminate the cutaneous disease, (iv) associated co-morbidities and immunocompromised patients'

Drug Discovery Series No. 60
Drug Discovery for Leishmaniasis
Edited by Luis Rivas and Carmen Gil

Published by the Royal Society of Chemistry, www.rsc.org

problems and finally, (v) malnutrition and limited access to medical care, both typically associated with poverty. Altogether, it is highly unlikely that a single new chemical entity will be enough to solve this therapeutic problem, but a combination of strategies, including an increased arsenal of safe and cheap drugs, will be required.

Although leishmaniasis has been neglected by pharmaceutical companies, a large number of scientific groups around the world, located in universities, research institutes and non-profit organizations such as Drugs for Neglected Diseases initiative (DNDi), make enormous efforts towards the development of new therapeutics. Unquestionably, the development of new drugs against leishmaniasis needs well-coordinated drug discovery projects, with strategic plans for fast and experienced decision making. It is worth mentioning, but also intriguing, that no new chemical entity has been introduced to the market for leishmaniasis. Interestingly, leishmaniasis profited from a therapeutic arsenal entirely discovered by the drug repositioning (or repurposing) approach, consisting of the search for new uses for approved or investigational drugs.[2] Among the drugs in clinical use, the antimonial therapy has origin in 16th century in France as a purgative agent, known as the "Paracelsian cure" by early 1560s. In 1912, the Brazilian researcher Gaspar de Oliveira Vianna used for the first time the emetic tartar (trivalent salt of antimony) in a patient with cutaneous leishmaniasis in Brazil.[3] Pentamidine, an aromatic diamine, was first synthesized at May and Baker pharmaceutical company during the preparation of anti-trypanosomal (*Trypanosoma brucei*) compounds.[4,5] Amphotericin B was first licensed in 1959 for the treatment of progressive and potentially life-threatening fungal infections, including aspergillosis, cryptococcosis, North American blastomycosis, systemic candidiasis and histoplasmosis.[6] Miltefosine, was synthesized as a part of an anti-inflammatory program in 1982 at the Burroughs Welcome (USA) pharmaceutical company.[7] Paromomycin was developed in 1959 as an oral broad-spectrum aminoglycoside antibiotic drug by the Carlo Erba (Italy) pharmaceutical company.[8–10] Through this brief summary, it is possible to understand that the discovery of new drugs against *Leishmania* is indeed a challenging journey.

Although a search in the literature for clinical trials identifies approximately 115 studies so far,[11] the vast majority of these studies investigate new protocols for treatment, using associations of available drugs as well as different drug regimens. This is because the parasite is, by itself, a cunning adversary, making the process more complicated and slow. Intracellular parasitism within host cells is a strategy used by some protozoan parasites, bacteria and viruses to evade host defenses and to allow survival. Usually, these are sophisticated strategies that engage specific pathogenic virulence factors and hijack host-cell components, such as the cytoskeleton and microtubule network, cellular machinery and signaling pathways. Intracellular amastigotes of *Leishmania* are able to replicate and persist in macrophages, causing long-term infections that can lead to reactivation of the disease years after the initial infection. *Leishmania* resides within a parasitophorous vacuole with efficient pathways for acquisition of host macromolecules, amino acids and sugars as carbon sources for its intravacuolar growth.[12]

Due to the intramacrophage nature of *Leishmania*, phenotypic drug screenings using intracellular amastigotes are considered the gold standard assays and a mandatory model for *in vitro* studies in drug discovery.[13] Drug candidates against *Leishmania* parasites, must inevitably be able to reach the intracellular amastigotes at the phagolysosome without harmful effects on host cells. Additionally, the drug candidate has another challenge: to resist a low-pH environment inside the phagolysosome,[14] in which some chemical compounds may be degraded.

9.2 The Interest in Natural Products as Scaffolds: Using a Rational Approach for Drug Discovery

Nature has provided more than 50% of scaffolds present in all FDA-approved drugs; for some therapeutic classes, such as antimicrobials, this ratio may be even higher. Since the early days of medicine, natural products have been used in folk medicine to treat ancient diseases. These substances may interfere with biological systems, blocking or activating receptors, developing a pharmacological response. These mainly include secondary metabolites isolated from plants and, to a lesser extent, marine invertebrates, animal venoms and microorganisms.[15]

The historical outlook of therapeutic drugs for leishmaniasis has strong connections to natural product chemistry. Amphotericin B is produced by the soil bacterium *Streptomyces nodosus;* despite its considerable toxicity, it has been a leading broad-spectrum antifungal antibiotic for more than 50 years.[16] Another leishmanicidal drug in clinical use is paromomycin, which was originally isolated from the bacteria *Streptomyces rimosus* in 1959.[8] Although academic drug discovery programs for neglected tropical diseases (NTDs) have increased the power and speed of screening, there are still important issues to be addressed. Increasing the screening capacity does not necessarily lead to an increased output in terms of new good hit compounds. Academic drug discovery programs need to be aligned with a rational approach to increase success. This includes the elimination of poor-quality compounds at early stages, these include toxic compounds with important structural alerts, poor absorption, distribution, metabolism, excretion and toxicity (ADMET) properties and the presence of interference compounds, named pan-assay interference compounds (PAINS). The PAINS concept has recently appeared as an important issue to be addressed at the beginning of drug discovery programs. Such compounds may demonstrate multiple behaviors that could interfere with assay readouts, such as metal chelation, redox cycling, and protein reactivity.[17] Many of them present appealing properties, such as high potency and "promising" *in vitro* selectivity, but researchers usually misperceive them and select such PAINS as new hit compounds.[18] So, large amounts of resources are expended in the lead optimization of poor-quality compounds; unsurprisingly, such projects have often a catastrophic end. Although despite problems like PAINS, natural products (NPs) have been considered promising scaffolds. Natural products enjoy a privileged position in drug discovery and few fell into compounds defined as PAINS.[15,19] How is it possible to administer this large quantity of information to make correct decisions? The answer for a successful drug discovery project could be the employment of a multidisciplinary team to make fast and adequate decisions, including parasitologists, biochemists, natural product chemists, synthetic organic chemists, medicinal chemists, pharmacologists and groups involved in drug product development among others.

9.2.1 Major Topics During Biological (*In Vitro* and *In Vivo*) Tests of Natural Products

9.2.1.1 Biological Testing of Crude Extracts or Pre-fractionated Samples for Screening Purposes

After collection and proper identification of material (plants, marine organisms, microorganisms as well as other samples), this material must be converted to extracts for drug screening purposes. These may be obtained using fresh or, more commonly, dried material. The resulting biomass may be extracted using different solvents by classical methods (percolation or maceration), or using automated systems such as accelerated solvent extraction (ASE) systems. After elimination of solvent under reduced pressure and/or lyophilisation, a crude extract is obtained. However, *in vitro* testing of crude extracts may result in false-negative results. To overcome this problem, some simple and rapid procedures should be implemented to enrich the active compounds in some specific fractions. As an example, the use of liquid–liquid partitioning allows concentration of the bioactive derivatives in a complex mixture of compounds based on solubility aspects.

9.2.1.2 Bioactivity-guided Fractionation and Concentration Test

In the quest for active anti-leishmanial compounds, crude extracts are usually submitted to bioactivity-guided fractionation. This technique consists of simultaneous chemical fractionation (frequently column chromatography over silica gel—normal and reversed phases, alumina, Sephadex LH-20 and other less common stationary phases) monitored by an *in vitro* assay (usually in 96-well microplates) using either extracellular forms (promastigotes or axenic amastigotes) or intracellular amastigotes. Despite the limitations of these techniques, both parasite forms could be used for bioactivity-guided fractionation, but the intracellular amastigote assay may not be the best approach. Some compounds present in a given fraction may be harmful to the macrophage, hence the fraction will be automatically rejected as "toxic", even when some anti-parasitic compounds (active compounds) may be also present.[20]

The selection of an appropriate cut-off concentration is important to avoid missing active compounds. For crude and pre-fractionated NP extracts active compounds are usually, but not always, present in a very modest quantities. Then, the selection of a high cut-off value will increase the number of active compounds. In a hypothetical situation, when a crude extract is tested at a low concentration (*e.g.* 10 $\mu g\,ml^{-1}$) and the active compound represents 0.1% of the crude extract, the active compound needs to eliminate the parasite at 0.01 $\mu g\,ml^{-1}$, which is, in practice, a highly unlikely event. Although this initial concentration depends on the method of parasite quantification (colorimetric, fluorimetric, light microscopy counting *etc.*), the top concentrations should not exceed 300 $\mu g\,ml^{-1}$ to avoid a high amount of insoluble material producing drawbacks.

9.2.1.3 Solutions and Dilution of Samples

Typically, few hydrophilic metabolites are found during bioactivity-guided fractionation procedures. Then a previous solubilisation step of the natural compound from the initial material using organic solvents, such as DMSO or methanol, before the *in vitro* studies is mandatory. Further serial dilutions in cell growth media are common strategies to obtain the 50% effective concentrations (EC_{50}), but organic solvents should never exceed 0.5% (v/v) in microplates.

9.2.1.4 Parasite Forms

Undoubtedly, the intracellular amastigote of *Leishmania* is the preferred form to be used in drug discovery.[21] *In vitro* macrophage infections need at least to be 50% of the population for a period of 48–120 h as this system is the most closely related to the pathological infection in mammals.

9.2.1.5 Potency, Efficacy and Selectivity

The selection of NPs as hit compounds with anti-leishmanial activity needs to follow criteria with increasing complexity from the hit to the lead phase. At the hit stage, the *potency* is defined as the amount of a drug to produce the desire effect. An EC_{50} value of less than 10 μM is suggested as an acceptable value. Furthermore, cytotoxicity studies are required to demonstrate safety of NPs on mammalian cells. Usually, the selectivity index (SI), defined as the ratio between the 50% cytotoxic concentration value (mammalian cells) and the 50% effective concentration value (parasites), is suggested to be more than 10. Finally, *in vivo efficacy* at the hit–to-lead stage for visceral leishmaniasis (VL) for the reduction of parasite burden in liver has been proposed to be 70% (after intraperitoneal administration) when compared with untreated animals.[21]

9.2.1.6 Chemical Tractability

This is an important issue for the selection of a new hit compound. Usually, NPs are used as scaffolds for lead optimization programs and ease of scaling up procedures needs to be considered when moving forward to the pre-clinical assays (amounts exceeding 1 g). The amenability of chemical synthesis must be considered at early stages of the research pipeline. Generally, for NTDs such as leishmaniasis, an acceptable synthetic pathway for a compound should involve no more than eight synthetic steps,[21] making the compound as cheap as possible.

9.2.1.7 Pharmacokinetics and Pharmacodynamics (PK–PD)

In the past, most PK–PD studies were conducted at late stages of the drug discovery programs. Recently, a result of the collaborative efforts of the distribution, metabolism and pharmacokinetic (DMPK) team from the pharmaceutical company Novartis recommended that PK–PD strategies should be implemented at early research phases of drug discovery. Effective PK–PD can undoubtedly help scientists to elucidate the relationship between PK and PD, to understand the mechanism of drug action, and to identify PK properties for further improvement and optimal compound design. The outcome allows reduction of the number of *in vivo* animal studies, and improves translation of findings from preclinical species into the clinical setting.[22]

9.3 From a Crude Extract to a Pure Compound—Aspects of Bioactive Compounds Isolation and Chemical Identification

Once a bioactive extract has been selected, the crude material must be chemically analysed before starting fractionation procedures aimed to isolate bioactive compounds. This approach, usually called *molecular dereplication*, employs the use of different combined techniques such as high performance liquid chromatography associated with mass spectrometry (LC–MS), nuclear magnetic resonance (LC–NMR) or both spectrometric techniques (LC/NMR–MS). These methods, together with the analysis of natural products databases, allow the detection, identification and quantification of different natural products in a complex mixture, such as a crude extract.[23] This procedure is useful to avoid the isolation of known bioactive compounds after exhaustive classical fractionation and permits detection of rare or even new natural products.

In a second step, the isolation of bioactive compound(s) in the crude extract must be conducted. A more rational approach involves the association of separation procedures with evaluation of the activity of each obtained fraction, usually called bioactivity-guided fractionation. After the use of several separation steps such as solvent–solvent partition, column chromatography using different stationary and mobile phases, counter-current chromatography or high-performance liquid chromatography (HPLC), spectrometric and/or spectroscopic methods could be used to complete the chemical characterization of the bioactive compound. This procedure essentially involves the use of NMR and MS, both crucial to the determination of the molecular mass of the isolated compound and, most importantly, the atoms' connectivity allowing the structural determination. In some cases, stereochemistry information may be inferred by using NMR experiments.[24,25] Therefore, this approach provides the structural characterization of a bioactive compound at high purity (over 99%) and is a crucial part of prospecting for natural products from different sources, such as marine organisms, plants, microorganisms, *etc.*

9.4 The Anti-leishmanial Potential of Natural Products in Animal Models

As reported in previous reviews,[26–29] the body of knowledge on the potential effects of natural products against *Leishmania* is quite large. Nevertheless, most of them were obtained by *in vitro* assays. Only few of these metabolites have been through tested in the *in vivo* models. Several secondary metabolites with *in vivo* anti-leishmanial potential have been isolated, essentially from plant species, although limited studies have been conducted with metabolites from marine organisms and endophytic fungi. These compounds represent potential prototypes for drug design studies against *Leishmania* and are classified as alkaloids (**1**–**5**) (Figure 9.1), terpenoids (**6**–**8**) (Figure 9.2), saponins (**9**–**16**) (Figure 9.3), phenylpropanoids (**17**–**20**) (Figure 9.4), flavonoids (**21**–**28**) (Figure 9.5), lignoid (**29**) (Figure 9.6), naphthoquinones (**30**–**33**) (Figure 9.7), and iridoid (**34**) (Figure 9.8), (Table 9.1).

9.4.1 Alkaloids

The alkaloid γ-fagarine (**1**), isolated from *Helietta apiculata* (Rutaceae), was studied in a cutaneous leishmaniasis (CL) murine model. BALB/c mice infected with *Leishmania amazonensis* and treated with compound **1** (orally, 10 mg kg^{-1}, 15 days) displayed a reduction of 97.4% of parasite load (in liver). Furthermore, when alkaloid **1** was tested at a higher dose (subcutaneous route at 100 mg kg^{-1}), this compound was able to reduce the lesional parasites (infected footpads) by 97.4%, a superior efficacy in comparison to Glucantime, with 66.9% reduction.[30] A related alkaloid, peganine hydrochloride (**2**) was isolated from the bioactive extract from seeds of *Peganum harmala* (Zygophyllaceae), a medicinal plant commonly used in India for the treatment of parasitic diseases. Compound **2** was orally administered in hamsters infected with *Leishmania donovani* at three different doses (50, 100 and 200 mg kg^{-1}, 5 days). At 200 mg kg^{-1} compound **2**

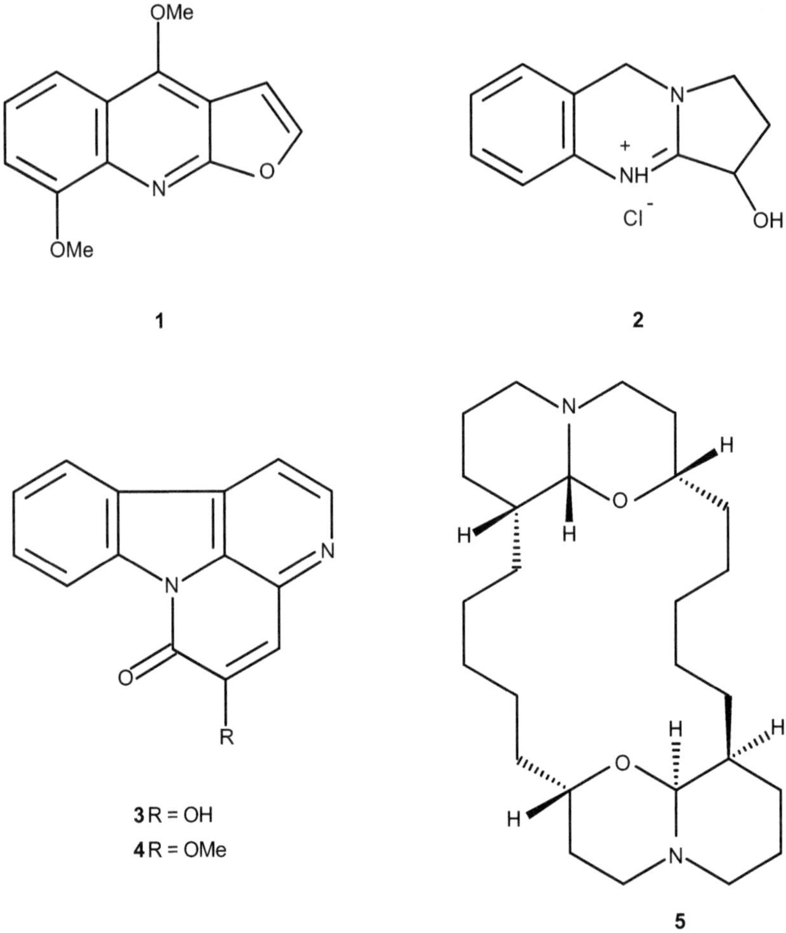

Figure 9.1 Structures of alkaloids with *in vivo* anti-leishmanial activity.

reduced the parasite splenic burden by 87.5% whilst at 100 mg kg^{-1} it inhibited only 79.6%. At 50 mg kg^{-1} compound **2** showed no effectiveness. The standard drug (miltefosine) (oral route, 40 mg kg^{-1}, 5 days) inhibited 95.5% of parasites.[31] Other bioactive alkaloids–ascanthin-6-one (**3**) and 5-methoxycanthin-6-one (**4**)–were isolated from extracts of stem bark of *Zanthoxylum chiloperone* var. *angustifolium* Engl. (Rutaceae). Treatment of BALB/c mice infected with *L. amazonensis* (oral route, 10 mg kg^{-1} day^{-1}, for 14 days) with compounds **3** and **4** resulted in a reduction of parasite load by 77.6 and 21.6%, respectively, after treatment.[32] Araguspongin C (**5**), an alkaloid isolated from the marine sponge *Haliclona exigua* (Porifera) was tested *in vivo* against *L. donovani* infection in hamster model and exhibited 38.7% reduction of parasitic burden (in spleen) at 100 mg kg^{-1} when administered orally for five days.[33]

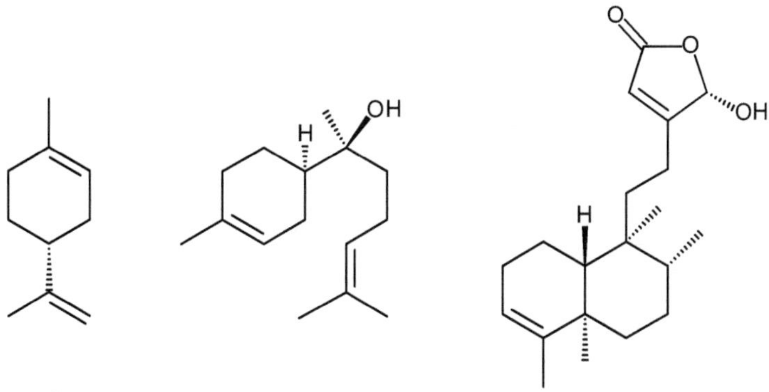

6 7 8

Figure 9.2 Structures of terpenoids with *in vivo* anti-leishmanial activity.

9.4.2 Terpenoids

The monoterpene limonene (**6**), isolated from *Citrus sinensis* (Rutaceae) as well as several other plant species, was tested in C57BL/6 mice against *L. amazonensis* infection. Compound **6** was administered topically (ointments vehicle at 10% by weight for 19 weeks) and intrarectally (100 mg kg^{-1} day^{-1}, 2 weeks). A reduction of more than 99.9% in the parasite load was observed in 80% of the intrarectally-treated mice. The average lesion size was significantly reduced in 80% of the intrarectally-treated animals. The topical treatment was effective (99.9% reduction of parasite loads) in 67–86% of animals and 13 weeks after treatment interruption very scarce or no traces of parasites were found, with no side effects detected in such cases.[34] BALB/c mice infected with *Leishmania infantum* promastigotes were treated (oral route, 200 mg kg^{-1}, 14 days) with the sesquiterpene (−)-α-bisabolol (**7**) isolated from chamomile (*Matricaria chamomilla*, Asteraceae) using Glucantime (104 mg Sb^{V} kg^{-1}) as a standard drug. In addition to the reduced toxicity, compound **7** improved the efficacy of the standard drug, decreasing the parasite burden in spleen (71.6%) and in liver (89.2%).[35] The diterpenoid 16α-hydroxycleroda-3,13(14)*Z*-dien-15,16-olide (**8**), isolated from *Polyalthia longifolia* (Annonaceae), demonstrated no toxicity to J774A.1 macrophages, and was administered to Golden hamsters infected with *L. donovani* (oral route, 25, 50, 100 and 250 mg kg^{-1}, 5 days). At the highest dose, the compound was effective for reducing the parasite burden in spleen by 87.5%.[36]

9.4.3 Saponins

Holothurin A (**9**), isolated from a coral reef sea cucumber *Actinopyga lecanora*, displayed moderate *in vivo* activity against *L. donovani*-infected hamsters, exhibiting efficacy of 44.6% (oral route, 100 mg kg^{-1} day^{-1}, five

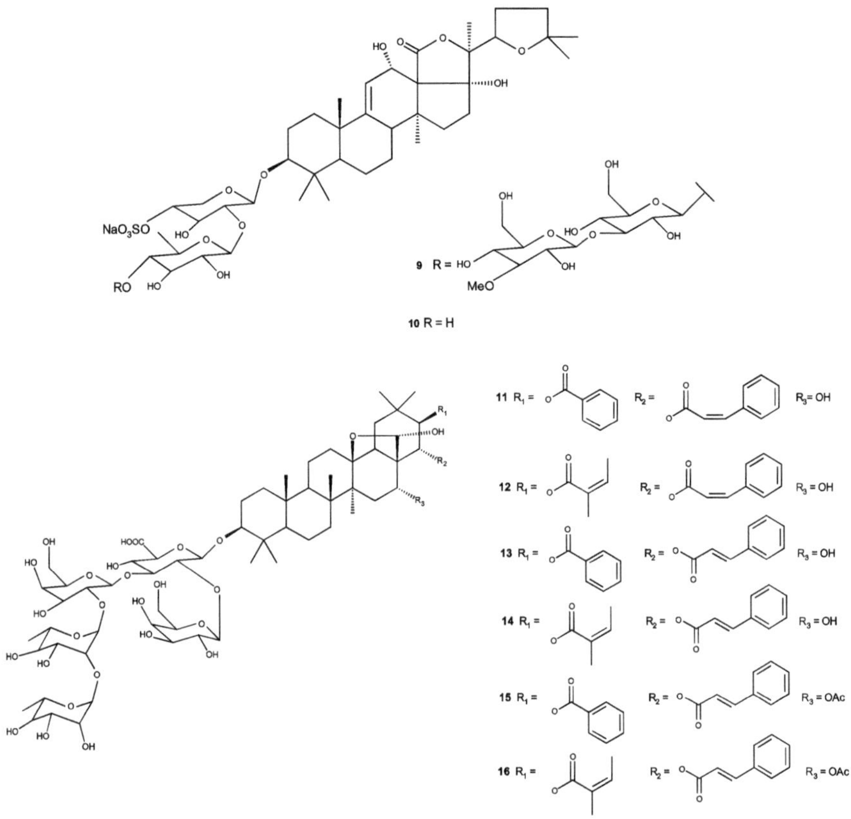

Figure 9.3 Structures of saponins with *in vivo* anti-leishmanial activity.

days treatment). A related compound, holothurin B (**10**) showed improved effectiveness (71.5%), using the same model and regimen of administration, even at 50 mg kg^{-1}.[37] A mixture of maesabalides I–VI (**11–16**) obtained from the leaves of *Maesa balansae* (Myrsinaceae), were evaluated against *L. donovani* in the BALB/c murine model. A single subcutaneous administration of 0.4 mg kg^{-1} at 1 day after infection reduced liver amastigote burdens by about 95% in all treated animals. If treatment was delayed until 14 days after infection, a dose of 1.6 mg kg^{-1} body weight was required to match the previous activity.[38] Prophylactic administration of purified compound **13** (subcutaneous dose at 0.2 mg kg^{-1}) reduced liver amastigote burdens by 99.8 and 83.0% within 7 and 56 days after treatment, respectively.[39]

9.4.4 Phenylpropanoids

Calceolarioside A (**17**), a glycoside phenylpropanoid derivative was isolated from leaves of *Nyctanthe sarbortristis* (Oleaceae). In golden hamsters infected

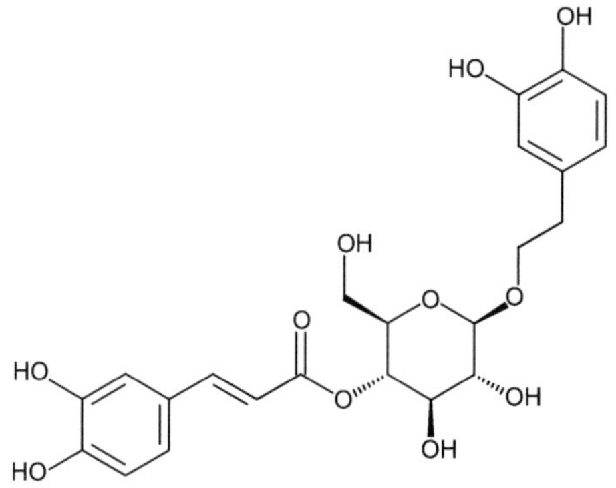

17

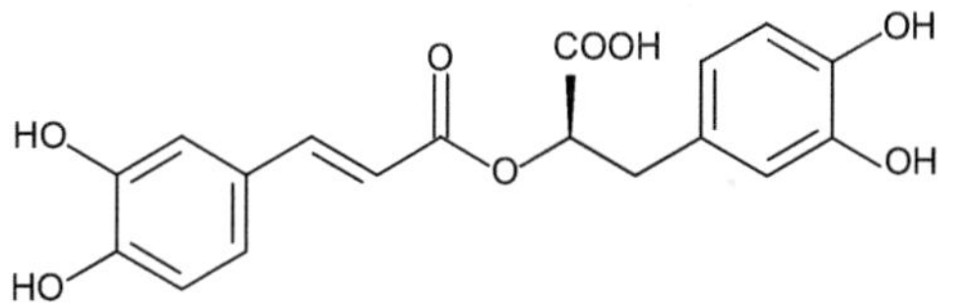

18

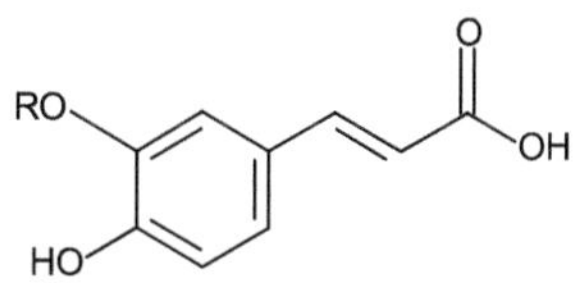

19 R = H

20 R = Me

Figure 9.4 Structures of phenylpropanoids with *in vivo* anti-leishmanial activity.

with *L. donovani*, compound **17** was able to reduce spleen (84%) and liver (79%) parasite burden (i.p. route, 20 mg kg^{-1}, 15 days).[40] The phenylpropanoids rosmarinic (**18**), caffeic (**19**) and ferulic (**20**) acids were isolated from *Pluchea carolinensis* (Asteraceae) and tested against *L. amazonensis* in the BALB/c murine model. After parasite inoculations, five doses (30 mg kg^{-1}) of compounds **18–20** were administered by the intralesional route every four days for 45 days. Compounds **18–20** reduced the lesion size and parasite load (52–65%) in treated groups. Additionally, the obtained results indicated that compound **20** displayed the highest efficacy in lesion size reduction and no parasites were found after six weeks of treatment.[41]

9.4.5 Flavonoids

The flavonoid (−)-epigallocatechin-3-*O*-gallate (**21**), isolated from the leaves of *Camellia sinensis* (Theaceae), was evaluated *in vivo* against *L. amazonensis*-infected BALB/c mice. The oral treatment (30 mg kg^{-1} day^{-1}, five times a week, 52 days), reduced the lesion size and parasite burden by 87.0%.[42] The flavonoid quercetin (**22**), obtained from several plant species, and flavone A (**23**), isolated from *Vitex negundo* (Lamiaceae) leaves,[43] were administered at 5, 10, 20, 30 and 40 mg kg^{-1} (orally, biweekly) and Syrian golden hamsters infected by *L. donovani* were sacrificed after 1 month of drug treatment.

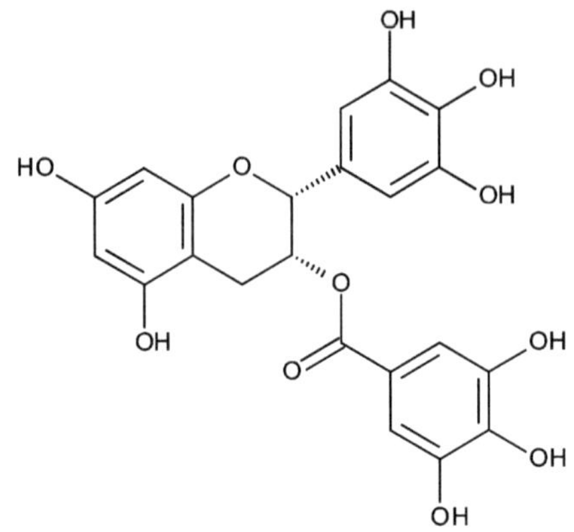

21

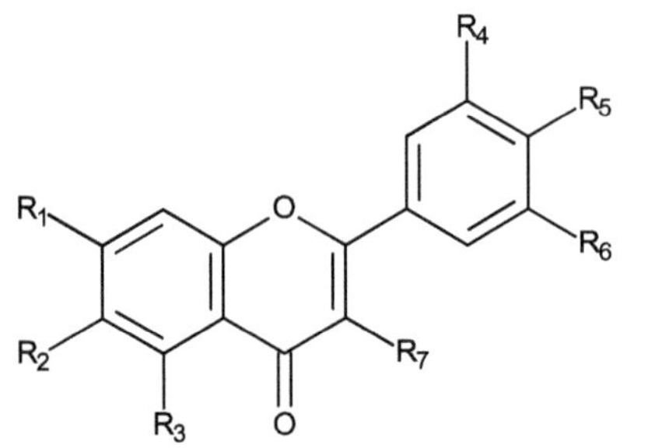

	R_1	R_2	R_3	R_4	R_5	R_6	R_7
22	R_1 = OH	R_2 = H	R_3 = OH	R_4 = OH	R_5 = OH	R_6 = H	R_7 = OH
23	R_1 = OMe	R_2 = OMe	R_3 = OH	R_4 = OMe	R_5 = Me	R_6 = H	R_7 = OMe
24	R_1 = OH	R_2 = H	R_3 = OH	R_4 = OH	R_5 = OH	R_6 = OH	R_7 = H
27	R_1 = OH	R_2 = H	R_3 = OH	R_4 = OH	R_5 = OH	R_6 = OH	R_7 = OAra-Rham
28	R_1 = OH	R_2 = H	R_3 = OH	R_4 = OH	R_5 = OH	R_6 = H	R_7 = ORham

Figure 9.5 Structures of flavonoids with *in vivo* anti-leishmanial activity.

Compound **22** displayed the highest performance in the subsequently *in vivo* tests, associated with a greater effect at prevention of lipid peroxidation, suppression of protein carbonyl increase, increased erythrocyte antioxidant capacity, correction of anemia levels during visceral leishmaniasis and reduction of splenic parasite burden. The combination of quercetin (**22**) with sodium stibogluconate (SAG), led to a higher prevention of reactive oxygen species (ROS) formation (23.7% SAG and 47.4% quercetin), lower proteolytic degradation, prolonged red cells life span and decreased spleen load of parasites.[44] However, taking into account that quercetin (**22**) is a well-known interference compound (see PAINS comments in Section 9.2), this compound could not be considered a promising hit.

The flavonoids quercetin (**22**) and luteolin (**24**) were also tested *in vivo* using Golden hamsters infected with *L. donovani*. In this assay, the animals were orally treated twice a week for four weeks with **24** at 3.5 mg kg^{-1} or **22** at 14 mg kg^{-1}. As results the flavonoids **24** and **22** were able to reduce the splenic parasite load at 80% and 90%, respectively.[44] Licochalcone A (**25**), a prenylated chalcone from *Glycyrrhiza ralensis* (Fabaceae), was tested *in vivo* against *Leishmania major* and *L. donovani*. Compound **25** (i.p. route, 2.5 and 5.0 mg kg^{-1} day^{-1}, 39 days) completely prevented disease development in *L. donovani*-infected BALB/c mice. In the Golden hamster model, this compound (i.p. route, 20 mg kg^{-1} day^{-1}, 6 days), showed efficacy, reducing the parasite burden in liver and spleen by 96.0%.[45] A related chalcone, flavokain B (**26**), isolated from *Piper rusbyi* (Piperaceae), displayed low activity against cutaneous leishmaniasis (*L. amazonensis*) in the BALB/c model (subcutaneous dose, 5 mg kg^{-1} day^{-1} for eight weeks) reducing the lesion sizes and parasite load by 32.0%.[46] Flavonoid glycosides quercetin-3-*O*-α-L-arabinopyranosyl-(1→2)-α-L-rhamnopyranoside (**27**) and quercetin-3-*O*-α-L-rhamnopyranoside (**28**) as well as quercetin (**22**), isolated from aqueous extract of *Kalanchoe pinnata* (Crassulaceae) were tested in *L. amazonensis* in a BALB/c model (oral route, 16 mg kg^{-1} day^{-1}, 30 days). Compounds **22**, **27** and **28** reduced the parasite burden by 76.0, 65.0 and 57.0%, respectively.[47]

9.4.6 Lignoid

The lignoid lyoniside (**29**), isolated from stem bark of *Saraca indica* (Fabaceae), was tested in the BALB/c mice infected with *L. donovani* model (i.p. route, 2.5 and 5.0 mg kg^{-1}, twice a week for three weeks). It was suggested that compound **29** elicited an immunomodulatory effect, reducing the parasite burden in liver and in spleen by 95.0% at the highest dose. Poor cytotoxic effects in murine peritoneal macrophages was observed even at 100 μM.[48]

9.4.7 Naphthoquinones

Bioactive naphthoquinone derivative 4-hydroxy-1-tetralone (**30**), isolated from stem barks of *Ampelocera edentula* (Ulmaceae), was tested in *L. amazonensis*- and *Leishmania venezuelensis*-infected BALB/c mice (intralesional, single dose at 50 mg kg^{-1}). After 14 days of treatment all mice treated with compound **30** developed a similar lesion size when compared with those treated with Glucantime® at 112 mg kg^{-1}, suggesting a greater effectivity than the antimony reference drug.[49] Other related naphthoquinones plumbagin (**31**), 3,3′-biplumbagin (**32**) and 8,8′-biplumbagin (**33**) were isolated from stem barks of *Pera benensis* (Euphorbiaceae). BALB/c mice infected with *L. amazonensis* were subcutaneously treated with compounds **31** (2.5 and 5.0 mg kg^{-1} day^{-1}) or **32**–**33** (25 mg kg^{-1} day^{-1}) for two weeks. Plumbagin (**31**) displayed superior efficacy (38.0% of reduction of parasite loads) with low toxicity.[50]

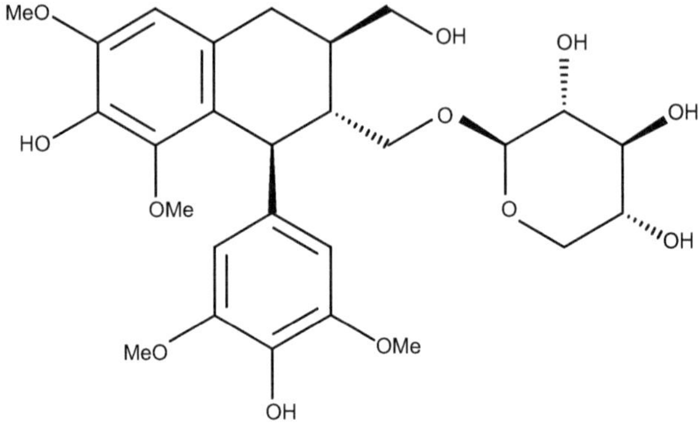

29

Figure 9.6 Structure of a lignoid with *in vivo* anti-leishmanial activity.

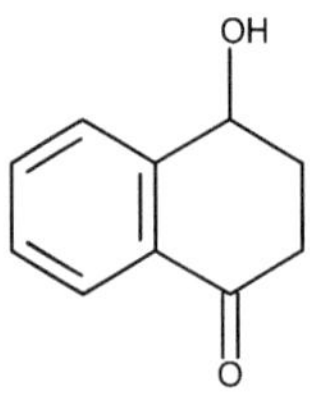

30

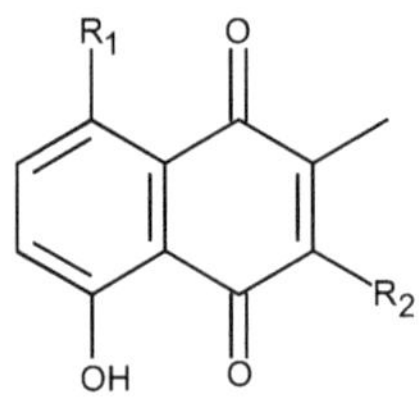

31 R_1 = H R_2 = H

32 R_1 = H R_2 = plumbagim

33 R_1 = plumbagin R_2 = H

Figure 9.7 Structure of naphtoquinones with *in vivo* anti-leishmanial activity.

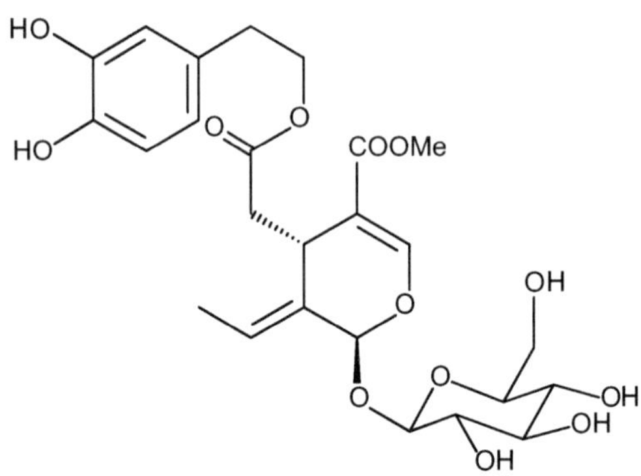

34

Figure 9.8 Structure of an iridoid with *in vivo* anti-leishmanial activity.

9.4.8 Iridoid

The iridoid oleuropein (**34**), isolated from *Olea europaea* (Oleaceae), was tested on BALB/c mice infected with *L. donovani* (i.p. route, 45, 15 and

Table 9.1 Natural products with *in vivo* anti-leishmanial activity.[a]

Compound	Species	Parasite	Route	Dose ($mg\,kg^{-1}\,day^{-1}$)	Frequency (days)	Reduction of parasite loads (%)
Alkaloids						
γ-Fagarine (**1**)	*Helietta apiculata*[30]	*L. amazonensis*	oral	10	15	97.4 (liver)
Peganine hydrochloride (**2**)	*Peganum harmala*[31]	*L. donovani*	oral	50	5	—
			oral	100	5	79.6 (spleen)
			oral	200	5	87.5 (spleen)
Canthin-6-one (**3**)	*Zanthoxylum chiloperone* var. *angustifolium*[32]	*L. amazonensis*	oral	10	14	77.6[b]
5-Methoxycanthin-6-one (**4**)	*Zanthoxylum chiloperone* var. *angustifolium*[32]	*L. amazonensis*	oral	10	14	21.6[b]
Araguspongin C (**5**)	*Haliclona exigua*[33]	*L. donovani*	oral	100	5	38.7 (spleen)
Terpenoids						
Limonene (**6**)	*Citrus sinensis*[34]	*L. amazonensis*	topical	10%[d]	135	>99.9[c]
			i.r.	100	14	>99.9[c]
(−)-α-Bisabolol (**7**)	*Matriarca chamoilla*[35]	*L. infantum*	oral	200	14	89.2 (liver)
						71.6 (spleen)
16α-Hydroxycleroda-3,13(14)*Z*-dien-15,16-olide (**8**)	*Polyalthia longifolia*[36]	*L. donovani*	oral	25	5	35.5 (spleen)
			oral	50	5	77.9 (spleen)
			oral	100	5	83.7 (spleen)
			oral	250	5	87.5 (spleen)
Saponins						
Holothurin A (**9**)	*Actinopyga lecanora*[37]	*L. donovani*	oral	100	5	44.6 (spleen)
Holothurin B (**10**)	*Actinopyga lecanora*[37]	*L. donovani*	oral	100	5	71.5 (spleen)
Maesabalide I (**11**)	*Maesa balansae*[38]	*L. donovani*	s.c.	0.4	1	95.0 (liver)
Maesabalide II (**12**)	*Maesa balansae*[38]	*L. donovani*	s.c.	0.4	1	95.0 (liver)
Maesabalide III (**13**)	*Maesa balansae*[38,39]	*L. donovani*	s.c.	0.4	1	95.0 (liver)
			s.c.	0.2	7	99.8 (liver)
			s.c.	0.2	56	83.0 (liver)
Maesabalide IV (**14**)	*Maesa balansae*[38]	*L. donovani*	s.c.	0.4	1	95.0 (liver)
Maesabalide V (**15**)	*Maesa balansae*[38]	*L. donovani*	s.c.	0.4	1	95.0 (liver)
Maesabalide VI (**16**)	*Maesa balansae*[38]	*L. donovani*	s.c.	0.4	1	95.0 (liver)

Table 9.1 (*Continued*)

Compound	Species	Parasite	Route	Dose ($mg\,kg^{-1}\,day^{-1}$)	Frequency (days)	Reduction of parasite loads (%)
Phenylpropanoids						
Calceolarioside A (**17**)	*Nyctanthe sarbortristis*[40]	*L. donovani*	i.p.	20	15	79.0 (liver) 84.0 (spleen)
Rosmarinic acid (**18**)	*Pluchea carolinensis*[41]	*L. amazonensis*	i.l.	30	45	65.0 (spleen)
Caffeic acid (**19**)	*Pluchea carolinensis*[41]	*L. amazonensis*	i.l.	30	45	52.0 (spleen)
Ferulic acid (**20**)	*Pluchea carolinensis*[41]	*L. amazonensis*	i.l.	30	45	65.0 (spleen)
Flavonoids						
(−)-Epi-gallocatechin-3-*O*-gallate (**21**)	*Camellia sinensis*[42]	*L. amazonensis*	oral	30	52	87.0 (liver)
Quercetin (**22**)	*Kalanchoe pinnata*[47]	*L. amazonensis*	oral	16	30	76.0 (spleen)
	Fagopyrum esculentum[44]	*L. donovani*	oral	14	28	90.0 (spleen)
	Myrica rubra[52]	*L. donovani*	i.p.	30	5	15.3 (liver)
	Pluchea carolinensis[41]	*L. amazonensis*	i.l.	30	45	28.0 (spleen)
	Vitex negundo[44]	*L. donovani*	oral	14	28	90.0 (spleen)
Flavone A (**23**)	*Vitex negundo*[43]	*L. donovani*	oral	20	30	94.4 (spleen)
			oral	40	30	91.4 (spleen)
Luteolin (**24**)	*Vitex negundo*[44]	*L. donovani*	oral	3.5	28	80.0 (spleen)
Licochalcone A (**25**)	*Glycyrrhizau ralensis*[45]	*L. donovani*	i.p.	2.5	39	96.0 (liver and spleen)
			i.p.	5.0	39	100.0 (liver and spleen)
Flavokain B (**26**)	*Piper rusbyi*[46]	*L. amazonensis*	s.c.	5.0	8	32.0 (spleen)
Quercetin 3-*O*-α-L-arabinopyranosyl-(1→2)-α-L-rhamnopyranoside (**27**)	*Kalanchoe pinnata*[47]	*L. amazonensis*	oral	16	30	65.0 (spleen)
Quercetin 3-*O*-α-L-rhamnopyranoside (**28**)	*Kalanchoe pinnata*[47]	*L. amazonensis*	oral	16	30	57.0 (spleen)

Lignoid						
Lyoniside (**29**)	*Saraca indica*[48]	*L. donovani*	i.p.	5.0	21	95.0 (liver and spleen)
Naphthoquinones						
4-Hydroxy-1-tetralone (**30**)	*Ampelocera edentula*[49]	*L. amazonensis*	i.l.	50	14	37.0[e]
		L. venezualensis	i.l.	50	14	70.0[e]
Plumbagin (**31**)	*Pera benensis*[50]	*L. amazonensis*	s.c.	2.5	14	38.0[c]
3,3′-Biplumbagin (**32**)	*Pera benensis*[50]	*L. amazonensis*	s.c.	25	14	36.6[c]
8,8′-Biplumbagin (**33**)	*Pera benensis*[50]	*L. amazonensis*	s.c.	25	14	22.0[c]
Iridoid						
Oleuropein (**34**)	*Olea europaea*[51]	*L. donovani*	i.p.	5	28	65.7 (liver) 57.2 (spleen)
			i.p.	5	42	96.6 (spleen)
			i.p.	15	28	62.0 (liver) 86.8 (spleen)
			i.p.	15	42	99.8 (spleen)
			i.p.	45	28	58.1 (liver) 79.5 (spleen)
			i.p.	45	42	99.7 (spleen)

[a]i.p., intraperitoneal; s.c., subcutaneous; i.l., intralesional; i.r., intrarectal.
[b]Suppression of parasite load in lesion.
[c]Values based in the reduction on the average lesion size.
[d]Treatment of mice infected at the basis of the tail and treated with a cream containing 10% (by weight) limonene.
[e]Based in the determination of Index of Leishmaniasis.[49]

5 mg kg^{-1} day^{-1} for 28 days). All tested dosages were demonstrated to be able to reduce the spleen (79.5, 86.8 and 57.2%, respectively) and liver (58.1, 62.0 and 65.7%, respectively) parasite burden after 28 days of treatment. Furthermore, spleen parasite burden diminished by 99.7, 99.8 and 96.9% (45, 15 and 5 mg kg^{-1}, respectively), six weeks after treatment completion, indicating an interesting *in vivo* effect of iridoid **34**.[51]

9.5 Conclusions

This chapter has presented a comprehensive discussion on the rational approaches to be used in drug discovery studies. This included the association of natural products with the treatment of leishmaniasis together with the efficacy of secondary metabolites isolated from different sources on animal models of *Leishmania* infections. These studies demonstrated the high potential of natural products for the discovery of new leads for the treatment of neglected parasitic diseases, such as leishmaniasis. Moreover, as only 10% of the biodiversity of the planet has been chemically and biologically explored, we may foresee that a huge number of new anti-leishmanial scaffolds could be discovered and evaluated in pre-clinical studies.

Acknowledgements

The authors thank the São Paulo Research Foundation—FAPESP (projects 2015/11936-2 and 2013/50228-8) and the Conselho Nacional de Desenvolvimento Cientifico e Tecnológico (CNPq) for the Scientific Research Award given to A.G.T. and J.H.G.L.

References

1. World Health Organization, Investing to overcome the global impact of neglected tropical diseases: third WHO report on neglected diseases. 6 Annual Reports, 2015.
2. R. A. Hodos, B. A. Kidd, K. Shameer, B. P. Readhead and J. T. Dudley, *Wiley Interdiscip. Rev. Syst. Biol. Med.*, 2016, **8**, 186.
3. S. Rath, L. A. Trivelin, R. T. Imbrunito, D. M. Tomazela, M. N. de Jeses, P. C. Marzal, H. D. De Andrade Jr. and A. G. Tempone, *Quim. Nova*, 2003, **26**, 550.
4. E. M. Lourie and W. Yorke, *Ann. Trop. Med. Parasitol.*, 1939, **33**, 289.
5. J. N. Ashley, H. J. Barber, A. J. Ewins and G. Newbery, *J. Chem. Soc.*, 1942, 103.
6. L. Ostrosky-Zeichner, K. A. Marr, J. H. Rex and S. H. Cohen, *Clin. Infect. Dis.*, 2003, **37**, 415.
7. S. L. Croft and J. Engel, *Trans. R. Soc. Trop. Med. Hyg.*, 2006, **100**, S4.
8. D. Botero, *Ann. Rev. Pharmacol. Toxicol.*, 1978, **18**, 1.
9. M. S. Wolfe, *Med. Clin. North. Am.*, 1982, **66**, 707.

10. W. Stauffer and J. I. Ravdin, *Curr. Opin. Infect. Dis.*, 2003, **16**, 479.
11. https://clinicaltrials.gov/ct2/home (last accessed October 2016).
12. V. Liévin-Le Moal and P. M. Loiseau, *FEBS J.*, 2016, **283**, 598.
13. K. Katsuno, J. N. Burrows, K. Duncan, R. H. Van Huijsduijnen, T. Kaneko, K. Kita, C. E. Mowbray, D. Schmatz, P. Warner and B. T. Slingsby, *Nat. Rev. Drug Discovery*, 2015, **14**, 751.
14. F. Real and R. A. Mortara, *PLoS Neglected Trop. Dis.*, 2012, **6**, e1518.
15. D. J. Newman and G. M. Cragg, *J. Nat. Prod.*, 2016, **79**, 629.
16. B. Murphy, K. Anderson, C. Borissow, P. Caffrey, G. Griffith, J. Hearn, O. Ibrahim, N. Khan, N. Lamburn, M. Lee, K. Pugh and B. Rawlings, *Org. Biomol. Chem.*, 2010, **21**, 3758.
17. J. B. Baell and G. A. Holloway, *J. Med. Chem.*, 2010, **8**, 2719.
18. J. B. Baell and M. A. Walters, *Nature*, 2014, **25**, 481.
19. M. S. Butler, A. A. Robertson and M. A. Cooper, *Nat. Prod. Rep.*, 2014, **31**, 1612.
20. A. G. Tempone, C. M. Oliveira and R. G. S. Berlinck, *Planta Med.*, 2011, 77, 572.
21. R. Don and J. R. Ioset, *Parasitology*, 2014, **141**, 140.
22. T. Tuntland, B. Ethell, T. Kosaka, F. Blasco, R. X. Zang, M. Jain, T. Gould and K. Hoffmaster, *Front. Pharmacol.*, 2014, **174**, 1.
23. S. P. Gaudêncio and F. Pereira, *Nat. Prod. Rep.*, 2015, **32**, 779.
24. M. H. Wright and S. A. Sieber, *Nat. Prod. Rep.*, 2016, **33**, 681.
25. N. Bross-Walch, T. Kühn, D. Moskau and O. Zerbe, *Chem. Biodiversity*, 2005, **2**, 147.
26. T. J. Schmidt, S. A. Khalid, A. J. Romanha, T. M. Alves, M. W. Biavatti, R. Brun, F. B. Costa, S. L. Castro, V. F. Ferreira, M. V. Lacerda, J. H. G. Lago, L. L. Leon, N. P. Lopes, R. C. N. Amorim, N. Niehues, I. V. Ogungbe, A. M. Pohlit, M. T. Scotti, W. N. Setzer, M. Soeiro, M. Steindel and A. G. Tempone, *Curr. Med. Chem.*, 2012, **19**, 2128.
27. T. J. Schmidt, S. A. Khalid, A. J. Romanha, T. M. Alves, M. W. Biavatti, R. Brun, F. B. Costa, S. L. Castro, V. F. Ferreira, M. V. Lacerda, J. H. G. Lago, L. L. Leon, N. P. Lopes, R. C. N. Amorim, N. Niehues, I. V. Ogungbe, A. M. Pohlit, M. T. Scotti, W. N. Setzer, M. Soeiro, M. Steindel and A. G. Tempone, *Curr. Med. Chem.*, 2012, **19**, 2176.
28. N. Singh, B. B. Mishra, S. Bajpai, R. K. Singh and V. K. Tiwari, *Bioorg. Med. Chem.*, 2014, **22**, 18.
29. I. V. Ogungbe and W. N. Setzer, *Molecules*, 2016, **21**, E1389.
30. M. E. Ferreira, A. R. de Arias, G. Yaluff, N. V. de Bilbao, H. Nakayama, S. Torres, A. Schinini, I. Guy, H. Heinzen and A. Fournet, *Phytomedicine*, 2010, **17**, 375.
31. T. Khaliq, P. Misra, S. Gupta, K. P. Reddy, R. Kant, P. R. Maulik, A. Dube and T. Narender, *Bioorg. Med. Chem. Lett.*, 2009, **19**, 2585.
32. M. E. Ferreira, A. R. de Arias, S. T. de Ortiz, A. Inchausti, H. Nakayama, C. Thouvenel, R. Hocquemiller and A. Fournet, *J. Ethnopharmacol.*, 2002, **80**, 199.

33. A. Dube, N. Singh, A. Saxena and V. Lakshmi, *Parasitol. Res.*, 2007, **101**, 317.
34. D. C. Arruda, D. C. Miguel, J. K. U. Yokoyama-Yasunaka, A. M. Katzin and S. R. B. Uliana, *Biomed. Pharmacother.*, 2009, **63**, 643.
35. V. Corpas-López, F. Morillas-Márquez, M. C. Navarro-Moll, G. Merino-Espinosa, V. Díaz-Sáez and J. Martín-Sánchez, *J. Nat. Prod.*, 2015, **78**, 1202.
36. P. Misra, K. V. Sashidhara, S. P. Singh, A. Kumar, R. Gupta, S. S. Chaudhaery, S. Gupta, H. K. Sen, H. K. Majumder, A. K. Saxena and A. Dube, *Br. J. Pharmacol.*, 2010, **159**, 1143.
37. N. Singh, R. Kumar, S. Gupta, A. Dube and V. Lakshmi, *Parasitol. Res.*, 2008, **103**, 351.
38. L. Maes, D. Vanden Berghe, N. Germonprez, L. Quirijnen, P. Cos, N. De Kimpe and L. Van Puyvelde, *Antimicrob. Agents Chemother.*, 2003, **48**, 130.
39. L. Maes, N. Germonprez, L. Quirijnen, L. Van Puyvelde, P. Cos and D. V. Berghe, *Antimicrob. Agents Chemother.*, 2004, **48**, 2056.
40. A. Poddar, A. Banerjee, S. Ghanta and S. Chattopadhyay, *Planta Med.*, 2008, **74**, 503.
41. E. Montrieux, W. H. Perera, M. García, L. Maes, P. Cos and L. Monzote, *Parasitol. Res.*, 2014, **113**, 2925.
42. J. D. F. Inacio, M. M. Canto-Cavalheiro and E. E. Almeida-Amaral, *J. Nat. Prod.*, 2013, **76**, 1993.
43. G. Sen, S. Mandal, S. S. Roy, S. Mukhopadhyay and T. Biswas, *Free Radical Biol. Med.*, 2005, **38**, 1257.
44. B. Mittra, A. Saha, A. R. Chowdhury, C. Pal, S. Mandal, S. Mukhopadhyay, S. Bandyopadhyay and H. K. Majumder, *Mol. Med.*, 2000, **6**, 527.
45. M. Chen, S. B. Christensen and T. G. Theander, *Antimicrob. Agents Chemother.*, 1994, **38**, 1339.
46. N. Flores, G. Cabrera, I. A. Jiménez, J. Piñero, A. Giménez, G. Bourdy, F. Cortés-Selva and I. L. Bazzocchi, *Planta Med.*, 2007, **73**, 206.
47. M. F. Muzitano, C. A. B. Falcão, E. A. Cruz, M. C. Bergonzi, A. R. Bilia, F. F. Vincieri, B. Rossi-Bergmann and S. S. Costa, *Planta Med.*, 2009, **75**, 307.
48. S. Saha, T. Mukherjee, S. Chowdhury, A. Mishra, S. R. Chowdhury, P. Jaisankar, S. Mukhopadhyay and H. K. Majumder, *Biochem. Pharmacol.*, 2013, **86**, 1673.
49. A. Fournet, A. A. Barrios, V. Muñoz, R. Hocquemiller, F. Roblot and A. Cavé, *Planta Med.*, 1994, **60**, 8.
50. A. Fournet and A. Barrios, *Trop. Med. Parasitol.*, 1992, **43**, 219.
51. J. D. Kyriazis, N. Aligiannis, P. Polychronopoulos, A.-L. Skaltsounis and E. Dotsika, *Phytomedicine*, 2013, **20**, 275.
52. D. Tasdemir, M. Kaiser, R. Brun, V. Yardley, T. J. Schmidt, F. Tosun and P. Rüedi, *Antimicrob. Agents Chemother.*, 2006, **50**, 1352.

CHAPTER 10

Organometallic Compounds in Chemotherapy Against Leishmania

BRUNO RODRIGUES DO PRADO,[b] ARSHAD ISLAM,[a,b] FRÉDÉRIC FRÉZARD[b] AND CYNTHIA DEMICHELI*[a]

[a] Department of Physiology and Biophysics, Institute of Biological Sciences, Federal University of Minas Gerais (UFMG), Av. Antônio Carlos 6627, Belo Horizonte 31270-901, Minas Gerais, Brazil; [b] Department of Chemistry, Institute of Exact Sciences, Federal University of Minas Gerais (UFMG), Av. Antônio Carlos 6627, Belo Horizonte 31270-901, Minas Gerais, Brazil

*Email: demichel@netuno.lcc.ufmg.br

10.1 Introduction

Organometallic compounds are molecules containing at least one metal–carbon bond.[1] The nature and character of the metal–carbon bond have an important role in chemical reactions, being either ionic, covalent or in between.[2] Today, most of the clinical drugs are purely organic compounds, while organometallic compounds are extensively being utilized in synthetic chemistry as catalysts. The organometallic class of compounds has not been widely considered for potential therapeutic applications until recently, due to the predominant consensus that the presence of a metal–carbon bond may lead to subsequent toxic properties and chemical instability, and

Drug Discovery Series No. 60
Drug Discovery for Leishmaniasis
Edited by Luis Rivas and Carmen Gil

Published by the Royal Society of Chemistry, www.rsc.org

therefore was generally considered to be non-compatible with biological systems.[3,4] Despite this perception, the medicinal properties of organometallic compounds have been investigated for a long time and in the last few years, interest in these metal complexes has grown substantially.[4]

The firstly documented and medicinally important organometallic compound was "salvarsan" (**1**) identified by Paul Ehrlich (1909). The compound contained arsenic bonded to a substituted aromatic ring and, thus, is an organometallic compound (Figure 10.1).

Salvarsan is active against syphilis, a bacterial infection, whose cure was not available until the beginning of the 20th century. The drug was used until about the end of the World War II, when modern antibiotics, primarily from the penicillin family, gradually replaced it.[5]

In 1934, Ernst Friedheim discovered another organoarsenical compound called melarsoprol (**2**) (Figure 10.1) for treatment of sleeping sickness, saving million of lives.[6] Nowadays, this compound remains the only drug available for stage 2 sleeping sickness, caused by *Trypanosoma brucei rhodesiense*.

Also in the 1970s, a few seminal papers reported on the anticancer activity of organometallic transition metal compounds. Köpf and Köpf-Maier reported that titanocene dichloride, and several other bent metallocene dihalides to a varying degree, were able to inhibit tumour growth in mice *in vivo*.[7]

Since then, the interest in the study of organometallic compounds has continued to increase and now they are being investigated for their potential medicinal properties. This chapter outlines the therapeutic potential of organometallic compounds that have been reported so far in the literature, in order to find novel, safer and more efficient treatments for leishmaniasis, a group of disorders with clinical manifestations that range from self-healing cutaneous and mucocutaneous skin ulcers to a fatal visceral form, caused by infection with different species of *Leishmania*, protozoan parasites belonging to the Trypanosomatidae family.

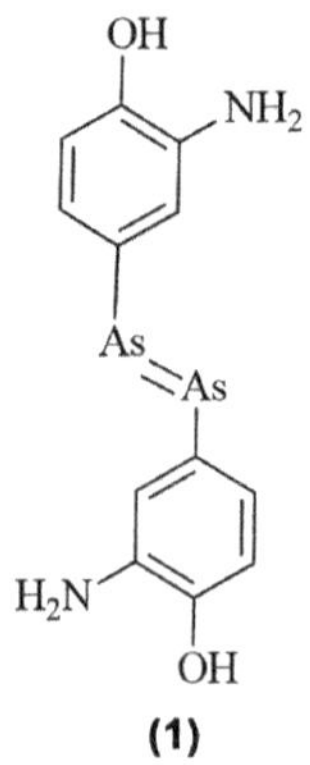

(1)

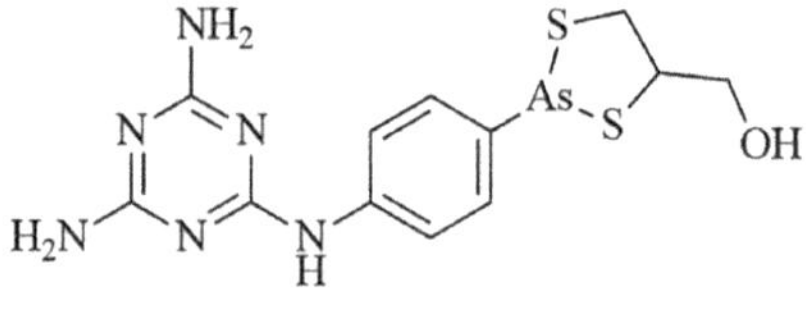

(2)

Figure 10.1 Chemical structures proposed for salvarsan (**1**) and melarsoprol (**2**).

10.2 History and Clinical Use of Antimony-based Drugs

The significant medicinal application of antimony in therapeutics can be traced back to the sixteenth century,[8] when Paracelsus used alcoholic solutions of nitrate and tartrate of antimony(III) to treat syphilis, rheumatism, fever, epilepsy and skin diseases.[9–11] Some authors have described that antimony had been utilized in cosmetics in early Egypt; however, this statement was shown to be false, based on a misreading of the ancient texts.[10]

The use of antimony in contemporary medicine commenced with the application of sodium and potassium antimony(III) tartrate (**3**) (tartar emetic) (Figure 10.2) against trypanosomes by Plimmer and Thompson in 1905, which were subsequently used in Africa for the treatment of human Trypanosomiasis for several decades.[11] Gaspar Vianna was the first to report, in 1912, the successful use of tartar emetic in the treatment of cutaneous leishmaniasis (CL).[11,12] The efficacy of tartar emetic drug in visceral

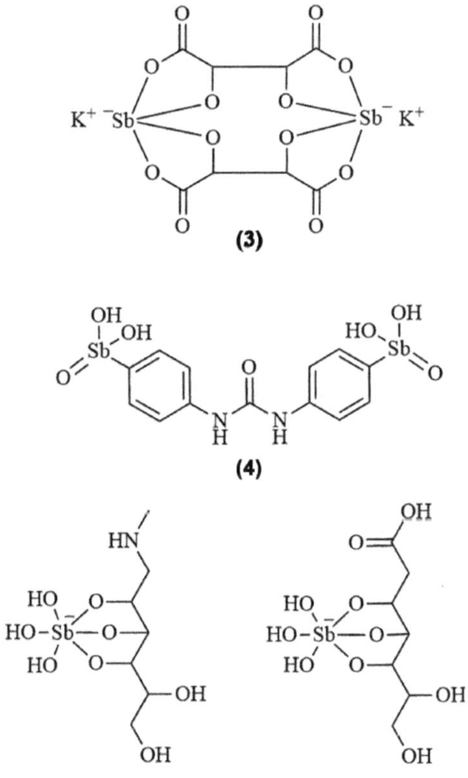

(3)

(4)

(5) **(6)**

Figure 10.2 Chemical structure of potassium antimony(III) tartrate (**3**) and of pentavalent antimonial urea stibamine (**4**), meglumine antimoniate (**5**) and stibogluconate (**6**).[21,33]

leishmaniasis (VL) was further confirmed by Rogers in India[13] and by Di Cristina and Caronia, in Sicily.[14] Due to the serious, sometimes fatal, side effects and the recurrence of the disease after treatment with tartar emetic drug, its prescription was banned later.[15,16] This led to the discovery and introduction of pentavalent antimonials due to the fact that antimony is less toxic in its pentavalent than its trivalent form.[17]

A major breakthrough in the discovery of pentavalent antimonial drugs against leishmaniasis was achieved in 1922 by Dr Upenderanath Brahmachari, who synthesized the organoantimony compound urea stibamine (**4**) (Figure 10.2), inspired by the idea that an antimonial having a constitution similar to atoxyl, which was found by Ehrlich to be effective against sleeping sickness, might prove to be useful in VL. In 1919, Brahmachari prepared *p*-stibanilic acid and various salts. In 1920, by heating stibanilic acid with urea, he produced the first organic antimonial and also demonstrated that it was an effective chemotherapeutic agent against VL.[18–20]

This discovery of urea stibamine saved millions of persons affected in areas with high VL epidemics in India. Brahmachari was nominated for Nobel Prize in 1929 for his tremendous efforts.[20] Due to its high toxicity, later, in the middle of the 20th century, urea stibamine was replaced by other less toxic and more effective pentavalent antimonials.

The pentavalent antimonials meglumine antimoniate (**5**) (Glucantime®) and sodium stibogluconate (**6**) (Pentostam®) (Figure 10.2), developed and first administered in the 1940s for the treatment of all types of leishmaniasis, are still the first-line drugs in most developing countries, and the most important and cost-effective drugs available to date.[22,23] They are effective in both VL and CL.[24] In the clinical setting, these inorganic Sb(v) compounds are less toxic than their Sb(III) analogues, though they are considered prodrugs, being reduced from Sb(v) to Sb(III), presumably by thiols inside the infected macrophages.[23,25] The trypanothione reductase (TR) is inhibited by Sb(III), disrupting the redox potential of the cell.[23,25] However, there are substantial risks; these pentavalent antimonials are still toxic drugs with severe adverse side effects.[26] Another major problem is the emergence of drug resistance that has reached epidemic proportions, particularly in some regions of India.[23,27] Although the mechanism responsible for the resistance of *Leishmania* to antimonial drugs is not fully understood, it often involves increased intracellular levels of thiols and the overexpression of ATP-binding cassette transporters, resulting in reduced intracellular concentration of the metal.[28] Liposomal amphotericin B (Ambisome®), Miltefosine and some other alternatives to Glucantime® and Pentostam®, have been used in many countries to treat leishmaniasis and showed excellent efficacy in some cases under trial. Besides severe side effects, Ambisome® is expensive and does not overcome the need for parenteral administration, while miltefosine, the only orally administered drug, remains in the bloodstream for up to five months after administration and has problems with teratogenicity.[29–31] Both drugs also present risks of resistance.[32] In this context, there is a great demand for new alternatives, safe and effective drugs that do not exhibit cross-resistance with conventional antimonial drugs or their derivatives. In this context, a considerable number of novel organometallic compounds have been evaluated for their anti-leishmanial potential in recent decades.

10.3 Organoantimonial Compounds

Organoantimony(v) and organoantimony(III) have gained extra attention due to the fact that antimony has been proven to have preeminent leishmanicidal activity for decades. Organoantimony(v) and organoantimony(III) carboxylates were extensively investigated in the literature, due to the versatile characteristics of the carboxylate group to act as unidentate, bidentate or bridging ligands.[34–37] The ligands studied include carboxylic acids, substituted benzoic acids, ferrocenyl benzoic acids and nitrogen donor heterocyclic ligands (2,2′-bipyridine and 1,10-phenanthroline). Regarding the aryl group, mainly phenyl and tolyl groups were investigated. Single-crystal X-ray diffraction analysis revealed that arylantimony(v) dicarboxylate complexes generally adopt five-coordinated trigonal bypiramidal geometries, binding with three aryl groups in the equatorial position and two carboxylic acid ligands (L) in the axial sites.

In one of the studies, carried out by Ali *et al.*,[38] an extensive range of well-characterized organoantimony(v) dicarboxylate complexes of the type [$R_3Sb(L)_2$] (7) (Figure 10.3), where R varies, being phenyl, *ortho*-, *meta*- or *para*-tolyl, or benzyl and L represents substituted benzoates, with some compounds incorporating acetate or cinnamate ligands, were evaluated for

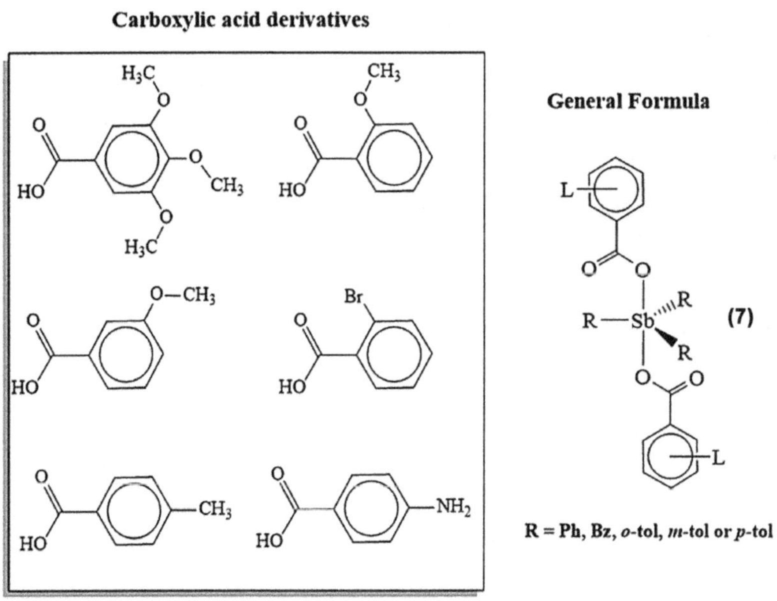

Figure 10.3 Chemical structures of some carboxylic acid derivative ligands and general formula for organoantimony(v) dicarboxylate complexes (7).

in vitro activity against the promastigote and intracellular amastigote forms of *Leishmania major*. Among the 26 compounds evaluated, 11 compounds, predominantly with *m*- or *p*-tolyl organometal moieties, showed high anti-leishmanial activity at concentrations lower than that of the reference drug, amphotericin B, in the range of 0.048–1.6 and 0.5–3.5 μM against *L. major* promastigotes and amastigotes, respectively, while remaining non-toxic towards the mammalian cells at concentrations below 25 μM. Complexes with the *o*-tolyl moiety were found to be comparatively less effective than *p*- and *m*-derivatives (Figure 10.4). It was also observed that compounds with phenyl- or benzyl-organometal moieties were generally more toxic and less selective. The substituted benzoic acids in free forms displayed negligible activity while organometal moieties were active both in promastigote and amastigote forms of *Leishmania* and in human skin fibroblast cells, evidencing that the organometal is responsible for the anti-leishmanial activity of the complexes. On the other hand, there was also no clear pattern with respect to which features of the carboxylate ligands make them more selective. As these organoantimony(v) dicarboxylates were effective against both promastigotes and amastigotes, it is suggested that mechanism of action of these complexes may differ from that proposed for standard inorganic antimony(v) drugs, because Glucantime and Pentostan® are primarily toxic to amastigotes.[39]

In another work, Islam *et al.*[40] synthesized similar $[R_3Sb(L)_2]$ (**7**) type of new organoantimony(v) dicarboxylate complexes (**8** and **9**) (Figure 10.5) through reaction of Ph_3SbCl_2 with acetylsalicylic acid or 3-acetoxybenzoic acid. These complexes were evaluated for *in vitro* activities against the promastigote form of *Leishmania* strains from two different New World

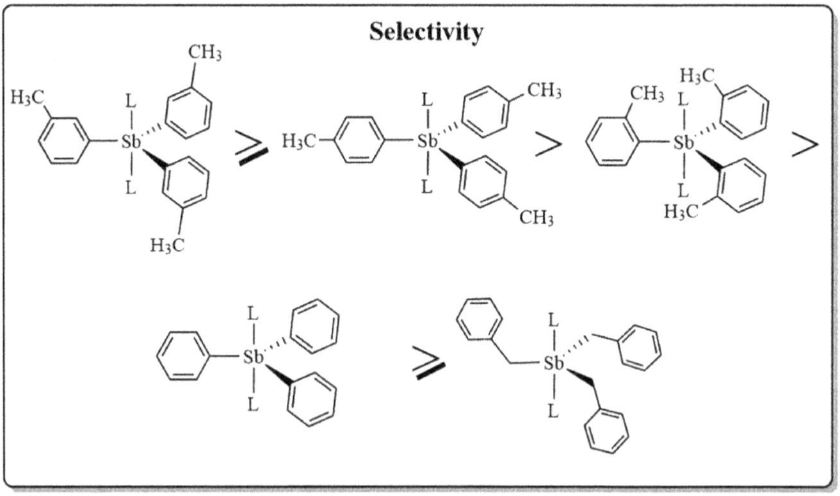

Figure 10.4 Selectivity order tendency for organoantimony(v) dicarboxylates, depending on the organometal moiety (considering their activity against *Leishmania* promastigotes and toxicity towards macrophages).

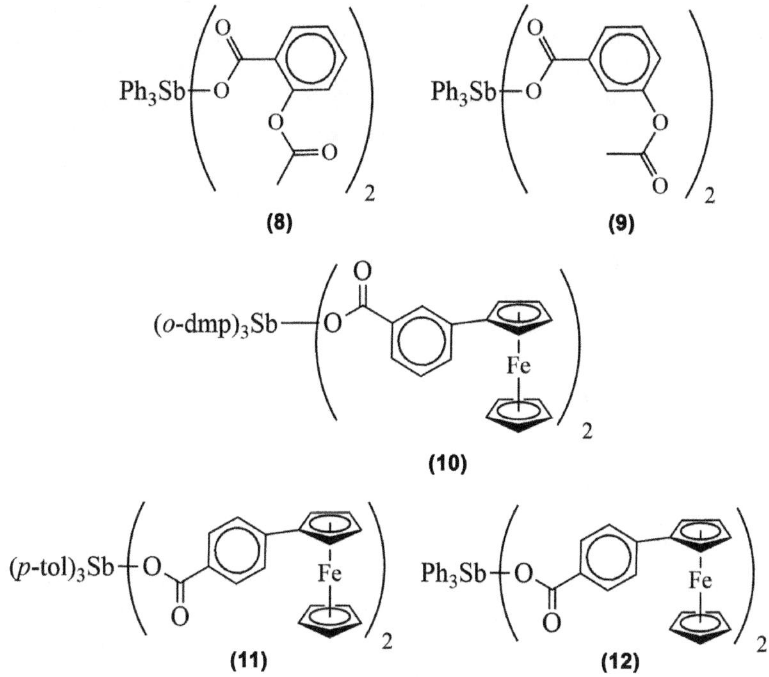

Figure 10.5 Chemical structures of triphenylatimony(v) complexes with acetylsalicylic acid (**8**) and 3-acetoxybenzoic acid (**9**) synthesized by Islam *et al.* and organoantimony(v) ferrocenyl (**10**), (**11**) and (**12**) synthesized by Rauf *et al.*[40,41] Dmp, dimethylphenyl.

Leishmania species: *Leishmania infantum*, the etiological agent for visceral leishmaniasis and *Leishmania amazonensis*, related to the cutaneous form. Both complexes were more active against *L. infantum* [50% inhibitory concentration (IC_{50}) 13.3 and 12.2 μM, respectively] and *L. amazonensis* (IC_{50} 30.7 and 8.9 μM, respectively) when compared with tartar emetic (IC_{50} 100 and 83 μM, respectively). Interestingly, the substituted benzoic acids showed no intrinsic anti-leishmanial activity, contrary to the organometal moiety, Ph_3SbCl_2 (IC_{50} 13.2 and 9.3 μM against *L. infantum* and *L. amazonensis*, respectively) that were at least as effective as their respective complexes. This data further supported the model that the organometallic moiety is responsible for the anti-leishmanial activity of the complexes and that the complexes must dissociate in order to exert their activity. It was observed that the complexes synthesized from 3-acetoxybenzoic acid also tended to be slightly more active than that synthesized from acetylsalicylic acid. When evaluated towards murine macrophages, both the complexes were more selective (SI range 2.0–4.5) than the organometal moiety (1.7–2.4), while the free ligands were non-toxic even at concentrations greater than 500 μM. This data indicates that substituted benzoic acid ligands may decrease the toxic effect of the organometal moiety upon complexation.

More recently, Rauf *et al.*[41] reported another interesting class of organoantimony(v) dicarboxylate complexes of similar [$R_3Sb(L)_2$] type (7), where *R* represents phenyl, 2,5-dimethylphenyl or *p*- or *o*-tolyl, and *L* represents *p*- or *o*-ferrocenyl benzoic acid (Figure 10.5). These ferrocenyl pentavalent antimonials were assessed for their leishmanicidal potential against *Leishmania tropica*, and also for their biocompatibility and membrane permeabilizing effect. Moreover, mechanistic studies using voltammetric analysis and computational docking were carried out to investigate the possible interaction of ferrocenyl antimonials with DNA and the leishmanial trypanothione reductase enzyme, respectively. The results of these studies indicated that the pentavalent antimonials were much more efficacious (IC_{50} range 0.0005–1.19 $\mu g\, ml^{-1}$) as compared with Glucantime® when screened for their activity in promastigote or axenic amastigote forms. Cytotoxicity assays in human macrophages isolated from human blood showed that these antimonials were not cytotoxic even at concentrations as high as 100 $\mu g\, ml^{-1}$ and exhibited selectivity indexes greater than 100. Membrane permeability assays conducted for ferrocenyl antimonials (**10**) and (**11**) established their permeabilising effect. Evidence was obtained for the interaction of complexes (**10**) and (**12**) with leishmanial DNA, in contrast to Glucantime. Furthermore, docking interactions highlighted the potential interactive binding of these antimonials within the trypanothione reductase active site, with van der Waals interactions contributing significantly to the process. Hence, considering their high anti-leishmanial activity, selectivity and potential interaction with multiple target sites in *Leishmania*, organoantimony(v) ferrocenyl dicarboxylates can be considered as promising anti-leishmanial drug candidates. Structure–activity relationships based on the results of anti-leishmanial assay of these complexes indicated that methyl from the tolyl group at the *para* position of antimony analogues promoted a higher anti-leishmanial activity, while the methyl at the *ortho* position has not shown significant anti-leishmanial activity, in agreement with the results described by Ali *et al.*[38] The results of both studies are also consistent with the idea that the nature of the R group attached to Sb(v) modulates the *in vitro* anti-leishmanial activity.

Meanwhile, another approach to enhance the anti-leishmanial activity of the organoantimony(v) complex was based on the complexation of the organometal with a bioactive ligand. de Oliveira *et al.*[42] synthesized and Rocha *et al.*[43] tested a new organoantimony(v) complex of lapachol (**13**) (Figure 10.6) against *L. amazonensis* intramacrophagic amastigotes. Lapachol (Lp), [2-hydroxy-3-(3′-methyl-2-butenyl)-1,4-napthoquinone], is a natural compound, extracted from the core of Bignonianceae trees, with medicinal properties, including antitumor, antibacterial, antimalarial, trypanocidal, leishmanicidal, anti-inflammatory and analgesic actions. The biological activity of the Lp molecule is generally attributed to the generation of reactive oxygen species (ROS) induced by the bioreduction of its quinonoid nucleus through specific enzymes and oxygen.[44–46] Lapachol, isolapachol and some of their derivatives showed activity against *Leishmania*

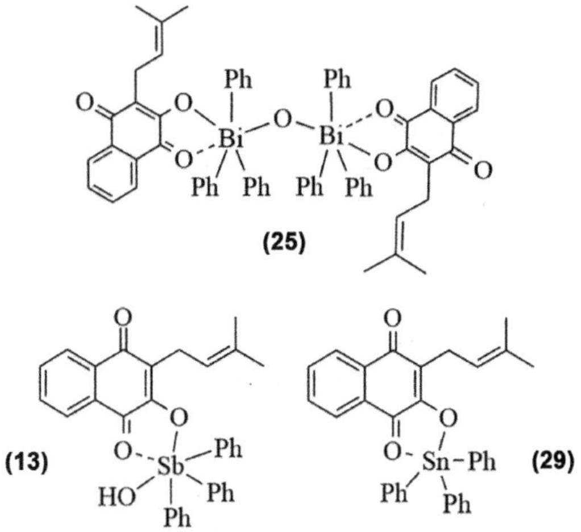

Figure 10.6 Chemical structures of triphenylantimony(v) (**13**), -bismuth(v) (**25**) and –tin(IV) (**29**) complexes with lapachol determined by Oliveira *et al.*[42]

brazilensis and *L. amazonensis* both *in vitro* and *in vivo*, respectively.[47] The organoantimony(v) lapachol complex [(Lp)(Ph_3Sb)OH] (IC_{50} approximately 30 μM) was found to be as effective as non-complexed organometal Ph_3SbCl_2 (IC_{50} approximately 27 μM) against *L. amazonensis* amastigotes, but more active than free ligand lapachol (Lp) (IC_{50} approximately 64 μM).[43] Although the anti-leishmanial activity of the complex could be attributed solely to the organoantimony(v) moiety, a contribution of Lp cannot be discarded. Interestingly, the organoantimony(v) complex of lapachol was threefold less cytotoxic against murine peritoneal macrophages than the organometal moiety, resulting in higher SI (approximately 7.1).

Following a similar approach, Lizarazo-Jaimes *et al.*[48] choose bioactive nitrogen donor heterocyclic ligands [2,2-bipyridine (bipy) and 1,10-phenanthroline (phen)] but differently, synthesized organoantimony(III) complexes through reaction with $PhSbCl_2$. The resulting complexes were characterized as [PhSb(bipy)Cl_2] (**14**) and [PhSb(phen)Cl_2]CH_3COOH (**15**) (Figure 10.7). The single-crystal X-ray diffraction data obtained for [PhSb(phen)Cl_2]CH_3COOH (**15**) showed the complex adopted a distorted square pyramid geometry with a five-coordinated Sb centre. The geometry adopted by this organoantimony(III) complex is different from those adopted by the organoantimony(v) complexes, where it is generally distorted bipyramidal trigonal geometry. These organoantimony(III) complexes showed high anti-leishmanial activities against *L. amazonensis* and *L. infantum* promastigotes, which could be attributed to both the ligand and the organometal moiety. Nevertheless, these complexes still showed relatively low SIs (in the range of 0.95–3.5). An important contribution of this work was the demonstration that the organoantimony(III) complexes and the non-complexed organometal were equally effective in Sb(III)-sensitive and -resistant *Leishmania* strains. These resistant

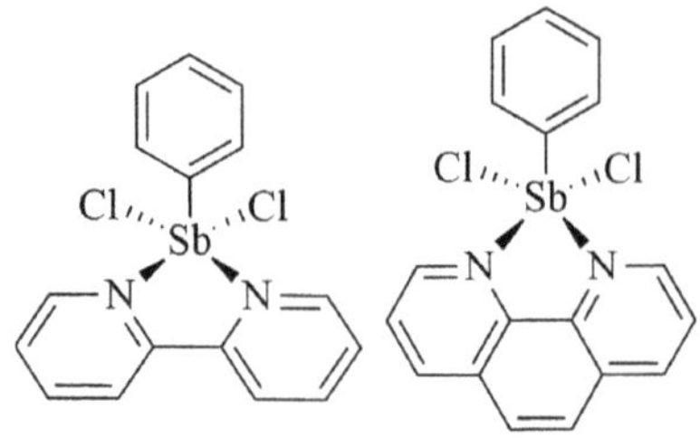

(14) **(15)**

Figure 10.7 Chemical structures of organoantimony(III) with bipyridine (**14**) and 1,10-phenanthroline (**15**) determined by Lizarazo-Jaimes *et al.*[48]

mutants were obtained through stepwise selection in the presence of increasing concentrations of potassium antimony(III) tartrate up to 2.7 mM Sb.[49] The lack of cross-resistance between organoantimony(III) and inorganic antimony(III) supports distinct mechanisms of action or, at least, distinct transport pathways across the parasite membranes. Interestingly, organoantimony(V) complexes also did not show cross-resistance with inorganic antimony(III) (Demicheli *et al.* unpublished results). This means that both organoantimony(III) and organoantimony(V) complexes are good candidate drugs to treat leishmaniasis resistant to conventional inorganic antimonial drugs.

In summary, several promising novel organoantimony compounds have been identified, mainly through *in vitro* assays of anti-leishmanial activity and cytotoxicity. It was found that *p*- and *m*-tolyl analogues were often more effective and selective than phenyl derivatives. A very interesting feature of these compounds is the lack of cross-resistance with inorganic antimony(III). In the future, efforts should be devoted to confirm their efficacy through *in vivo* assays in experimental models of leishmaniasis. There is some evidence that the mechanism of action of this class of compounds differs from that of conventional inorganic antimonial drugs. However, the mode of action of these compounds still needs to be elucidated.

10.4 Organobismuth Compounds

Bismuth(III) carboxylate derivatives, such as bismuth subsalicylate and colloidal bismuth subcitrate, are commonly used to treat a variety of gastrointestinal disorders, including gastric diseases caused by *Helicobacter pylori*.[50]

Bismuth is positioned below antimony in the periodic table and therefore displays many similar chemical and electronic properties. Unexpectedly for a heavy metal, bismuth and its compounds are considered to be of low toxicity. Therefore, it has been postulated that the toxic effects associated with antimonial drugs might be minimized by replacing antimony(V), which has the ability to get reduced to antimony(III), by less toxic bismuth(III).

Andrews *et al.*[51] evaluated four bis-carboxylate phenylbismuth(III) complexes (**16** to **19**) (Figure 10.8), against *L. major* promastigotes and human primary fibroblast cells. The complexes were synthesized by the reaction of $BiPh_3$ with carboxylate derivative ligands (HL) *o*-nitrobenzoic acid,[52] *o*-methoxybenzoic acid, *m*-methoxybenzoic acid and 5-[(*R*/*S*)-2,3-dihydroxypropyl carbamoyl]-2-pyridine carboxylic acid, with a 1 : 2 molar ratio. Solid-state structures of two complexes (**16** and **17**) were determined by single-crystal X-ray diffraction and were shown to adopt a six-coordinated distorted pentagonal pyramidal geometry. The substituted benzoic acids were essentially considered non-toxic towards the *L. major* parasites. The corresponding organobismuth(III) complexes showed near-quantitative anti-leishmanial activity at very low concentrations (1.9–30.8 $\mu g\ ml^{-1}$) but were still less active than amphotericin B (IC_{50} 0.46 $\mu g\ ml^{-1}$). When evaluated in human fibroblast cells, the substituted benzoic acids showed poor toxicity at concentration as high as 500 $\mu g\ ml^{-1}$, but their respective complexes were toxic at concentration below 10 $\mu g\ ml^{-1}$. From the results of this study, it was concluded that it is the presence of Bi(III) which accounts for the observed increase in toxicity. This work also indicates that these Bi(III) complexes are unsuitable anti-leishmanial drug candidates due to their low selectivity indexes.

Ong *et al.*[53] prepared a different series of organobismuth complexes (**20**) through reaction of pentavalent triphenylbismuth, in the presence of H_2O_2, with *o*-methoxybenzoic acid, *p*-nitrobenzoic acid and new benzoic acid derivatives, some of which are classified as non-steroidal anti-inflammatory

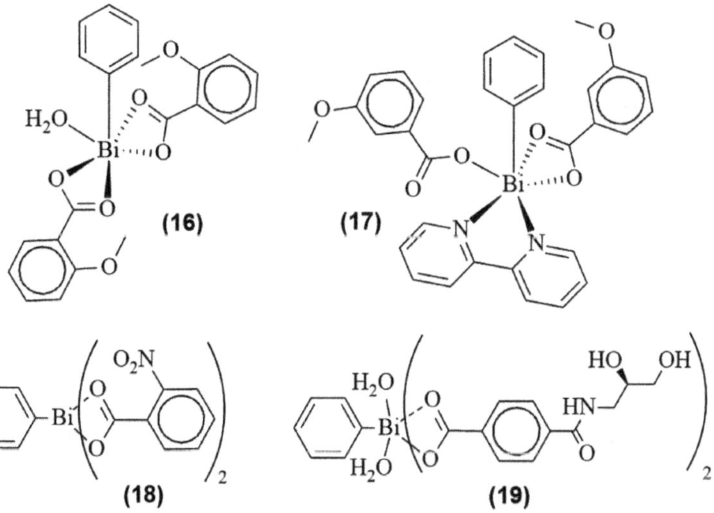

Figure 10.8 Chemical structures of organobismuth(III) complexes with carboxylate derivative ligands *o*-methoxybenzoic acid (**16**), *m*-methoxybenzoic acid (**17**), *o*-nitrobenzoic acid (**18**) and 5-[(*R*/*S*)-2,3-dihydroxypropyl carbamoyl]-2-pyridine carboxylic acid (**19**), respectively.

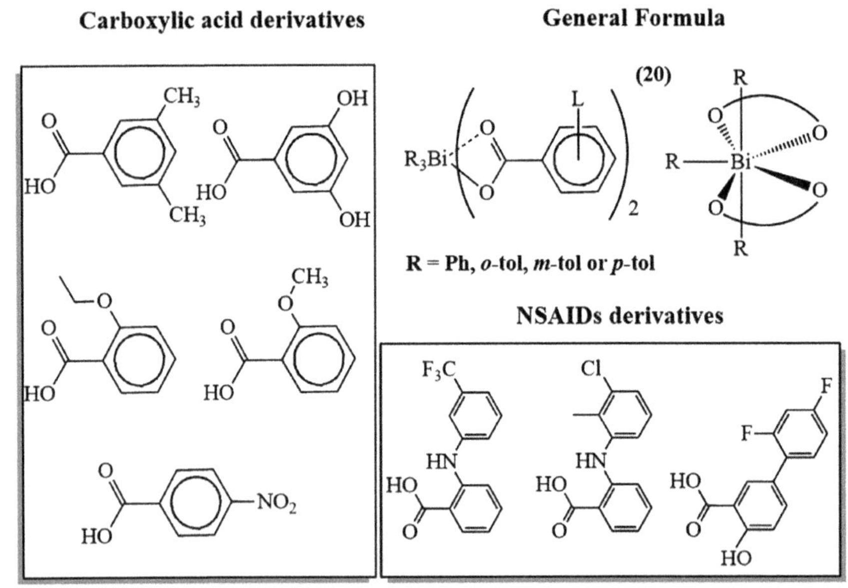

Figure 10.9 Chemical structures of some carboxylic acid ligands, flufenamic acid, tolfenamic acid and difunisal, respectively. Representation of the general formula of organobismuth(v) dicarboxylate complexes (**20**).

drugs (NSAIDs) (Figure 10.9). Single-crystal X-ray diffraction analysis of eight of these complexes revealed that they adopt seven-coordinated typical pentagonal bipyramidal geometry, where the carboxylate groups are coordinated to the central metal as bidentate ligand. When evaluated against *L. major* promastigotes, the complexes were more active (IC_{50} range 0.7–2.5 μM) than the organometal precursor $BiPh_3$ (IC_{50} approximately 45.8 μM) and the free carboxylic acids, however, all of them were non-selectively toxic towards fibroblasts with SI in the range of 1.6–4.2. The stability of one of the complexes was confirmed by ^{1}H-NMR analysis in DMSO and water. On the other hand, it underwent a slow decomposition (with a half-life greater than 6 hours) in culture medium undergoing reduction to give $BiPh_3$ and the free carboxylic acid. It could be concluded that these triphenylbismuth(v) dicarboxylates are stable for long enough to have a significant *in vitro* leishmanicidal activity.

In a subsequent report, Ong *et al.*[54] synthesized novel tris-tolyl bismuth(v) dicarboxylates, of the type $[R_3Bi(L)_2]$ (**20**), where $R = o$-, m- or p-tolyl and $L =$ substituted benzoic acid (Figures 10.9 and 10.10). Single-crystal X-ray diffraction analysis showed seven-coordinated complexes adopting typical pentagonal bipyramidal geometry, where aryl ligands occupy the equatorial plane. Although stable in D_2O and DMSO, the complexes underwent rapid transformation in the culture medium (half-life approximately 1.2 hours), with glucose being a contributing factor in the reduction to $Bi(Tol)_3$. Despite their short lifetimes, many complexes showed significant activity towards

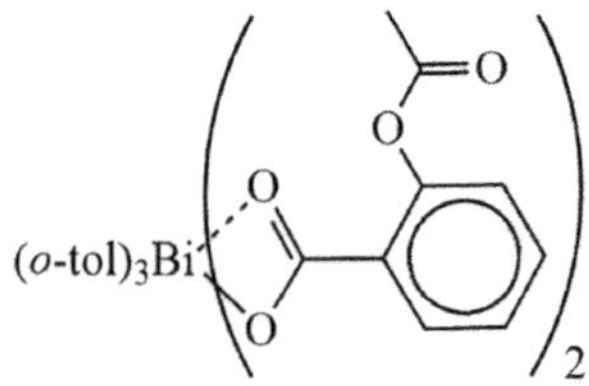

(21)

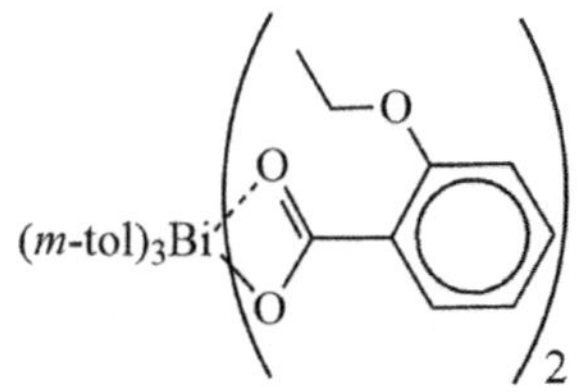

(22)

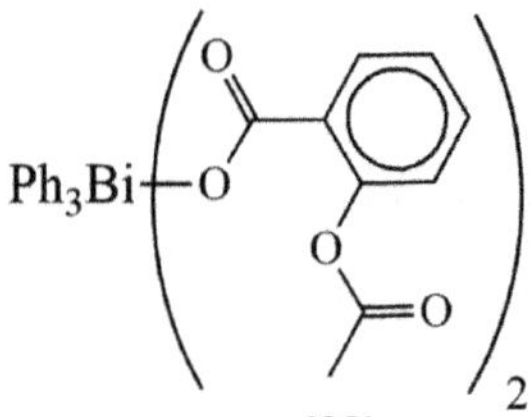

(23)

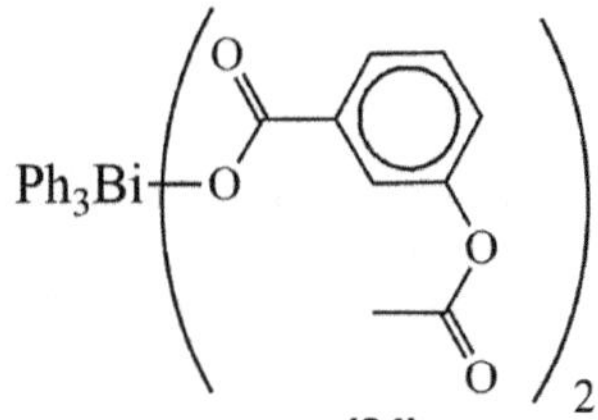

(24)

Figure 10.10 Chemical structures for tristolylbismuth(v) complexes (**21**) and (**22**) synthesized by Ong *et al.* and triphenylatimony(v) complexes with acetylsalicylic acid (**23**) and 3-acetoxybenzoic acid (**24**) synthesized by Islam *et al.*[40,54]

promastigotes. Evaluation of their toxicity against *L. major* promastigotes and human fibroblast cells revealed anti-leishmanial activity at low concentrations (IC_{50} approximately 0.57–1.77 μM) and acceptable SI (4.4–29.0). In contrast, the organometal (*m*-Tol)$_3$Bi precursor showed low toxicity towards both *Leishmania* (IC_{50} approximately 21 μM) and fibroblasts (IC_{50} approximately 460 μM) and the free ligands remained essentially non toxic. Surprisingly, non-complexed (*o*-Tol)$_3$Bi and (*p*-Tol)$_3$Bi compounds displayed significant anti-leishmanial activities (0.85 μM and 0.98 μM) and selectivity indexes (13.0 and 29.0). The results taken together indicated that the toxic effect on human cells appeared to be mainly based on the aryl group, while the anti promastigote activity was influenced by the nature of the carboxylate ligand. From the results presented in this and the previous[53] reports, it was suggested that the *ortho* and *meta* analogues may be more selective and promising for further development (Figure 10.11), while the *p*-Tol and the Ph series would be similar in terms of activity on both parasites and fibroblasts, and would not be as selective. When further tested against intracellular amastigotes, the tris-tolyl bismuth(v) dicarboxylates complexes were found to be less effective. Nevertheless, a significant reduction of the percentage of infected macrophages was observed after exposition to [Bi(*o*-Tol)$_3$($O_2CC_6H_4$(2-EtO))$_2$] (**21**) and [Bi(*m*-Tol)$_3$($O_2CC_6H_4$(2-OAc))$_2$] (**22**) at 1 μM.

Islam *et al.*[40] have reported the synthesis of triphenylbismuth(v) dicarboxylate complexes (**23** and **24**) through reaction of Ph_3BiCl_2 with acetylsalicylic acid or 3-acetoxybenzoic acid (Figure 10.10). A significant contribution of this study was the comparison of these complexes to their triphenylantimony(v) dicarboxylates counterparts, regarding their activity against *Leishmania* promastigotes and cytotoxicity towards murine

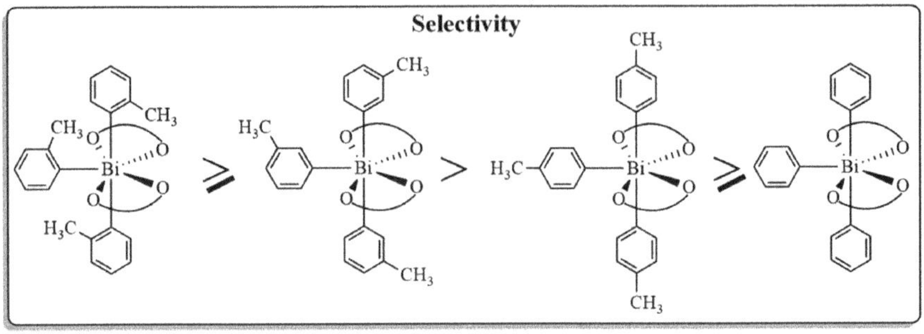

Figure 10.11 Selectivity order tendency for organobismuth(v) dicarboxylates, depending on the organometal moiety (considering their activity against *Leishmania* promastigotes and toxicity towards macrophages).

macrophages. Both the organobismuth(v) dicarboxylate complexes were found to be 1.5- to 3.6-fold more effective against *L. infantum* and *L. amazonensis* (IC_{50} 2.9–8.6 μM), when compared with their antimony(v) analogues. Moreover, bismuth(v) complexes were 9- to 15-fold more cytotoxic towards macrophages and so were considered to be less selective. Interestingly, the organobismuth(v) salt, Ph_3BiCO_3, was more active against *Leishmania* promastigotes (IC_{50} 1.1–2.7 μM) than the organobismuth(v) complexes, indicating that the organometallic salt in its free form is responsible for the anti-leishmanial activity of the complexes and that the complexes dissociate in order to exert their activity.

Oliveira *et al.*[42] synthesized and Rocha *et al.*[43] evaluated a new triphenylbismuth(v) complex of lapachol (**25**) against *L. amazonensis* intramacrophagic amastigotes (Figure 10.6).[42] The complex showed modest anti-leishmanial activity (IC_{50} approximately 21 μM), close to the cytotoxic concentration towards macrophages. Interestingly, the Ph_3SbCl_2 metal precursor was found to be more active (IC_{50} approximately 10 μM) and less cytotoxic (CC_{50}, compound concentration that kills 50% of viable cells in the host, approximately 49 μM) than the complex, indicating that it contributed significantly to the anti-leishmanial activity of the complex and that Lp might be responsible for the cytotoxicity of the complex.

Collectively, the results of these investigations indicated a general trend of anti-leishmanial activity where the organobismuth(v) complexes were found to be more active than organoantimony(v) ones. The current data also indicates that the toxic effect on human cells may depend on the aryl group, while the anti-promastigote activity would be influenced by the nature of carboxylate ligand. In the future, efforts should be devoted to confirming the efficacy of these complexes through *in vivo* assays in experimental models of leishmaniasis. The *in vivo* reduction of Bi(v) to Bi(III) seems to be important in the mode of action of these complexes, however, it is not clear whether the reduced species contribute or not to the biological activity of the complexes. Furthermore, the molecular targets of these compounds have not yet been identified.

10.5 Organotin(IV) Compounds

Organotin(IV) compounds are being extensively studied with special reference to their methods of synthesis, structural elucidation and biological activities.[55,56] Tributyltin and triphenyltin derivatives have been used as industrial and agricultural biocides, respectively, acting mainly as antifungal agents.[57] In this context, organotin(IV) compounds have been investigated as bactericides, fungicides, acaricides and wood preservatives.[58] During the last decades, interesting results regarding their antitumor[59] and anti-leishmanial activities have also been reported.

Raychaudhury *et al.*[60] screened the organometallic complex of triphenyltin(IV) with salicylanilide thiosemicarbazone (**26**) (Figure 10.12) against *L. donovani* both in *in vitro* and *in vivo* experimental models. Treatment of infected mice with organotin(IV) (10 mg kg^{-1}) promoted reduction of parasite burden in spleen by 87%, compared with 65% following Pentostam® (20 mg kg^{-1}). To understand its mechanism of action, a study with superoxide dismutase (SOD) was carried out. SOD is one of the key enzymes of the oxygen defense system[61] and is highly active during the infection of *Leishmania*.[62] The organotin(IV) was found to act through inhibition of superoxide dismutase and simultaneous increase of toxic oxygen metabolites.

According to the literature, organotin(IV) complexes which contain carboxylic acids as ligands are among the most effective biologically active compounds. It has been proposed that the formation of Sn–O bonds increases the hydrolysability of the organotin(IV) precursors,[63–65] allowing the attack of the hydrolysed $[R_2Sn(IV)]^{2+}$ moiety at the target cell, thereby enhancing the anti-leishmanial activity.[66] Khan *et al.*[66,67] synthesized organotin(IV) complexes with maleimido derivatives substituted with carboxylic acids (Figure 10.13). The complexes have general formula $[R_3Sn(L)_1]$ (**27**) or $[R_2Sn(L)_2]$ (**28**), where R = Me, Et, *n*-Pr, *n*-Bu, Ph or Bz and L = maleimido derivative. Solid state FTIR and ^{119m}Sn Mössbauer spectroscopy revealed bidentate coordination through carboxylic acid function. *In vitro* anti-leishmanial screenings have been conducted on the promastigote forms of *L. major*, *L. tropica*, *L. infantum*, *L. mexicana* and *L. donovani*. For some complexes, the IC_{50} values obtained (IC_{50} range 0.3–101 μg ml^{-1}) were lower

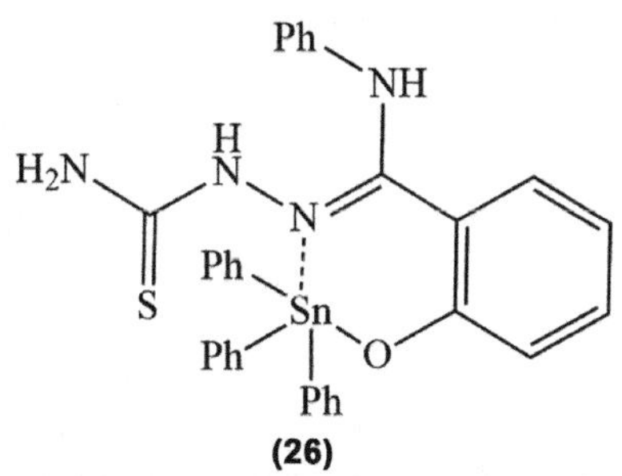

(26)

Figure 10.12 Chemical structure of organotin(IV) complex (**26**).

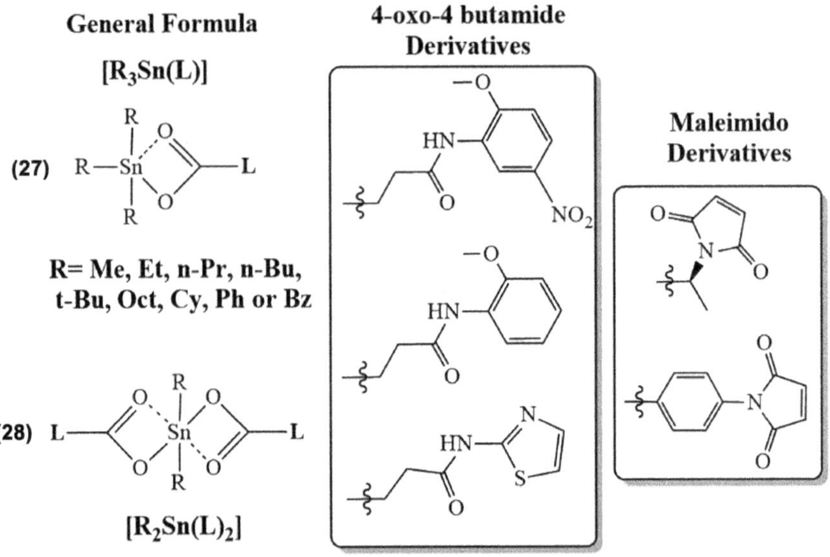

Figure 10.13 Chemical structures of 4-oxo-4 butamide and maleimido derivative ligands. Representation of the general formula of organotin(IV) carboxylate complexes (**27** and **28**).

than those for amphotericin B and pentamidine. The results also indicate that the nature and size of the R group attached to Sn(IV) affect the *in vitro* anti-leishmanial activity. The increasing size of the R groups, Bz > Ph > *n*-Bu > *n*-Pr > Et > Me, leads to increased lipophilicity of the corresponding moieties, resulting in an increase in their activity.[66]

Later, a group of organotin(IV) carboxylates, with 4-oxo-4-butanamide derivatives as ligands (Figure 10.13), was assayed by Ali and colleagues[68–71] for their anti-leishmanial activity. The complexes with general formula $[R_3Sn(L)_1]$ (**27**) or $[R_2Sn(L)_2]$ (**28**), where R = alkyl or aryl group and L = 4-oxo-4-butanamide derivative, showed significant activities and the ability to lead *Leishmania* parasites to apoptotic death. Cyclic voltammetry and molecular docking support the hypothesis that the investigated compounds interact with the *Leishmania* DNA. The compounds also showed both oxidation and reduction behaviour in the presence of the DNA strand, which may cause the degradation of the genetic material.

A triphenyltin(IV) complex using lapachol as ligand (**29**) (Figure 10.6) was also synthesized by Oliveira *et al.*[42] and evaluated by Rocha *et al.*[43] against *L. amazonensis* intramacrophagic amastigotes. The complex and Ph_3SnCl were equally active against *Leishmania* (IC_{50} approximately 0.29 μM and 0.26 μM, respectively) and exhibited close cytotoxic activities towards murine macrophages (CC_{50} approximately 2.7 μM and 1.9 μM, respectively). Interestingly, when compared with triphenylantimony(V)- and triphenylbismuth(V)-lapachol complexes, as described in previous sections, the organotin(IV)-lapachol was 70- to 100-fold more active and tended to be slightly more selective.

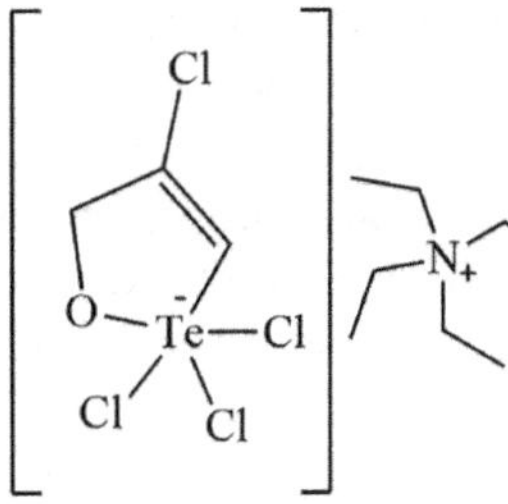

(30)

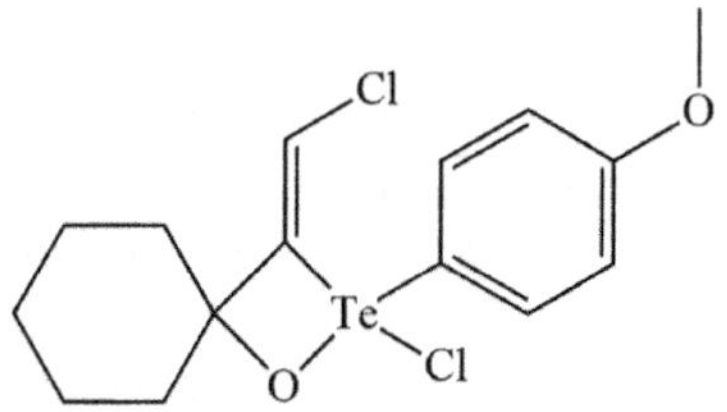

(31)

Figure 10.14 Chemical structures of organotelluranes(IV) complexes (**30**) and (**31**) proposed by Lima *et al.*[75] and Pimentel *et al.*,[76] respectively.

10.6 Organotellurium(IV) Compounds

During the last decade, organotellurium compounds have been studied as chemotherapeutic agents. Organotellurane compounds proved to be involved in the inhibition of proteases, with specificity towards cysteine proteases, like papain and cathepsin B.[72,73] Tellurium(IV)-based complexes were claimed to be low-toxicity anti-leishmanial agents.[74]

Lima *et al.*[75] synthesized the organotellurane (**30**) (Figure 10.14) by the reaction of tellurium tetrachloride with propargyl alcohol and triethylbenzylammonium chloride. Compound (**30**) was screened for *in vitro* and *in vivo* activity against *L. amazonensis*, showing anti-leishmanial activity in both cases. A detailed study revealed that compound (**30**) administered intralesionally produced a similar effect in the lesions of mice, as Glucantime®. According to a previous report, this complex caused inhibition of the extracellular form of cathepsin B, a proteolytic enzyme.[73] Based on this fact, it was proposed that (**30**) inhibits the parasite cysteine proteases, which is an important factor for *Leishmania* infection process and survival.

Another organotellurane(IV) (**31**) (Figure 10.14) was synthesized by Pimentel *et al.*[76] through reaction of *p*-methoxyphenyl tellurium with 1-ethynyl-1-cyclohexanol. The compound was screened against the amastigote form of *L. infantum* where it showed significant activity at low concentrations (IC_{50} approximately 0.5 μM) and displayed a tenfold lower cytotoxic concentration towards macrophages. A 99.6% reduction of parasite burden of hamsters infected with *L. infantum* was reported following multiple-dose treatment by the intraperitoneal route. Compound (**31**) also was found to inhibit cathepsin B.

10.7 Iron- and Ruthenium-based Organometallic Compounds

Studies conducted in the last decade have generated promising and potential organometal-based chemotherapeutic agents, mostly emerging from the structural modification of an existing drug, and all demonstrating the advantageous properties unique to organometallic chemistry.

Vale-Costa *et al.*[77] synthesized ferrocenyl primaquine derivatives and screened these against *L. infantum.* Primaquine (**32**) is an antimalarial that contains a quinolone heterocycle and presents low bioavailability due to its premature oxidative deamination (Figure 10.15). It has been reported that primaquine exhibits activity against VL, though it has no applicability in a clinical setting for this disease.[78] In order to prevent the metabolism of primaquine, the primaquine-derivative ligands were linked to an aliphatic amine fragment. The resulting ferrocenyl primaquine compounds demonstrated significant activity against *L. infantum*. The organometal complex (**33**) exhibited good *in vitro* activity against the intracellular amastigote form, while being less cytotoxic to mammalian macrophages than the reference anti-leishmanial sitamaquine and miltefosine drugs.

Iniguez *et al.*[79] synthesized organoruthenium(II) complexes with ketoconazole (**34**) (Figure 10.16) and tested for their anti-leishmanial activity. Ketoconazole is a well-established sterol biosynthesis inhibitor used primarily to treat fungal infections. The organometallic portion is occupied by a *p*-cymene ligand. The complexes have displayed activity and selectivity similar to free ketoconazole. It is suggested that there is a dual-target mechanism, which involves the sterol biosynthesis inhibitory action by the released ketoconazole in the cellular medium and the covalent binding of the remaining ruthenium-containing moiety to parasite DNA.

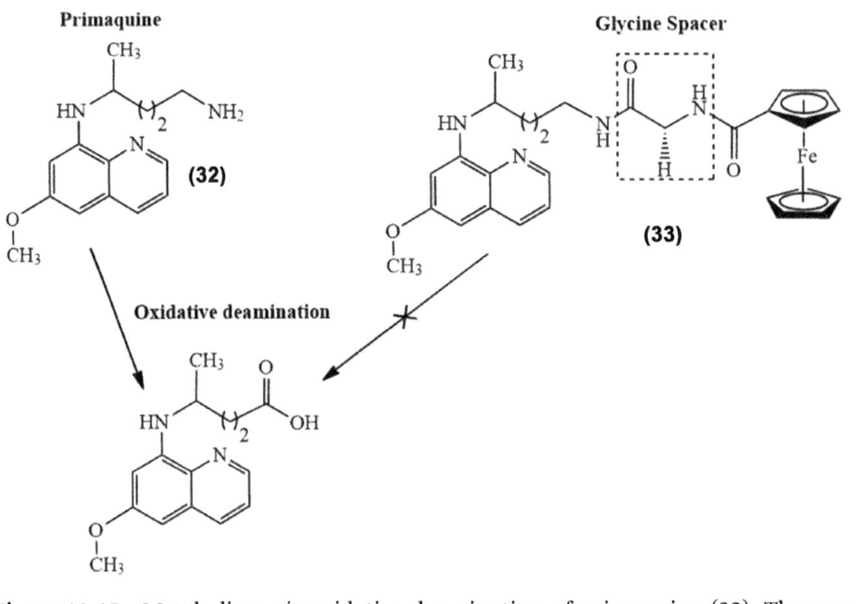

Figure 10.15 Metabolism *via* oxidative deamination of primaquine (**32**). The presence of a spacer on primaquine-derivative ligands avoids the deactivation of primaquine.

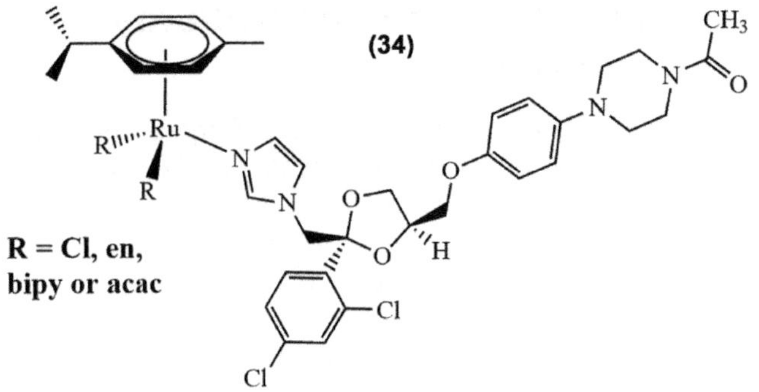

Figure 10.16 Representation of chemical structure for organoruthenium(II) complexes with ketoconazole (**34**) synthesized by Iniguez *et al.*[79]

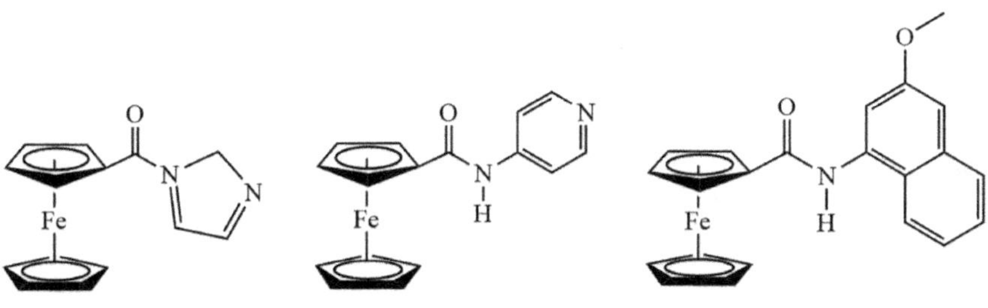

Figure 10.17 Chemical structures of *N*-heterocyclic ferrocenyl complexes (**35**), (**36**) and (**37**) proposed by Quintal *et al.*[80]

Since heterocyclic compounds play important roles in biology and pharmacology, Quintal *et al.*[80] have investigated ferrocene organometallics linked to various *N*-heterocyclic compounds (Figure 10.17) such as quinolone, benznidazole, imidazole and pyridine derivatives, where the coordination occurred *via* either amide or ester bridging. Among all the complexes, only three (**35**, **36** and **37**) showed a high selectivity based on their anti-leishmanial activity in *L. infantum* promastigotes and cytotoxicity towards the human Caucasian histiocytic lymphoma U-937 cell line, and were therefore further analysed for their activity on the *Leishmania* amastigote form. These compounds exhibited a high activity against intracellular amastigotes (IC_{50} approximately 5 μM) and higher SI (16–89), compared with the reference anti-leishmanial drug miltefosine (SI 6). The electrochemical data indicated that Fe(II) in the ferrocenyl moiety oxidized easily to Fe(III) at acidic conditions, which might be the reason for the increased anti-leishmanial activity of the complexes, possibly due to the production of the toxic ferrocenium ion. It was also suggested that the production of ferrocenium inside the host cell may be facilitated by the intracellular conditions.

10.8 Palladium, Rhodium and Iridium-based Organometallic Compounds

Mbongo *et al.*[81] have evaluated another series of organometallic complexes based on rhodium and iridium (Figure 10.18) for *in vitro* activity against *L. donovani*. The complexes have as the general formula $[M_2(R)_2(L)]X_2$ where M = rhodium(I) or iridium(I); R = 1,5-cyclooctadiene (**38**), 1,3-1,5-cyclooctatetraene (**39**) or CO; L = pentamidine and X = tetraphenylborate, NO^{3-}, ethylfumarate, alizarin red, mordant orange or tetraiodophenolphthalein. The authors have suggested that the nature of the salts could interfere with transport of the active agent through the parasite membrane, hence significantly affecting the anti-leishmanial activity. Some of the complexes exhibited better selectivity indices than pentamidine.

Organopalladium compounds are often used as a catalyst for several reactions in chemistry. In another context, Franco *et al.*[82] synthesized organopalladium derivatives based on two types of imine ligands (Figure 10.19)

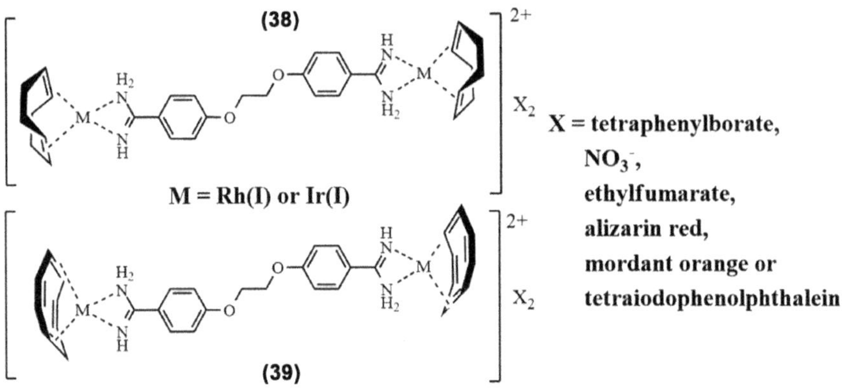

Figure 10.18 Chemical structures of pentamidine organo-rhodium(I) and -iridium(I) complexes (**38** and **39**) proposed by Mbongo *et al.*[81]

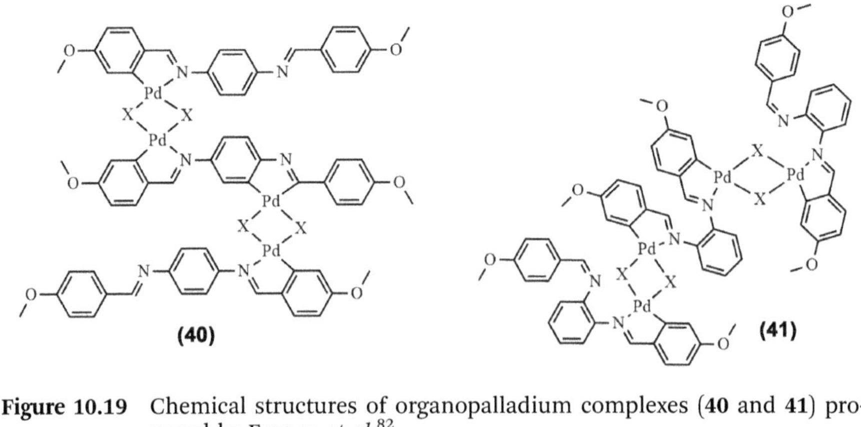

Figure 10.19 Chemical structures of organopalladium complexes (**40** and **41**) proposed by Franco *et al.*[82]

and tested them against the promastigote and amastigote forms of *L. amazonensis*. The complexes have the general formula $[Pd_4(LA)n(X)_4]$ (**40**) or $[Pd_4(LB)n(X)_4]$ (**41**), where LA = *N*,*N*′-bis[(1*E*)-(4-methoxyphenyl)methylene]-benzene-1,4-diamine, LB = *N*,*N*′-bis[(1*E*)-(4-methoxyphenyl)methylene]benzene-1,2-diamine and X = Cl or SCN. The ligands' activities were found to be improved by the coordination of Pd(II). Complexes with thiocyanate were the most active (IC_{50} range 2.2–20.8 $\mu g\,ml^{-1}$) and showed better SI (>1), even higher than for the reference drug pentamidine.

10.9 Conclusion

During the last two decades, several new antimony-, bismuth-, tin-, tellurium-, palladium-, rhodium-, iridium-, ruthenium- and iron-based organometallic compounds have been synthesized and evaluated for their activity in experimental models of leishmaniasis. Most of these complexes showed high *in vitro* anti-leishmanial activities in the micromolar range that could be attributed mainly to the organometal moiety. Some of these complexes also exhibited relatively high toxicity against the mammalian host cell. The selectivity index was found to depend on the nature of the metal, the aryl group and the labile ligand. So far, only organotellurane(IV) and organotin(IV) have been tested *in vivo*, showing promising results. The development of new organometallic drugs still relies on future progress regarding the understanding of the metabolism, cellular processing and mechanism of toxicity of these complexes and the organometal moiety and the identification of the pharmacological target(s).

Acknowledgements

The authors are grateful to the Brazilian agency CNPq, FAPEMIG, CAPES for financial support.

References

1. A. Salzer, *Pure Appl. Chem.*, 1999, **71**, 1557.
2. J. F. Hartwig, *Organotransition Metal Chemistry, from Bonding to Catalysis*, University Science Books, New York, 2010, ISBN 978-1-891389-53-5.
3. R. W. Brown and C. J. T. Hyland, *Med. Chem. Commun.*, 2015, **6**, 1230.
4. C. S. Allardyce and P. J. Dyson, in *Bioorganometallic Chemistry (Topics in Organometallic Chemistry)*, ed. G. Simonneaux, Springer-Verlag, Berlin, Heidelberg, 2006, vol. 17, p. 177.
5. V. F. Vecki and M. Ottinger, *Cal. State J. Med.*, 1921, **19**, 438.
6. C. Chesterman, *Trans. Royal Soc. Tropical Med. Hyg.*, 1979, **73**, 597.
7. H. Köpf and P. Köpf-Maier, *Angew. Chem.*, 1979, **91**, 509.
8. S. Yan, L. Jin and H. Sun, *Metallotherapeutic Drugs and Metal-Based Diagnostic Agents*, ed. M. Gielen and E. R. T. Tiekink, JohnWiley & Sons, Ltd., Chichester, 2005, p. 441.
9. J. Duffin and P. René, *J. Hist. Med. Allied Sci.*, 1991, **46**, 440.

10. J. W. Estes, *The Medical Skills of Ancient Egypt*, Science History Publications, Canton, MA, 1989.
11. G. Vianna, *Tratamento da leishmaniose tegumentar por injecçoes intravenosas de tartaro emetic*, In: 7 Congresso Brasileiro de Medicina Tropical de Sao Paulo, 1912, vol. 4, p. 426.
12. R. C. Palenik, K. A. Abboud and G. J. Palenik, *Inorg. Chim. Acta*, 2005, **358**, 1034.
13. G. C. Cook, *Notes Rec. R. Soc.*, 2006, **60**, 171.
14. G. D. Cristina and G. Caronia, *Pathologica*, 1915, **7**, 82.
15. U. N. Brahmachari, *Indian J. Med. Res.*, 1989, **89**, 393.
16. H. E. Shortt, *Trans. Royal Soc. Tropical Med. Hyg.*, 1945, **39**, 13.
17. S. Sundar and J. Chakrvarty, *Int. J. Env. Res. Public Health*, 2010, **7**, 4267.
18. P. D. Marsden, *Rev. Soc. Bras. Med. Trop.*, 1986, **19**, 115.
19. U. N. Brahmachari, *Indian J. Med. Res.*, 1922, **10**, 492.
20. S. Singh and R. Sivakumar, *J. Inf. Chemother.*, 2004, **10**, 307.
21. S. B. Mahato, B. C. Pal, S. Khan, A. K. Ghosh, K. Mukerji and D. K. Ghosh, *Biochem. Med. Metab. Biol.*, 1987, **38**, 47.
22. H. Veeken, K. Ritmeijer, J. Seaman and R. A. Davidson, *Trop. Med. Int. Health*, 2000, **5**, 312.
23. F. Frézard, C. Demicheli and R. R. Ribeiro, *Molecules*, 2009, **14**, 2317.
24. S. Singh and R. Sivakumar, *J. Infect. Chemother.*, 2004, **10**, 307.
25. A. K. Haldar, P. Sen and S. Roy, *Mol. Biol. Int.*, 2011, **2011**, 571242.
26. S. Coolin, R. Davidson, K. Ritmeijer, K. Keus, Y. Melaku, S. Kipngetich and C. Davies, *Clin. Infect. Dis.*, 2004, **38**, 612.
27. S. Sundar, D. K. More, M. K. Singh, V. P. Singh, S. Sharma, A. Makharia, P. C. K. Kumar and H. W. Murray, *Clin. Infect. Dis.*, 2000, **31**, 1104.
28. F. Frézard, R. Monte-Neto and P. G. Reis, *Biophys. Rev.*, 2014, **6**, 119.
29. P. K. Sinha, J. G. Van, K. Pandey, N. Kumar, N. Verma, R. Mahajan, P. Kumar, R. Kumar, P. Das and G. Mitra, *Clin. Inf. Dis.*, 2011, **53**, e91–e98.
30. J. Berman, *Expert Opin. Pharmacother.*, 2005, **6**, 1381.
31. T. Jha, S. Sundar, C. P. Thakur, J. M. Felton, A. J. Sabin and J. Horton, *Am. J. Trop. Med. Hyg.*, 2005, **73**, 1005.
32. T. P. C. Dorlo, M. Balasegaram, J. H. Beijnen and P. J. J. de Vries, *J. Antimicrob. Chemother.*, 2012, **67**, 2576.
33. F. Frézard, P. S. Martins, M. Barbosa, A. Pimenta, W. A. Ferreira, J. E. de Melo, J. B. Mangrum and C. Demicheli, *J. Inorg. Biochem.*, 2008, **102**, 656.
34. V. V. Sharutin, O. K. Sharutina, A. P. Pakusina, T. P. Platonova, G. K. Fukin and L. N. Zakharov, *Russ. J. Coord. Chem.*, 2001, **27**, 368.
35. H. Barucki, S. J. Coles, J. F. Costello, T. Gelbrich and M. B. Hursthouse, *J. Chem. Soc. Dalton Trans.*, 2000, 2319.
36. V. V. Sharutin, O. K. Sharutina, E. A. Bonsae, A. P. Pakusina, Y. V. Gatilov, N. Y. Adonin and V. F. Starichenko, *Russ. J. Coord. Chem.*, 2000, **28**, 333.
37. N. C. Kasuga, K. Onodera, S. Nakano, K. Hayashi and K. Nomiya, *J. Inorg. Biochem.*, 2006, **100**, 1176.

38. M. I. Ali, M. K. Rauf, A. Badshah, I. Kumar, C. M. Forsyth, P. C. Junk, L. Kedzierski and P. C. Andrews, *Dalton Trans.*, 2013, **42**, 16733.
39. H. Barucki, Ashutosh, S. Sundar and N. Goyal, *J. Med. Microbiol.*, 2007, **56**, 143.
40. A. Islam, J. G. da Silva, F. M. Berbet, S. M. da Silva, B. L. Rodrigues, H. Beraldo, M. N. Melo, F. Frézard and C. Demicheli, *Molecule*, 2014, **19**, 6009.
41. M. K. Rauf, U. Shaheen, F. Asghar, A. Badshah, A. Nadhman, S. Azam, M. I. Ali, G. Shahnaz and M. Yasinzai, *Arch. Pharm. Chem. Life Sci.*, 2016, **349**, 50.
42. L. G. de Oliveira, M. M. Silva, F. C. S. de Paula, E. C. Pereira-Maia, C. L. Donnici, C. A. de Simone, F. Frézard, E. N. da, S. Júnior and C. Demicheli, *Molecules*, 2011, **16**, 10314.
43. M. N. Rocha, P. M. Nogueira, C. Demicheli, L. G. de Oliveira, M. M. da Silva, F. Frézard, M. N. Melo and R. P. Soares, *Bioinorg. Chem. Appl.*, 2013, **2013**, Article ID 961783.
44. J. Tonholo, L. R. Freitas, F. C. De Abreu, D. C. Azevedo, C. L. Zani, A. B. de Oliveira and M. O. F. Goulart, *J. Braz. Chem. Soc.*, 1998, **9**, 163.
45. J. Benites, J. A. Valderrama, F. Rivera, L. Rojo, N. Campos, M. Pedro and M. S. J. Nacsimento, *Bioorg. Med. Chem.*, 2008, **16**, 862.
46. E. A. Hillard, F. C. de Abreu, D. C. M. Ferreira, G. Jaouen, M. O. F. Foulart and C. Amatore, *Chem. Comm.*, 2008, **23**, 2612.
47. N. M. F. Lima, C. S. Correia, L. L. Leon, G. M. C. Machado, M. F. Madeira, A. E. G. Santana and M. O. F. Goulart, *Mem. Inst. Oswaldo Cruz*, 2004, **99**, 757.
48. E. H. Lizarazo-Jaimes, P. G. Reis, F. M. Bezerra, B. L. Rodrigues, R. L. Monte-Neto, M. N. Melo, F. Frézard and C. Demicheli, *J. Inorg. Biochem.*, 2014, **132**, 30.
49. R. L. Monte-Neto, A. C. Coelho, F. Raymond, D. Légaré, J. Corbeil, F. Frézard and M. Ouellette, *PLoS Neglected Trop. Dis.*, 2011, **5**, e1167.
50. J. A. R. Salvador, R. M. A. Pinto and S. M. Silvestre, *Curr. Org. Synth.*, 2009, **6**, 426.
51. P. C. Andrews, R. Frank, P. C. Junk, L. Kedzierski, I. Kumar and J. G. MacLellan, *J. Inorg. Biochem.*, 2011, **105**, 454.
52. P. C. Andrews, G. B. Deacon, P. C. Junk, I. Kumar and J. G. MacLellan, *Organometallics*, 2009, **28**, 3999.
53. Y. C. Ong, V. L. Blair, L. Kedzierski and P. C. Andrews, *Dalton Trans.*, 2014, **43**, 12904.
54. Y. C. Ong, V. L. Blair, L. Kedzierski, K. L. Tuck and P. C. Andrews, *Dalton Trans.*, 2015, **44**, 18215.
55. M. S. Ahmad, M. Hussain, M. Hanif, S. Ali, M. Qayyum and B. Mirza, *Chem. Biol. Drug Design*, 2008, **71**, 568.
56. M. Sirajuddin, Ali, F. A. Shah, M. Ahmad and M. N. Tahir, *J. Iran. Chem. Soc.*, 2014, **11**, 297.
57. A. G. Davies, *Organotin Chemistry*, Wiley-VCH, Weinheim, 2nd edn, 2004, ISBN 978-3-527-31023-4.

58. A. K. Saxena, S. Ram, P. C. Jain and G. K. Patnaik, *Appl. Organomet. Chem.*, 1985, **1**, 16.
59. A. K. Saxena and F. Huber, *Coord. Chem. Rev.*, 1989, **95**, 109.
60. B. Raychaudhury, S. Banerjee, S. Gupta, R. V. Singh and S. C. Datta, *Acta Tropica*, 2005, **95**, 1.
61. C. L. Fattman, L. M. Schaefer and T. D. Oury, *Free Radic. Biol. Med.*, 2003, **35**, 236.
62. R. Dey, S. Mitra and S. C. Datta, *J. Parasitol.*, 1995, **81**, 683.
63. P. Apodaca, F. Lee, F. Cervantes and H. K. Pannell, *Main Group Met. Chem.*, 2001, **24**, 597.
64. M. Gielen, A. E. Khloufi, M. Biesemans, F. Kayser and R. Willem, *Appl. Organomet. Chem.*, 1993, **7**, 201.
65. N. M. Brown, PhD thesis, Clemsom University, 1972.
66. M. I. Khan, M. K. Baloch, M. Ashfaq and S. Gul, *J. Braz. Chem. Soc.*, 2009, **20**, 341.
67. M. I. Khan, M. K. Baloch and M. Ashfaq, *J. Enzyme Inhib. Med. Chem.*, 2007, **22**, 343.
68. Q. H. Qureshi, A. Nadhman, M. Sirajuddin, G. Shahnaz, S. Ali, A. Shah and M. M. Yasinzai, *Inorg. Chim. Acta*, 2014, **423**, 220.
69. M. Sirajuddin, S. Ali, V. Mckee, S. Zaib and J. Iqbal, *RSC Adv.*, 2014, **4**, 57505.
70. M. Sirajuddin, S. Ali, V. Mckee, M. Sohail and H. Pasha, *Eur. J. Med. Chem.*, 2014, **84**, 343.
71. F. Javed, S. Ali, S. Shahzadi, S. K. Sharma, K. Qanungo, M. N. Tahir, N. A. Shah, M. R. Khan and N. Khalid, *J. Inorg. Organomet. Polym.*, 2016, **26**, 48.
72. T. S. Abondanza, C. R. Oliveira, C. M. Barbosa, F. E. Pereira, R. L. Cunha, A. C. Caires, J. V. Comasseto, M. L. Queiroz, M. C. Valadares and C. Bincoletto, *Food Chem. Toxicol.*, 2008, **46**, 2540.
73. R. L. Cunha, M. E. Urano, J. R. Chagas, P. C. Almeida, C. Bincoletto, I. L. Tersariol and J. V. Comasseto, *Bioorg. Med. Chem. Lett.*, 2005, **15**, 755.
74. D. Sredni-Kenigsbuch, B. Shohat, B. Shohat, D. Ben-Amitai, C. C. Chan and M. David, *J. Dermatol. Sci.*, 2008, **50**, 232.
75. C. B. C. Lima, W. W. Arrais-Silva, R. L. O. R. Cunha and S. Giorgio, *Korean J. Parasitol.*, 2009, **47**, 213.
76. I. A. S. Pimentel, C. S. Paladi, S. Katz, W. A. S. Júdice, R. L. O. R. Cunha and C. L. Barbiéri, *PLoS One*, 2012, **7**, e48780.
77. S. Vale-Costa, N. Vale, J. Matos, A. Tomás, R. Moreira, P. Gomes and M. S. Gomes, *Antimicrob. Agents Chemother.*, 2012, **56**, 5774.
78. R. Gaspar, F. R. Opperdoes, V. Preat and M. Roland, *Ann. Trop. Med. Parasitol.*, 1992, **86**, 41.
79. E. Iniguez, A. Sánchez, M. A. Vasquez, A. Martínez, J. Olivas, A. Sattler, R. A. Sánchez-Delgado and R. A. Maldonado, *J. Biol. Inorg. Chem.*, 2013, **18**, 779.

80. S. Quintal, T. S. Morais, C. P. Matos, M. P. Robalo, M. F. M. Piedade, M. J. V. de Brito, M. H. Garcia, M. Marques, C. Maia, L. Campino and J. Madureira, *J. Organomet. Chem.*, 2013, **745**, 299.
81. N. Mbongo, P. M. Loiseau, F. Lawrence, C. Bories, D. G. Craciunescu and M. Robert-Gero, *Parasitol Res.*, 1997, **83**, 515.
82. L. P. Franco, E. P. de Góis, B. S. Codonho, A. L. R. Pavan, I. O. Pereira, M. J. Marques and E. T. de Almeida, *Med. Chem. Res.*, 2013, **22**, 1049.

CHAPTER 11

New Avenues for Drug Delivery in Leishmania*: Using Treatment of Visceral Leishmaniasis with Amphotericin B as an Exemplar*

KATHARINE C. CARTER* AND ALEXANDER B. MULLEN

Strathclyde Institute for Pharmacy and Biomedical Sciences, University of Strathclyde, 161 Cathedral Street, Glasgow G4 0RE, Scotland, UK
*Email: k.carter@strath.ac.uk

11.1 Introduction

Leishmaniasis is a disease caused by infection with the protozoan parasite *Leishmania.* The disease can take three main forms, depending on the species infecting the individual and their inherent susceptibility, these are cutaneous leishmaniasis (CL), caused by infection with species such as *Leishmania major* or *Leishmania mexicana,* mucocutaneous leishmaniasis, caused by infection with *Leishmania braziliensis*, or visceral leishmaniasis (VL), caused by infection with *Leishmania donovani*, *Leishmania infantum* or *Leishmania chagasi.* Leishmaniasis is endemic in a number of countries and in 2012 a World Health Organisation study determining the incidence of leishmaniasis in 102 countries found that 90% of visceral leishmaniasis (VL) cases occurred in six countries and over 70% of cutaneous leishmaniasis

Drug Discovery Series No. 60
Drug Discovery for Leishmaniasis
Edited by Luis Rivas and Carmen Gil

Published by the Royal Society of Chemistry, www.rsc.org

(CL) cases occurred in ten countries.[1] A more recent survey looking at incidence in 25 countries with high levels of leishmaniasis showed that in 2014, approximately 556 million people were at risk of VL and 399 million people were at risk of CL in high burden countries.[2] Currently there is no vaccine to protect against infection therefore the only way to control disease is to target the vector or any reservoir host, use practical methods to prevent individuals becoming infected *e.g.* use bed nets or insect repellents to prevent infection, or use drug treatment to cure infected individuals and reduce the number of parasites available to infect the vector. There are a limited number of drugs available for the treatment of leishmaniasis and the emergence of drug resistance within the parasite population is limiting their use in certain geographical locations.[3,4] Drug resistance has led to the development of combination drug treatments in an attempt to extend the clinical life of existing drugs. A recent study in Sudan showed that joint treatment with sodium stibogluconate (SSG, 20 $mg\,Sb^{v}\,kg^{-1}\,day^{-1}$ and paromomycin (15 $mg\,kg^{-1}\,day^{-1}$) for 17 days was as safe and effective as the standard 30 day SSG treatment.[5] Another way to improve the treatment with existing drugs is to reformulate them into a drug delivery system (DDS) that targets entrapped drug to the amastigote, increases the drug's *in vivo* half-life, ameliorates the drug's side effect profile or facilitates delivery by use of a non-invasive route of administration. This approach has been used by pharma companies as a way of extending the patent life of a drug, as generic equivalents can come on to the market when patent protection expires. This approach can ensure that a clinically approved drug can still secure revenues, as development of a new entity as a drug can cost a pharma company \$2.78 billion,[6] which is probably why repurposing existing clinical drugs for new indications is now important to pharma companies. Most anti-leishmanial drugs are given by an injectable route of administration[3–5] but the most practical route for clinical use is oral administration as this allows the patient to self-treat and reduces medical costs, an important consideration in developing countries. Currently miltefosine (MIL) is the only oral drug treatment for leishmaniasis. It was originally developed as an anti-cancer drug but was repurposed for the treatment of VL when it was found to be effective against *L. donovani* in murine studies.[7] However, there are drawbacks to developing an oral drug formulation for the treatment of leishmaniasis. The experience from using MIL for the treatment of VL have indicated that problems with patient compliance may be partly responsible for the emergence of MIL-resistant strains of *L. donovani* in India and it is now recommended that the drug should be given to patients under direct observation to encourage patient adherence.[8]

11.2 Drug Delivery Systems (DDS)

There are a wide variety of DDS currently being investigated and the system selected depends upon the condition being treated and the most advantageous route of administration. Even within a particular category of DDS

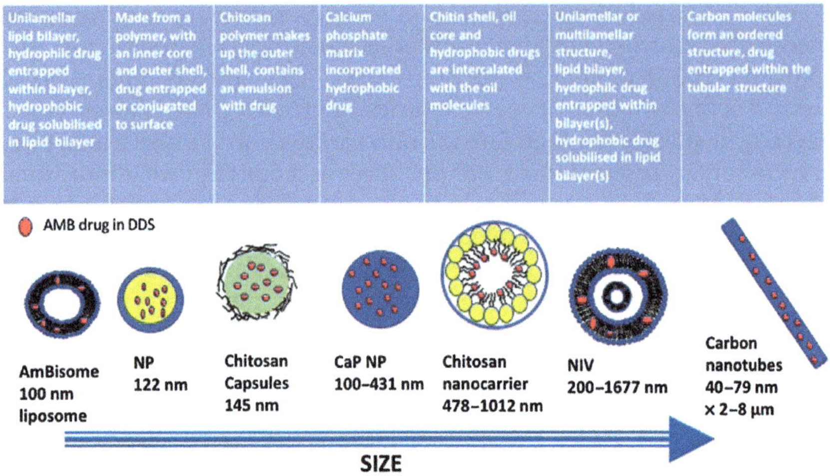

Figure 11.1 Comparison of different types of DDS, loaded with AMB, used in the treatment of VL.

there are a variety of types and subtypes (examples are shown in Figure 11.1). For example, lipid based systems[9] can be divided into three main categories (*i.e.* emulsions, vesicular systems and lipid particles), and further subdivided into subtypes. The properties of the drug *e.g.* its aqueous solubility and compatibility with the components of the DDS have a profound influence on which DDS is most suitable for a particular drug. This review will focus on different DDS designed for amphotericin B (AMB) to illustrate the advantages of using a DDS for the treatment of VL. AMB, a polyene macrolide antibiotic, is derived from the actinomycete *Streptomyces nodosus*. It was originally introduced as a broad-spectrum intravenous antifungal agent in the 1950s. Its water insolubility and physicochemical stability necessitated formulation as a lyophilised micellar complex with desoxycholic acid (Fungizone®). AMB binds to sterols and it has a greater affinity for ergosterol, the predominant sterol in fungal and *Leishmania* cell membranes, compared with cholesterol, the prominent sterol in mammalian cell membranes. The drug increases the permeability of cell membranes by causing the formation of channels, which compromises the integrity of the organism's cell membrane. Although, studies using *Candida* found that binding to ergosterol alone, without channel formation was sufficient to induce cell killing.[10] This simple mechanism may explain why AMB resistance is surprising rare. AMB use is associated with a number of adverse effects, with nephrotoxicity being the most severe. It can cause acute kidney failure, and incidences as high as 65% have been reported in patients.[11] In studies, the effect of treatment on kidney function is assessed by measuring changes in serum creatinine levels. Using a DDS improves AMB targeting to tissues, resulting in lower kidney levels or allows treatment with a lower drug dose compared with treatment with the original micellar drug formulation, or more aggressive treatment with a formulation with lower inherent toxicity.

11.3 Parenteral Administration

The intravenous route is an excellent way of screening the anti-leishmanial activity of drugs and their novel formulations, as it allows rapid access to the systemic circulation and gives reproducible drug levels after administration. A single randomised blind study was carried out to compare the pharmacokinetics of Fungizone® (0.6 mg AMB kg^{-1}) and a liposomal formulation of AMB (AmBisome®, 2 mg AMB kg^{-1}) in healthy volunteers.[12] AmBisome® contains spherical unilamellar liposomes with a mean diameter of less than 100 nm. Both formulations had similar plasma drug half-lives but the terminal elimination phase contained 80% of the total area under the curve (AUC) for Fungizone® but only 47% of the total AUC for liposomal AMB, and the volume of distribution for AmBisome® was significantly lower than that of Fungizone® ($p<0.05$). Thus, indicating that more of the drug remains in the central compartment and less is in peripheral tissues for AmBisome® treatment. The urinary recovery of AMB was significantly higher in Fungizone®-treated individuals (20.6% versus 4.5%), which explains the lower renal clearance value in AmBisome®-treated individuals (0.05 ± 0.01 versus 0.32 ± 0.06, $p<0.01$).[12] These data clearly indicate that using the liposomal formulation, where AMB is incorporated into the lipophilic bilayer of the liposome, increased tissue targeting of the drug to specific sites within the body, such as the reticuloendothelial system, where the infection resides. This would explain the higher drug levels in *L. infantum*-infected mice, treated by intravenous injection on days 14, 16 and 18 post-infection (0.8 mg AMB kg^{-1}) with Fungizone® or AmBisome® (mean AMB levels $\pm$ SE, $\mu g\,g^{-1}$, liver, 6.46 ± 0.80 versus 0.06 ± 0.02, spleen, 2.05 ± 0.51 versus 0.24 ± 0.05)[13]. Most experimental rodent studies in leishmaniasis use intravenous injection in murine studies, or intravenous or intraperitoneal injection in hamsters. Intravenous treatment is difficult in hamsters, even though researchers have indicated that non invasive injection is possible using the cephalic[14] or sub-lingual veins.[15] However, in our hands this has proved problematic and we use the jugular vein for intravenous administration in hamsters.[16] The Centre for Replacement, Refinement and Reduction of Animals in Research guidelines advise that cardiac puncture with recovery can only be carried out once on an animal, indicating that this route is not suitable for infection and drug treatment.[17]

In our studies, we have shown that five doses of 1 mg kg^{-1} of a non-ionic surfactant vesicle (NIV) formulation of AMB, a lipid based vesicular DDS, was as effective as the same dose of AmBisome® against bone marrow parasites ($p<0.001$), but was less effective against splenic and liver parasites (mean suppression $\pm$ SE, AMB-NIV, spleen 70 ± 4, liver 89 ± 2, bone marrow 74 ± 7; AmBisome®, spleen 86 ± 3, liver 99 ± 1, bone marrow 73 ± 6) in

L. donovani-infected mice.[18] In more recent studies nanocapsules, which have an oil-based hydrophobic central cavity, surrounded by chitosan have been used to deliver AMB. Chitosan is a naturally occurring polymer, is biocompatible, biodegradable, has generally regarded as safe (GRAS) approval and has immunostimulatory effects, which could boost local immune responses as well as acting as a constituent of the DDS.[19] AMB-containing nanocapsules (CNC–AMB) had a mean size of 145 nm, an AMB loading of 9.1% and an entrapment efficiency of 97%, and a net positive zeta potential of 29 mV. Formulations with a net positive surface charge are generally associated with higher cytotoxicity.[20] However, this study showed that the CNC–AMB formulation was less toxic than Fungizone® against the J774A cell line, and had similar cytotoxicity to AmBisome® (mean IC_{50} dose $\mu g\,ml^{-1} \pm$ SD: Fungizone® 0.4 ± 0.1; CNC-AMB 10.7 ± 0.1; AmBisome® 9.8 ± 0.2). Five doses of CNC–AMB to *L. donovani* infected hamsters by the intraperitoneal route were significantly more effective at reducing splenic parasites burden than Fungizone® or AmBisome®, administered in the same way ($p < 0.01$, mean suppression $\pm$ SE: CNC-AMB 86 ± 2; Fungizone®, 56 ± 4; AmBisome®, 70 ± 3).[19]

More recently, nanoparticles (NP) produced from calcium phosphate (CaP) have been produced. This compound is present in bone and teeth enamel, and has the advantages of being biocompatible and being metabolized to a non-toxic degradation product. However, the NP produced are unstable in the systemic circulation and they have a tendency to cluster together, which could block a blood vessel.[21] CaP–NP containing AMB (CaP–AMB–NP) physical characteristics depended on the production method, but they had a size range of 100–431 nm, a zeta potential of 17–31 mV and an entrapment efficiency of 65–71%.[21] Treatment of *L. donovani*-infected hamsters with five consecutive doses of AMB formulations (1 mg AMB kg^{-1}) by intraperitoneal injection showed that CaP–AMB–NP had a similar activity to liposomal AMB (mean percentage suppression in splenic parasite numbers compared with controls $\pm$ SD, CaP–AMB–NP 78 ± 15, liposomal AMB 66 ± 9). There was no sign of nephrotoxicity in mice given five consecutive doses of CaP–AMB–NP by the intravenous route (3 or 5 mg AMB kg^{-1}), based on blood urea nitrogen or plasma creatine levels compared with controls.[21]

11.4 Oral Administration

Oral administration of anti-leishmanial drugs would allow patients to self-treat, a major advantage for the patient and countries with limited resources to fund health services. Oral delivery, however, presents a number of problems, for example, a drug may be inactivated or damaged by stomach acid, enzymes present in the host intestinal tract or enzymes in gut flora, or liver metabolism before the drug enters the systemic circulation. A formulation must protect against these challenges and allow the drug to reach its site of action.[22] AMB is poorly absorbed after oral administration, it is unstable at gastric pH, and it is unable to penetrate the brush border membrane of the small intestine. Therefore, the drug needs to be incorporated into a formulation that allows effective absorption from the gastrointestinal tract. Oral treatment of *L. donovani*-infected mice with five doses of a lipid-based formulation of AMB (iCo-009: 2.5, 5, 10 or 20 mg kg^{-1}, twice daily for five consecutive days), produced using Peceol® and distearoylphosphatidylethanolamine poly(ethylene glycol)2000, resulted in dose dependent killing of liver parasites.[23] The highest drug dose caused a 99.8 ± 0.2% reduction in liver burdens compared with controls. In contrast, single-dose treatment by intravenous injection with AmBisome® (2 mg AMB kg^{-1}) completely cleared liver parasites, showing that this formulation was significantly more effective than iCo-009.[23] Subsequent studies using a multiple-dosing regimen showed that oral treatment with iCo-009 resulted in detectable AMB levels in the spleen and liver of uninfected mice or *L. donovani*-infected mice, but the amount present was lower than that achieved with AmBisome® treatment. Interestingly, significantly higher drug levels were present in uninfected mice compared with VL-infected mice for both AMB formulations ($p < 0.01$),[24] indicating that infection affects drug uptake, metabolism, distribution or elimination and that pharmacokinetic studies should be completed in uninfected and infected individuals for leishmaniasis. The formulation of iCo-009 was subsequently adapted so that it was stable at tropical temperatures.[25] Some drug formulations require storage at ambient (*e.g.* 25 °C) or refrigerated temperatures (*e.g.* 4 °C or −20 °C) for long-term stability. For example, chitosan, a natural polymer used to formulate DDS, is susceptible to degradation at high or low temperatures, which can affect the methods used to produce a chitosan-containing DDS and storage conditions post-production.[26] Liposomal formulations are susceptible to hydrolysis or peroxidation, which is why lyophilised products are often produced as these are more stable on storage compared with hydrated formulations.[27] Development of a formulation that is stable without special storage conditions could give a drug formulation a competitive advantage, especially in countries where there is difficulty in maintaining a constant electrical supply. The iCo-010 formulation[25] contains AMB with 60:40 (v:v) mono- and diglycerides with D-α-tocopheryl polyethylene glycol 1000 succinate (Vitamin E TPGS), an FDA-approved component which is a GRAS-listed excipient. Vitamin E–TPGS is often used as a drug solubilizer in oral, parenteral, topical, nasal and rectal or vaginal therapies. It is synthesised from α-tocopherol (vitamin E), which has important anti-oxidant properties.[28,29] The results of stability studies over 30 days at different temperatures indicated that iCo-010 lost more AMB content at 43 °C compared with storage at 30 °C (mean AMB content lost, μg ml^{-1} day^{-1} ± SD, 43 °C, 16.3 ± 7.3; 30 °C 13.8 ± 4.5), leading to 80% drug stability after 60 days at 30 °C and 75% after 60 days at 43 °C.[30] This formulation caused a significant dose-dependent reduction in *L. donovani* in parasite numbers (mean percentage suppression in liver parasite burdens, five-day treatment, 3 mg kg^{-1} day^{-1}, 48%; 5 mg kg^{-1} day^{-1}, 98%; 10 mg kg^{-1} day^{-1}, 99%).[25]

Other types of AMB formulations have been developed for oral treatment of leishmaniasis, including NP, functionalized carbon nanotubes and mannose-anchored thiolated nanocarriers. A NP formulation of AMB, suitable for large-scale manufacture, was produced using polylactic-co-glycolic acid (PLGA, a biodegradable polymer) and Vit–TPGS (AMB–NP).[31] The resultant formulation was produced using an ultrafiltration step to remove unentrapped drug, gave a formulation with a mean particle size of 122 nm, an entrapment efficiency of 71% and a nanoparticle recovery of 98%. Six rounds of ultrafiltration reduced residual dimethyl sulphoxide levels to below the limits of detection. Single-dose oral treatment (2.5 mg AMB kg^{-1}) with this formulation resulted in a reduction in *L. donovani* liver parasite burdens. The formulation was more effective by the intravenous route but its activity was significantly lower than that of AmBisome® given by the intravenous route ($p < 0.01$, mean suppression compared with control ± SE: AMB–NP oral 27 ± 13%; AMB–NP intravenous 51 ± 3; AmBisome® intravenous, 78 ± 4). Increasing the dose of AMB–NP given to animals increased the efficacy of drug treatment but did not result in parasite clearance in the liver (mean suppression ± SE: two doses AMB–NP 50 mg kg^{-1} $dose^{-1}$, 51 ± 14; five doses AMB–NP 50 mg kg^{-1} $dose^{-1}$, 50 ± 7).[31] Functionalized carbon nanotubes (f-CNTs) are produced by chemical covalent or non-covalent bonding between the CNTs and the material of interest, where amino groups act as chemical bridges. The CNTs need to be functionalized as carbon nanotubes have a low solubility in aqueous or organic solvents.[32,33] Functionalized-CNTs of amphotericin B (f-CNT–AMB) had a drug loading of 72.4%, a diameter of 40–70 nm and a length of 2–8 μm.[34] Oral treatment of *L. donovani*-infected hamsters with five doses of f-CNT–AMB gave a significant reduction in splenic parasite burdens (99% reduction compared with controls), 5 mg kg^{-1} does being as active as the 10 or 15 mg kg^{-1} doses. This formulation was as active as oral dosing with MIL (5 mg kg^{-1}, five doses) or intraperitoneal injection with AmBisome® (5 mg kg^{-1}, one dose).[34] However, there is a lot of concern about the potential toxicity of these types of formulations.[35] Mannose-anchored thiolated chitosan nanocarriers containing AMB (MTC–AMB) were produced to improve cell uptake *via* the mannose receptor on macrophages.[36] MTC–AMB particles had a mean size of 478–1012 nm, a net positive charge (19–29 mV) and a drug entrapment of 26–76%. The results of *in vitro* studies indicated that significantly greater amounts of drug ($p < 0.01$) were delivered to uninfected and *L. donovani*-infected macrophages compared with unmodified chitosan nanocarriers containing AMB (UCC–AMB, mean AMB uptake *L. donovani*-infected macrophages, μg AMB 10^{-6} cells, UCC–AMB, 6.3 ± 0.7; MTC–AMB 27–28, depending on the strain used to infect cells). This implied that surface modification enhanced cellular uptake. An oral dose of MTC–AMB was significantly more effective ($p < 0.001$) than similar treatment with an equivalent dose of the carrier alone, AMB solution or AmBisome® (mean percentage inhibition splenic parasite burden compared with control ± SD, carrier alone, 36 ± 2, MTC–AMB, 89 ± 7, AMB solution, 17 ± 4).[36]

11.5 Pulmonary Administration

Pulmonary administration is a non-invasive route and has been used for systemic drug delivery. It allows rapid absorption due to the huge surface area of the respiratory epithelium and drugs avoid metabolism associated with first passage through the liver. However, pulmonary drug delivery is associated with a number of challenges. A formulation must protect a drug against mucociliary clearance, damage by enzymes present within the lungs and favour systemic absorption.[37] A lyophilised NIV–AMB formulation for treatment by the pulmonary route had an AMB concentration of 0.93–1.19 mg AMB ml^{-1} after rehydration, and a mean vesicle size of 1677 nm, and a surface charge of −72 mV. The treatment efficacy of a single intravenous dose of AmBisome® (2.5 mg AMB kg^{-1}) or AMB–NIV (2.5 mg AMB kg^{-1}, 150 mM lipid. 0.93 mg AMB ml^{-1}) was compared with multiple dosing with inhaled AMB–NIV (five doses, 150 mM lipid. 0.93 mg AMB ml^{-1}) in a murine model of VL. Intravenous therapy with the two AMB formulations was equally active in the liver (mean suppression ± SE, AmBisome 90 ± 3, AMB–NIV, 85 ± 3), whereas the NIV formulation delivered by the pulmonary route was less effective (mean suppression ± SE, AMB–NIV, 46 ± 3).[38]

11.6 Summary

Studies have shown that various DDS can be used to increase the therapeutic efficacy and reduce the nephrotoxicity of AMB. Although there is a clinical need for new drugs and better drug formulations for leishmaniasis, there is not a great commercial drive for their development. The only way to drive drug development is through specific partnerships with academic institutions, health services, philanthropic organizations and pharma.[39] AMB has the advantage of being a drug used for both leishmaniasis and fungal treatment, and the fungal drug market is of commercial interest to pharma. It was worth US$ 10.7 billion in 2015 and may be valued at US$ 12.6 billion by 2025.[40] Using *Leishmania* as an initial screening model has many advantages as it provides (i) well characterized animal models, (ii) the ability to screen delivery to multiple sites of infection (spleen, liver, bone marrow, VL; lesions in different cutaneous sites, CL), (iii) higher throughput screening assays and the ability to monitor parasite levels in the same animal over the course of an experiment by assessing bioluminescence or fluorescence, (v) availability of drug-resistant lines. Oral treatment is arguably the ideal route of administration for leishmaniasis, and most studies have used batch manufacture methods, whereas future studies will probably move to continuous manufacturing methods, a smaller manufacturing footprint, and the ability to produce personalised medicines. This will, however, depend on having a good understanding of the factors which result in a uniform product, and the availability of in-process sensors to give strict control over processing variables.[41] For leishmaniasis, monotherapy is probably not the best treatment option, and future formulations should incorporate drug combinations, in an attempt to combat emergence of parasite drug resistance.

References

1. J. Alvar, I. D. Vélez, C. Bern, M. Herrero, P. Desjeux, J. Cano, J. Jannin and M. den Boer, *PLoS One*, 2012, **7**, e35671.
2. WHO/Department of Control of Neglected Tropical Diseases, *Wkly. Epidemiol. Rec.*, 2016, **91**, 287.
3. S. Sundar and A. Singh, *Ther. Adv. Infect. Dis.*, 2016, **s3**, 98.
4. H. J. de Vries, S. H. Reedijk and H. D. Schallig, *Am. J. Clin. Dermatol.*, 2015, **16**, 99.
5. A. Musa, E. Khalil, A. Hailu, J. Olobo, M. Balasegaram, R. Omollo, T. Edwards, J. Rashid, J. Mbui, B. Musa, A. A. Abuzaid, O. Ahmed, A. Fadlalla, A. El-Hassan, M. Mueller, G. Mucee, S. Njoroge, V. Manduku, G. Mutuma, L. Apadet, H. Lodenyo, D. Mutea, G. Kirigi, S. Yifru, G. Mengistu, Z. Hurissa, W. Hailu, T. Weldegebreal, H. Tafes, Y. Mekonnen, E. Makonnen, S. Ndegwa, P. Sagaki, R. Kimutai, J. Kesusu, R. Owiti, S. Ellis and M. Wasunna, *PLoS Neglected Trop. Dis.*, 2012, **6**, e1674.
6. J. A. DiMasi, H. G. Grabowski and R. W. Hansen, *J. Health Econ.*, 2016, **47**, 20.
7. S. L. Croft, R. A. Neal, W. Pendergastand and J. H. Chan, *Biochem. Pharmacol.*, 1987, **36**, 2633.
8. S. Sundar and P. L. Olliaro, *Ther. Clin. Risk Manag.*, 2007, **3**, 733.
9. H. Shrestha, R. Bala and S. Arora, *J. Pharm. (Cairo)*, 2014, **2014**, ID801820.
10. K. C. Gray, D. S. Palacios, I. Dailey, M. M. End, B. E. Uno, B. C. Wilcock and M. D. Burke, *Proc. Natl. Acad. Sci. U.S.A.*, 2012, **109**, 2234.
11. G. Delay, *J. Antimicrobiol. Chemother.*, 2002, **49**(Suppl. S1), 37.
12. I. Bekersky, R. M. Fielding, D. E. Dressler, J. W. Lee, D. N. Buell and T. J. Walsh, *Antimicrob. Agents Chemother.*, 2002, **46**, 828.
13. M. Paul, R. Durand, H. Fessi, D. Rivollet, R. Houin, A. Astier and M. Deniau, *Antimicrob. Agents Chemother.*, 1987, **41**, 1731.
14. J. H. Ransom, *Lab. Anim. Sci.*, 1984, **34**, 200.
15. H. B. Waynforth and R. Parkin, *Lab. Anim.*, 1969, **3**, 35.
16. K. C. Carter, J. O'Grady, T. F. Dolan, A. J. Baillie, J. Alexander and J. A. Keys, *Int. J. Pharm.*, 1989, **53**, 129.
17. https://www.nc3rs.org.uk/hamster-cardiac-puncture-recovery-surgical, accessed 12/12/16.
18. A. B. Mullen, K. C. Carter and A. J. Baillie, *Antimicrob. Agents Chemother.*, 1997, **41**, 2089.
19. A. Asthana, A. K. Jaiswal, P. K. Gupta, K. P. Vivek, A. Dube and M. Chourasia, *Antimicrob. Agents Chemother.*, 2013, **57**, 1714.
20. M. L. Immordino, F. Dosio and L. Cattel, *Int. J. Nanomed.*, 2006, **1**, 297.
21. M. Chaurasia, P. K. Singh, A. K. Jaiswal, A. Kumar, V. K. Pawar, A. Dube, S. K. Paliwal and M. K. Chourasia, *Pharm. Res.*, 2016, **33**, 2617.
22. H. Wen, H. Jung and X. Li, *AAPS J.*, 2015, **17**, 1327.

23. K. M. Wasan, E. K. Wasan, P. Gershkovich, X. Zhu, R. R. Tidwell, K. A. Werbovetz, J. G. Clement and S. J. Thornton, *J. Infect. Dis.*, 2009, **200**, 357.
24. P. Gershkovich, E. K. Wasan, O. Sivak, R. Li, X. Zhu, K. A. Werbovetz, R. R. Tidwell, J. G. Clement, S. J. Thornton and K. M. Wasan, *J. Antimicrob. Chemother.*, 2010, **65**, 535.
25. E. K. Wasan, G. Gershkovich, J. Zhao, X. Zhu, K. Werbovetz, R. R. Tidwell, J. G. Clement, S. J. Thornton and K. M. Wasan, *PLoS Neglected Trop. Dis.*, 2010, **4**, e913.
26. E. Szymańska and K. Winnicka, *Mar. Drugs*, 2015, **13**, 1819.
27. N. M. Payton, M. F. Wempe, Y. Xu and T. J. Anchordoquya, *J. Pharm. Sci.*, 2014, **103**, 3869.
28. P. P. Constantinides, J. Han and S. S. Davis, *Pharm. Res.*, 2006, **23**, 243.
29. M. G. Traber and J. Atkinson, *Free Radic. Biol. Med.*, 2007, **43**, 4.
30. F. Ibrahim, O. Sivak, E. K. Wasan, K. Bartlett and K. M. Wasan, *Lipids Health Dis.*, 2013, **29**, 158.
31. J. L. Italia, M. N. Kumar and K. C. Carter, *J. Biomed. Nanotechnol.*, 2012, **8**, 695.
32. A. Di Crescenzo, V. Ettorre and A. Fontana, *Beilstein J. Nanotechnol.*, 2014, **5**, 1675.
33. M. Karimi, N. Solati, A. Ghasemi, M. A. Estiar, M. Hashemkhani, P. Kiani, E. Mohamed, A. Saeidi, M. Taheri, P. Avci, A. R. Aref, M. Amiri, F. Baniasadi and M. R. Hamblin, *Expert Opin. Drug Delivery*, 2015, **12**, 1089.
34. V. K. Prajapati, K. Awasthi, S. Gautam, T. P. Yadav, M. Rai, O. N. Srivastava and S. Sundar, *J. Antimicrob. Chemother.*, 2011, **66**, 874.
35. Y. Liu, Y. Zhao, B. Sun and C. Chen, *Acc. Chem. Res.*, 2013, **46**, 702.
36. G. Shahnaz, B. J. Edagwa, J. McMillan, S. Akhtar, A. Raza, N. A. Quresh, M. Yasinzai and H. E. Gendelman, *Nanomedicine (Lond.)*, 2017, **12**, 99.
37. K. C. Carter and M. Puig-Sellart, *Curr. Pharm. Des.*, 2016, **22**, 3324.
38. M. Alsaadi, J. L. Italia, A. B. Mullen, M. N. Ravi Kumar, A. A. Candlish, R. A. Williams, C. D. Shaw, F. Al Gawhari, G. H. Coombs, M. Wiese, A. H. Thomson, M. Puig-Sellart, J. Wallace, A. Sharp, L. Wheeler, P. Warn and K. C. Carter, *J. Control Release*, 2012, **160**, 685.
39. L. Lezaun and C. M. Montgomery, *Sci. Technol. Hum. Values*, 2015, **40**, 3.
40. http://www.prnewswire.com/news-releases/antifungal-drugs-market-size-projected-to-reach-126-billion-by-2025-grand-view-research-inc-602833686.html, accessed 12/12/16.
41. J. Rantanen and J. Khinast, *J. Pharm. Sci.*, 2015, **104**, 3612.

III. The Quest for Achille's Heel of Leishmania. Singular Targets as New Avenues for Drug Development

CHAPTER 12

Addressing the Molecular Biology of Leishmania *for Drug Development*

BRIANNA NORRIS-MULLINS AND MIGUEL A. MORALES*

Eck Institute for Global Health, Department of Biological Sciences, University of Notre Dame, Notre Dame, IN 46556, USA
*Email: miguel.morales@nd.edu

12.1 Introduction

Kinetoplastid parasites of the genus *Leishmania* are the causative agents of leishmaniasis, a vector-borne disease that affects millions of people worldwide.[1–3] Globally, approximately 310 million people are at risk of acquiring the disease, which is endemic in over 90 countries spanning four continents.[4] In its digenetic life cycle *Leishmania* parasites cycle between the sand fly vector, where they persist as promastigotes, and phagolysosomes of mammalian macrophages, where they survive as amastigotes. The ability of the parasites to survive these distinct environments is dependent on the developmental regulation of numerous genes, whose identification may result in therapeutic targets.[5] Though there is currently no human safe vaccine for leishmaniasis, some treatments are available and are, typically, species-specific. Current existing therapies for leishmaniasis are inadequate due to resistance, safety and cost, underscoring the necessity for safer therapies with alternate modes of action. Leishmaniasis affects the poorest sectors of low-income countries, therefore, public health therapies to combat the disease have little prospect of major economic return for private industry. Consequently, there are few

Drug Discovery Series No. 60
Drug Discovery for Leishmaniasis
Edited by Luis Rivas and Carmen Gil

Published by the Royal Society of Chemistry, www.rsc.org

large-scale drug discovery programs to identify new chemical entities to combat leishmaniasis.[4,6,7] In implementing effective control strategies against leishmaniasis, it is vital to have a thorough understanding of the host–pathogen–vector relationship and, more importantly, of their interactions at a molecular level.[8,9] Recent advances in 'omics' technologies, including genomics, transcriptomics and proteomics, make available the tools necessary for advancing our current understanding of the molecular biology of *Leishmania*[9] with the purpose of exploiting this information for the development of new alternatives for anti-leishmanial therapeutics.

12.2 The *Leishmania* Genome

In 2005, the first ever sequenced genomes of *Leishmania major*[10] and two distantly related kinetoplastid protozoa, *Trypanosoma brucei*[11] and *Trypanosoma cruzi*,[12] revealed large-scale genetic preservation.[13] Despite a conserved core of genes, more than 1000 *Leishmania*-specific genes were identified; many of which remain uncharacterized.[14] In 2007, genomic studies were extended to two additional *Leishmania* species: *Leishmania infantum* and *Leishmania braziliensis*. Upon bioinformatics analysis, more than 99% gene conservation between the three *Leishmania* genomes (*L. major*, *L. infantum* and *L. braziliensis*) was identified, yielding roughly 200 differences at the gene or pseudogene level with only a fraction of these being species-specific.[14] The genomes of *Leishmania amazonensis*, *Leishmania mexicana*, *Leishmania donovani* and *Leishmania tarentolae* have since been sequenced, and comparative analyses have shown similar results.[9,15] While findings indicate that minimal species-specific genes are important for pathogenesis, the underlying profound and unexpected conclusion is that the parasite genome plays a very small part in determining disease clinical presentation;[14] this makes it nearly impossible to use species-specific genes for the development of therapeutic agents aimed at treating distinct clinical manifestations caused by *Leishmania* spp.

Leishmania parasites exhibit unique biological features ranging from gene organization, through transcription and mRNA processing, to post-translational modifications.[5,16] For example, at the genomic level, many *Leishmania* genes are arranged into long, strand-specific, polycistronic clusters (Figure 12.1).[10,17,18] This major difference between *Leishmania* spp. and higher eukaryotes has been extensively studied, has potential therapeutic implications for drug discovery and will be discussed in detail in the following section.

12.3 The *Leishmania* Transcriptome

Although comparing the sequenced genomes of *Leishmania* species provides a solid basis for beginning to understand mechanisms of host–parasite interactions, it seems likely that a better understanding of disease pathogenesis could be gained by studying the regulation of gene expression during different parasite life cycle stages.[9] The advent of high-throughput

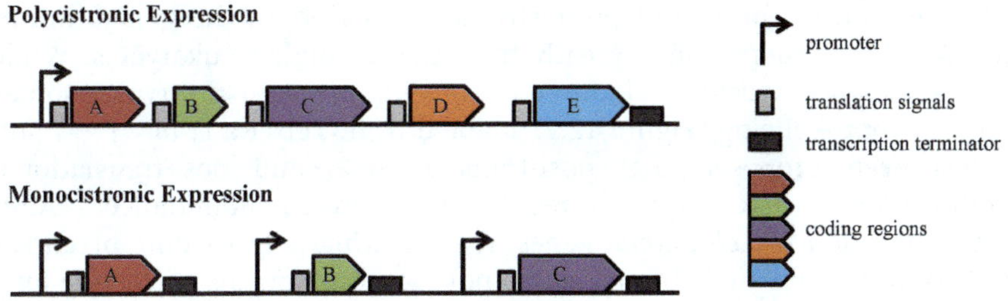

Figure 12.1 Gene Expression in Trypanosomatids *vs*. Higher Eukaryotes. In trypanosomatid parasites, large clusters of unrelated genes (coding regions) are organized as polycistronic transcription units (PTUs) under the control of a single promoter and termination codon (top). This differs from higher eukaryotes, where each individual gene has its own set of transcription signals (bottom).

transcriptomic technologies, such as RNA sequencing, have facilitated such studies, however, the peculiar mechanisms of *Leishmania* gene expression partially dampen the enthusiasm for large-scale transcriptomic analyses. Indeed, most *Leishmania* genes have been shown to be constitutively expressed throughout the life cycle. It is important to note, though, that the noticeable variation in chromosome and gene copy numbers among *L. infantum*, *L. braziliensis* and *L. major* indicates that genome plasticity, rather than differential expression of single genes, could be the key to the different tissue tropism of *Leishmania* spp.[9]

12.3.1 Polycistronic Transcription and *Trans*-splicing Mechanisms

Similar to other eukaryotic organisms, *Leishmania* parasites use three different types of RNA polymerases (pol) to transcribe their genes. RNA pol I and RNA pol III are responsible for transcription of ribosomal RNA (rRNA) and transfer RNA (tRNA) genes, respectively, while RNA pol II transcribes protein-coding genes.[19] It has been suggested, however, that in *Leishmania*, RNA pol II does not regulate transcription; this is due to the lack of transcription initiation sites for protein-coding genes within the organism.[20] Furthermore, common transcription factors typically present in eukaryotes, such as the TATA-binding protein and TFIIB, are highly divergent in trypanosomatids.[21] The parasites navigate their lack of control of transcription initiation by means of polycistronic transcription, where polycistronic clusters of tens to hundreds of functionally unrelated genes are transcribed similarly across the genome.[22] This means that the same polycistronic transcriptional units that make the parasite genomes so similar are also what are responsible for a lack of transcriptional gene expression regulation.[23] Indeed, over 94% of genes are constitutively expressed in both promastigote and amastigote life cycle stages, indicating that the parasites are well adapted for survival and replication in both their vector and host.[24]

In addition to the use of polycistronic transcription, the generation of mRNA in *Leishmania* differs greatly from that of higher eukaryotes. While many eukaryotic species maintain a *cis*-splicing mechanism, trypanosomes undergo *trans*-splicing (Figure 12.2, adapted from Teixeira *et al.*[25])[25–27] and it has been proposed that post-transcriptional and post-translational modifications are required for regulation of protein abundance.[28] After co-transcription of *Leishmania* genes, the resulting polycistronic precursor RNA is processed in a bi-modal fashion. A common 39-nucleotide-long spliced leader (SL) sequence containing a 'cap' structure is *trans*-spliced to the 5′ end of every mRNA. Meanwhile, polyadenylation of the 3′ end of each mRNA occurs at a fixed distance (100–400 nucleotides) upstream of the splicing signals.[5,16] Together, *trans*-splicing coupled with polyadenylation play a role in mRNA maturation and their post-transcriptional RNA processing reactions are thought to help control gene expression.[5]

To date, there has been no indication that identical *trans*-splicing mechanisms occur in the hosts of trypanosomatid parasites, rendering this process as parasite-specific.[29] The requirement of SL *trans*-splicing for mRNA maturation has drawn a lot of attention, specifically with the aims of identifying protein factors or factor domains required only for *trans*-splicing with attempts to determine the potential of these features to be inactivated

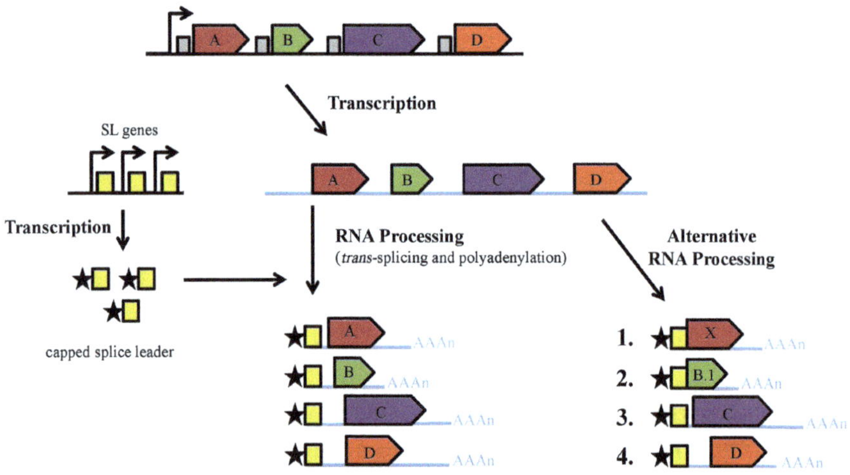

Figure 12.2 *Trans*-splicing and RNA Editing Processes. Parasite coding and spliced leader (SL) genes are transcribed simultaneously. During RNA processing, the polycistronic RNAs (pre-mRNAs) are individualized into monocistronic mRNAs after the addition of a 39-nucleotide capped splice leader RNA through a *trans*-splicing reaction coupled with polyadenylation. Polycistronic pre-mRNAs may undergo alternative RNA processing where the alteration of protein translation (1), targeting and/or function (2) may occur. Additionally, alternative splicing and polyadenylation can alter gene expression by the inclusion or exclusion of regulatory elements present in 5′ untranslated regions (UTRs) (3) or 3′ UTRs (4). Adapted from Teixeira *et al.*[25]

in a parasite-specific manner. Recently, it was shown that the inhibition of cdc2-related kinases implicated in the phosphorylation of RPB1 (one of the largest RNA Pol II subunits) led to a block in SL *trans*-splicing,[30] suggesting the use of parasite-specific kinase inhibitors to inhibit this fundamental process. This is challenging, however, due to the multifarious parasite splicing machinery, dubbed the spliceosome.[31]

12.3.2 Spliceosome

The spliceosome is a large, dynamic complex comprised of both structural RNAs and proteins. The spliceosome consists of U1, U2, U4, U5 and U6 small nuclear ribonucleic acids (snRNAs) and, in the human system, upwards of 170 spliceosome-associated protein factors.[32] *Leishmania* parasites possess all five spliceosomal U snRNAs, however, only U2, U4, U5 and U6 have been suggested to function in *trans*-splicing.[33,34] Though still not as significant as progress on the well-characterized human system, major progress has been made in recent years on the identification of spliceosomal proteins and the characterization of U small nuclear ribonucleoprotein particles (U snRNPs) in trypanosomes.[31]

Complete spliceosome development undergoes numerous distinct assembly steps as well as dynamic changes, where both the RNA and protein composition are altered before, during and after splicing. Despite few publications on functional characterizations of trypanosomal RNA splicing factors, many trypanosome-specific characteristics of the splicing machinery have been identified. Parasite-specific features include, but are not limited to, U snRNP elements, like Sm core variation and the architectural makeup of the U2 RNP core.[31] Tandem affinity purification (TAP) followed by liquid chromatography–tandem mas spectrometry (LC–MS/MS) has been particularly useful for identifying spliceosome proteins, revealing the presence of trypanosomatid-specific splicing factors and several conserved snRNPs with parasite-specific functions.[35] Such differences may be focused on for anti-trypanosomal drug targets, an idea not altogether uncommon as the spliceosome has been validated as a drug target for anticancer therapy.[36]

12.3.3 Mitochondrial mRNA Editing

Once mitochondrial mRNA has been transcribed it has the possibility of undergoing a post-transcriptional maturation process, termed RNA editing.[37] Like many processes in *Leishmania*, RNA editing is developmentally regulated and is necessary for parasite survival through its cycle between insect vectors and mammalian hosts.[38] During traditional RNA editing, mRNAs are produced by site-specific U insertion or deletion processing of pre-mRNAs.[39] Briefly, RNA editing requires small, non-coding RNAs, termed guide RNAs (gRNAs). The editing reaction is initiated by the binding of gRNA to pre-mRNA molecules, forming a three-way helix junction topology. Included in the structure is an ‘anchor’ duplex element that borders the sequence to be edited. It is at this point that U insertion or deletion occurs: unpaired gRNA nucleotides next to the anchor region use free UTP for a U insertion and further, non-base-paired uridylates in the pre-mRNA are deleted.[40] Importantly, and in addition to traditional RNA editing, the utilization of atypical gRNAs can cause alternative editing events to occur.[40] The progression of all RNA editing processes has been suggested to be coordinated by important multiprotein complexes, termed editosomes, the importance of which is highlighted later in this chapter.[39–41]

12.3.3.1 Alternative mRNA Editing

It has been proposed that alternative RNA editing events can be described as the addition or deletion of any uridine that henceforth produces open reading frame (ORF) changes that differ from anticipated coding sequences.[42] The utilization of different gRNAs for alternative editing events contributes greatly to protein diversity within the parasite mitochondria by producing functionally discrete protein isoforms that have different enzymatic activities, expression levels, substrate specificities, localization characteristics or interaction abilities with other proteins, DNA and/or RNA (Figure 12.2).[40,42] Additionally, it is thought that alternative editing may have importance in producing mitochondrial proteins whose functions are specific to trypanosomes.[42]

12.3.3.2 Editosome

Functional editosome proteins have been identified by several methods, including mass spectrometry, tandem affinity purification (TAP) and electron microscopy in combination with single-particle reconstruction techniques.[40,43] Although the number of proteins in the fully functional editosome is not yet known, recent studies have identified upwards of 20 proteins as being highly dynamic components of the stable 20S editosome core complex.[41,44]

The identification, characterization and subsequent knockdown or knockout of proteins that complex with the editosome has shown that RNA editing is essential in both promastigote and amastigote life cycle stages. The essentiality of this process to the survival of the parasites in addition to its specificity to these organisms makes the editosome viable for drug target development. High-throughput virtual and chemical screening of libraries is one approach that has been proposed to identify chemical compounds that specifically target this complex. It is anticipated that any chemical compounds generated from the screening will block or, at minimum, affect one or more steps in the editing cycle, including editosome assembly and termination. It is likely that by interfering with the editosome function and assembly *via* a chemical genetics approach, rather than RNAi or gene replacement strategies, we will be more informed of likely functions of specific components of editosome proteins and their order of assembly. Therefore, the benefit of developing and characterizing editosome inhibitors encompasses opportunities for potential therapeutics as well as intimate details of editosome assembly and function.[43]

12.4 Post-transcriptional Regulation of Gene Expression

As previously discussed, the *Leishmania* genome encodes few potential regulatory transcription factors, making post-transcriptional modifications central to gene expression regulation. Apart from *trans*-splicing and RNA editing mechanisms, additional factors contribute to gene expression variation including mRNA degradation and untranslated region (UTR) control.[45] However, as stated above, it is worth mentioning that variations in chromosome and gene copy numbers among strains of *L. major*, *L. infantum* and *L. braziliensis* have unveiled a degree of aneuploidy for the first time in these parasites.[46] Therefore, rather than relying on differential expression of single genes and gene products, 'genome plasticity' and 'gene dosage' are being increasingly considered as important factors for disease progression in *Leishmania*;[47] this makes it challenging to employ transcriptomics as the sole tool to identify potential drug targets.

12.4.1 3′ UTR Control and mRNA Degradation

The regulation of genes post-transcriptionally in trypanosomatid parasites is well documented within the *Leishmania* community. In depth analysis has demonstrated direct involvement of untranslated regions in such regulation, where sequences within both[48] or either 5′ and 3′ UTRs play central roles in the molecular machinery responsible for mRNA stability and translation.[49,50] For example, the completion of trypanosomatid whole-genome sequencing allowed the discovery of "extinct" transposable elements [short interspersed degenerated retroposons (SIDERs)] located in the 3′ UTR where they, despite previously being thought of as "junk" DNA, are actually functionally important in gene expression regulation.[50] To be more specific, gene expression may be altered by the degradation of individual mRNAs, the process of which is initiated by endonuclease cleavage by sequence-specific endonucleases binding within the 3′ UTR.[51] Recently, an Alba-domain protein has been identified in *Leishmania* as a specific RNA-binding factor to the 3′ UTR of the amastin gene and its role in mRNA destabilization, specifically in the amastigote life cycle stage of the parasite, has been shown.[52] Currently, however, not much is known about the mechanisms underlying mRNA degradation in *Leishmania* parasites.[53]

12.4 Perspectives

Until recently, genomic and transcriptomic analyses were unmatched in their ability to enable highly parallel insight into the biology of cells and organisms.[54] Proteomics, and more recently metabolomics, have since evolved and are now contributing to expanding this knowledge[21] (see Chapter 6 of this volume). Because *Leishmania* parasites regulate gene expression *via* post-transcriptional modifications, proteomics is thought to yield additional, yet essential, insights into the mechanisms of protein regulation and function, including stage differentiation, species differences, virulence and drug resistance. Altogether, it is crucial that we take into consideration every fundamental aspect of the molecular biology of these parasites in our endeavors to develop successful anti-leishmanial therapeutics.

References

1. R. W. Ashford, P. Desjeux and P. Deraadt, *Parasitol. Today*, 1992, **8**, 104.
2. A. L. Banuls, M. Hide and F. Prugnolle, *Adv. Parasitol.*, 2007, **64**, 1.
3. A. Ambit, K. L. Woods, B. Cull, G. H. Coombs and J. C. Mottram, *Eukaryotic Cell*, 2011, **10**, 1429.
4. B. L. Herwaldt, *Lancet*, 1999, **354**, 1191.
5. B. Papadopoulou, X. Huang, N. Boucher and F. McNicoll, *ASM News*, 2003, **69**, 282.
6. S. D. Kruchten, K. M. Bacon and B. Y. Lee, *Am. J. Trop. Med. Hyg.*, 2014, **91**, 520.
7. O. P. Singh and S. Sundar, *Front. Immunol.*, 2014, **5**, 296.
8. D. Otranto and F. Dantas-Torres, *Trends Parasitol.*, 2013, **29**, 339.
9. C. Cantacessi, F. Dantas-Torres, M. J. Nolan and D. Otranto, *Trends Parasitol.*, 2015, **31**, 100.
10. A. C. Ivens, C. S. Peacock, E. A. Worthey, L. Murphy, G. Aggarwal, M. Berriman, E. Sisk, M. A. Rajandream, E. Adlem, R. Aert, A. Anupama, Z. Apostolou, P. Attipoe, N. Bason, C. Bauser, A. Beck, S. M. Beverley, G. Bianchettin, K. Borzym, G. Bothe, C. V. Bruschi, M. Collins, E. Cadag, L. Ciarloni, C. Clayton, R. M. Coulson, A. Cronin, A. K. Cruz, R. M. Davies, J. De Gaudenzi, D. E. Dobson, A. Duesterhoeft, G. Fazelina, N. Fosker, A. C. Frasch, A. Fraser, M. Fuchs, C. Gabel, A. Goble, A. Goffeau, D. Harris, C. Hertz-Fowler, H. Hilbert, D. Horn, Y. Huang, S. Klages, A. Knights, M. Kube, N. Larke, L. Litvin, A. Lord, T. Louie, M. Marra, D. Masuy, K. Matthews, S. Michaeli, J. C. Mottram, S. Muller-Auer, H. Munden, S. Nelson, H. Norbertczak, K. Oliver, S. O'Neil, M. Pentony, T. M. Pohl, C. Price, B. Purnelle, M. A. Quail, E. Rabbinowitsch, R. Reinhardt, M. Rieger, J. Rinta, J. Robben, L. Robertson, J. C. Ruiz, S. Rutter, D. Saunders, M. Schafer, J. Schein, D. C. Schwartz, K. Seeger, A. Seyler, S. Sharp, H. Shin, D. Sivam, R. Squares, S. Squares, V. Tosato, C. Vogt, G. Volckaert, R. Wambutt, T. Warren, H. Wedler, J. Woodward, S. Zhou, W. Zimmermann, D. F. Smith, J. M. Blackwell, K. D. Stuart, B. Barrell and P. J. Myler, *Science*, 2005, **309**, 436.
11. M. Berriman, E. Ghedin, C. Hertz-Fowler, G. Blandin, H. Renauld, D. C. Bartholomeu, N. J. Lennard, E. Caler, N. E. Hamlin, B. Haas, U. Bohme, L. Hannick, M. A. Aslett, J. Shallom, L. Marcello, L. Hou,

B. Wickstead, U. C. Alsmark, C. Arrowsmith, R. J. Atkin, A. J. Barron, F. Bringaud, K. Brooks, M. Carrington, I. Cherevach, T. J. Chillingworth, C. Churcher, L. N. Clark, C. H. Corton, A. Cronin, R. M. Davies, J. Doggett, A. Djikeng, T. Feldblyum, M. C. Field, A. Fraser, I. Goodhead, Z. Hance, D. Harper, B. R. Harris, H. Hauser, J. Hostetler, A. Ivens, K. Jagels, D. Johnson, J. Johnson, K. Jones, A. X. Kerhornou, H. Koo, N. Larke, S. Landfear, C. Larkin, V. Leech, A. Line, A. Lord, A. Macleod, P. J. Mooney, S. Moule, D. M. Martin, G. W. Morgan, K. Mungall, H. Norbertczak, D. Ormond, G. Pai, C. S. Peacock, J. Peterson, M. A. Quail, E. Rabbinowitsch, M. A. Rajandream, C. Reitter, S. L. Salzberg, M. Sanders, S. Schobel, S. Sharp, M. Simmonds, A. J. Simpson, L. Tallon, C. M. Turner, A. Tait, A. R. Tivey, S. Van Aken, D. Walker, D. Wanless, S. Wang, B. White, O. White, S. Whitehead, J. Woodward, J. Wortman, M. D. Adams, T. M. Embley, K. Gull, E. Ullu, J. D. Barry, A. H. Fairlamb, F. Opperdoes, B. G. Barrell, J. E. Donelson, N. Hall, C. M. Fraser, S. E. Melville and N. M. El-Sayed, *Science*, 2005, **309**, 416.

12. N. M. El-Sayed, P. J. Myler, D. C. Bartholomeu, D. Nilsson, G. Aggarwal, A. N. Tran, E. Ghedin, E. A. Worthey, A. L. Delcher, G. Blandin, S. J. Westenberger, E. Caler, G. C. Cerqueira, C. Branche, B. Haas, A. Anupama, E. Arner, L. Aslund, P. Attipoe, E. Bontempi, F. Bringaud, P. Burton, E. Cadag, D. A. Campbell, M. Carrington, J. Crabtree, H. Darban, J. F. da Silveira, P. de Jong, K. Edwards, P. T. Englund, G. Fazelina, T. Feldblyum, M. Ferella, A. C. Frasch, K. Gull, D. Horn, L. Hou, Y. Huang, E. Kindlund, M. Klingbeil, S. Kluge, H. Koo, D. Lacerda, M. J. Levin, H. Lorenzi, T. Louie, C. R. Machado, R. McCulloch, A. McKenna, Y. Mizuno, J. C. Mottram, S. Nelson, S. Ochaya, K. Osoegawa, G. Pai, M. Parsons, M. Pentony, U. Pettersson, M. Pop, J. L. Ramirez, J. Rinta, L. Robertson, S. L. Salzberg, D. O. Sanchez, A. Seyler, R. Sharma, J. Shetty, A. J. Simpson, E. Sisk, M. T. Tammi, R. Tarleton, S. Teixeira, S. Van Aken, C. Vogt, P. N. Ward, B. Wickstead, J. Wortman, O. White, C. M. Fraser, K. D. Stuart and B. Andersson, *Science*, 2005, **309**, 409.
13. N. M. El-Sayed, P. J. Myler, G. Blandin, M. Berriman, J. Crabtree, G. Aggarwal, E. Caler, H. Renauld, E. A. Worthey, C. Hertz-Fowler, E. Ghedin, C. Peacock, D. C. Bartholomeu, B. J. Haas, A. N. Tran, J. R. Wortman, U. C. Alsmark, S. Angiuoli, A. Anupama, J. Badger, F. Bringaud, E. Cadag, J. M. Carlton, G. C. Cerqueira, T. Creasy, A. L. Delcher, A. Djikeng, T. M. Embley, C. Hauser, A. C. Ivens, S. K. Kummerfeld, J. B. Pereira-Leal, D. Nilsson, J. Peterson, S. L. Salzberg, J. Shallom, J. C. Silva, J. Sundaram, S. Westenberger, O. White, S. E. Melville, J. E. Donelson, B. Andersson, K. D. Stuart and N. Hall, *Science*, 2005, **309**, 404.
14. C. S. Peacock, K. Seeger, D. Harris, L. Murphy, J. C. Ruiz, M. A. Quail, N. Peters, E. Adlem, A. Tivey, M. Aslett, A. Kerhornou, A. Ivens, A. Fraser, M. A. Rajandream, T. Carver, H. Norbertczak, T. Chillingworth,

Z. Hance, K. Jagels, S. Moule, D. Ormond, S. Rutter, R. Squares, S. Whitehead, E. Rabbinowitsch, C. Arrowsmith, B. White, S. Thurston, F. Bringaud, S. L. Baldauf, A. Faulconbridge, D. Jeffares, D. P. Depledge, S. O. Oyola, J. D. Hilley, L. O. Brito, L. R. Tosi, B. Barrell, A. K. Cruz, J. C. Mottram, D. F. Smith and M. Berriman, *Nat. Genet.*, 2007, **39**, 839.
15. F. Real, R. O. Vidal, M. F. Carazzolle, J. M. Mondego, G. G. Costa, R. H. Herai, M. Wurtele, L. M. de Carvalho, R. Carmona e Ferreira, R. A. Mortara, C. L. Barbieri, P. Mieczkowski, J. F. da Silveira, M. R. Briones, G. A. Pereira and D. Bahia, *DNA Res.*, 2013, **20**, 567.
16. J. K. Stiles, P. I. Hicock, P. H. Shah and J. C. Meade, *Ann. Trop. Med. Parasitol.*, 1999, **93**, 781.
17. P. J. Myler, L. Audleman, T. deVos, G. Hixson, P. Kiser, C. Lemley, C. Magness, E. Rickel, E. Sisk, S. Sunkin, S. Swartzell, T. Westlake, P. Bastien, G. Fu, A. Ivens and K. Stuart, *Proc. Natl. Acad. Sci. U. S. A.*, 1999, **96**, 2902.
18. E. A. Worthey, S. Martinez-Calvillo, A. Schnaufer, G. Aggarwal, J. Cawthra, G. Fazelinia, C. Fong, G. Fu, M. Hassebrock, G. Hixson, A. C. Ivens, P. Kiser, F. Marsolini, E. Rickel, R. Salavati, E. Sisk, S. M. Sunkin, K. D. Stuart and P. J. Myler, *Nucleic Acids Res.*, 2003, **31**, 4201.
19. N. E. Padilla-Mejia, L. E. Florencio-Martinez, R. Moreno-Campos, J. C. Vizuet-de-Rueda, A. M. Cevallos, R. Hernandez-Rivas, R. Manning-Cela and S. Martinez-Calvillo, *Eukaryotic Cell*, 2015, **14**, 216.
20. M. G. Mayer and L. M. Floeter-Winter, *Mem. Inst. Oswaldo Cruz*, 2012, **107**, 1070.
21. *Leishmania: After the Genome*, ed. P. Myler and N. Fasel, Caister Academic Press, United Kingdom, 2008.
22. L. A. Dillon, K. Okrah, V. K. Hughitt, R. Suresh, Y. Li, M. C. Fernandes, A. T. Belew, H. Corrada Bravo, D. M. Mosser and N. M. El-Sayed, *Nucleic Acids Res.*, 2015, **43**, 6799.
23. R. Soysa, N. S. Carter and P. A. Yates, *Mol. Biochem. Parasitol.*, 2014, **195**, 1.
24. K. Leifso, G. Cohen-Freue, N. Dogra, A. Murray and W. R. McMaster, *Mol. Biochem. Parasitol.*, 2007, **152**, 35.
25. S. M. Teixeira, R. M. de Paiva, M. M. Kangussu-Marcolino and W. D. Darocha, *Genet. Mol. Biol.*, 2012, **35**, 1.
26. G. Di Segni, S. Gastaldi and G. P. Tocchini-Valentini, *Proc. Natl. Acad. Sci. U. S. A.*, 2008, **105**, 6864.
27. S. Michaeli, *Future Microbiol.*, 2011, **6**, 459.
28. J. M. Requena, *Front. Biosci., Landmark Ed.*, 2011, **16**, 2069.
29. V. Douris, M. J. Telford and M. Averof, *Mol. Biol. Evol.*, 2010, **27**, 684.
30. N. Badjatia, D. L. Ambrosio, J. H. Lee and A. Gunzl, *Mol. Cell. Biol.*, 2013, **33**, 1965.
31. A. Gunzl, *Eukaryotic Cell*, 2010, **9**, 1159.
32. E. M. Makarov, O. V. Makarova, H. Urlaub, M. Gentzel, C. L. Will, M. Wilm and R. Luhrmann, *Science*, 2002, **298**, 2205.

33. X. H. Liang, A. Haritan, S. Uliel and S. Michaeli, *Eukaryotic Cell*, 2003, **2**, 830–840.
34. C. Tschudi and E. Ullu, *Cell*, 1990, **61**, 459.
35. I. D. Tkacz, S. K. Gupta, V. Volkov, M. Romano, T. Haham, P. Tulinski, I. Lebenthal and S. Michaeli, *J. Biol. Chem.*, 2010, **285**, 27982.
36. R. J. van Alphen, E. A. Wiemer, H. Burger and F. A. Eskens, *Br. J. Cancer*, 2009, **100**, 228.
37. A. K. Panigrahi, A. Schnaufer and K. D. Stuart, *Methods Enzymol.*, 2007, **424**, 3.
38. A. Schnaufer, G. J. Domingo and K. Stuart, *Int. J. Parasitol.*, 2002, **32**, 1071.
39. A. K. Panigrahi, A. Schnaufer, N. L. Ernst, B. Wang, N. Carmean, R. Salavati and K. Stuart, *RNA*, 2003, **9**, 484.
40. H. U. Goringer, V. S. Katari and C. Bohm, *Wiley Interdiscip. Rev. RNA*, 2011, **2**, 395.
41. J. Carnes, J. R. Trotter, A. Peltan, M. Fleck and K. Stuart, *Mol. Cell. Biol.*, 2008, **28**, 122.
42. T. Ochsenreiter, M. Cipriano and S. L. Hajduk, *PLoS One*, 2008, **3**, e1566.
43. R. Salavati, H. Moshiri, S. Kala and H. Shateri Najafabadi, *Int. J. Parasitol. Drugs Drug Resist.*, 2012, **2**, 36.
44. A. K. Panigrahi, N. L. Ernst, G. J. Domingo, M. Fleck, R. Salavati and K. D. Stuart, *Ribonucleic Acid*, 2006, **12**, 1038.
45. C. Clayton and M. Shapira, *Mol. Biochem. Parasitol.*, 2007, **156**, 93.
46. M. B. Rogers, J. D. Hilley, N. J. Dickens, J. Wilkes, P. A. Bates, D. P. Depledge, D. Harris, Y. Her, P. Herzyk, H. Imamura, T. D. Otto, M. Sanders, K. Seeger, J. C. Dujardin, M. Berriman, D. F. Smith, C. Hertz-Fowler and J. C. Mottram, *Genome Res.*, 2011, **21**, 2129.
47. Y. Sterkers, L. Lachaud, N. Bourgeois, L. Crobu, P. Bastien and M. Pages, *Mol. Microbiol.*, 2012, **86**, 15.
48. R. Aly, M. Argaman, S. Halman and M. Shapira, *Nucleic Acids Res.*, 1994, **22**, 2922.
49. L. Quijada, M. Soto, C. Alonso and J. M. Requena, *Mol. Biochem. Parasitol.*, 2000, **110**, 79.
50. F. Bringaud, M. Muller, G. C. Cerqueira, M. Smith, A. Rochette, N. M. El-Sayed, B. Papadopoulou and E. Ghedin, *PLoS Pathog.*, 2007, **3**, 1291.
51. R. Parker and H. Song, *Nat. Struct. Mol. Biol.*, 2004, **11**, 121.
52. A. Dupe, C. Dumas and B. Papadopoulou, *Mol. Microbiol.*, 2014, **91**, 548.
53. S. Haile, A. Dupe and B. Papadopoulou, *Nucleic Acids Res.*, 2008, **36**, 1634.
54. D. Paape and T. Aebischer, *J. Proteomics*, 2011, **74**, 1614.

CHAPTER 13

The Physical Matrix of the Plasma Membrane as a Target: The Charm of Drugs with Low Specificity

LUIS RIVAS,*[a] MONTSERRAT NÁCHER-VÁZQUEZ[a] AND DAVID ANDREU[b]

[a] Department of Chemical and Physical Biology, Centro de Investigaciones Biológicas (CSIC), Ramiro de Maeztu 9, 28040 Madrid, Spain;
[b] Department of Experimental and Health Sciences, Pompeu Fabra University, Barcelona Biomedical Research Park, 08003 Barcelona, Spain
*Email: luis.rivas@cib.csic.es

13.1 Outline for an Antimicrobial Peptide-based Chemotherapy against Leishmaniasis

Of the so called neglected tropical diseases (NTDs), leishmaniasis is positioned among those with higher importance for human health. As mentioned recurrently throughout this volume, chemotherapy is nowadays the only effective treatment for this disease, aside from those treatments based on physical methods, as surgical laser ablation of the ulcer,[1] or thermotherapy or cryotherapy,[2] limited to some forms of cutaneous leishmaniasis (CL).

This chapter is focused on the interaction of antimicrobial peptides (AMPs) and peptide-like agents with the plasma membrane of *Leishmania*. Peptide-based chemotherapy is a rather heterodox approach in chemotherapy; much

Drug Discovery Series No. 60
Drug Discovery for Leishmaniasis
Edited by Luis Rivas and Carmen Gil

Published by the Royal Society of Chemistry, www.rsc.org

more so when intended for leishmaniasis. Membrane is the ultimate and preformed target for the peptides. As such, AMPs either cause disruption of the structure of the phospholipid matrix of the plasma membrane, leading to the loss of intracellular homeostasis, or reach their intracellular targets by translocation throughout the phospholipid bilayer in the absence of a cognate protein transporter. In any event, AMPs utilize a new and scarcely exploited target.

13.2 General Appraisal of Peptide-based Therapies

Compared with drugs fulfilling the Lipinsky rules, AMPs have poor bioavailability (being prone to proteolytic degradation and with limited tissue penetration), and their production costs largely surpass those for classical antibiotic therapy. On the other hand, peptides offer almost endless diversity.

According to the Global Peptide Therapeutics Market, 136 peptides are currently under clinical use, 308 in preclinical use and 722 in the pipeline.[3] Nowadays there is a positive appraisal of peptide therapies by the pharma industry.[4,5]

The field of anti-infective peptides has been recently reviewed.[6,7] Some of the reviews address AMPs on parasites or are even specifically focused on *Leishmania*.[8–14]

13.3 Natural History of AMPs

The astounding variety of AMPs runs parallel to biodiversity. Each organism has its own set of AMPs, increased by the proteolytic trimming of mature AMPs,[15,16] unmasking new antimicrobial cryptopeptides, as well as by the addition of proteins or peptides with antimicrobial activities that were formerly described for other functionalities such as chemokines,[17] nucleic acid binding proteins[18] or neuropeptides,[19] among others.

AMPs are regularly compiled in different updated databases (APD3;[20] CAMP[21]). The relatively uncomplicated rules for AMP identification, underlie the definition of putative AMPs by genome interrogation through dedicated algorithms.[22]

AMPs are components of the innate immunity, and, as such, deployed at the anatomical locations in first contact with the invading pathogen, such as mucosae,[23–26] biological fluids,[27,28] or professional phagocytes.[29] They are endowed with a broad specificity of pathogen recognition, which encompasses viruses,[30] bacteria,[31,32] fungi,[33] and even tumoural cells.[34] Protozoans as AMP targets have been scarcely addressed,[8,12,35,36] with anti-protozoan activity described for only 2.1% of more than 10 000 AMPs compiled in the DRAMP database.[37]

In addition, some AMPs have additional functionalities not related to microbicidal activity. For instance, they can prime antigen-specific immunity, thus establishing a cross-talk between the two branches of immunity.[38,39] Also, specific AMPs are involved in wound healing,[40,41] anti-endotoxic activities[42] and even angiogenesis.[43]

Targeting of the pathogen cell membrane by AMPs is effected through structure-derived physico-chemical interactions which do not require a defined sequential motif.[44] In this regard, AMPs have been structurally classified into four major groups: (i) α-helical AMPs, (ii) AMPs enriched in certain residues poorly represented in average protein composition (His, Pro, Gly, Trp or Phe), (iii) cyclic peptides and (iv) circular AMPs with N- and C-terminal ends fused by means of a peptide bond.

13.4 Mechanism of Action of Antimicrobial Peptides

AMP–membrane interaction is mandatory for functionality, regardless of whether membranes are the ultimate AMP target, or the interaction is limited to a translocation process in order to access intracellular targets.

13.4.1 Molecular Characteristics of AMPs

Most structure–activity relationships (SAR) studies for AMPs have understandably been performed on α-helical peptides whose easy manipulation allows straightforward structural parametrization.[45,46]

AMP lengths vary from 8 up to 50 residues. Most AMPs described to date are cationic and amphipathic, with net charge between +2 and +9. This feature underlies their preferential recognition of anionic membranes, typical of pathogens, as the basis of their specificity (see Section 13.4.2). While most AMPs are unstructured in aqueous solution, interaction with the membrane can drive α-helix formation, and as about 50% of the total residues are hydrophobic, the acquisition of an amphipathic (helical) structure is a strong driving force for membrane insertion that facilitates the interaction of AMP with acyl chain regions of the bilayer and disruption of phospholipid packing.[47,48] Nevertheless, helicity is not essential for antimicrobial activity, while hydrophobicity is associated with cytotoxicity.[47,49,50] Similarly, a positive charge is no longer viewed as a must for antimicrobial activity, as a growing number of AMPs discovered in recent years are anionic.[51,52]

13.4.2 Antimicrobial Peptide–Membrane Interaction

Membrane is a universal structure preserved throughout evolution. Thus, the self *vs.* non-self cell membrane differentiation by AMPs is mostly achieved by the phospholipid charge and orientation at both sides of the membrane.[53,54] The net electric charge of the external leaflet is the main determinant of AMP specificity. Prokaryotic and lower eukaryotic cells have a higher percentage of anionic phospholipids in their membranes and, even more relevantly, these phospholipids face the external medium, thus favouring privileged interaction with cationic AMPs. In contrast, higher eukaryotes display zwitterionic phospholipids to the extracellular space, thus impairing interaction with AMPs.

Other factors decreasing AMP–membrane interaction include: (i) The geometrical shape of the phospholipids near the inserted AMP, phosphatidylethanolamine being of special relevance due to its truncated cone shape.[55] (ii) Lower membrane fluidity, associated with phospholipids with long and saturated acyl chains,[56] or the presence of sterols. In this respect, ergosterol is less effective than cholesterol.[57,58] (iii) Lower membrane potential values, negative at the cytoplasmic side the membrane. Metabolically quiescent cells are more resistant to AMPs due to their lower membrane potential.[59] (iv) Biological barriers beyond the plasma membrane (outer membrane, peptidoglycan layer, capsule, glycocalyx, *etc.*). These structures may trap or sterically hinder the access of AMPs to the cell membrane.[60] (v) Highly proteolytic environments that may degrade AMPs.[61] (vi) AMP removal by efflux pumps.[62]

13.4.3 Models for Antimicrobial Peptide–Membrane Interaction

Several models have been proposed to account for the permeabilisation of the membrane by AMPs.[53,63] The same AMP may be allocated into different models according to its concentration and composition of the targeted membrane.

Barrel-stave Model. The pore is formed exclusively by peptides lying perpendicular to the membrane plane like the staves of a barrel. The hydrophilic surface area of the AMP (helix) faces the aqueous lumen of the pore. Pore assembly is driven by high affinity between peptide monomers relative to peptide–phospholipid affinity and displays low membrane discrimination, typical of toxins.

Leaky slit Model. This is a variant of the barrel-stave. In this case AMP monomers aggregate into a fibrilar amyloid amphipathic structure instead of forming a circular pore.

Carpet-like Model. AMPs act as a sophisticated biological detergent, leading to micellisation and catastrophic membrane disruption once the density of peptides inserted into the membrane reaches a threshold.

Toroidal Pore, Sinking Raft, and Molecular Electroporation Models. These are variants of the aforementioned models, accounting for special pore geometries or for mass or electric imbalance created by the initial asymmetrical binding of AMPs on the external membrane layer.

Phospholipid Clustering Model. This is based on preferential binding of membrane-inserted cationic AMP to the anionic phospholipids, leading to a spatial rearrangement of phospholipids, with creation of new microdomain boundaries, that cause faulty phospholipid packing and, hence, membrane leakage.[64]

13.4.4 The Plasma Membrane of *Leishmania* as a Target for Antimicrobial Peptides

Like other trypanosomatids, *Leishmania* has a subpellicular layer of microtubules running longitudinally underneath its plasma membrane.[65] Only the flagellar pocket, a specialized area of the plasma membrane from which the flagellum emerges, is devoid of such a network.[66] This limits all endo- and exocytic membrane traffic into this area,[67] as well as feasible membrane repair mechanisms active in *Leishmania*. The major components of the plasma membrane facing the extracellular medium are anchored to the membrane through glycosylphosphatidylinositol structures (GPI), to avoid their restrained diffusion on the plane of the membrane by this microtubular network.[68]

Leishmania has a well-developed glycocalyx, formed by different glycoconjugates and heavily phosphoglycosylated proteins. Lipophosphoglycan (LPG) is mainly responsible for the highly anionic character of the promastigote surface.[69,70] LPG expression has been estimated at $1.3–6\times10^6$ copies per cell, depending on the species, and accounting for a surface coverage of 20 to 80%.[71] LPG expression scarcely reaches the detection threshold in the amastigote.[70,72,73]

The basic structure of LPG encompasses four different domains.[69,73] The anionicity of LPG is due to the repeating phosphorylated unit, formed by the polymerization of the phosphodiester disaccharide $[6Gal\beta(1\text{-}4)Man\alpha1\text{-}PO_4]_{n=14-30}$. During metacyclogenesis LPG undergoes changes in the substituent sugars, and for species belonging to the *Leishmania donovani* and *Leishmania major* complexes, duplication in the number of its repetitive units.[74,75]

The glycosylinositol phospholipids (GIPLs)[76–78] are small heterogeneous phosphatidylinositol oligosaccharides, of up to eight saccharide units, anchored to the membrane through alkyl–acyl phosphatidylinositol or lyso–alkyl phosphatidylinositol GPI motifs. In molar terms, GIPLs are the most abundant components of *Leishmania* membrane, both for promastigote ($1–4\times10^7$ copies per cell) and amastigotes ($1–1.8\times10^7$ copies per cell).[79] The biological relevance of GIPLs is higher for the amastigote, almost fully devoid of LPG molecules. Ablation of GIPL expression jeopardizes parasite survival in the macrophage, but not inside the sandfly.[68]

Additionally, glycosphingolipids from the host have been detected in *Leishmania mexicana* amastigotes.[79]

Proteophosphoglycans (PPGs) are proteins extremely rich in serine and heavily phosphoglycosylated with oligosaccharides made of $[6Gal\beta(1\text{-}4)\text{-}Man\alpha1\text{-}PO_4]_n$, the repetitive LPG unit.[80,81] PPGs are expressed by both promastigotes and amastigotes, although with different molecular patterns.[80] PPGs have been found in the extracellular media in *in vitro* culture, at the lumen of the sandfly gut, bound to the plasma membrane of the parasite[82] and inside the parasitophorus vacuole of *L. mexicana*-infected macrophages,[80] forming a protective barrier against proteolytic and AMP attack.

The 63 kDa Zn^{2+}-metalloproteinase leishmaniolysin (Gp63) is the most abundant protein in the promastigote (5×10^5 copies per cell), and it is present in all *Leishmania* species. This enzyme has a broad substrate activity and loose pH requirements.[83] Leishmaniolysin has been extensively reviewed from both biological and biochemical perspectives,[84,85] as well as for its functional role in virulence, strongly associated with its level of expression.[86]

Finally, protein surface antigen 2 (PSA-2 or Gp46), the second most abundant protein on the promastigote surface,[87] is ubiquitous in all *Leishmania* species except those of the *Leishmania braziliensis* complex. It may function as a barrier to prevent damage from host proteinases,[87] or to complement lysis.[88]

The lipid composition of different *Leishmania* species has been reported.[89–92] These studies were carried out on promastigotes and without addressing lipid asymmetry.

The lipid composition for *L. major* promastigotes was 33% phosphatidylcholine (PC), 10% phosphatidylinositol (PI), 10% phosphatidylethanolamine (PE), 7% other phospholipids, 10% sphingolipids and 33% sterol plus other lipid species.[93] In *L. donovani* promastigotes 188 species have been detected, PCs being the most diverse, with 59 species.[91] The composition disclosed for the plasma membrane of promastigotes was as 15% PC, 37% PE, 18% PI and 10% PS.[94]

The most abundant phospholipid in *Leishmania* is PC, with long and highly unsaturated acyl chains.[95] In *L. major*; PE appears mostly as the plasmalogen 1-*O*-alk-1′-enyl-2-acyl-*sn*-glycero-3-phosphoethanolamine or plasmenylethanolamine (PME), accounting for 80–90% of the total PE.[96] The free-form of PI appears under the diacyl form. Most of the PI is committed to the formation of GPI anchors.[93] Finally, the presence of PS in *Leishmania* is controversial.[97]

Sphingolipids (SL) are second in importance after glycerophospholipids among *Leishmania* lipids. *Leishmania* lacks sphingomyelin; the functionality of sphingolipids in *Leishmania* is to supply ethanolamine and phosphoethanolamine as a building block in aminophospholipid biosynthesis.[98]

Similarly to fungi, the main SL species in *Leishmania* is inositolphosphorylceramide (IPC),[99] that accounts for 5–10% of the total lipid in *Leishmania*, d16:1/18:0-IPC being the most abundant.[93]

The sterol composition of *Leishmania* consists of species biosynthesized by the parasite through the ergosterol pathway, plus cholesterol acquired from the host. For *L. donovani* promastigotes, ergosterol isomers are the most abundant, followed by ergosta-7,22-dien-3β-ol, stigmasta-7,24 (28)-dien-3β-ol and cholesterol.[100] Ergosterol and cholesterol account for 70% and 15%, respectively, of the total sterol of the *Leishmania infantum* procyclic promastigotes, whereas the percentages for stationary promastigotes are 40% and 28%, respectively.[100]

The high sterol and SL content in *Leishmania* favours the existence of lipid rafts or detergent-resistant membrane (DRM), enriched in IPC and GPI-anchored proteins, such as PSA-2 and leishmaniolysin, and required for virulence.[101]

Furthermore, through the life cycle of *Leishmania*, plasma membrane architecture undergoes substantial changes needed to survive in different environments and to maintain virulence.

13.5 Natural AMPs as Leishmanicidal Agents

AMPs with leishmanicidal activity are listed in Tables 13.1 to 13.4. No bias with respect to biological origin has been introduced. AMPs from invertebrates (Table 13.1) and amphibians (Table 13.2) are the most thoroughly studied.[8,102] Of special relevance for chemotherapy are AMPs of mammalian origin (Table 13.3). It is worth noting the growing number of AMPs of marine origin, especially from cyanobacteria endosymbionts from a wide variety of marine invertebrates (Table 13.4 and Figure 13.1).[103,104]

Unfortunately, comparison of activities among substances listed in the different tables is hampered by the different species and stages of *Leishmania*, as well as the variety of incubation media and cellular densities among the references.

Most AMPs are leishmanicidal at low micromolar concentrations. A few natural AMPs (melittin and phylloseptin 1) display a 50% inhibitory concentration (IC_{50}) at a submicromolar range on axenic parasites. The subnanomolar IC_{50}s reported for indolicidin and two fragments of seminal plasmin would appear most likely to be an inadvertent error.[136]

In general, there are no sharp differences in AMP susceptibility among the different species of *Leishmania*, according to the few reports where a given AMP has been compared in several species. For instance, urocortin II has been tested on promastigotes of different *Leishmania* species,[137] with *L. mexicana* and *L. donovani* showing lower susceptibility while IC_{50}s for other species (*L. major*, *L. infantum* and *L. tropica*) were rather similar.

The vasoactive intestinal peptide VIP51(6–30) analogue has also been tested on different *Leishmania* species.[150] Only *L. major* and *L. tropica* showed IC_{50}s under 24 μM, whereas those of *L. mexicana*, *L. donovani* and *L. infantum* were higher.

Susceptibility to AMP among *Leishmania* species does not provide clues to visceralization. For instance, cathelin-related antimicrobial peptide (CRAMP), the murine cathelicidin, is highly expressed on the skin and macrophages, and is active on *L. amazonensis*. Knock-out (KO) CRAMP mice infected with *L. amazonensis* showed larger cutaneous ulcers and, most importantly, visceralization into spleen and liver not observed in normal mice.[130] In contrast, other AMPs such as phylloseptin PLS-8b[126] or temporin-SHd[151] showed rather similar activities on different *Leishmania* species, regardless of whether they cause cutaneous or visceral leishmaniasis.

Axenic amastigotes are more resistant than promastigotes to AMPs, as illustrated by temporins A, B, F, L and 1Sa on *L. mexicana*,[115,145] and ocellatins PT-1 to PT-8 on *L. infantum*.[128] For SPYY, a slower killing of

Table 13.1 Representative examples of leishmanicidal peptides from invertebrates.

Peptide	Sequence	*Leishmania* species and stage tested[a]	Organism
Melittin	GIGAVLKVLTTGLPALISWIKRKRQQ-NH_2	*L. maj.* [**p.**[105] and **a.a.**[106]]; *L. pan.* [**p.**[106]]; *L. inf.* [**p.** and **i.a.**[107]]; *L. don.* [**p.**[108]]	Honey bee: *Apis mellifica*
Decoralin	SLLSLIRKLIT	*L. maj.* [**p.**[109]]	Solitary wasp: *Oreumenes decoratus*
Spinigerin	HVDKKVADKVLLLKQLRIMRLLTRL	*L. don.* [**p.** and **i.a.**[110]]	Termite: *Pseudacanthotermes spiniger*
Cecropin A	KWKLFKKIEKVGQNIRDGIIKAGPAVAVVGQATQIAK-NH_2	*L. aethiop.* [**p.** and **i.a.**[111]]; *L. pan.* [**p.** and **i.a.**[105]]; *L. maj.* [**p.**[105]]; *L. don.* [**p.**[108]]	American silk moth: *Hyalophora cecropia*
Andropin	VFIDILDKMENAIHKAAQAGIGIAKPIEKMILPK	*L. pan.* [**p.** and **i.a.**[105]]; *L. maj.* [**p.**[105]]	Fruit fly: *Drosophila melanogaster*
Sandfly defensin[b]	$ATC^1DLLSAFGVGHAAC^2AAHC^3IGHGYRGGYC^1NSKAVC^2TC^3RR$	*L. maj.* [**p.**[112]]	Sand fly: *Phlebotomus duboscqi*
Gomesin[b]	$QC^1RRLC^2YKQRC^1VTYC^2RGR$-$NH_2$	*L. amaz.* [**p.**[113]]	Spider: *Acanthoscurria gomesiana*
Mytilin-A[b]	$GC^1ASRC^2KAKC^3AGRRC^4KGWASASFRGRC^1YC^2KC^3FRC^4$	*L. braz.* [**p.**[114]]	Mussel: *Mytilus edulis*

[a]*Leishmania* species: *L. aethiop.*, *L. aethiopica*; *L. amaz.*, *L. amazonensis*; *L. braz.*, *L. braziliensis*; *L. don.*, *L. donovani*; *L. inf.*, *L. infantum*; *L. maj.*, *L. major*; *L. pan.*, *L. panamensis*. *Leishmania* stage: **p.**: promastigote; **a.a.**: axenic amastigote; **i.a.**: intracellular amastigote.
[b]Superindex refers to the arrangement of disulfide bonds.

Table 13.2 Representative examples of leishmanicidal peptides from amphibians.

Peptide	Sequence	*Leishmania* species and stage tested[a]	Organism
Temporin B	LLPIVGNLLKSLL-NH_2	*L. mex.* [**p.** and **a.a.**[115,116]]; *L. pif.* [**a.a.**[115,117]]; *L. don.* [**p.**[117]]	European frog: *Rana temporaria*
Temporin-SHd	FLPAALAGIGGILGKLF-NH_2	*L. maj.*, *L. trop.*, *L. amaz.*, *L. braz.*: [**p.**[118]]; *L. inf.* [**p.** and **a.a.** and **i.a.**[118]]	Sahara frog: *Pelophylax saharicus*
Bombinin H2	IIGPVLGLVGSALGGLLKKI-NH_2	*L. pif.* [**a.a.**[119]]; *L. don.* [**p.**[119]]	Yellow-bellied toad: *Bombina variegata*
SPYY	YPPKPESPGEDASPEEMNKYLTALRHYINLVTRQRY	*L. maj.* [**p.** and **i.a.**[120]]	Giant leaf frog: *Phyllomedusa bicolor*
Dermaseptin DRS-1, DS1	ALWKTMLKKLGTMALHAGKAALGAAADTISQGTQ	*L. maj.* [**p.**[105,121,122] and **i.a.**[105]]; *L. mex.* [**p.**[123]]; *L. pan.* [**p.** and **i.a.**[105]]	Painted-belly leaf frog: *Phyllomedusa sauvagii*
Dermaseptin DRS-4, DS4	ALWMTLLKKVLKAAAKAALNAVLVGANA	*L. maj.* [**p.**[121,124,125]]	Painted-belly leaf frog: *Phyllomedusa sauvagii*
Phylloseptin PLS-H8	FLSLLPSLVSGAVSLVKKL	*L. amaz.* [**p.** and **i.a.**[126]]; *L. inf.* [**p.** and **i.a.**[126]]	Frog: *Phyllomedusa nordestina*
Phylloseptin PLS-S1	FLSLIPHIVSGVASIAKHF-NH_2	*L. inf.*, *L. maj.*, *L. braz.*: [**p.**[127]]	Painted-belly leaf frog: *Phyllomedusa sauvagii*
Ocellatin PT-6	GVFDIIKGAGKQLIAHAMEKIAEKVGLNKDGN-NH_2	*L. inf.* [**p.** and **a.a.**[128]]	Frog: *Lectodactylus pustulatus*

[a]*Leishmania* species: *L. amaz.*, *L. amazonensis*; *L. braz.*, *L. braziliensis*; *L. don.*, *L. donovani*; *L. inf.*, *L. infantum*; *L. maj.*, *L. major*; *L. mex.*, *L. mexicana*; *L. pan.*, *L. panamensis*; *L. pif.*, *L. pifanoi*; *L. trop.*, *L. tropica*. *Leishmania* stage: **p.**: promastigote; **a.a.**: axenic amastigote; **i.a.**: intracellular amastigote.

Table 13.3 Representative leishmanicidal peptides of mammalian origin.

Peptide	Sequence	*Leishmania* species and stage tested[a]	Organism
Bovine AMP (BAMP-28)	GGLRSLGRKILRAWKKYGPIIVPIIRIGLDRI	*L. maj.* [**p.** and **i.a.**[129]]	Cattle: *Bos taurus*
CRAMP (cathelicidin)	GLLRKGGEKIGEKLKKIGQKIKNFFQKLVPQPEQ	*L. maj.* [**p.**[130]]	Mouse: *Mus musculus*
LL-37 (cathelicidin)	LLGDFFRKSKEKIGKEFKRIVQRIKDFLRNLVPRTES-NH_2	*L. maj.* [**p.** and **i.a.**[131]]; *L. don.* [**p.** and **i.a.**[131]]	Man: *Homo sapiens*
Histatin 5	DSHAKRHHGYKRKFHEKHHSHRGY	*L. pif.* [**a.a.**[132]]; *L. don.* [**p.**[132]]	Man: *Homo sapiens*
Chemokyne CXCL9[b]	Acc. Number: Q07325	*L. mex.* [**p.**[133]]	Man: *Homo sapiens*
HNP-1 (α-defensin)[c]	AC^{1}YC^{2}RIPAC^{3}IAGERRYGTC^{2}IYQGRLWAFC^{5}C^1	*L. maj.* [**p.** and **i.a.**[134]]	Man: *Homo sapiens*
Θ-defensin 2[c] (θ-defensin)	GVC^{1}RC^{2}LC^{3}RRGVC^{3}RC^{2}LC^{1}RR (cyclic)	*L. maj.* [**p.**[135]]	Rhesus macaque: *Macaccus mulatta*
Protegrin-1[c] (cathelicidin)	RGGRLC^{1}YC^{2}RRRFC^{2}VC^{1}VGR	*L. maj.* [**p.**[135]]	Pig: *Sus scrofa*
Indolicidin (cathelicidin)	ILPWKWPWWPWRR-NH_2	*L. don.* [**p.**[136]]	Cattle: *Bos taurus*
Urocortin II (neuropeptide)	VILSLDVPIGLLRILLEQARYKAARNQAATHAQILAHV	*L. maj.* [**p.** and **i.a.**[137]]; *L. trop.*, *L. mex.*, *L. inf.*, *L. don.*: [**p.**[137]]	Cattle: *Bos taurus*
H2A (histone)[b]	Acc. Number: Q7L7L0	*L. amaz.* [**p.** and **i.a.**[138]]; *L. mex.* [**i.a.**[138]]; *L. braz.* [**p.**[138]]	Man: *Homo sapiens*

[a]*Leishmania* species: *L. amaz.*, *L. amazonensis*; *L. braz.*, *L. braziliensis*; *L. don.*, *L. donovani*; *L. inf.*, *L. infantum*; *L. maj.*, *L. major*; *L. mex.*, *L. mexicana*; *L. pif.*, *L. pifanoi*; *L. trop.*, *L. tropica*. *Leishmania* stage: **p.**: promastigote; **a.a.**: axenic amastigote; **i.a.**: intracellular amastigote.
[b]Entry reference for Uniprot database.
[c]Superindex refers to the arrangement of disulfide bonds.

Table 13.4 Leishmanicidal AMPs from miscellaneous origin: leishmanicidal peptoids and AMP-mimicking agents.

Peptide	Sequence	*Leishmania* species and stage tested[a]	Organism/type
Plant AMPs			
Thionin α-1[b]	$KSC^1C^2RTTLGRNC^3YNLC^4RSRGAQKLC^4STVC^3RC^2KLTSGLSC^1PKGFPK$	*L. pif.* [**a.a.**[139]]; *L. don.* [**p.**[139]]	Wheat: *Triticum aestivum*
Bacteriocins (bacterial AMPs encoded by genes)			
AS-48[c]	Acc. Number: Q47765 (circular peptide)	*L. pif.* [**a.a.**[140]]; *L. don.* [**p.**[140]]	Bacteria: *Enterococcus faecalis*
Non-ribosomally synthesized peptides from cyanobacterial or fungal origin			
IB-01212	2(Ser3,Ser3), ester (Figure 13.1A)	*L. pif.* [**a.a.**[141]]; *L. don.* [**p.**[141]]	Marine fungus: *Clonostachys sp.*
Kahalalide-F(KF)	(Figure 13.1B)	*L. pif.* [**a.a.**[142]]; *L. don.* [**p.**[142]]	Mollusc & green alga: *Elysia rufescens* and *Bryopsis. sp.*
Viridamide A	(Figure 13.1C)	*L. mex.* [**p.**[143]]	Cyanobacteria: *Oscillatoria nigro-viridis*
Almiramide B	(Figure 13.1D)	*L. don.* [**p.**[144]]	Cyanobacteria: *Lyngbya majuscula*
Peptoids and leishmanicidal amphipatic reagents			
Peptoid 16[d]	$(NaeNspeNspe)_4$ (Figure 13.1E)	*L. mex.* [**p.** and **i.a.**[145,146]]	Peptoid
CSA-13	Ceragenin (Figure 13.1F)	*L. maj.* [**p.**[147]]	Cationic steroid peptide
AA1	Arylalkylamide (Figure 13.1G)	*L. maj.* [**p.**[148]]	Arylalkylamide
1212RAc	1,2-dilauroyl-*rac*-glycero-3-*O*-(N^{α}-acetyl-L-arginine) (Figure 13.1H)	*L. pif.* [**a.a.**[149]]; *L. don.* [**p.**[149]]	Diacylglycerol arginine

[a] *Leishmania* species: *L. don.*, *L. donovani*; *L. maj.*, *L. major*; *L. mex.*, *L. mexicana*; *L. pif.*, *L. pifanoi*. *Leishmania* stage: **p.**: promastigote; **a.a.**: axenic amastigote; **i.a.**: intracellular amastigote.
[b] Superindex refers to the arrangement of disulfide bonds.
[c] Entry reference for Uniprot database.
[d] Peptoid nomenclature: *N*ae, *N*-(2-aminoethyl) glycine; *N*spe, (*S*)-*N*-(1-phenylethyl)glycine.

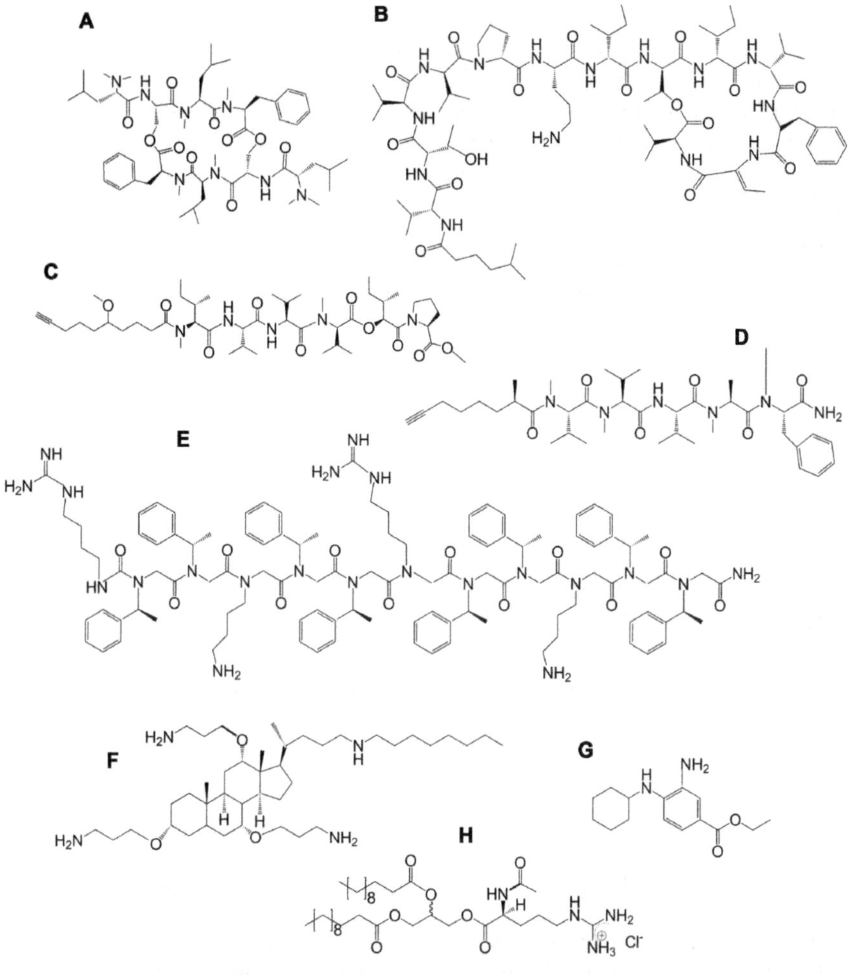

Figure 13.1 Structures mentioned in Table 13.4. (A) IB-01212; (B) Kahalalide F; (C) Viridamide A; (D) Almiramide B; (E) Peptoid NaeNspeNspe; (F) Ceragenin CSA-13; (G) Arylalkylamide AA-1; (H) 1212RAc.

amastigotes compared with promastigotes was observed.[120] Human histones 2A and 2B were exclusively active on the promastigote stage.[138]

The reason for the higher AMP resistance of amastigotes is puzzling. For one thing, they have smaller surfaces, hence higher molar AMP : phospholipid ratios, and express many fewer copies of leishmaniolysin, a resistance factor against AMPs (see Section 13.12.1); plus their membrane potential is even slightly higher than that of promastigotes.[152] Amastigotes from the *mexicana* complex are rich in cysteine proteinases partially secreted into the medium, but the feature is exclusive to this complex.[153] The resilience to AMPs occurs even with peptoids, impervious to proteolytic degradation,[145,154] and with diacyl glycerol arginines, minimalistic mimetics of AMPs.[149] Altogether, the most likely cause of AMP resistance in the amastigote might be lipid composition, but, unfortunately, no report has addressed the subject.

In contrast, intracellular amastigotes are often, though not always, more susceptible to AMPs than promastigotes.[105,111,131,155] Similar observations on *Leishmania* host cells other than macrophages, such as dendritic cells,[105] or neutrophils,[134] have been made. Other AMPs showed similar activities on both stages (phylloseptin PLS-S1[126]), or even an inverted rank of AMP susceptibility for the protamine-derived peptides Pr-1 and Pr-2.[105]

Studies of AMP effectiveness on axenic *vs.* intracellular amastigotes are scarce. Results are available for temporin Shd on *L. infantum*,[151] AS-48 on *L. pifanoi*,[140] and for peptoid 47 on *L. amazonensis*,[154] in all cases intracellular amastigotes being more AMP-susceptible.

All this highlights the importance of macrophages in the final outcome of AMPs. The AMP either concentrates inside the parasitophorus vacuole, where it may work synergistically with other macrophage AMPs, or triggers some leishmanicidal mechanism of the macrophage. AMP visualization inside the parasitophorus vacuole has been barely reported.[132,137,140] One may surmise that a distended parasitophorus vacuole, such as that housing *Leishmania* of the *mexicana* complex, or tightly opposed to the surface of the parasite for the other species, will be relevant to the result.[156] Finally, the acidic pH of the parasitophorus vacuole[157] will increase the cationic character of AMPs, including full protonation of His. A third option is the inhibition of amastigote reinvasion into new macrophages by AMPs, as described for HNP-1[134] or the human histone H2A.[138]

13.6 Assessment of Plasma Membrane Permeabilisation by AMPs in *Leishmania*

Permeabilisation of the *Leishmania* plasma membrane is a key step in the lethal mechanism of AMPs. It is assessed mostly by the entrance of vital dyes, by membrane depolarization, or by the different electron microscopy techniques.

Different vital probes, each with its own fluorescent properties, have been used for a variety of AMPs, SYTOX green, propidium iodide and ethidium homodimer being most frequently used.[117,119,138,158–160]

Membrane depolarization is exclusively dependent on ionic gradients across the plasma membrane; as such, it is more sensitive for monitoring slight changes in membrane permeabilisation. Bisoxonol is the probe commonly selected for this task on *Leishmania*.[117,119,141,150,158–162] Other authors measured the collapse of pH gradient across the plasma membrane due to AMPs.[108,110,139]

The bioenergetic collapse of the parasite is a direct consequence of membrane permeabilisation, and can be easily monitored by the decrease of ATP upon AMP incubation. Total ATP can be evaluated after extraction from AMP-treated parasites. Another alternative is kinetic assessment of the variation of free-cytoplasmic ATP levels after AMP addition on a real-time basis on living parasites.[137,150,159] For this dynamic appraisal, parasites transfected with a cytoplasmic form of firefly luciferase were incubated with AMPs in the presence of a free-membrane permeable caged luciferin analogue. Under these conditions, free ATP is the limiting substrate for the luminescence output.[117,119,139,141,142,158–162]

In other cases, the release of intracellular GFP[136] or lactate dehydrogenase[110] from the parasite after AMP addition was measured.

Transmission electron microscopy (TEM) allows the visualization of membrane disruption caused by AMP, as well as the structural modification of intracellular organelles, robust evidence for the existence of intracellular targets.[132,163]

The formation of blebs at the plasma membrane, with separation of the membrane from the subpellicular layer of microtubules, occurs at sublethal concentrations of AMP. At higher AMP concentrations, parasite ghosts with large membrane disruption, loss of intracellular material and blurred definition of internal organelles were observed by TEM (Figure 13.2).[117,119,137,139,141,142,150,159–162] TEM also defined the autophagic death of *L. donovani* promastigotes due to indolicidin, according to the strong internal vacuolization of the parasite and absence of large lesions at its plasma membrane.[136]

Scanning electron microscopy can also document membrane damage by AMPs.[128,133,138] AMPs increase the roughness of the promastigote surface and induce pits and circular features, assimilated to blisters. The same

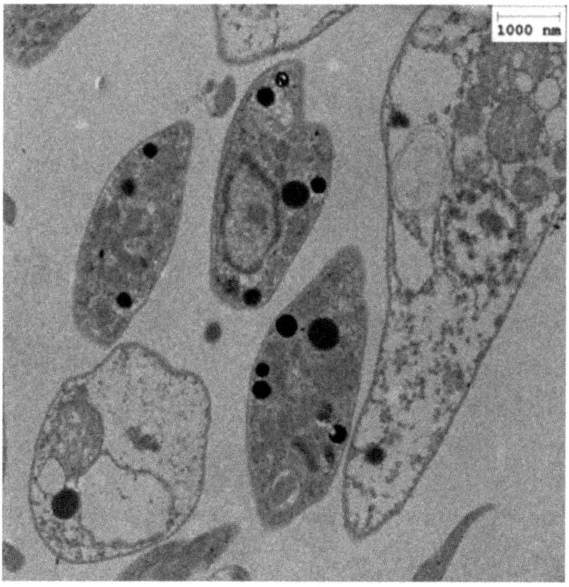

Figure 13.2 Electron micrograph of *Leishmania donovani* promastigotes incubated with the hybrid peptide lactoferrampin (265–284)-lactoferricin (17–30). Incubation conditions: 4 h, 26 °C; peptide concentration: 1.7 μM (IC_{50}).

conclusions have been drawn by the use of atomic force microscopy.[128,164] Freeze–fracture techniques visualized deep structural damage on the *Leishmania* membrane.[123]

The cellular distribution of AMP on *Leishmania* has been documented using either antibodies against the respective AMP[123,138] or fluorescent versions.[130,132,135,137,140,150]

Despite its intrinsic caveats (see Section 13.4.2), the use of artificial membranes remains the standard method to study AMP–membrane interactions. Thus, the interaction of dermaseptin DS 01 on membranes from a lipid extract of *L. amazonensis* promastigotes,[165] or of bombinins H2 and H4 with (PE : PC : PI : PS : ergosterol, 4 : 2 : 2 : 1 : 3) lipid composition,[119] are the only reported attempts to simulate the plasma membrane of *Leishmania*. The biophysical parameters disclosed by these studies highlight some differential features of the interaction of AMP with *Leishmania* membrane with respect to those from mammalian or bacterial origin.

13.7 Intracellular Targets

Identification of intracellular targets for AMPs follows similar rules to those for small-molecule drugs. Phenotypic inhibition assays for the suspected targets, pull-down from a parasite lysate by capture with immobilized peptides followed by proteomic analysis, or two-hybrid analysis, are the standard methodologies in this regard.[166–168]

Some clues for the involvement of intracellular targets are the visualization of the AMP inside the parasite at its IC_{50},[163] or a higher leishmanicidal effect at long kinetic points in the absence of strong membrane permeabilisation, as described for the potato defensin PTH1,[139] tachyplesin,[114] or FW5-magainin and its analogues.[161]

In rare cases a defined intracellular target has been identified. The mitochondrial F_0F_1 ATP synthase for histatin 5,[132] or down-regulation of expression of ascorbate peroxidase and trypanothione reductase, together with ROS production by spinigerin,[110] are cases in point.

13.8 Induction of Programmed Cell Death of *Leishmania* by Antimicrobial Peptides

Programmed cell death can be triggered on *Leishmania* by AMPs. The mechanism in Trypanosomatidae has not been yet fully unveiled, but it differs significantly from mammalian apoptosis,[169,170] due to the absence of the extrinsic pathway, and the functional replacement of caspases by cysteine proteinases, including metacaspases with different substrate specificity.

AMPs are classified with respect to their mode of leishmanicidal mechanisms into those that proceed through non-apoptotic (Class I) or apoptotic-like death (Class II). The apoptotic-like process is characterized by an increase in intracellular Ca^{2+}, PS exposure, mitochondrial depolarization, release of cytochrome *c* and of the endonuclease EndoG from mitochondria, and in some cases activation of specific cysteine proteinases and DNA nucleosomal degradation.[171]

Apoptosis without caspase activation has been described for spinigerin,[110] and a subtle modification of a single residue in kinocidins changed an apoptotic death of *L. donovani* promastigotes into necrosis.[172] Finally, killing of *L. donovani* promastigotes by indolicidin occurs through an autophagic pathway.[136]

The importance of the class of death is based on the deactivation of the inflammatory response of the macrophage by apoptotic parasites, presumably protective for the host. This deactivation is not produced by necrotic parasites.

13.9 Structure–Activity Relationship of Leishmanicidal Antimicrobial Peptides

Improvement of the natural structures of AMPs is a frequent approach to ensure their clinical implementation. It involves a finely tuned trade-off between leishmanicidal and cytotoxic activities within a set of AMP analogues, obtained either through chemical synthesis or genetic engineering.[173] Some of these strategies have been described in detail in other reviews.[174–176] Figure 13.3 shows some common approaches to this goal, and Table 13.5 summarizes some optimization studies carried out on leishmanicidal peptides.

Shortening of the active sequence. The definition of the shortest active analogue is usually carried out at the very onset of the optimization process, to minimize cost and immunogenicity, while preserving bioavailability. Sequence shortening has been performed on dermaseptins DSR1[122] and DRS4,[124] and on cecropin A–melittin hybrid peptides.[159] This rationale can be extended into the definition of internal sequence stretches of a protein, endowed with leishmanicidal activities. Successful leishmanicidal crypto-peptides have been found in lactoferrin,[162] protamine,[105] chemokines (kinocidins),[172] mussel defensin[177] and phage lysins.[180]

Taken to the limit, this rationale leads to defining amphipathic leishmanicidal non-peptide structures, such as cationic steroids (ceragenins),[147] arylalkylamides[148] or arginine-based detergents.[149]

Charge optimization. A strong cationic character underlies AMP effectiveness. Increase of leishmanicidal activity after replacement of neutral by basic residues has been described for dermaseptin DR-S4,[124] for peptides derived from loop 3 of mussel defensin,[177] and for cationic variants of VIP, VIP51 and VIP51(6–30).[150] Likewise, pexiganan, a magainin-2 analogue optimized for clinical use against bacteria[181] is more effective on *Leishmania* than its parent structure.[135]

Excessive cationicity may cause the AMP to stick into the membrane, preventing further insertion. This may explain the innocuous binding of histones H2A and H2B to *Leishmania* amastigotes.[138] In contrast, acid

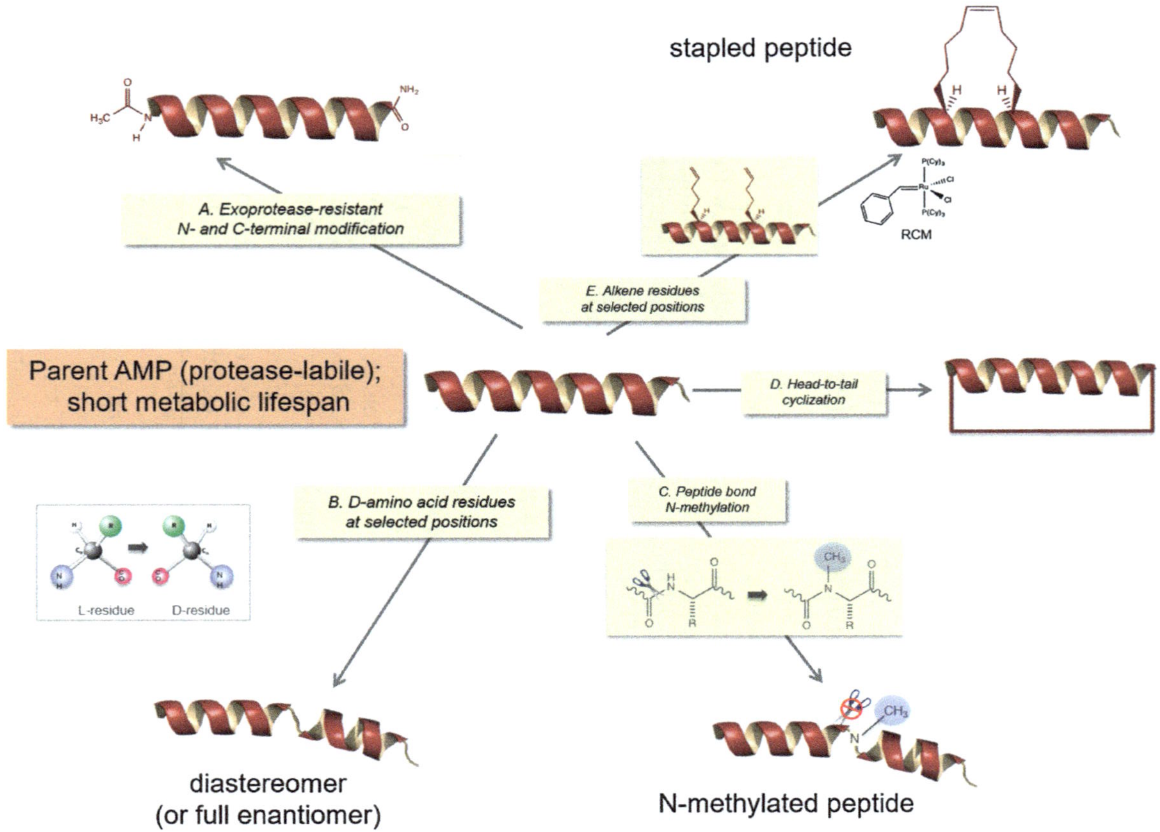

Figure 13.3 Representative approaches employed for the optimization of AMP activity.

Table 13.5 Some optimization strategies to improved the activity of leishmanicidal AMPs.

Peptide	Sequence	*Leishmania* species and stage tested[a]	Comments
Hybrid peptides			
CA(1–8)M(1–18)	KWKLFKKIGIGAVLKVLTTGLPALIS-NH_2	*L. don.* [**p.**[108]]	Cecropin A and melittin hybridization.
LF-chimera	FKCRRWQWRMKKLG-K-$CONH_2$ (ε) DLIWKLLSKAQEKFGKNKSR	*L. pif.* [**a.a.**[162]]; *L. don.* [**p.**[162]]	Sequence hybridization simulating lactoferrin topology.
Increase of cationicity			
Fragment P. Mussel defensin[b,c]	$C^1GGYC^2G\underline{K}W\underline{K}RLRC^2TSYRC^1G$	*L. maj.* [**p.**[177]]	Two residues replaced by Lys.
K_4K_{20}-S4[c]	ALW**K**TLLKKVLKAAAKAAL**K**AVLVGANA	*L. maj.* [**p.**[124]]	Shortening and increased cationicity. Dermaseptin DRS-4.
VIP51	HSDAVFTANYTRLRRQLAVRRYLAAILGR	*L. maj., L. trop., L. mex., L. don., L. inf.*: [**p.**[137]]	Increase of charge, biological stability and leishmanicidal activity over the parental VIP.
Pexiganan, MSI-94, Lys-Pex	GIGKFLKKAKKFGKAFVKMKK-NH_2	*L. maj.* [**p.**[178]]; *L. amaz.* [**p.** and **a.a.**[178]]; *L. braz.* [**p.**[114]]	Amidated analogue of magainin 2.
Arg-Pex	GIGRPLRRARRPGARPVRILRR-NH_2	*L. maj.* [**p.**[178]]; *L. amaz.* [**p.** and **a.a.**[178]]	Pexiganan, Lys by Arg replacement. Improved proteolytic stability.
K^6 (Me3)–CA(1–7)M(2–9)	KWKLFKK$(Me)_3$IGAVLKVL-NH_2	*L. pif.* [**a.a.**[160]]; *L. don.* [**p.**[160]]	Lys–trimethylation. Improved proteolytic stability. Reduced cytotoxicity.

Table 13.5 (*Continued*)

Peptide	Sequence	*Leishmania* species and stage tested[a]	Comments
Shortening of active sequence (parental peptide)			
DS1(1–15)a	ALWKTMLKKLGTMAL-NH_2	*L. maj.* [**p.**[122]]	Dermaseptin DS1.
CA(1–7)M(2–9)	KWKLFKKIGAVLKVL-NH_2	*L. pif.* [**a.a.**[158]]; *L. don.* [**p.**[158]]	CA(1–8)M(1–18).
K4-S4(1–13)a	ALWKTLLKKVLKA-NH_2	*L. maj.* [**p.**[125]]	Shortening and increased cationicity. (Dermaseptin DRS-4).
Increase of hydrophobicity			
CXCL4 AA-RP-1	AL(anthryl)YKKFKKKLLKSLKRLG	*L. maj., L. braz., L. inf.*: [**p.**[172]]	Substitution of Leu2 by Leu(anthryl)
N1$^{\varepsilon}$ Oct-CA(1–7)M(2–9)	C_7H_{15}-CO-KWKLFKKIGAVLKVL-NH_2	*L. pif.* [**a.a.**[158]]; *L. don.* [**p.**[158]]	*N*-terminal fatty acid acylation.
Lau-[K4-S4(1–13)a]	($C_{11}H_{23}$-CO)-ALWKTLLKKVLKA-NH_2	*L. maj.* [**p.**[179]]	*N*-terminal fatty acid acylation.
Aminolau-[K4-S4(1–13)a]	(NH_2-$C_{11}H_{22}$-CO)-ALWKTLLKKVLKA-NH_2	*L. maj.* [**p.**[179]]	*N*-terminal aminoacylation DRS4
Enantiomer versions			
Histatin 5 (L and D)[d]	DSHAKRHHGYKRKFHEKHHSHRGY	*L. pif.* [**a.a.**[132]]; *L. don.* [**p.**[132]]	Improved activity of D-peptide.
BAMP-28 (L and D and RI)[d]	GGLRSLGRKILRAWKKYGPIIVPIIRIGLDRI	*L. maj.* [**p.** and **a.a.** and **i.a.**[129]]	Improved activity of D- and RI peptides.

[a]*Leishmania* species: *L. amaz.*, *L. amazonensis*; *L. braz.*, *L. braziliensis*; *L. don.*, *L. donovani*; *L. inf.*, *L. infantum*; *L. maj.*, *L. major*; *L. mex.*, *L. mexicana*; *L. pif.*, *L. pifanoi*; *L.trop.*, *L. tropica*. *Leishmania* stage: **p.**: promastigote; **a.a.**: axenic amastigote; **i.a.**: intracellular amastigote.
[b]Superindex refers to the arrangement of disulfide bonds.
[c]The substituted residues are in bold type and underlined.
[d]Enantiomer versions: L, L-amino acid peptide; D, D-amino acid peptide; RI, Retroinverso peptide (sequence of the L-version read from the C- to the N-terminus and synthesized with D-amino acids.

pH increased by 30% the leishmanicidal activity of histatin 5, most probably by extensive protonation of histidine residues.[132] No leishmanicidal activity has yet been reported for anionic peptides. It is likely that the strong anionic glycocalyx of the promastigote acts as a deterrent for this class of AMPs.

Hydrophobicity. Few studies have addressed the role of hydrophobicity in AMP leishmanicidal activity. Positional hydrophobicity was studied for FW5-magainin and their analogues MG-H1 and MG-H2. Leishmanicidal activity increased when hydrophobicity was evenly distributed along the sequence, as in analogue MG-H2, rather than concentrated at a specific stretch, where it made it prone to aggregation in aqueous medium and more cytotoxic (analogue MG-H1).[161]

Acylation or aminoacylation of a single amino group were tested on the cecropin A–melittin hybrid CA(1–7)M(2–9)[158] and on the dermaseptin analogue K4–S4(1–13).[179] A significant increase in leishmanicidal activity was obtained by acylation with medium-size fatty acids. Long fatty acids increased the haemolytic effect and AMP aggregation, in part offset in aminoacylated analogues.[179]

Resistance to proteolytic degradation. A reasonable lifespan under physiological conditions is essential for AMP-based therapies. C-terminal amidation shields the peptide against carboxypeptidases, sometimes with significant improvement of leishmanicidal activity as found for decoralin,[109] or for AMPs derived from phage D3.[180] Amidation also increases by one the positive charge and creates an additional hydrogen bond, that can enhance structuration if the AMP is an α-helix.[182]

Lysine residues are especially prone to degradation by trypsin-like and other *Leishmania* proteinases. Full replacement of lysine residues by arginine in pexiganan (Arg-pex) improved the leishmanicidal activity and extended half-life compared with the parental peptide, severely fragmented by leishmaniolysin.[178] Likewise, trimethylation of the ε-NH_2 group of lysine residues increases notably the biological stability of CA(1–7)M(2–9). The effectiveness of the substitution depends on the number and position of the substituted lysines, with a decreased peptide toxicity and prolonged lifespan of the AMPs, although leishmanicidal activity was totally lost in the fully Lys-trimethylated analogue.[160]

Other strategies to render AMPs impervious to proteases have used all-D-enantiomer and retro-inverso (RI) versions (sequence read from C- to N-terminus and synthesized with D-amino acids). The D-versions of CA(1–8)M(1–18),[108] BMAP-28,[129] or histatin 5,[132] have threefold to sevenfold higher activity than the respective L-versions, regardless of the stage of the parasite assayed. The RI versions of BMAP-18 or AMP1018 performed rather similarly to the D-versions.[131]

Amphibian bombinins H2 and H4 are examples of native epimerization within an AMP sequence, in this case the Ile2 residue.[119] The activity of bombinin H4 is four times higher than that of H2, probably due to the lower aggregation tendency of bombinin H4.

Hybrid antimicrobial peptides. The combination of sequence stretches from different AMPs into a single sequence can improve microbicidal activity, or reduce toxicity relative to parental AMPs.[183] The two most representative examples in *Leishmania* are the cecropin A-melittin hybrids, such as CA(1–8)M(1–18),[108] or lactoferricin–lactoferrampin hybrid peptides.[162]

Analysis by molecular modelling and quantitative structure-activity relationships (QSAR) of 23 α-helical AMPs revealed charge, solvation of the backbone, solvent accessible area and volume of the peptide as the most relevant descriptors for leishmanicidal activity.[184]

13.10 Antimicrobial Peptides in Animal Models of Leishmaniasis

The toxicity of dermaseptins DS-01, DD-L and DD-K was studied in Swiss mice (i.v. single dose: 5.0 mg kg^{-1}, 1.9 nmol kg^{-1}).[185] No variation was observed in leukocyte counts, or in liver, spleen and kidney histology in mice sacrificed two weeks after AMP inoculation.

BALB/c mice infected by *L. major* were treated with VIP analogues,[150] or urocortin II.[137] Two weeks after *Leishmania* inoculation, the respective AMPs were injected s.c. into the ulcer (1.5 nmol kg^{-1} every other day for 7 weeks), and parasite load was evaluated. VIP and its two analogues VIP51 and VIP51(6–30) diminished pad swelling, but only VIP51(6–30) reduced by 13 log units the parasite load, and reverted the granulomatous lesion into a normal state. Furthermore, these AMPs prevented visceralization.[150] Urocortin II treatment followed a very similar pattern, *i.e.*, strong reduction of parasite load at the hind pad (20 log units), recovery of a normal histology at the lesion and prevention of visceralization.[137]

Two AMPs were modelled on the helical domain of the human chemokine CXCL4 (PF4), present in platelets.[172] These peptides differed exclusively at position 2 (alanine in RP-1, anthryl-alanine in AA-RP-1). Both AMPs showed a significant leishmanicidal activity on *in vitro* infections by a variety of *Leishmania* species. One week after infection of BALB/c mice with *L. infantum chagasi*, mice were treated with 12.5 mg kg^{-1} every other day for two weeks. Both peptides decreased the parasite load of the liver between 60 and 65%, but the decrease in the spleen was only 42% for AA-RP-1 and 10% for RP-1.

Dogs naturally infected with *L. infantum* received i.v. injections of Oct-*N*-CA(1–7)M(2–9) [5 mg (2.6 nmol) at days 0, 2 and 4] at the onset of clinical symptomatology and in the absence of any prior treatment.[186] Parasitaemia in peripheral blood decreased by 80% one week after the end of the treatment. Haematological and biochemical parameters did not vary significantly.

13.11 Immunomodulation by Leishmanicidal AMPs

The influence of mammalian AMPs on the macrophage has been specifically addressed in a recent review.[14] Thus, we exclusively refer to immunomodulation described for peptides mentioned in this chapter.

Melittin decreases the production of TNF-α of *Leishmania chagasi*-infected macrophages, with simultaneous decrease of IL-10 and increase of IL-12, a clue for the induction of a moderate inflammatory response. Nevertheless, melittin also abrogates almost completely H_2O_2 and NO production.[107] This calls into question the role of immunomodulation in the global leishmanicidal effect of melittin. HNP-1 induced TNF-α in macrophages regardless of their infection or not with *L. major*.[134]

HNP-1 induced the release of IL-8 from *L. major*-infected neutrophils, but with nil effect on the uninfected cells.[134] In human dendritic cells, a decrease of IL-8 levels was reported for andropin, but not for cecropin A or dermaseptin DR-S1.[105]

CA (1–8) M (1–18) induces partial depolarization of the plasma membrane of the macrophage in a concentration-dependent manner, assessed by patch-clamp.[187] This induces a transitory rise of intracellular Ca^{2+} and induction of nitric oxide synthase expression without prior priming.[188]

13.12 The Outlook for AMP-based Therapies for Leishmaniasis

13.12.1 Intrinsic Resistance of *Leishmania* to Antimicrobial Peptides

A repeated concern regarding the therapeutic use of AMPs is the induction of cross-resistance, as they have phospholipid interaction as a shared target. Furthermore, the alleged loss of fitness associated with AMP resistance was disproved in bacteria.[189] Consistent with this, improved fitness of *Leishmania* parasites associated with drug resistance has been described.[190] Nevertheless, AMPs modulate *Leishmania* infection, as described for CRAMP on murine CL.[130]

Studies on resistance to AMPs have almost exclusively focused on the promastigote. LPG and PPG are mostly responsible for the anionicity of the promastigote glycocalyx (see Section 13.4.4), however only simultaneous abrogation of LPG and PPG results in significant protection of the parasite against VIP51,[150] H2A,[138] or temporins A, L, and 1Sa.[116]

Leishmaniolysin is a strong deterrent to AMP leishmanicidal activity. Thus, it alone hydrolyses pexiganan into 14 fragments,[135] and promastigotes with low expression or devoid of leishmaniolysin are more susceptible towards cryptdin 4, pexiganan, defensin θ-II,[135] and CRAMP.[130]

13.12.2 Role of Enviromental Conditions for AMPs on CL Lesions

Since insights into the efficacy of AMPs as leishmanicidal agents are mostly obtained from *in vitro* studies rather than from the scarce studies on animal infections, a strong disparity exists between the former type of studies and the environments where AMPs are expected to be active.

Ulcers from CL are frequently infected by bacteria and less frequently by fungi. Bacterial infection in the skin promotes a hypertonic Na^+ rich environment, which favours the clearance of *Escherichia coli* and *L. major*,[191] but it will condition the outcome of AMPs. Presumably, salinity would be much less relevant for the AMP–intracellular amastigote interaction.

In general, poor AMP bioavailability is an important caveat for topical administration.[192] Even so, tissue penetrability is much higher in ulcers than in the surrounding healthy tissue.[193] Furthermore, melittin penetration into the skin is slow but high, provided that proteolytic degradation is inhibited.[194] In addition, skin penetration of AMPs can be improved by strategies commonly used for other cutaneous drugs.[195]

13.12.3 Synergism and AMPs for Leishmaniasis

Synergism is a common phenomenon for natural AMPs acting at the same anatomical or subcellular location.[196,197] Artificial synergism has also been described for a wide variety of AMP combinations, including peptidomimetics.[198,199]

AMPs also synergize with small-molecule antibacterial[200] and antifungal[201] drugs. Again, no study addressing AMP synergism with small-molecule leishmanicidal drugs exists. Since AMPs and amphotericin B share the same mechanism of action, although respectively achieved by interaction with phospholipids and ergosterol, one would at least expect additive permeation effects, as already demonstrated on fungal targets.[202]

The best documented effect of miltefosine is the inversion of the PC : PE ratio due to the increment of PE in the overall phospholipid composition.[203] The increase in PE may hamper membrane permeabilisation by AMPs, as this phospholipid is a typical inducer of negative curvature at the membrane, in contrast to the positive one induced by many AMPs.

Many leishmanicidal drugs induce reactive oxygen species (ROS) production in the parasites,[204–206] so one might envision synergism between these drugs and AMPs that induce ROS production, such as spinigerin.[110] Furthermore, oxidized phospholipids may serve as receptors for AMPs.[207] If this could be demonstrated for *Leishmania*, it would amplify the lethal effects of AMPs by increasing the density of AMP bound to the parasite surface.

Finally, the bioenergetic collapse created by AMPs can be exploited to revert drug resistance (*e.g.*, antimonials, miltefosine and paromomycin) due to efflux pumps.[208,209]

13.12.4 Antimicrobial Peptide Production. Scale-up Production or *In situ* Expression

As stated in Section 13.10, substantial amounts of AMPs are required to cure or alleviate leishmaniasis in animal models,[172,186] even when locally administered.[137,150] The production costs associated with AMP treatment are presently too high for a massive implementation into populations of largely impoverished patients.

Several alternatives to this gloomy situation include: (i) Synergy among AMPs, improving their final global effects; up to 100-fold increases in the microbicidal activities of dermaseptins were obtained with a judicious choice of partners.[196] (ii) Scale-up of AMP production, with special focus on biofactories.[210] In this respect, bacterial AMPs obtained from fermentation broths are easy to scale-up, with low to moderate cost and amenable to improvement by genetic engineering. These AMP sources remain practically unexplored for leishmaniasis to date.[140] (iii) Direct targeting of AMPs by fusing the AMP with a specific recognition sequence. Filamentous phage peptide libraries are especially useful in this regard. By means of this technology, hexapeptides against metacyclic promastigotes from *L. major* were obtained after several biopanning cycles. The peptide obtained was by itself effective against intracellular amastigotes, and provided significant protection in animal models.[211] AMPs can also be preferentially delivered to macrophages after incorporation into vehicles through a large variety of nanotechnological strategies.[212–214]

Under natural conditions, AMP expression is induced at the site of infection in a quite limited temporal frame, and at high local concentrations, to ensure the elimination of the invading pathogen. This landscape can be mimicked by several strategies, some of them outlined here.

Gene therapy is envisaged to eliminate intracellular pathogens by expression of foreign AMPs, but has not been assayed on *Leishmania* yet. Its efficacy for other intracellular pathogens has been demonstrated against *Mycoplasma humanis* and *Chlamydia trichomonatis*, even in animal models of infection by these pathogens,[215,216] on *Salmonella typhimurium*,[217] and on *Histoplasma capsulatum*.[218]

The induction of AMP expression was observed for a set of metabolites such as isoleucine, butyrate or vitamin D_3 among others, mostly identified from the gut microbiome,[176] but also induced by synthetic aroylated phenylendiamines.[219] Many of these inducers are inhibitors of histone deacetylase,[219] although the scarce studies on this approach for leishmaniasis are nowadays controversial.[220,221]

13.13 Conclusions

Despite the scarcity of studies of AMPs on *Leishmania*, a proof-of mechanism for their use in future therapy has been consistently proven. The big challenge for AMPs is to get a leishmanicidal concentration at the location of the parasite without toxicity to surrounding tissues. This concern is also shared for other AMP-based treatments of bacterial and fungal infections. This and the cost of peptide-based therapies are important hurdles that must be urgently addressed if implementation of an AMP-based therapy against *Leishmania* is intended. Non-disseminated ulcers of cutaneous leishmaniasis are perhaps the most appealing and easiest application for AMPs.

Development of an AMP-based therapy must assay a larger number of AMPs, as well as their optimization by peptide-engineering strategies. To this end, those leishmanicidal AMPs of bacterial origin are quite enticing, as their production can be easily scaled up at affordable cost, and previous clinical experience against bacteria is available. Furthermore, synergism and antagonism of foreign AMPs with those from the host must be addressed, as well as AMP combination with current clinical leishmanicidal drugs, with those targeting the membrane (*e.g.* amphotericin B) or its lipid composition being of special relevance.

Finally, the local production of AMPs by gene therapy, by modification of the microbiome of the skin or by administration of inducers for AMP production, are promising strategies to be studied in the future.

Acknowledgements

LR is supported by grants from Plan Estatal de Investigación Científica y Técnica y de Innovación 2013-2016 (SAF2015-65740-R), Subdirección General de Redes y Centros de Investigación Cooperativa-FEDER (RICET RD12/0018/0007, and RD16/0027/0010) and CSIC (PIE 201620E038). DA acknowledges support from MINECO, Spain (grant AGL2014-52395-C2-R).

References

1. H. C. Stahl, F. Ahmadi, U. Schleicher, R. Sauerborn, J. L. Bermejo, M. L. Amirih, I. Sakhayee, C. Bogdan and K. W. Stahl, *BMC Infect. Dis.*, 2014, **14**, 619.
2. B. Kunzler, *J. Spec. Oper. Med.*, 2013, **13**, 40.
3. Transparency Market Research, *Peptide Therapeutics Market: Global Industry Analysis, Size, Share, Growth, Trends and Forecast 2020*, Albany (USA) 2016. pp. 1–236.
4. B. Gevaert, S. Stalmans, E. Wynendaele, L. Taevernier, N. Bracke, M. D'Hondt and B. De Spiegeleer, *Protein Pept. Lett.*, 2016, **23**, 324.
5. E. Sancho-Vaello and K. Zeth, *Future Microbiol.*, 2015, **10**, 1103.
6. J. Lakshmaiah Narayana and J. Y. Chen, *Peptides*, 2015, **72**, 88.
7. F. Rabanal and Y. Cajal, *New Weapons to Control Bacterial Growth*, ed. T. G. Villa and M. Viñas. Springer, Berlin, 2016, ch. 16, pp. 433–455.
8. L. Rivas, J. R. Luque-Ortega and D. Andreu, *Biochim. Biophys. Acta*, 2009, **1788**, 1570.
9. S. L. Cobb and P. W. Denny, *Curr. Opin. Investig. Drugs*, 2010, **11**, 868.
10. B. S. McGwire and M. M. Kulkarni, *Exp. Parasitol.*, 2010, **126**, 397.
11. A. K. Marr, B. S. McGwire and W. R. McMaster, *Future Microbiol.*, 2012, **7**, 1047.
12. M. Torrent, D. Pulido, L. Rivas and D. Andreu, *Curr. Drug Targets*, 2012, **13**, 1138.
13. P. Mendez-Samperio and J. L. de-la-Rosa-Arana, *J. Egypt. Soc. Parasitol.*, 2013, **43**, 195.

14. Z. Abdossamadi, N. Seyed and S. Rafati, *Cell. Immunol.*, 2016, **309**, 23.
15. X. Sun, E. Salih, F. G. Oppenheim and E. J. Helmerhorst, *FASEB J.*, 2009, **23**, 2691.
16. M. Murakami, B. López-García, M. Braff, R. A. Dorschner and R. L. Gallo, *J. Immunol.*, 2004, **172**, 3070.
17. M. Wolf and B. Moser, *Front. Immunol.*, 2012, **3**, 213.
18. H. Kawasaki and S. Iwamuro, *Infect. Disord. Drug Targets*, 2008, **8**, 195.
19. I. A. El Karim, G. J. Linden, D. F. Orr and F. T. Lundy, *J. Neuroimmunol.*, 2008, **200**, 11.
20. G. Wang, X. Li and Z. Wang, *Nucleic Acids Res.*, 2016, **44**, D1087.
21. F. H. Waghu, R. S. Barai, P. Gurung and S. Idicula-Thomas, *Nucleic Acids Res.*, 2016, **44**, D1094.
22. S. Liu, L. Fan, J. Sun, X. Lao and H. Zheng, *J. Peptide Sci.*, 2017, **23**, 4.
23. R. L. Gallo and L. V. Hooper, *Nat. Rev. Immunol.*, 2012, **12**, 503.
24. M. Hans and V. Madaan Hans, *Int. J. Peptides*, 2014, 370297.
25. D. Parker and A. Prince, *Am. J. Respir. Cell. Mol. Biol.*, 2011, **45**, 189.
26. J. Schröder, *Allergol. Int.*, 2011, **60**, 17.
27. T. K. Fábián, P. Hermann, A. Beck, P. Fejérdy and G. Fábián, *Int. J. Mol. Sci.*, 2012, **13**, 4295.
28. A. M. McDermott, *Exp. Eye Res.*, 2013, **117**, 53.
29. Å. Björstad, K. L. Brown, H. Forsman, C. Dahlgren, A. Karlsson and J. Bylund, *Anti-Infect. Agents Med. Chem.*, 2008, **7**, 155.
30. P. G. Barlow, E. G. Findlay, S. M. Currie and D. J. Davidson, *Future Microbiol.*, 2014, **9**, 55.
31. F. Findlay, L. Proudfoot, C. Stevens and P. G. Barlow, *Pathog. Glob. Health*, 2016, **110**, 137.
32. M. García-Quintanilla, M. R. Pulido, R. López-Rojas, J. Pachón and M. J. McConnell, *Trends Microbiol.*, 2013, **21**, 157.
33. N. L. Van Der Weerden, M. R. Bleackley and M. A. Anderson, *Cell. Mol. Life Sci.*, 2013, **70**, 3545.
34. C. Oelkrug, M. Hartke and A. Schubert, *Anticancer Res.*, 2015, **35**, 635.
35. L. Angélique, W. J. Frederik, J. Garmi and D. P. L. Hester, *Molecules*, 2015, **20**, 15392.
36. N. Vale, L. Aguiar and P. Gomes, *Front. Pharmacol.*, 2014, **5**, 275.
37. L. Fan, J. Sun, M. Zhou, J. Zhou, X. Lao, H. Zheng and H. Xu, *Sci. Rep.*, 2016, **6**, 24482.
38. A. Fabisiak, N. Murawska and J. Fichna, *Pharmacol. Rep.*, 2016, **68**, 802.
39. R. E. W. Hancock, E. F. Haney and E. E. Gill, *Nat. Rev. Immunol.*, 2016, **16**, 321.
40. M. L. Mangoni, A. M. McDermott and M. Zasloff, *Exp. Dermatol.*, 2016, **25**, 167.
41. L. Otvos Jr. and E. Ostorhazi, *Expert Rev. Anti-Infect. Ther.*, 2015, **13**, 871.
42. L. Martin, A. van Meegern, S. Doemming and T. Schuerholz, *Front. Immunol.*, 2015, **6**, 404.
43. D. Xhindoli, S. Pacor, M. Benincasa, M. Scocchi, R. Gennaro and A. Tossi, *Biochim. Biophys. Acta*, 2016, **1858**, 546.

44. H. G. Boman, *Annu. Rev. Immunol.*, 1995, **13**, 61.
45. D. Juretić, D. Vukičević and A. Tossi, *Methods Mol. Biol.*, 2017, **1548**, 23.
46. I. Zelezetsky and A. Tossi, *Biochim. Biophys. Acta*, 2006, **1758**, 1436.
47. M. Dathe and T. Wieprecht, *Biochim. Biophys. Acta*, 1999, **1462**, 71.
48. Y. Shai, *Biopolymers*, 2002, **66**, 236.
49. K. Matsuzaki, *Biochim. Biophys. Acta*, 2009, **1788**, 1687.
50. N. Papo and Y. Shai, *Peptides*, 2003, **24**, 1693.
51. F. Harris, S. R. Dennison and D. A. Phoenix, *Curr. Prot. Pept. Sci.*, 2009, **10**, 585.
52. S. Prabhu, S. R. Dennison, B. Lea, T. J. Snape, I. D. Nicholl, I. Radecka and F. Harris, *Crit. Rev. Plant Sci.*, 2013, **32**, 303.
53. V. Teixeira, M. J. Feio and M. Bastos, *Prog. Lipid Res.*, 2012, **51**, 149.
54. M. R. Krause and S. L. Regen, *Acc. Chem. Res.*, 2014, **47**, 3512.
55. B. S. Perrin Jr., A. J. Sodt, M. L. Cotten and R. W. Pastor, *J. Membr. Biol.*, 2015, **248**, 455.
56. M. A. Sani, T. C. Whitwell and F. Separovic, *Biochim. Biophys. Acta*, 2012, **1818**, 205.
57. A. J. Mason, A. Marquette and B. Bechinger, *Biophys. J.*, 2007, **93**, 4289.
58. R. Sood and P. K. J. Kinnunen, *Biochim. Biophys. Acta*, 2008, **1778**, 1460.
59. M. R. Yeaman and N. Y. Yount, *Pharmacol. Rev.*, 2003, **55**, 27.
60. S. Sánchez-Gómez, R. Conde-Alvarez, J. Antonio, I. Moriyón and G. Martínez-de-Tejada, in *Antimicrobial Peptides: Properties, Functions and Role in Immune Response*, ed. M.-D. Seong and Y.-I. I. Hak. Nova Science Publishers, New York, USA, 2013, ch. 1, pp. 1–30.
61. A. Schmidtchen, I. M. Frick, E. Andersson, H. Tapper and L. Björck, *Mol. Microbiol.*, 2002, **46**, 157.
62. M. V. Buri, T. M. Domingues, E. J. Paredes-Gamero, R. L. Casaes-Rodrigues, E. G. Rodrigues and A. Miranda, *PLoS One*, 2013, **8**, e80924.
63. L. T. Nguyen, E. F. Haney and H. J. Vogel, *Trends Biotechnol.*, 2011, **29**, 464.
64. R. M. Epand and R. F. Epand, *Mol. BioSyst.*, 2009, **5**, 580.
65. K. Gull, *Annu. Rev. Microbiol.*, 1999, **53**, 629.
66. S. M. Landfear and M. Ignatushchenko, *Mol. Biochem. Parasitol.*, 2001, **115**, 1.
67. M. C. Field and M. Carrington, *Nat. Rev. Microbiol.*, 2009, **7**, 775.
68. S. C. Ilgoutz and M. J. McConville, *Int. J. Parasitol.*, 2001, **31**, 899.
69. S. J. Turco and A. Descoteaux, *Annu Rev. Microbiol.*, 1992, **46**, 65.
70. S. J. Turco and D. L. Sacks, *Mol. Biochem. Parasitol.*, 1991, **45**, 91.
71. S. W. Homans, A. Mehlert and S. J. Turco, *Biochemistry*, 1992, **31**, 654.
72. P. Schneider, J. P. Rosat, J. Bouvier, J. Louis and C. Bordier, *Exp. Parasitol.*, 1992, **75**, 196.
73. V. Bahr, Y. D. Stierhof, T. Ilg, M. Demar, M. Quinten and P. Overath, *Mol. Biochem. Parasitol.*, 1993, **58**, 107.
74. D. L. Sacks, P. F. P. Pimenta, M. J. McConville, P. Schneider and S. J. Turco, *J. Exp. Med.*, 1995, **181**, 685.

75. D. L. Sacks, T. N. Brodin and S. J. Turco, *Mol. Biochem. Parasitol.*, 1990, **42**, 225.
76. M. J. McConville and M. A. J. Ferguson, *Biochem. J.*, 1993, **294**, 305.
77. M. J. McConville and J. E. Ralton, *Behring Inst. Mitt.*, 1997, **99**, 34.
78. M. J. McConville, L. F. Schnur, C. Jaffe and P. Schneider, *Biochem. J.*, 1995, **310**, 807.
79. G. Winter, M. Fuchs, M. J. McConville, Y. D. Stierhof and P. Overath, *J. Cell Sci.*, 1994, **107**, 2471.
80. T. Ilg, *Parasitol. Today*, 2000, **16**, 489.
81. N. M. Novozhilova and N. V. Bovin, *Biochemistry (Moscow)*, 2010, **75**, 686.
82. B. Foth, A. Piani, J. M. Curtis, T. Ilg, M. McConville and E. Handman, *Int. J. Parasitol.*, 2002, **32**, 1701.
83. C. Yao, J. E. Donelson and M. E. Wilson, *Mol. Biochem. Parasitol.*, 2003, **132**, 1.
84. J. Bouvier, R. Etges and C. Bordier, *Mol. Biochem. Parasitol.*, 1987, **24**, 73.
85. K. P. Chang, G. Chaudhuri and D. Fong, *Annu. Rev. Microbiol.*, 1990, **44**, 499.
86. M. Olivier, V. D. Atayde, A. Isnard, K. Hassani and M. T. Shio, *Microbes Infect.*, 2012, **14**, 1377.
87. L. P. Kahl and D. McMahon-Pratt, *J. Immunol.*, 1987, **138**, 1587.
88. L. M. Lincoln, M. Ozaki, J. E. Donelson and J. K. Beetham, *Mol. Biochem. Parasitol.*, 2004, **137**, 185.
89. D. H. Beach, G. G. Holz Jr. and G. E. Anekwe, *J. Parasitol.*, 1979, **65**, 201.
90. R. K. Adosraku, M. M. Anderson, G. J. Anderson, G. Choi, S. L. Croft, V. Yardley, J. D. Phillipson and W. A. Gibbons, *Mol. Biochem. Parasitol.*, 1993, **62**, 251.
91. L. Zheng, R. T'Kind, S. Decuypere, S. J. von Freyend, G. H. Coombs and D. G. Watson, *Rapid Commun. Mass Spectrom.*, 2010, **24**, 2074.
92. L. Imbert, M. Gaudin, D. Libong, D. Touboul, S. Abreu, P. M. Loiseau, O. Laprevote and P. Chaminade, *J. Chromatogr. A*, 2012, **1242**, 75.
93. K. Zhang and S. M. Beverley, *Mol. Biochem. Parasitol.*, 2010, **170**, 55.
94. M. K. Wassef, T. B. Fioretti and D. M. Dwyer, *Lipids*, 1985, **20**, 108.
95. R. Zufferey, S. Allen, T. Barron, D. R. Sullivan, P. W. Denny, I. C. Almeida, D. F. Smith, S. J. Turco, M. A. J. Ferguson and S. M. Beverley, *J. Biol. Chem.*, 2003, **278**, 44708.
96. M. C. Pawlowic, F. F. Hsu, S. Moitra, N. Biyani and K. Zhang, *Mol. Microbiol.*, 2016, **101**, 238.
97. A. Weingartner, G. Kemmer, F. D. Muller, R. A. Zampieri, M. Gonzaga dos Santos, J. Schiller and T. G. Pomorski, *PLoS One*, 2012, 7, e42070.
98. K. Zhang, J. M. Pompey, F. F. Hsu, P. Key, P. Bandhuvula, J. D. Saba, J. Turk and S. M. Beverley, *EMBO J.*, 2007, **26**, 1094.
99. F. F. Hsu, J. Turk, K. Zhang and S. M. Beverley, *J. Am. Soc. Mass Spectrom.*, 2007, **18**, 1591.
100. C. Yao and M. E. Wilson, *Parasit. Vectors*, 2016, **9**, 200.

101. C. Yao, U. Gaur Dixit, J. H. Barker, L. M. Teesch, L. Love-Homan, J. E. Donelson and M. E. Wilson, *Infect. Immun.*, 2013, **81**, 2507.
102. J. M. Conlon, *Cell. Mol. Life Sci.*, 2011, **68**, 2303.
103. A. Falanga, L. Lombardi, G. Franci, M. Vitiello, M. R. Iovene, G. Morelli, M. Galdiero and S. Galdiero, *Int. J. Mol. Sci.*, 2016, **17**, E785.
104. J. Fotie, *Bioactive Natural Products*, ed. G. Brahmachari, Wiley-VCH Verlag GmbH & Co., 2014, ch. 10, pp. 279–320.
105. J. J. Pérez-Cordero, J. M. Lozano, J. Cortés and G. Delgado, *Peptides*, 2011, **32**, 683.
106. J. M. Lozano, B. S. Cuadrado, G. Delgado and M. E. Patarroyo, *J. Microb. Biochem. Technol.*, 2014, **6**, 54.
107. A. V. Pereira, G. de Barros, E. G. Pinto, A. G. Tempone, O. Orsi Rde, L. D. Dos Santos, S. Calvi, R. S. Ferreira Jr., D. C. Pimenta and B. Barraviera, *J. Venom Anim. Toxins Incl. Trop. Dis.*, 2016, **22**, 1.
108. P. Diaz-Achirica, J. Ubach, A. Guinea, D. Andreu and L. Rivas, *Biochem. J.*, 1998, **330**, 453.
109. K. Konno, M. Rangel, J. S. Oliveira, M. P. Dos Santos Cabrera, R. Fontana, I. Y. Hirata, I. Hide, Y. Nakata, K. Mori, M. Kawano, H. Fuchino, S. Sekita and J. R. Neto, *Peptides*, 2007, **28**, 2320.
110. A. H. Sardar, S. Das, S. Agnihorti, M. Kumar, A. K. Ghosh, K. Abhishek, A. Kumar, B. Purkait, M. Y. Ansari and P. Das, *Exp. Parasitol.*, 2013, **135**, 715.
111. H. Akuffo, D. Hultmark, A. Engstom, D. Frohlich and D. Kimbrell, *Int. J. Mol. Med.*, 1998, **1**, 77.
112. N. Boulanger, C. Lowenberger, P. Volf, R. Ursic, L. Sigutova, L. Sabatier, M. Svobodova, S. M. Beverley, G. Spath, R. Brun, B. Pesson and P. Bulet, *Infect. Immun.*, 2004, **72**, 7140.
113. P. I. Silva Jr., S. Daffre and P. Bulet, *J. Biol. Chem.*, 2000, **275**, 33464.
114. S. E. Lofgren, L. C. Miletti, M. Steindel, E. Bachere and M. A. Barracco, *Exp. Parasitol.*, 2008, **118**, 197.
115. F. L. Chadbourne, C. Raleigh, H. Z. Ali, P. W. Denny and S. L. Cobb, *J. Pept. Sci.*, 2011, **17**, 751.
116. G. A. Eggimann, K. Sweeney, H. L. Bolt, N. Rozatian, S. L. Cobb and P. W. Denny, *Molecules*, 2015, **20**, 2775.
117. M. L. Mangoni, J. M. Saugar, M. Dellisanti, D. Barra, M. Simmaco and L. Rivas, *J. Biol. Chem.*, 2005, **280**, 984.
118. F. Abbassi, Z. Raja, B. Oury, E. Gazanion, C. Piesse, D. Sereno, P. Nicolas, T. Foulon and A. Ladram, *Biochimie*, 2013, **95**, 388.
119. M. L. Mangoni, N. Papo, J. M. Saugar, D. Barra, Y. Shai, M. Simmaco and L. Rivas, *Biochemistry*, 2006, **45**, 4266.
120. I. Vouldoukis, Y. Shai, P. Nicolas and A. Mor, *FEBS Lett.*, 1996, **380**, 237.
121. L. Gaidukov, A. Fish and A. Mor, *Biochemistry*, 2003, **42**, 12866.
122. D. Savoia, R. Guerrini, E. Marzola and S. Salvadori, *Bioorg. Med. Chem.*, 2008, **16**, 8205.
123. C. Hernández, A. Mor, F. Dagger, P. Nicolas, A. Hernandez, E. L. Benedetti and I. Dunia, *Eur. J. Cell. Biol.*, 1992, **59**, 414.

124. R. Feder, A. Dagan and A. Mor, *J. Biol. Chem.*, 2000, **275**, 4230.
125. I. Kustanovich, D. E. Shalev, M. Mikhlin, L. Gaidukov and A. Mor, *J. Biol. Chem.*, 2002, **277**, 16941.
126. G. D. Brand, R. C. Santos, L. M. Arake, V. G. Silva, L. M. Veras, V. Costa, C. H. Costa, S. S. Kuckelhaus, J. G. Alexandre, M. J. Feio and J. R. Leite, *Molecules*, 2013, **18**, 7058.
127. Z. Raja, S. Andre, C. Piesse, D. Sereno, P. Nicolas, T. Foulon, B. Oury and A. Ladram, *PLoS One*, 2013, **8**, e70782.
128. M. Oliveira, A. G. Gomes-Alves, C. Sousa, M. Mirta Marani, A. Plácido, N. Vale, C. Delerue-Matos, P. Gameiro, S. A. S. Kückelhaus, A. M. Tomas, J. R. S. A. Leite and P. Eaton, *Biopolymers*, 2016, **105**, 873.
129. M. A. Lynn, J. Kindrachuk, A. K. Marr, H. Jenssen, N. Pante, M. R. Elliott, S. Napper, R. E. Hancock and W. R. McMaster, *PLoS Neglected Trop. Dis.*, 2011, **5**, e1141.
130. M. M. Kulkarni, J. Barbi, W. R. McMaster, R. L. Gallo, A. R. Satoskar and B. S. McGwire, *Cell. Microbiol.*, 2011, **13**, 913.
131. A. K. Marr, S. Cen, R. E. W. Hancock and W. R. McMaster, *Antimicrob. Agents Chemother.*, 2016, **60**, 2484.
132. J. R. Luque-Ortega, W. van't Hof, E. C. Veerman, J. M. Saugar and L. Rivas, *FASEB J.*, 2008, **22**, 1817.
133. S. K. Sobirk, M. Morgelin, A. Egesten, P. Bates, O. Shannon and M. Collin, *PLoS One*, 2013, **8**, e58129.
134. S. Dabirian, Y. Taslimi, F. Zahedifard, E. Gholami, F. Doustdari, M. Motamedirad, S. Khatami, K. Azadmanesh, S. Nylen and S. Rafati, *PLoS Neglected Trop. Dis.*, 2013, **7**, e2491.
135. M. M. Kulkarni, W. R. McMaster, E. Kamysz, W. Kamysz, D. M. Engman and B. S. McGwire, *Mol. Microbiol.*, 2006, **62**, 1484.
136. A. Bera, S. Singh, R. Nagaraj and T. Vaidya, *Mol. Biochem. Parasitol.*, 2003, **127**, 23.
137. J. Campos-Salinas, M. Caro, A. Cavazzuti, I. Forte-Lago, S. M. Beverley, F. O'Valle and E. Gonzalez-Rey, *J. Immunol.*, 2013, **191**, 6040.
138. Y. Wang, Y. Chen, L. Xin, S. M. Beverley, E. D. Carlsen, V. Popov, K. P. Chang, M. Wang and L. Soong, *Infect. Immun.*, 2011, **79**, 1124.
139. M. Berrocal-Lobo, A. Molina, P. Rodriguez-Palenzuela, F. Garcia-Olmedo and L. Rivas, *Exp. Parasitol.*, 2009, **122**, 247.
140. M. A. Abéngozar, R. Cebrián, J. M. Saugar, T. Gárate, E. Valdivia, M. Martínez-Bueno, M. Maqueda and L. Rivas, *Antimicrob. Agents Chemother.*, 2017, **61**, e02288.
141. J. R. Luque-Ortega, L. J. Cruz, F. Albericio and L. Rivas, *Mol. Pharm.*, 2010, **7**, 1608.
142. L. J. Cruz, J. R. Luque-Ortega, L. Rivas and F. Albericio, *Mol. Pharm.*, 2009, **6**, 813.
143. T. L. Simmons, N. Engene, L. D. Urena, L. I. Romero, E. Ortega-Barria, L. Gerwick and W. H. Gerwick, *J. Nat. Prod.*, 2008, **71**, 1544.
144. L. M. Sánchez, D. López, B. A. Vesely, G. Della Togna, W. H. Gerwick, D. E. Kyle and R. G. Linington, *J. Med. Chem.*, 2010, **53**, 4187.

145. G. A. Eggimann, H. L. Bolt, P. W. Denny and S. L. Cobb, *Chem. Med. Chem.*, 2015, **10**, 233.
146. H. L. Bolt, G. A. Eggimann, P. W. Denny and S. L. Cobb, *Med. Chem. Commun.*, 2016, **7**, 799.
147. D. Lara, Y. Feng, J. Bader, P. B. Savage and R. A. Maldonado, *J. Parasitol.*, 2010, **96**, 638.
148. E. A. Iniguez, A. Pérez, R. A. Maldonado and R. Skouta, *Bioorg. Med. Chem. Lett.*, 2015, **25**, 5315.
149. N. Lozano, L. Pérez, R. Pons, J. R. Luque-Ortega, M. Fernández-Reyes, L. Rivas and A. Pinazo, *Colloids Surf. A Physicochem. Eng. Asp.*, 2008, **319**, 196.
150. J. Campos-Salinas, A. Cavazzuti, F. O'Valle, I. Forte-Lago, M. Caro, S. M. Beverley, M. Delgado and E. Gonzalez-Rey, *J. Biol. Chem.*, 2014, **289**, 14583.
151. F. Abbassi, B. Oury, T. Blasco, D. Sereno, G. Bolbach, P. Nicolas, K. Hani, M. Amiche and A. Ladram, *Peptides*, 2008, **29**, 1526.
152. T. A. Glaser, G. L. Utz and A. J. Mukkada, *Mol. Biochem. Parasitol.*, 1992, **51**, 9.
153. S. Monroe Duboise, M. A. Vannier-Santos, D. Costa-Pinto, L. Rivas, A. A. Pan, Y. Traub-Cseko, W. De Souza and D. McMahon-Pratt, *Mol. Biochem. Parasitol.*, 1994, **68**, 119.
154. H. L. Bolt, P. W. Denny and S. L. Cobb, *J. Vis. Exp.*, 2016, DOI: 10.3791/54750.
155. F. Abbassi, C. Galanth, M. Amiche, K. Saito, C. Piesse, L. Zargarian, K. Hani, P. Nicolas, O. Lequin and A. Ladram, *Biochemistry*, 2008, **47**, 10513.
156. C. Matte, J. Mallégol and A. Descoteaux, *Intracellular Niches of Microbes: A Pathogens Guide Through the Host Cell*, eds. U. E. Schaible and A. Haas. Wiley-VCH Verlag GmbH & Co. KGaA, 2009, ch. 32, pp. 583–595.
157. J. C. Antoine, E. Prina, C. Jouanne and P. Bongrand, *Infect. Immun.*, 1990, **58**, 779.
158. C. Chicharro, C. Granata, R. Lozano, D. Andreu and L. Rivas, *Antimicrob. Agents Chemother.*, 2001, **45**, 2441.
159. J. R. Luque-Ortega, J. M. Saugar, C. Chiva, D. Andreu and L. Rivas, *Biochem. J.*, 2003, **375**, 221.
160. M. Fernández-Reyes, D. Diaz, B. G. de la Torre, A. Cabrales-Rico, M. Valles-Miret, J. Jimenez-Barbero, D. Andreu and L. Rivas, *J. Med. Chem.*, 2010, **53**, 5587.
161. E. Guerrero, J. M. Saugar, K. Matsuzaki and L. Rivas, *Antimicrob. Agents Chemother.*, 2004, **48**, 2980.
162. T. Silva, M. A. Abengózar, M. Fernández-Reyes, D. Andreu, K. Nazmi, J. G. Bolscher, M. Bastos and L. Rivas, *Amino Acids*, 2012, **43**, 2265.
163. V. V. do Nascimento, O. Mello Ede, L. P. Carvalho, E. J. de Melo, O. Carvalho Ade, K. V. Fernandes and V. M. Gomes, *Biosci. Rep.*, 2015, **35**, e00248.

164. P. Eaton, C. R. Bittencourt, V. Costa Silva, L. M. Veras, C. H. Costa, M. J. Feio and J. R. Leite, *Nanomedicine*, 2014, **10**, 483.
165. L. C. Salay, T. M. Nobre, M. C. Colhone, M. E. D. Zaniquelli, P. Ciancaglini, R. G. Stabeli, J. R. S. A. Leite and V. Zucolotto, *J. Pept. Sci.*, 2011, **17**, 700.
166. P. Nicolas, *FEBS J.*, 2009, **276**, 6483.
167. M. Scocchi, M. Mardirossian, G. Runti and M. Benincasa, *Curr. Top. Med. Chem.*, 2016, **16**, 76.
168. M. Benincasa, G. Runti, M. Mardirossian, R. Gennaro and M. Scocchi, *Methods Mol. Biol.*, 2017, **1548**, 283.
169. A. Jimenez-Ruiz, J. F. Alzate, E. T. Macleod, C. G. Luder, N. Fasel and H. Hurd, *Parasit. Vectors*, 2010, **3**, 104.
170. S. Gannavaram and A. Debrabant, *Front. Cell. Infect. Microbiol.*, 2012, **2**, 95.
171. M. M. Kulkarni, W. R. McMaster, W. Kamysz and B. S. McGwire, *J. Biol. Chem.*, 2009, **284**, 15496.
172. M. C. Erfe, C. V. David, C. Huang, V. Lu, A. C. Maretti-Mira, J. Haskell, K. W. Bruhn, M. R. Yeaman and N. Craft, *Antimicrob. Agents Chemother.*, 2012, **56**, 658.
173. S. Wang, X. Zeng, Q. Yang and S. Qiao, *Int. J. Mol. Sci.*, 2016, **17**, E603.
174. S. E. Blondelle and K. Lohner, *Curr. Pharm. Des.*, 2010, **16**, 3204.
175. C. D. Fjell, R. E. W. Hancock and A. Cherkasov, *Bioinformatics*, 2007, **23**, 1148.
176. B. Mishra, S. Reiling, D. Zarena and G. Wang, *Curr. Opin. Chem. Biol.*, 2017, **38**, 87.
177. P. Roch, A. Beschin and E. Bernard, *J. Evidence-Based Complementary Altern. Med.*, 2004, **1**, 167.
178. M. M. Kulkarni, A. Karafova, W. Kamysz and B. S. McGwire, *Parasitol. Res.*, 2014, **113**, 1971.
179. I. S. Radzishevsky, S. Rotem, F. Zaknoon, L. Gaidukov, A. Dagan and A. Mor, *Antimicrob. Agents Chemother.*, 2005, **49**, 2412.
180. S. Rotem, I. Radzishevsky, R. T. Inouye, M. Samore and A. Mor, *Peptides*, 2006, **27**, 18.
181. L. M. Gottler and A. Ramamoorthy, *Biochim. Biophys. Acta*, 2009, **1788**, 1680.
182. M. Mura, J. Wang, Y. Zhou, M. Pinna, A. V. Zvelindovsky, S. R. Dennison and D. A. Phoenix, *Eur. Biophys. J.*, 2016, **45**, 195.
183. H. G. Boman, D. Wade, I. A. Boman, B. Wahlin and R. B. Merrifield, *FEBS Lett.*, 1989, **259**, 103.
184. M. E. Guerra, V. Fadel, V. G. Maltarollo, G. Baldissera, K. M. Honorio, J. R. Ruggiero and M. P. Dos Santos Cabrera, *Chem. Biol. Drug Des.*, 2017, DOI: 10.1111/cbdd.12970.
185. J. R. S. A. Leite, G. D. Brand, L. P. Silva, S. A. S. Kückelhaus, W. R. C. Bento, A. L. T. Araújo, G. R. Martins, A. M. Lazzari and C. Bloch Jr., *Comp. Biochem. Physiol. A Mol. Integr. Physiol.*, 2008, **151**, 336.

186. J. Alberola, A. Rodríguez, O. Francino, X. Roura, L. Rivas and D. Andreu, *Antimicrob. Agents Chemother.*, 2004, **48**, 641.
187. C. Arias, M. Guizy, J. R. Luque-Ortega, E. Guerrero, B. G. De La Torre, D. Andreu, L. Rivas and C. Valenzuela, *Biochim. Biophys. Acta*, 2006, **1763**, 110.
188. M. Velasco, M. J. M. Díaz-Guerra, P. Díaz-Achirica, D. Andreu, L. Rivas and L. Boscá, *J. Immunol*, 1997, **158**, 4437.
189. D. I. Andersson, D. Hughes and J. Z. Kubicek-Sutherland, *Drug Resist. Updates*, 2016, **26**, 43.
190. M. Vanaerschot, S. Decuypere, M. Berg, S. Roy and J. C. Dujardin, *Crit. Rev. Microbiol.*, 2013, **39**, 384.
191. J. Jantsch, V. Schatz, D. Friedrich, A. Schröder, C. Kopp, I. Siegert, A. Maronna, D. Wendelborn, P. Linz, K. J. Binger, M. Gebhardt, M. Heinig, P. Neubert, F. Fischer, S. Teufel, J. P. David, C. Neufert, A. Cavallaro, N. Rakova, C. Küper, F. X. Beck, W. Neuhofer, D. N. Muller, G. Schuler, M. Uder, C. Bogdan, F. C. Luft and J. Titze, *Cell Metab.*, 2015, **21**, 493.
192. M. L. Clausen and T. Agner, *Curr. Probl. Dermatol.*, 2016, **49**, 38.
193. K. Van Bocxlaer, V. Yardley, S. Murdan and S. L. Croft, *J. Antimicrob. Chemother.*, 2016, **71**, 1578.
194. N. Do, G. Weindl, L. Grohmann, M. Salwiczek, B. Koksch, H. C. Korting and M. Schafer-Korting, *Exp. Dermatol.*, 2014, **23**, 326.
195. M. R. Hamblin, P. Avci and T. W. Prow, *Nanoscience in Dermatology*, Elsevier Inc., 2016.
196. A. Mor, K. Hani and P. Nicolas, *J. Biol Chem.*, 1994, **269**, 31635.
197. A. Patrzykat and S. E. Douglas, *Protein Pept. Lett.*, 2005, **12**, 19.
198. N. P. Chongsiriwatana, M. Wetzler and A. E. Barron, *Antimicrob. Agents Chemother.*, 2011, **55**, 5399.
199. D. Knappe, N. Kabankov, N. Herth and R. Hoffmann, *Future Med. Chem.*, 2016, **8**, 1035.
200. O. Cirioni, C. Silvestri, R. Ghiselli, F. Orlando, A. Riva, F. Mocchegiani, L. Chiodi, S. Castelletti, E. Gabrielli, V. Saba, G. Scalise and A. Giacometti, *J. Antimicrob. Chemother.*, 2008, **62**, 1332.
201. S. C. Park, Y. M. Kim, J. K. Lee, N. H. Kim, E. J. Kim, H. Heo, M. Y. Lee, J. R. Lee and M. K. Jang, *J. Control. Release*, 2017, **256**, 46.
202. C. S. Cavalcante, C. B. Falcao, R. O. Fontenelle, D. Andreu and G. Radis-Baptista, *J. Antibiot.*, 2017, **70**, 231.
203. M. Rakotomanga, S. Blanc, K. Gaudin, P. Chaminade and P. M. Loiseau, *Antimicrob. Agents Chemother.*, 2007, **51**, 1425.
204. J. M. Basu, A. Mookerjee, P. Sen, S. Bhaumik, P. Sen, S. Banerjee, K. Naskar, S. K. Choudhuri, B. Saha, S. Raha and S. Roy, *Antimicrob. Agents Chemother.*, 2006, **50**, 1788.
205. L. Carvalho, J. R. Luque-Ortega, J. I. Manzano, S. Castanys, L. Rivas and F. Gamarro, *Antimicrob. Agents Chemother.*, 2010, **54**, 5344.
206. E. F. Cunha-Júnior, V. V. Andrade-Neto, M. L. Lima, T. A. da Costa-Silva, A. J. Galisteo Junior, M. A. Abengózar, C. Barbas, L. Rivas,

E. E. Almeida-Amaral, A. G. Tempone, E. C. and Torres-Santos, *PLoS. Neglected Trop. Dis.*, 2017, **11**, e0005281.
207. J. P. Mattila, K. Sabatini and P. K. J. Kinnunen, *Biochim. Biophys. Acta*, 2008, **1778**, 2041.
208. A. C. Coelho and P. C. Cotrim, *Drug Resistance in Leishmania Parasites: Consequences, Molecular Mechanisms and Possible Treatments*, ed. A. Ponte-Sucre, E. Diaz, M. Padrón-Nieves, Springer-Verlag, Wien, 2013, ch. 11, pp. 237–258.
209. F. Gamarro, M. P. Sánchez-Cañete and S. Castanys, *Drug Resistance in Leishmania Parasites: Consequences, Molecular Mechanisms and Possible Treatments*, ed. A. Ponte-Sucre, E. Diaz, M. Padrón-Nieves, Springer-Verlag, Wien, 2013, ch. 16, pp. 351–380.
210. N. S. Parachin, K. C. Mulder, A. A. B. Viana, S. C. Dias and O. L. Franco, *Peptides*, 2012, **38**, 446.
211. R. B. Rhaiem and M. Houimel, *Acta Trop.*, 2016, **159**, 11.
212. E. Moreno, J. Schwartz, C. Fernández, C. Sanmartín, P. Nguewa, J. M. Irache and S. Espuelas, *Expert Opin. Drug Deliv.*, 2014, **11**, 579.
213. A. Reinhardt and I. Neundorf, *Int. J. Mol. Sci.*, 2016, **17**, E701.
214. P. Urbán, J. J. Valle-Delgado, E. Moles, J. Marques, C. Díez and X. Fernàndez-Busquets, *Curr. Drug Targets*, 2012, **13**, 1158.
215. V. N. Lazarev, M. M. Shkarupeta, E. S. Kostryukova, S. A. Levitskii, G. A. Titova, T. A. Akopian and V. M. Govorun, *Bull. Exp. Biol. Med.*, 2007, **144**, 452.
216. V. N. Lazarev, M. M. Shkarupeta, G. A. Titova, E. S. Kostrjukova, T. A. Akopian and V. M. Govorun, *Biochem. Biophys. Res. Commun.*, 2005, **338**, 946.
217. Y. A. Wen, X. L. Yu, Q. F. Xia, D. Cen, J. B. Liu and Z. G. Tu, *Int. J. Antimicrob. Agents*, 2009, **34**, 315.
218. M. A. Couto, L. Liu, R. I. Lehrer and T. Ganz, *Infect. Immun.*, 1994, **62**, 2375.
219. H. Ottosson, F. Nylén, P. Sarker, E. Miraglia, P. Bergman, G. H. Gudmundsson, R. Raqib, B. Agerberth and R. Strömberg, *Sci. Rep.*, 2016, **6**, 36692.
220. E. Ramos-Martínez, M. I. Villaseñor-Cardoso, M. R. López-Vancell, F. J. García-Vázquez, A. Pérez-Torres, N. Salaiza-Suazo and R. Pérez-Tamayo, *Exp. Parasitol.*, 2013, **134**, 413.
221. P. V. Chang, L. Hao, S. Offermanns and R. Medzhitov, *Proc. Natl. Acad. Sci. U. S. A.*, 2014, **111**, 2247.

CHAPTER 14

Nutrient Transport and Sensing as Pharmacological Targets for Leishmaniasis

DAN ZILBERSTEIN

Faculty of Biology, Technion-Israel Institute of Technology, Haifa 320003, Israel
Email: danz@technion.ac.il

14.1 Introduction

A key feature in cellular adaptation of *Leishmania* parasites to the distinct environments of the insect vector and the mammalian host is nutrient transport in and out of the cells. Membrane transporters are the major means used by cells to translocate nutrients across their plasma membranes. Together with metabolic enzymes, cells maintain homeostatic levels of nutrients. An example of this is the cellular pool of proline that promastigotes maintain at 10 mM. Parasites that lack the gene encoding the proline transporter protein (LdAAP24) lack the cellular pool of proline as well.

This chapter summarizes the up-to-date biochemical and molecular knowledge on transport of amino acids, sugars, nucleosides and purines, as well as on nutrient sensing in *Leishmania*. It also discusses the potential use of *Leishmania* nutrient transporters as drug delivery systems for increased drug efficacy.

Drug Discovery Series No. 60
Drug Discovery for Leishmaniasis
Edited by Luis Rivas and Carmen Gil

Published by the Royal Society of Chemistry, www.rsc.org

14.2 The Metabolic Environment Inside the Vector and Host

14.2.1 The Insect Vector

During their life cycle, *Leishmania* parasites alternate between two hydrolytic but distinct environments: mid-gut of female sand flies and phagolysosome of mammalian macrophages. The first is relatively alkaline and rich in sugar, as insects of the phlebotomine family feed on plant juice.[1] Occasionally (every 10 days), the female sand fly that carries fertilized eggs sucks blood from an animal or human to enable egg maturation. This blood meal temporarily enriches the mid-gut with proteins and amino acids. Furthermore, insects utilize alanine and proline as sources of metabolic energy for their muscles, including those used for flying.[1–4] For example, hemolymphs of tsetse flies accumulate up to 60 mM alanine, which is converted by oxidative deamination to α-ketoglutarate and then *via* the Krebs cycle contributes electrons for energy production through oxidative phosphorylation. *Leishmania* cells are highly flexible in their ability to utilize proline, alanine and glucose as metabolic energy precursors.[5] In the absence of sugar, *Leishmania* can grow on proline as an alternative carbon source. However, glucose represses amino-acid catabolism through an unknown mechanism. If both glucose and proline are available in the medium, glucose will always be used by the promastigotes. The results of studies conducted in the 1970s indicated that in culture, promastigotes prefer proline at log phase and shift to glucose towards the stationary phase.[6] However, the relevance of this phenomenon remained unknown, as no further research was performed on this preliminary observation. Unlike African and American trypanosomes, during development in the insect vector, *Leishmania* parasites are not directly exposed to the alanine-rich hemolymph. Nonetheless, procyclics of both parasites (*e.g.* epimastigotes and promastigotes, respectively) show almost identical catabolic preferences; both prefer glucose but can grow equally well on proline and alanine.

It is likely that the vector hemolymph influences catabolic preferences in all *Trypanosoma cruzi*, *Trypanosoma brucei* and *Leishmania* (TriTryps) regardless of the lack of physical contact in *Leishmania*. The parasites in the vector's mid gut face unstable osmotic conditions due to dramatic changes between blood meals. Consequently, osmoregulatory mechanisms developed that help promastigotes survive in the constantly changing environment. Two features of promastigotes support osmotic resistance. The first is the strong and rigid sub-pelicular microtubule that mechanically protects parasite cells from lysis upon exposure to hypotonic environments. The second is the use of osmolytes (*e.g.* proline and alanine) to minimize swelling.[7,8] Upon encountering a hypo-osmotic environment, promastigotes rapidly release proline and alanine from a large cellular pool (10 and 25 mM, respectively) thereby reducing internal osmolarity. The high concentrations of both amino acids are essential for a normal osmoregulatory response.[7] The pool of proline is essential for the ability of promastigotes to recover from swelling, whereas that of alanine is essential for their ability to minimize the size of swelling. Transport of proline and alanine is thus critical for maintaining the cellular pools of these amino acids.

14.2.2 Development Inside the Host

The environment inside the mammalian phagolysosome is acidic (pH 4.5–5.5)[9] and the temperature is approximately 10 °C higher than that in the insect vector (33–37 °C). This creates a unique situation in which invading parasites encounter two stresses: acidic pH and high temperature. Normally, they respond to each stress separately using a mechanism resembling that of other lower eukaryotes.[10] However, *Leishmania* parasites respond to concomitant exposure to 37 °C and pH 5.5 as a signal for initiating differentiation from promastigote to amastigote.[11,12] The kinetics of this signal-derived differentiation have been extensively investigated, the results indicating that differentiation is a regulated process involving post-transcriptional, translational and post-translational processes.

The nutrient content of phagolysosome before and during infection is still an enigma. To date, only a few studies have determined the amino acid content of lysosomes, and none (to the best of my knowledge) determined the content with respect to sugar or fatty acids. Harms *et al.*[13] developed a method to purify intact lysosomes from liver tissue and determined their amino acid content. These lysosomes contained 95.24 nmoles free amino acids mg^{-1} protein. The major amino acids were glutamate and/or glutamine (15 nmoles mg^{-1} protein), lysine (11 nmoles mg^{-1} protein) and leucine (8.4 nmoles mg^{-1} protein). Interestingly, the amino acid content in the lysosomes was different from that of total cell extracts (Figure 14.1). Over half of the amino acids were more abundant in lysosomes than in liver tissue cells. Notably, arginine and cystine were found to be 34 and 12 times more abundant in lysosomes than in tissue extracts, respectively (Figure 14.1). The second most abundant nutrients in lysosomes are probably fatty acids, as they are byproducts of lipid degradation, whereas the sugar content in lysosomes is expected to be very low. Naderer *et al.*[14] found that, in contrast to free-living promastigotes, gluconeogenesis in amastigotes is an essential pathway, further supporting the notion that glucose is scarce in lysosomes and that gluconeogenesis supplies the missing free glucose. Naderer *et al.* proposed that amino glycans replace glucose as the energy carbon source.[14] Biochemical, proteomics and metabolomics analyses demonstrated major changes in catabolic preferences between promastigotes and amastigotes. For example, Rosenzweig *et al.*[15] used a differentiation time course analysis and found a decrease in the abundance of glycolytic enzymes but an increase in catabolism of amino acids and β-oxidation.

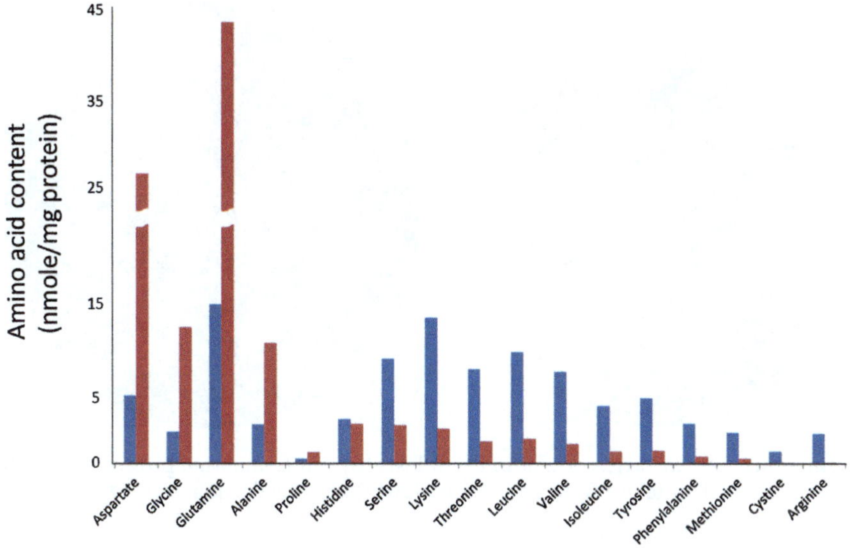

Figure 14.1 Amino acid content of liver lysosomes (blue bars) compared with whole-tissue extract (red bars).
Data taken from Harms *et al.* (1981).[13]

14.3 Amino Acid Transport and Homeostasis

14.3.1 The Amino Acid Pool in *Leishmania* Cells

Comparative analysis of cellular content of free amino acids revealed a few interesting differences between the parasite and its host. The major amino acids in the mammalian cellular pool are glutamine (33%), glutamate (19%), and threonine (11%) (Figure 14.2A).[16] In *Leishmania* promastigotes, the major amino acids are alanine (32%), glutamate (12%) and proline (10%) (Figure 14.2B). Similarly to the other Trypanosomatids, *Leishmania* parasites utilize the uncharged alanine and proline as major osmolytes,[7] secreting them in large amounts upon exposure to hypotonic solutions. In contrast, mammalian cells secrete KCl and non-essential organic compounds to minimize swelling, and set new regulated volume decrease (RVD) upon exposure to hypotonic environment. Inbar *et al.*[7] showed that the cellular pool of proline is totally dependent on the presence and activity of the *Leishmania* proline-alanine transporter LdAAP24. Abrogating this gene cascade in the *Leishmania* genome emptied the proline pool and reduced the alanine pool by half. They also demonstrated that alanine and proline have distinct roles in the response to hypo-osmotic shock, alanine helps to minimize the magnitude of swelling, whereas proline is responsible for RVD that recovers cells from swelling.

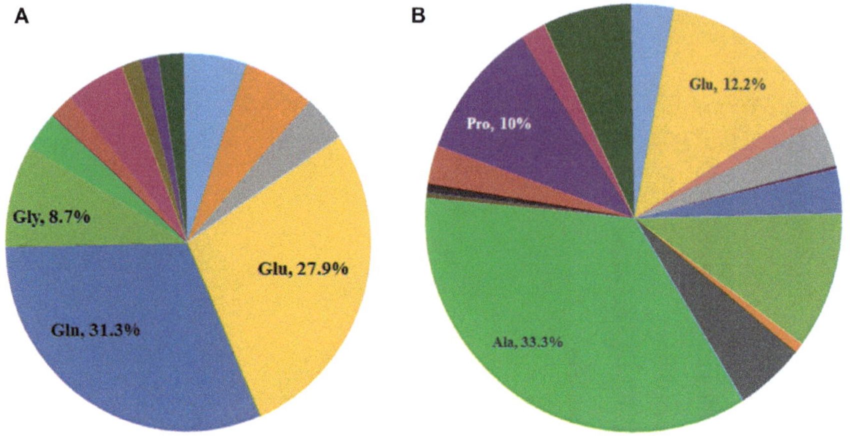

Figure 14.2 Amino acid content in fibroblasts (A) and *L. donovani* promastigotes (B). Data taken from Inbar *et al.* (2013) and Bevilacqua *et al.* (2005).[7,16]

14.3.2 LdAAP24: A New Role for Permease N-termini in Transport Regulation

Proline transport was first reported in *Leishmania tropica* promastigotes and amastigotes as broad-specificity active transport systems,[17,18] driven by the proton motive force. Further biochemical analysis in *Leishmania donovani* revealed three proline transport systems: A) cation-dependent, promastigote-specific; B) cation-independent, promastigote-specific; and C) cation-dependent, amastigote-specific.[19] The active transport of proline in promastigotes (systems A and B) is energized by the proton motive force,[20] as reviewed by Bringaud *et al.*[21] Moreover, as predicted by Law and Mukkada in 1979,[17] only 10% of the accumulated proline is incorporated into proteins, while the remaining proline forms a large pool (approximately 10 mM).

Two identical genes on the reverse strand of *L. donovani* chromosome 10 encode system A transporters (LinJ.10.0760 and LinJ.10.0770).[7] These genes differ in splice sites location, resulting in two protein variants: a full-size (488 amino acids) and an 18 amino acids (at the N-terminus) shorter protein, LdAAP24.2 and LdAAP24.1, respectively.[22] LdAAP24.2 transports both proline and alanine, whereas LdAAP24.1 transports only proline. Interestingly, Schlisselberg *et al.*[22] demonstrated that the 18 amino acids at the tip of the LdAAP24 N-terminus form an element essential for alanine transport activity. Co-expression of LdAAP24.1 with the full-length soluble 89-amino-acid N-terminus peptide recovered the alanine transport activity in this variant. Recent, as yet unpublished, results indicated that phosphorylation of two adjacent sites in the 18 amino acid element (Ser6 and Thr7) is essential for activating alanine transport in LdAAP24.2. As shown biochemically, both *LdAAP24* genes are expressed only in promastigotes. Once they are exposed to conditions that initiate axenic differentiation into amastigotes, both LdAAP24 protein variants rapidly and completely degrade [half life ($t_{1/2}$) = 35 min; Liburkin-Dan, T. *et al.*, personal communication]. Stage-specific expression of LdAAP24 also occurs *in vivo*, in macrophage- and hamster-derived parasites.

Most *Leishmania* amino acid transporters have long hydrophilic N-termini (over 70 amino acids),[23] which probably control transport activity. To date, three such transporters have been cloned and functionally characterized: LdAAP24 (proline–alanine), LdAAP7 (lysine), and LdAAP3 (arginine). More studies are warranted to elucidate the significance of the long N-termini phenomenon in *Leishmania* transporters.

14.3.3 A *Leishmania*-Specific Gene Encodes the Arginine Transporter AAP3 Family

Arginine is an essential amino acid for *Leishmania* but semi essential for the host.[5] *L. donovani* promastigotes take up arginine *via* a high-affinity and specificity transport system [Michaelis constant (K_m) = 14.20 μM, maximum reaction rate (V_{max}) = 166 pmol min^{-1} 10^{-8} cells].[24] None of the classical analogs of arginine (*e.g.* agmatine, homoarginine), nor any of the other amino acids was found to inhibit transport in promastigotes[24,25] or amastigotes (Goldmane-Pinkovich *et al.*, unpublished). Interestingly, diamidines, such as pentamidine, berenil and dibrompropamidine, were demonstrated to be efficient competitive inhibitors of arginine transport [50% inhibitory concentration (IC_{50}) = 38 μM for pentamidine].[26] The concentrations of pentamidine that inhibited arginine transport were more than 30 times higher than the concentrations that killed parasites. To date, no experimental evidence has been provided on the nature of arginine transport inhibition by pentamidine. Pentamidine transport assays indicated that arginine did not inhibit its transport, strongly indicating that LdAAP3 is not a diamidine transporter.[27]

In *L. donovani*, two adjacent genes on the reverse strand of chromosome 31 encode the protein that transports arginine (LdAAP3). This is a high-affinity (K_m = 1.9 μM) transporter whose sole substrate is L-arginine. Orthologues of AAP3 have been identified in *Leishmania amazonensis* (LaAAP3)[28] and *Leishmania mexicana* (LmxAAP3).[29] These genes are syntenic but genera-specific. Biochemical analyses of LdAAP3 strongly support the notion that LdAAP3 is the transport system described by Kandpal *et al.*[24]

Interestingly, LdAAP3 localized in promastigotes to the surface of the flagellum and to glycosome membranes.[29] Protein abundance in both locations increased in response to arginine deprivation. The mechanism that determines the dual localization of LdAAP3 is as yet unknown, although the biological rationale for it is clear. In *Leishmania*, arginine is the sole precursor for polyamine biosynthesis.[5,30] Arginase, the enzyme that converts arginine to ornithine localizes to parasite lysosomes,[31,32] a localization that is essential for parasite virulence. An interplay between LdAAP3 activity and/or abundance and the polyamine pathway has been demonstrated.[33] In ornithine decarboxylase (ODC) or spermidine synthase null mutants, the cellular level of arginine is reduced and LdAAP3 activity is decreased by half.[33]

14.3.4 Lysine Transport

In mammalian cells, lysine and arginine are translocated *via* the same transporters, for example, the cation amino acid (CAT). In contrast, in *Leishmania* parasites, there are separate transporters for arginine and lysine (LdAAP3 and LdAAP7, respectively).[25,34] The *L. donovani* lysine transporter, like the arginine transporter, is a high-affinity and specific transporter. The kinetic parameters of LdAAP7 resemble those of LdAAP3 ($K_m = 3$ μM, $V_{max} = 0.27$ nmol min^{-1} 10^{-8} cells). LdAAP7 localizes to the surface of promastigotes. The orthologous gene for LdAAP7 in *T. brucei* has recently been identified and characterized. RNAi assays indicated that TbAAP7 and TbAAP3 are both essential for the procyclics as well as for the bloodstream forms.[35]

14.3.5 Transporters of Other Amino Acids

L. tropica promastigotes take up methionine *via* a high-affinity and low-specificity system.[36,37] The gene(s) that encode the methionine transporter have not yet been identified. On the other hand, the gene encoding the *S*-adenosylmethionine transporter has been identified in *Leishmania infantum* promastigotes.[38] This surface membrane protein is a high-affinity transporter that belongs to the folate–biopterine transporter family.

Transport of polyamines was biochemically investigated. Putrescine and spermidine are taken up *via* high-affinity and high-specificity transport systems.[39] Polyamine transport enables ablation of upstream pathway enzymes such as arginase, ODC and spermidine synthase.[27,40]

14.4 Glucose Transport

As indicated earlier in this chapter, procyclic promastigotes feed mostly on sugar provided by the vector's diet. In phagolysosomes, the situation with respect to nutrients is different, although not entirely clear. It is widely agreed that amino acids [either free or bound to sugar (aminoglycosides)] and fatty acids are the major metabolic energy providers in phagolysosomes.[41,42] Regardless, glucose transporters are active in both intracellular amastigotes and extracellular promastigotes. Glucose transport in *Leishmania* spp. has been biochemically described a long time ago using the non-metabolized analogue 2-deoxy D-glucose.[43–46] These studies revealed high-affinity systems with K_m values of 24 and 29 μM in promastigotes and amastigotes, respectively.[43–46] Transport was active in nature, driven by the proton motive force.[20] Of all the traditional inhibitors of mammalian erythrocyte glucose transport, only cytochalasin B efficiently blocked glucose transport in *Leishmania*. A protein of 20–30 kDa was identified as the *Leishmania* glucose transporter in experiments using radiolabeled cytochalasin B chemically crosslinked to *L. donovani* surface proteins.[47] However, the results of later studies indicated that this size was approximately half that predicted from the genes encoding *Leishmania* glucose transporters proteins.

The first glucose transporter gene was identified and cloned by Cairns *et al.*[48] as a promastigote-specific gene. This gene encoded a protein of 61.4 kDa that showed a high level of similarity to the conserved human erythrocyte transporter (HGT)[49] and to *Escherichia coli* arabinose transporter.[50] Both erythrocyte and bacterial transporters translocate glucose passively using facilitated diffusion. This contradicted previous observations on the active nature of glucose transport in *Leishmania*.[20] Further analyses in *Leishmania enriettii* revealed two isoforms that differ mostly in their N-termini sequences and are expressed mostly in promastigotes.[51] The first (isoform 1, or iso-1) localizes at the surface of the flagellum and the second (iso-2) is expressed mostly in the surface membrane. Interestingly, the flagellar localization of iso-1 was determined by the N-terminus association with the cortex subpelicular microtubule.[52,53]

The *L. mexicana* genome includes a gene family (LmGT) that encodes three isoforms of glucose transporters: LmGT1, LmGT2 and LmGT3.[54] These transporters differ from each other in the sequences at their N- and C-termini. LmGT1 and LmGT3 are expressed constitutively throughout the parasite life cycle; whereas LmGT2 is expressed exclusively in promastigotes.[54] Interestingly, ablation of LmGT isoforms from the *L. mexicana* genome was not toxic to parasite growth but did attenuate the ability of promastigote to differentiate into amastigotes. Of the three isoforms, only LmGT3 expression in Δ*LmGT* strains rescued axenic and *in vivo* differentiation from promastigotes into amastigotes.[55] The observation that *Leishmania* promastigotes can grow without importing glucose from the outside raises a question about the metabolic changes in these promastigotes that enable growth.[56] Gluconeogenesis is probably the alternative pathway providing parasites with the missing glucose. Moreover, the inability of Δ*LmGT* null mutants to grow as amastigotes was surprising as it was assumed that amastigotes rely less on glucose. Furthermore, whereas ablating the phosphofructokinase gene had no effect on promastigotes, such an ablation stopped amastigotes from developing, indicating that this enzyme was essential for their survival.[14] The strongest phenotype of LmGT is loss of virulence, or more specifically, the inability to differentiate into amastigotes. Feng *et al.*[57] identified LmGT4 (an additional hexose transporter that is part of the LmGT cascade), which is amplified in the Δ*LmGT* null mutants and rescues virulence.

14.5 Purine, Pyrimidine and Nucleoside Transport

Many metabolic pathways are similar in parasites and their hosts. One of the exceptions is the purine metabolism. All the parasitic protozoa characterized to date are unable to synthesize the purine ring *de novo*. In each group, there is a unique series of purine salvage pathways that enable the parasites to scavenge host purines.[59] Both promastigote and amastigote forms of the *Leishmania* species are likely to use the nucleotides of the host organism as their major source of purines, since they are present in vast excess over nucleosides or bases. *L. donovani* and *Leishmania major* promastigotes possess four independent purine nucleoside transporters with no overlapping substrate specificity. These include: NT1, a high-affinity transporter that translocates adenosine and pyrimidines; NT2, a high-affinity transporter that translocates inosine and guanosine; NT3, a transporter that translocates hypoxanthine, xanthine, adenosine and guanosine; and NT4, that transports purine nucleobases at acidic pH.[58,59]

14.6 Nutrient Sensing Has a Role in Virulence

Nutrient sensing is an intracellular function that supports metabolic homeostasis, largely through controlled uptake or release of nutrients.

14.6.1 Intracellular Sensing: Response to Purine Starvation

Depletion of purine nucleosides, nucleotides or nucleobases from the growth medium for at least six hours induced a significant increase of more than tenfold in transport of nucleosides, such as inosine, hypoxanthine and others.[60,61] Long-term starvation (*e.g.* over 48 hours) halted cell growth, with most of the cells stopping at G_1–G_0. Starvation was purine-specific, as it affected only the uptake of purines, not amino acids (arginine) or glucose. The increased transport *via* LdNT1, LdNT2 and LdNT3 was due to a change in mRNA-independent *de novo* transporter protein synthesis.[60] Interestingly, kinetic analysis of nucleoside transporter before and after starvation indicated that under the latter conditions, only the rate (V_{max}) of transport increased significantly, with no effect on the affinity of the transporters. In order to understand how *Leishmania* parasites sense and respond to changes in their purine environment, Martin *et al.*[62] have exploited several purine pathway mutants, including some in which adenine and guanine nucleotide metabolism was uncoupled from transport. They found that adaptation of the parasites arises from surveillance of the intracellular purine nucleotide pools rather than from direct sensing of the extracellular purine content.[62] The response to purine starvation is slow and parasites can stay alive for a few days before purine is salvaged, thereby resuming growth. This is probably because the starved cells arrest growth and minimize their metabolic activities, thereby enabling survival until purines are available again.

14.6.2 Extracellular Sensing: Arginine-deprivation Response

Unlike purine starvation, which leads to long-term growth arrest, lack of arginine in the growth medium results in cell death within 48–72 hours. In *Leishmania* parasites, arginine is the sole precursor for the polyamine synthesis pathway that provides spermidine for the biosynthesis of the essential trypanothione.[5] During host infection, the macrophage arginine pool is depleted due to its use in producing metabolites (NO and polyamines) that constitute part of the host defense response and its suppression. Hence, there are mechanisms by which parasites sense critical changes in the environmental pool of arginine in both host and vector. Indeed, depriving the growth medium of *L. donovani* promastigotes of amino acids reduced the cellular pool of several amino acids, including arginine and intermediates of the polyamine pathway.[33] In *L. donovani*, the parasites respond within minutes by increasing the abundance of the arginine transporter LdAAP3.[29] This intriguing observation led Goldman-Pinkovich *et al.*[29] to identify a new pathway in *Leishmania* that specifically senses the extracellular levels of arginine. They found that both promastigotes and amastigotes, in axenic culture and inside macrophages, respond to shortage of arginine by upregulating expression and activity of the LdAAP3, as well as several other transporters.[29] The results of the analysis indicated that parasites monitor arginine levels in the environment rather than in the intracellular pools. Results of phosphoproteomics and genetic assays indicated that this arginine-deprivation response (ADR) is mediated through a mitogen-activated protein kinase-2-dependent signaling cascade.[29]

Interesting, as yet unpublished, experiments demonstrated that the minimal arginine concentration that activates the ADR is similar to the K_m of LdAAP3 for arginine. In addition, pentamidine, which competitively inhibits LdAAP3 transport activity,[26] also attenuated ADR activation. These preliminary data support the notion that LdAAP3 is a transceptor, *i.e.* it both translocates arginine and senses its absence.

14.6.3 Proteins Involved in Sensing are Localized in the Flagella Membrane

An interesting phenomenon in Trypanosomatids is the localization of several membrane proteins, including transporters and channels, in the flagella membrane. Some of these proteins are involved in sensing, including LdAAP3[29] and aquaglyceroporin 1.[63] Rodrigues-Contreras and Landfear[64] have recently suggested that the flagellar localization is linked to metabolic sensing. They observed that glucose deprivation in the growth medium of *L. mexicana* promastigotes induced 50-fold increases in the expression and activity of LmGT1 and that LmGT1-null mutants died when subjected to glucose deprivation.[65] All in all, the investigators concluded that GT1, one of three isoforms in *L. mexicana*, is a glucose sensor due to its localization in parasite flagella.

14.7 Can Nutrient Transporters Deliver Drugs into *Leishmania* Cells?

14.7.1 Amino Acid Transporters

Leishmania amino acid transporters differ from those of the host (for example see Inbar *et al.*);[34] therefore, these transporter might be excellent targets for delivering drugs into the parasite cells. Prati *et al.*[66] investigated this hypothesis using appropriately-designed chemical probes. As indicated earlier in this chapter, *L. donovani*, the parasite that causes visceral leishmaniasis, is auxotrophic for arginine and lysine and has specific transporters (LdAAP3 and LdAAP7) to import these nutrients. The investigators conjugated cytotoxic quinone fragments to arginine, lysine and gamma amino isobutyric acid (GABA) with amide linkage and investigated the toxicity of these conjugates against promastigotes and amastigotes as well their ability to inhibit the relevant amino acid transporters. Indeed, some conjugates were found to be toxic to the parasites.[66] In particular, the conjugate with GABA was identified as the most potent derivative. Notably, the conjugate with lysine retained the cytotoxic activity of quinone II and displayed no toxicity against mammalian THP1 macrophages. Results of transport assays indicated that the novel conjugates inhibited transport activity of lysine, arginine and proline transporters. The results of the analysis indicated that the toxic conjugates might be translocated by the transporters into the cells. The non-toxic probes that inhibited transport competed with the natural substrates for binding to the transporters without being translocated. Thus, it is likely that the lysine conjugated at position 6, can selectively deliver its toxic effects to *L. donovani* cells by exploiting amino acid transporters.[66] This work provides a proof-of-concept for modulation of *Leishmania* amino acid transporters with small molecules and warrants their further investigation from drug discovery and chemical biology perspectives.

14.7.2 Pentamidine Transport in *Leishmania*

Pentavalent antimonial (SbV) drugs are the standard first-line treatment for leishmaniasis, albeit there is increasing resistance in the highly endemic regions.[67] The diamidine pentamidine is often used as an alternative to SbV in leishmaniasis treatment.[67,68] A definitive mode-of-action has yet to be elucidated for pentamidine, despite studies on its uptake and distribution in drug-sensitive and -resistant parasites (*Leishmania* and the closely related parasite *T. brucei*). Pentamidine is a competitive inhibitor of arginine transport *via* the high-affinity *L. donovani* AAP3.[26,29] It is a non-competitive inhibitor of putrescine and spermidine transport in *L. infantum*,[69] *L. donovani* and *L. mexicana*.[27] The availability of radiolabeled pentamidine helped to determine whether it is a substrate of AAP3, putrescine or spermidine transporters.[70] Pentamidine inhibited each of these metabolites; however, this did not inhibit [^{3}H]-pentamidine uptake, indicating that pentamidine is not a substrate of these specific transporters.[70] Pentamidine translocates into both promastigotes and amastigotes of *L. mexicana via* a low-affinity carrier-mediated transport system ($K_m = 7.4$ and 13.9 mM, respectively). Interestingly, three inhibitors of P-glycoproteins (trifluoperazine, prochlorperazine and verapamil) competitively inhibited pentamidine transport. Even though *Leishmania* parasites take up pentamidine, the nature of the transporter is an enigma. Pentamidine transport and mode-of-action has been the focus of research in African trypanosomes.[71] One of these studies revealed that aquaglyceroporin 2 (TbAQP2) transports pentamidine and is one of the proteins responsible for its toxicity. Expression of *TbAQP2* in *L. mexicana* introduced a pentamidine transport activity indistinguishable from that of HAPT1.

14.8 Concluding Remarks

Publication of the genome sequence of *Leishmania*, first for *L. major* and then for other species opened a new era in parasite molecular physiology. As noted in this chapter, to date, a considerable number of genes that encode membrane transporters have been identified and functionally characterized, resulting in better understanding of the molecular mechanisms involved in development, metabolism and host–parasite interactions.[72] This chapter has summarized new advances in this field and the evidence for their significance in *Leishmania* development, including the most important aspect, namely, the increasing evidence that nutrient transporters can be exploited for drug delivery into parasite cells.

Acknowledgements

This work was supported by grant number 1766-16 from the Israel Science Foundation (ISF) and grant number 2316/15 from ISF as part of the ISF–University Grant Commission of India (UGC) joint program.

References

1. I. Cunningham and J. S. Slater, *Acta Trop.*, 1974, **31**, 83.
2. E. Brusell, *Comp. Biochem. Physiol.*, 1966, **10**, 809.
3. E. Brusell, *An Introduction to Insect Physiology*, Academic Press, London, 1970, pp. 31–34.
4. S. R. Raghupathi Reddy and J. W. Campbell, *Biochem. J.*, 1969, **115**, 495.
5. F. R. Opperdoes and G. H. Coombs, *Trends Parasitol.*, 2007, **23**, 149.
6. A. J. Mukkada, F. W. Schaefer, III, M. W. Simon and C. Neu, *J. Protozool.*, 1974, **21**, 393.
7. E. Inbar, D. Schlisselberg, M. Suter Grotemeyer, D. Rentsch and D. Zilberstein, *Biochem. J.*, 2013, **449**, 555.
8. L. L. Vieira, E. Lafuente, J. Blum and Z. I. Cabantchik, *Mol. Biochem. Parasitol.*, 1997, **90**, 449.

9. J. C. Antoine, E. Prina, T. Lang and N. Courret, *Trends Microbiol.*, 1998, **6**, 392.
10. D. Zilberstein and M. Shapira, *Annu. Rev. Microbiol.*, 1994, **48**, 449.
11. D. Zilberstein, in *Leishmania After the Genome: Biology and Control*, ed. P. J. Myler and N. Fasel, Horizon Scientific Press and Caiser Academic Press, New York, 2008, pp. 107–122.
12. T. Naderer and M. J. McConville, *Essays Biochem.*, 2011, **51**, 81.
13. E. Harms, N. Gochman and J. A. Schneider, *Biochem. Biophys. Res. Commun.*, 1981, **99**, 830.
14. T. Naderer, M. A. Ellis, M. F. Sernee, D. P. De Souza, J. Curtis, E. Handman and M. J. McConville, *Proc. Natl. Acad. Sci. U. S. A.*, 2006, **103**, 5502.
15. D. Rosenzweig, D. Smith, F. Opperdoes, S. Stern, R. W. Olafson and D. Zilberstein, *FASEB J.*, 2008, **22**, 590.
16. E. Bevilacqua, O. Bussolati, V. Dall'Asta, F. Gaccioli, R. Sala, G. C. Gazzola and R. Franchi-Gazzola, *FEBS Lett.*, 2005, **579**, 3376.
17. S. S. Law and A. J. Mukkada, *J. Protozool.*, 1979, **26**, 295.
18. T. A. Glaser and A. J. Mukkada, *Mol. Biochem. Parasitol.*, 1992, **51**, 1.
19. S. Mazareb, Z. Y. Fu and D. Zilberstein, *Exp. Parasitol.*, 1999, **91**, 341.
20. D. Zilberstein and D. M. Dwyer, *Proc. Natl. Acad. Sci. U. S. A.*, 1985, **82**, 1716.
21. F. Bringaud, M. P. Barrett and D. Zilberstein, *Front. Biosci.*, 2012, **17**, 349.
22. D. Schlisselberg, E. Mazarib, E. Inbar, D. Rentsch, P. J. Myler and D. Zilberstein, *Sci. Rep.*, 2015, **5**, 16289.
23. M. Akerman, P. Shaked-Mishan, S. Mazareb, H. Volpin and D. Zilberstein, *Biochem. Biophys. Res. Commun.*, 2004, **325**, 353.
24. M. Kandpal, R. B. Fouce, A. Pal, P. Y. Guru and B. L. Tekwani, *Mol. Biochem. Parasitol.*, 1995, **71**, 193.
25. P. Shaked-Mishan, M. Suter-Grotemeyer, T. Yoel-Almagor, N. Holland, D. Zilberstein and D. Rentsch, *Mol. Microbiol.*, 2006, **60**, 30.
26. M. Kandpal, B. L. Tekwani, P. M. Chauhan and A. P. Bhaduri, *Life Sci.*, 1996, **59**, PL75.
27. M. Basselin, G. H. Coombs and M. P. Barrett, *Mol. Biochem. Parasitol.*, 2000, **109**, 37.
28. E. A. Castilho-Martins, M. F. Laranjeira da Silva, M. G. dos Santos, S. M. Muxel and L. M. Floeter-Winter, *PLoS One*, 2011, **6**, e27818.
29. A. Goldman-Pinkovich, C. Balno, R. Strasser, M. Zeituni-Molad, K. Bendelak, D. Rentsch, M. Ephros, M. Wiese, A. Jardim, P. J. Myler and D. Zilberstein, *PLoS Pathog.*, 2016, **12**, e1005494.
30. M. F. da Silva and L. M. Floeter-Winter, *Subcell. Biochem.*, 2014, **74**, 103.
31. S. C. Roberts, M. J. Tancer, M. R. Polinsky, K. M. Gibson, O. Heby and B. Ullman, *J. Biol. Chem.*, 2004, **279**, 23668.
32. M. F. da Silva, R. A. Zampieri, S. M. Muxel, S. M. Beverley and L. M. Floeter-Winter, *PLoS One*, 2012, **7**, e34022.
33. I. Darlyuk, A. Goldman, S. C. Roberts, B. Ullman, D. Rentsch and D. Zilberstein, *J. Biol. Chem.*, 2009, **284**, 19800.

34. E. Inbar, G. E. Canepa, C. Carrillo, F. Glaser, M. Suter Grotemeyer, D. Rentsch, D. Zilberstein and C. A. Pereira, *Amino Acids*, 2012, **42**, 347.
35. C. Mathieu, J. P. De Macêdo, D. Hürlimann, C. Wirdnam, A. Gonzalez-Salgado, M. Suter Grotemeyer, E. Inbar, P. Bütikofer, D. Zilberstein and D. Rentsch, *PLoS One*, 2017, **12**, e0168775.
36. A. J. Mukkada and M. W. Simon, *Exp. Parasitol.*, 1977, **42**, 87.
37. M. W. Simon and A. J. Mukkada, *Exp. Parasitol.*, 1977, **42**, 97.
38. L. Dridi, A. Haimeur and M. Ouellette, *J. Biol. Chem.*, 2010, **285**, 19767.
39. M. Kandpal and B. L. Tekwani, *Life Sci.*, 1997, **60**, 1793.
40. S. C. Roberts, C. Kline, W. Liu and B. Ullman, *Exp. Parasitol*, 2011, **128**, 166.
41. R. J. Burchmore and M. P. Barrett, *Int. J. Parasitol.*, 2001, **31**, 1311.
42. M. J. McConville, E. C. Saunders, J. Kloehn and M. J. Dagley, *F1000-Research*, 2015, **4**, 938.
43. D. Zilberstein and D. M. Dwyer, *Mol. Biochem. Parasitol.*, 1984, **12**, 327.
44. R. J. Burchmore and D. T. Hart, *Mol. Biochem. Parasitol.*, 1995, **74**, 77.
45. F. W. Schaefer, 3rd and A. J. Mukkada, *J. Protozool.*, 1976, **23**, 446.
46. F. W. Schaefer, 3rd, E. Martin and A. J. Mukkada, *J. Protozool.*, 1974, **21**, 592.
47. D. Zilberstein, D. M. Dwyer, S. Matthaei and R. Horuk, *J. Biol. Chem.*, 1986, **261**, 15053.
48. B. R. Cairns, M. W. Collard and S. M. Landfear, *Proc. Natl. Acad. Sci. U. S. A.*, 1989, **86**, 7682.
49. M. Mueckler, C. Caruso, S. A. Baldwin, M. Panico, I. Blench, H. R. Morris, W. J. Allard, G. E. Lienhard and H. F. Lodish, *Science*, 1985, **229**, 941.
50. M. C. Maiden, E. O. Davis, S. A. Baldwin, D. C. Moore and P. J. Henderson, *Nature*, 1987, **325**, 641.
51. S. P. Stack, D. A. Stein and S. M. Landfear, *Mol. Cell Biol.*, 1990, **10**, 6785.
52. E. L. Snapp and S. M. Landfear, *J. Cell Biol.*, 1997, **139**, 1775.
53. M. I. Nasser and S. M. Landfear, *Mol. Biochem. Parasitol.*, 2004, **135**, 89.
54. R. J. Burchmore and S. M. Landfear, *J. Biol. Chem.*, 1998, **273**, 29118.
55. R. J. Burchmore, D. Rodriguez-Contreras, K. McBride, M. P. Barrett, G. Modi, D. Sacks and S. M. Landfear, *Proc. Natl. Acad. Sci. U. S. A.*, 2003, **100**, 3901.
56. D. Rodriguez-Contreras and S. M. Landfear, *J. Biol. Chem.*, 2006, **281**, 20068.
57. X. Feng, D. Rodriguez-Contreras, C. Buffalo, H. G. Bouwer, E. Kruvand, S. M. Beverley and S. M. Landfear, *Mol. Microbiol.*, 2009, **71**, 369.
58. B. Aronow, K. Kaur, K. McCartan and B. Ullman, *Mol. Biochem. Parasitol.*, 1987, **22**, 29.
59. J. M. Boitz, B. Ullman, A. Jardim and N. S. Carter, *Trends Parasitol.*, 2012, **28**, 345.
60. N. S. Carter, P. A. Yates, S. K. Gessford, S. R. Galagan, S. M. Landfear and B. Ullman, *Mol. Microbiol.*, 2010, **78**, 92.
61. A. Seyfang and S. M. Landfear, *Mol. Biochem. Parasitol.*, 1999, **104**, 121.

62. J. L. Martin, P. A. Yates, J. M. Boitz, D. R. Koop, A. L. Fulwiler, M. B. Cassera, B. Ullman and N. S. Carter, *Mol. Microbiol.*, 2016, **101**, 299.
63. K. Figarella, N. L. Uzcategui, Y. Zhou, A. LeFurgey, M. Ouellette, H. Bhattacharjee and R. Mukhopadhyay, *Mol. Microbiol.*, 2007, **65**, 1006.
64. D. Rodriguez-Contreras and S. M. Landfear, *Channels*, 2014, **8**, 477.
65. D. Rodriguez-Contreras, H. Aslan, X. Feng, K. Tran, P. A. Yates, S. Kamhawi and S. M. Landfear, *FASEB J.*, 2015, **29**, 11.
66. F. Prati, A. Goldman-Pinkovich, F. Lizzi, F. Belluti, R. Koren, D. Zilberstein and M. L. Bolognesi, *PLoS One*, 2014, **9**, e107994.
67. G. Matlashewski, B. Arana, A. Kroeger, S. Battacharya, S. Sundar, P. Das, P. K. Sinha, S. Rijal, D. Mondal, D. Zilberstein and J. Alvar, *Lancet Infect. Dis.*, 2011, **11**, 322.
68. M. Sands, M. A. Kron and R. B. Brown, *Rev. Infect. Dis.*, 1985, **7**, 625.
69. R. Reguera, R. Balaña Fouce, J. C. Cubria, M. L. Alvarez Bujidos and D. Ordoñez, *Biochem. Pharmacol.*, 1994, **47**, 1859.
70. M. Basselin, H. Denise, G. H. Coombs and M. P. Barrett, *Antimicrob. Agents Chemother.*, 2002, **46**, 3731.
71. J. C. Munday, L. Settimo and H. P. de Koning, *Front. Pharmacol.*, 2015, **6**, 32.
72. M. J. McConville, *Trends Parasitol.*, 2016, **32**, 666.

CHAPTER 15

Carbon Metabolism as a Drug Target in Leishmania

HÉCTOR ACOSTA,[a] MENG YUAN,[b] ANA J. CÁCERES,[a] WILFREDO A. QUIÑONES,[a] JUAN LUIS CONCEPCIÓN[a] AND PAUL A. M. MICHELS*[b,c]

[a] Laboratorio de Enzimología de Parásitos, Departamento de Biología, Facultad de Ciencias, Universidad de Los Andes, Mérida 5101, Venezuela; [b] Centre for Translational and Chemical Biology, School of Biological Sciences, The University of Edinburgh, The King's Buildings, Michael Swann Building, Max Born Crescent, Edinburgh EH9 3BF, Scotland, UK; [c] Centre for Immunity, Infection and Evolution, School of Biological Sciences, The University of Edinburgh, The King's Buildings, Ashworth Building, Charlotte Auerbach Road, Edinburgh EH9 3FL, Scotland, UK
*Email: paul.michels@ed.ac.uk

15.1 Introduction

During its life cycle, *Leishmania* encounters environments that are very varied with respect to temperature, host defence mechanisms, chemical composition and pH. As an adaptation to the different environments the parasite undergoes drastic morphological changes, notably in its size, the composition of its surface and, as reflected in the names of the different developmental forms, the presence of the flagellum. However, the nutrient availability for the different forms also differs and has a major effect on the metabolism of the parasites. Although the overall network of carbon and energy metabolism between different forms does not differ importantly, in

Drug Discovery Series No. 60
Drug Discovery for Leishmaniasis
Edited by Luis Rivas and Carmen Gil

Published by the Royal Society of Chemistry, www.rsc.org

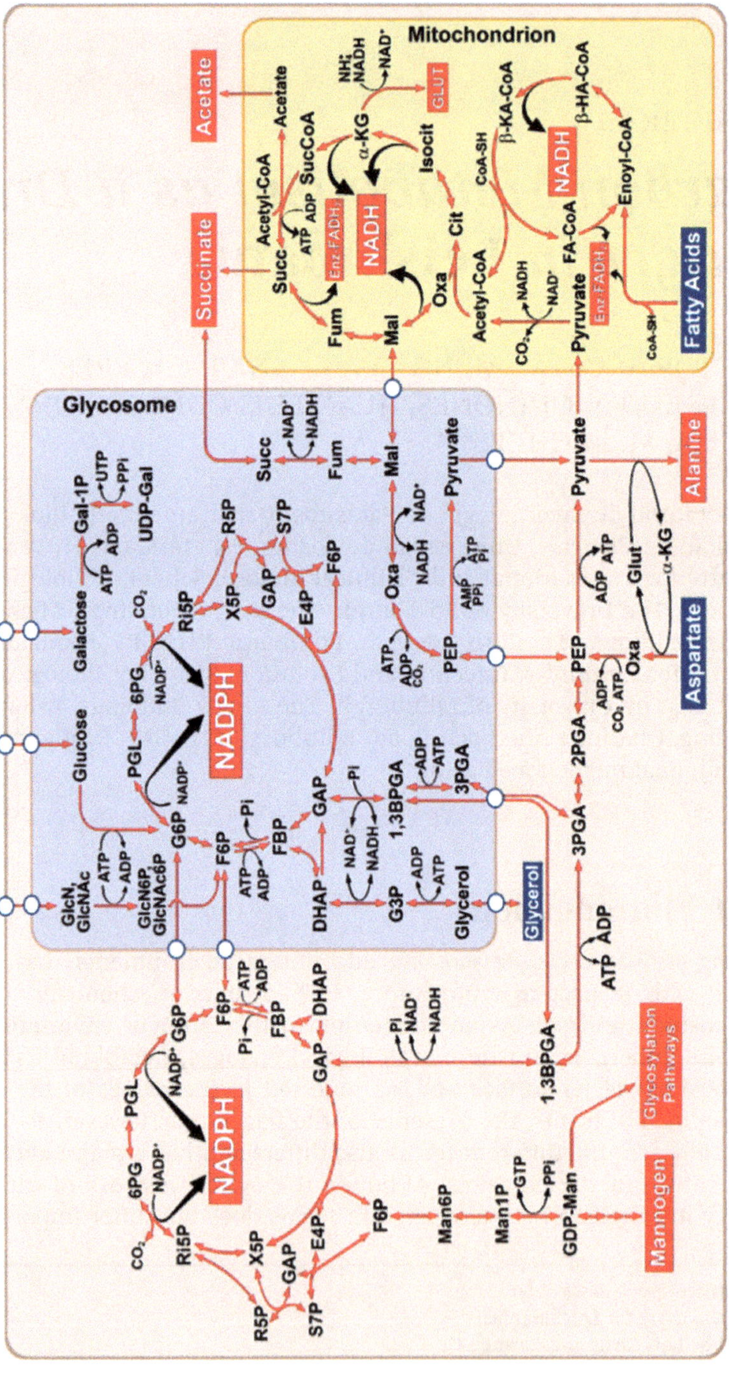

GlcN, GlcNAc
Glucose
Galactose
Glycosome
Mitochondrion
Acetate
Succinate
Glycerol
Alanine
Aspartate
Fatty Acids
Glycosylation Pathways
Mannogen
NADPH
NADPH
NADH
NADH
GLUT

contrast to other trypanosomatid parasites such as *Trypanosoma brucei* and *Trypanosoma cruzi*,[1,2] major quantitative metabolic changes occur upon differentiation.[3,4]

The flagellated promastigotes live extracellularly in the intestinal tract of the sand fly, where they find an abundant supply of saccharides and amino acids, which can be taken up *via* specific transporters. The non-flagellated amastigotes, the human pathogenic forms living in the phagolysosomes of macrophages, also have access to a large variety of carbon substrates, such as sugars, lipids, peptides and amino acids, which are obtained by uptake of macromolecules by the macrophages *via* endocytosis followed by degradation in the lysosomes by the various hydrolases. Amastigotes have transporters to import these small solutes, and additionally they can acquire carbon nutrients by taking up host macromolecules and hydrolysing them in their own lysosome.

Leishmania spp. possess the different pathways of carbon metabolism to deal with the different nutrients available in the different stages (Figure 15.1): the catabolic glycolytic pathway, with its extension for succinate fermentation, the mitochondrial tricarboxylic acid (TCA) cycle with the machinery for oxidative phosphorylation, and the anabolic gluconeogenic route of sugar-phosphate synthesis from amino acids and fatty acids. A third crucial route of carbon metabolism is the pentose-phosphate pathway (PPP). This sugar catabolic process plays a crucial role in cells of all organisms by sustaining various biosynthetic processes through the production of NADPH and precursors for the synthesis of nucleotides and aromatic amino acids. In addition, NAPDH is essential for dealing with oxidative stress, very important for parasites that are combatted by

Figure 15.1 The organisation of pathways of carbon and energy metabolism in *Leishmania* spp. The relative abundance of the different processes within carbon metabolism differs between promastigote and amastigote forms, with different ratios of substrates used and resulting in different end products, as discussed in the text. Red boxes indicate metabolic products, blue boxes substrates. Abbreviations: 1,3BPGA, 1,3-bisphosphoglycerate; Cit, citrate; CoA-SH, coenzyme A; DHAP, dihydroxyacetone phosphate; E4P, erythrose 4-phosphate; FA-CoA, fatty-acyl-CoA; FBP, fructose 1,6-bisphosphate; F6P, fructose 6-phosphate; Fum, fumarate; Gal-1P, galactose 1-phosphate; GDP-Man, GDP-mannose; GlcN, D-glucosamine; GlcN6P, D-glucosamine 6-phosphate; Glut, glutamate; GAP, glyceraldehyde 3-phosphate; G3P, glycerol 3-phosphate; G6P, glucose 6-phosphate; β-HA-CoA, β-hydroxyacyl-CoA; Pi, inorganic phosphate; PPi, inorganic pyrophosphate; Isocit, isocitrate; β-KA-CoA, β-ketoacyl-CoA; α-KG, α-ketoglutarate; Mal, malate; Man6P, mannose 6-phosphate; Man1P, mannose 1-phosphate; GlcNAc, *N*-acetylglucosamine; GlcNAc6P, *N*-acetylglucosamine 6-phosphate; Oxa, oxaloacetate; PEP, phosphoenolpyruvate; 6PG, 6-phosphogluconate; PGL, 6-phosphogluconolactone; 3PGA, 3-phosphoglycerate; 2PGA, 2-phosphoglycerate; R5P, ribose 5-phosphate; Ri5P, ribulose 5-phosphate; S7P, sedoheptulose 7-phosphate; Succ, succinate; SucCoA, succinyl-CoA; UDP-Gal, UDP-galactose; X5P, xylulose 5-phosphate.

their hosts by different oxidative attack mechanisms. This is the case for trypanosomatid parasites in general, and particularly for *Leishmania* amastigotes living in the very hostile environment within the phagolysosomes of their host's macrophages.[5]

In this chapter, we discuss the potential of enzymes of these various processes of carbon metabolism as drug targets and the current state of drug discovery using some of the validated or potential targets.

15.2 Carbon Metabolic Network of *Leishmania*

Organisms of the genus *Leishmania* are able to take up and metabolise glucose and other sugars in each of their life-cycle stages.[6,7] Like in other trypanosomatids, with the exception of the bloodstream-form of *T. brucei*, which produces predominantly pyruvate, the major excreted end product of sugar metabolism by *Leishmania* promastigotes, both in aerobic and anoxic conditions, is succinate, while smaller amounts of products like CO_2, acetate and alanine are also formed (Figure 15.1). The parasites thus exhibit a so-called overflow metabolism, despite the presence of a mitochondrion with well-developed capacity for complete oxidation of the substrates and oxidative phosphorylation *via* the TCA cycle and respiratory chain. In addition to the consumption of glucose, considerable uptake of amino acids from the culture medium has been observed during exponential growth of promastigotes.[8] While the rates of glucose and amino-acid uptake decreased with increasing culture age, that of fatty acids increased when the parasites reached the stationary phase.[8–11] These data were initially interpreted as an ability of promastigotes to use multiple substrates as sources of free energy, and also a preference to catabolise amino acids, overlooking the possibility that some substrates could preferentially serve for biomass production. Indeed, more recent metabolite analysis by nuclear magnetic resonance (NMR) of cells provided with a variety of ^{13}C-labelled carbon sources revealed that fermentation of glucose to succinate served both as the preferred ATP-generating process in promastigotes and to produce substrates to the mitochondrial TCA cycle for anabolic processes;[12,13] in contrast, the amino acids serve primarily for biosynthesis. In addition, ^{13}C glucose was readily incorporated in PPP intermediates and in the reserve material, the oligosaccharide mannogen.

Results from similar isotope studies indicated that differentiation of promastigotes to amastigotes is associated with a drastic decrease in metabolic activity; the parasites enter a quiescent state, exhibiting a stringent metabolic response. They show an important reduction in uptake and utilisation of hexoses and amino acids, a decreased secretion of organic acids and increased fatty-acid β-oxidation.[4,14] The reduced hexose uptake is due to the parasites decreasing the level of their transporters by routing them to the lysosome for degradation.[15] In addition, the differentiation is associated with important changes in post-translational modification of many proteins, such as phosphorylation and dephosphorylation, that may affect the *in vivo* activity of many proteins, including metabolic enzymes. Possible explanations for the metabolic quiescence of amastigotes have been discussed by McConville *et al.*[4] Nonetheless, the energy supply remains largely dependent on the catabolism of hexoses, whereas amino-acid uptake still serves for biomass. Hexose catabolism *via* extensions of the glycolytic pathway provides succinate, malate and acetyl-CoA—the latter also from β-oxidation—for the synthesis of anabolic precursors, such as glutamate, *via* the TCA cycle (Figure 15.1). In addition, hexoses continue to be catabolised by the PPP for NADPH production and the synthesis of mannogen and surface glycoconjugates. Lesion-derived and *in vitro* differentiated amastigotes exhibit a similar metabolism, indicating that the decreased metabolic activity is not due to limited exogenous nutrient availability, but is part of the differentiation program.

Leishmania promastigotes are able to synthesise sugar-phosphates *de novo*, *via* gluconeogenesis from metabolites such as alanine, aspartate and glycerol, under sugar starvation.[16,17] In contrast, *Leishmania* amastigotes carry out gluconeogenesis even in the presence of external sugar.[16] Glucose uptake alone, followed by phosphorylation by hexokinase (HK) and/or glucokinase (GlcK) seems insufficient for amastigotes to produce all the glucose 6-phosphate (G6P) for the cellular processes in which it is required: glycolysis and the PPP to produce NADPH, cell-surface glycoconjugates and mannogen (Figure 15.1).[16] Mannogen has been shown to be required for a normal establishment of infection and survival in macrophages.[18,19] The key gluconeogenic enzyme fructose-1,6-bisphosphatase (FBPase) has been shown to be important for the virulence of *Leishmania major*; mutant parasites lacking this enzyme were unable to replicate in the macrophage's phagolysosome and to generate normal lesions in mice.[20] *Leishmania* spp. also have the ability to take up other sugars besides glucose, such as D-fructose, D-mannose, D-galactose and the amino sugars D-glucosamine and *N*-acetylglucosamine, some of them primarily being used for the synthesis of glycoconjugates and mannogen.

A major part of carbon metabolism of Kinetoplastea takes place within glycosomes, peroxisome-related organelles so named upon the detection in *T. brucei* that they contain the first seven enzymes of glycolysis.[21] Indeed, such compartmentalisation has also been found in promastigotes and amastigotes of *Leishmania* spp.[22,23] The main aspects of this organelle and the role of the compartmentalised pathways in Kinetoplastea are discussed in detail elsewhere.[24]

The glycolytic pathway is compartmentalised in glycosomes from its first enzymes, HK, responsible for phosphorylation of hexoses such as glucose, fructose and mannose, GlcK, specific for glucose phosphorylation, and phosphoglycerate kinase (PGK) (Figure 15.1). PGK is represented in *Leishmania* by two almost identical (99%) differently localised isoenzymes, PGKC in glycosomes and PGKB in the cytosol, which are simultaneously expressed in both the promastigote and amastigote stages.[25] The last three enzymes of the glycolytic pathway, phosphoglycerate mutase (PGAM), enolase (ENO) and pyruvate kinase (PYK), are exclusively present in the cytosol.

The glycolytic pathway is extended with enzymes for succinate formation from phosphoenolpyruvate (PEP), including phosphoenolpyruvate carboxykinase (PEPCK), malate dehydrogenase (MDH) and fumarate reductase (FR) and with pyruvate–phosphate dikinase (PPDK), all present inside glycosomes. In addition, the organelles contain enzymes of glycerol metabolism, an NADH-dependent glycerol-3-phosphate dehydrogenase (GPDH) and glycerol kinase (GK), as well as an adenylate kinase (AK). The glycosomal membrane is impermeable to NAD(H), ATP, ADP and AMP, and other molecules with a molecular mass greater than 400–500 Da. However, the different glycosomally located kinases and dehydrogenases act together in such a way that the consumption and production of ATP and the reduction and oxidation of NAD^+ and NADH in the organelles are each in balance. No transporters have been identified in the glycosomal membrane for any metabolites of carbon metabolism. However, studies with *T. brucei* have indicated the presence of glycosomal membrane proteins capable of forming non-selective pores, which may allow the passage of such molecules.[26]

Most enzymes of glycolysis catalyse equilibrium reactions and therefore can also participate in the reverse process of gluconeogenesis. Only the reaction by PFK is virtually irreversible under physiological conditions. Therefore, the reverse reaction, the dephosphorylation of fructose 1,6-bisphosphate (F16BP) to fructose 6-phosphate (F6P) occurs by a different enzyme, FBPase, but without concomitant formation of ATP. Interestingly, this enzyme has also been located in glycosomes,[20,27] indicating that both glycolysis and gluconeogenesis occur inside the organelles. How the activities of PFK and FBPase are reciprocally regulated, to avoid ATP loss by futile cycling, remains to be determined. AMP may participate in such regulation, because it is an allosteric activator of PFK and inhibitor of FBPase (see Section 15.4).[28,29] In contrast to their homologues in most other eukaryotes, these enzymes are not susceptible to the reciprocally acting effector fructose 2,6-bisphosphate (F26BP),[30,31] which is, however, in trypanosomatids uniquely a potent allosteric activator of PYK (see also Section 15.4). It is also feasible that futile cycling occurs to some extent in order to regulate fluxes and/or to maintain a metabolic homeostasis within the glycosomes.

The PPP plays an important role in all life-cycle stages of *Leishmania* but has been particularly studied in promastigotes.[32] Like in other kinetoplastids, this route has a dual distribution over two compartments, cytosol and glycosomes (Figure 15.1).[5] The pathway is divided into an oxidative phase and a non-oxidative phase. The first, oxidative, phase converts the G6P into the pentose phosphate ribulose 5-phosphate, with the concomitant production of two NADPH molecules and CO_2, and involves three enzymes: glucose-6-phosphate dehydrogenase (G6PDH), 6-phosphogluconolactonase and 6-phosphogluconate dehydrogenase (6PGL). The subsequent non-oxidative phase is responsible for rearrangements of the carbon skeleton and comprises an epimerase (ribulose-5-phosphate 3′-epimerase), an isomerase (ribose 5-phosphate isomerase) and two transferases (transketolase and transaldolase).

The synthesis of the essential reserve material mannogen is unique to members of the genus *Leishmania*. It is a homo-oligomer or polymer of β-1,2-linked mannose residues, present in non-dividing promastigotes and lesion-derived amastigotes, where it can constitute up to 80–90% of the cellular carbohydrate.[19] The molecule is synthesised in the cytosol from hexose-phosphates or hexosamines like glucosamine or *N*-acetylglucosamine (Figure 15.1).[33]

Leishmania also takes up galactose from the environment using the hexose transporters to incorporate it into surface glycoconjugates in the form of galactopyranose (Gal_p) and UDP-galactofuranose (Gal_f).[34,35] Gal_f is an uncommon, highly immunogenic monosaccharide, originating from the action of UDP–galactopyranose mutase (UGM). UDP-Gal is synthesised by two routes; *de novo* by the interconversion of UDP-α-D-glucose (UDP–Glc) by UDP-glucose-4-epimerase, and by a salvage pathway. This latter route involves the phosphorylation of α-D-galactose to α-D-galactose 1-phosphate (Gal-1P) by a galactokinase, followed by activation of the Gal-1P by UDP–sugar pyrophosphorylase (USP).[35] However, upon deletion of the USP gene, *L. major* lost the ability to synthesise UDP–galactose from galactose 1-phosphate but its ability to convert glucose 1-phosphate into UDP–glucose was fully maintained. Accordingly, USP appeared to be dispensable for growth and glycoconjugate biosynthesis under standard *in vitro* growth conditions, but it may restrict growth if galactose is the only monosaccharide source.[36] This galactose salvage pathway seems to be active only in *Leishmania*,[35] but recently another one has been identified in *T. cruzi*, involving a novel UDP–sugar pyrophosphorylase and two glycosomal galactokinases.[37,38]

15.3 Drug Targets of Carbon Metabolism in *Leishmania*

Contrary to the situation in *T. brucei*, where RNAi is available, little genetic validation of enzymes of carbon and energy metabolism of *Leishmania* has been performed so far. However, there is ample support for the notion that major parts of this metabolism are essential for growth and/or virulence of the pathogenic forms of *Leishmania* spp. For example, both the consumption of sugar and FBPase activity are required for proliferation and virulence of intracellular amastigotes, indicating that the formation of G6P by sugar uptake and phosphorylation by HK and/or GlcK, and gluconeogenesis offers likely targets for therapy. Together with the finding that lesion-derived amastigotes are dependent on glucose catabolism, even when the infected macrophages are supplied with excess amino acids render all enzymes of both glycolysis and gluconeogenesis candidate drug targets.[14] Since mannogen has also been shown to be essential for the virulence of *Leishmania*,[18,19] the successive enzymes responsible for its synthesis from F6P could also be considered as targets. Also galactose metabolism offers prospects as a drug target, because it is incorporated as Gal_f in molecules that play an important role in host-specific recognition of *Leishmania* and its growth and pathogenesis.[39] Gal_f does not occur in humans and is produced by UGM whose deletion leads to attenuated virulence.[40]

TCA cycle enzymes play important roles in providing intermediates for the synthesis of amino acids, such as glutamate, glutamine and aspartate (and from there other important products such as thiols, pyrimidine and amino sugars), from acetyl-CoA. This latter metabolite results only partially from glycolysis but mainly from fatty-acid β-oxidation that thus also plays an important role.[4,14]

With the amastigotes being exposed to high oxidative stress in the macrophages, the PPP is of crucial importance for the production of NADPH required to neutralise the oxidative attacks. However not only the NADPH-producing oxidative branch of the pathway is crucial, but also the non-oxidative branch as illustrated by the essentiality of the enzyme ribose-5-phosphate isomerase (RPI). Viable double-knockout mutants could only be obtained when complemented with an episomal copy encoding a metabolically active RPI.[41] Single-knockout promastigotes exhibited no defect in *in vitro* growth, metacyclogenesis or macrophage infection, but in intracellular amastigotes replication was impaired. These observations indicate that glucose and ribose uptake from the phagolysosome is a limiting factor for intracellular growth. From a drug discovery perspective, it is also relevant that trypanosomatids contain a B-type RPI whereas humans have the structurally unrelated RPI-A.

Together, all these data reveal the importance or essentiality of (parts of) the pathways of carbon and energy metabolism (glycolysis, gluconeogenesis, TCA cycle, PPP, β-oxidation) for the pathogenic forms of *Leishmania*. In addition, for many of the enzymes of such pathways, the sequence divergence between parasites and humans is considerable, due to the long, independent evolutionary paths. These differences often even involve residues for substrate binding, and, in many cases, parasite and host enzymes use different allosteric effectors. For some reactions parasites and humans even use non-homologous enzymes. In the next section, examples of enzymes as specific drug targets will be discussed and the druggability of some enzymes explored.

15.4 Functional and Structural Properties of Carbon-metabolic Enzymes that are Potential Drug Targets in *Leishmania*

Several enzymes of *Leishmania* spp. involved in the carbon metabolic pathways that are potential drug targets have been structurally and functionally characterised in great detail. High-resolution crystal structures are available for the exclusively glycolytic enzyme PYK, enzymes shared by glycolysis and gluconeogenesis, such as PGI, ALD, TIM, GAPDH and PGAM, as well as

G3PDH of glycerol metabolism that also plays a role in the reoxidation of the glycolytically reduced cofactor NADH. Also resolved are structures of the gluconeogenic enzyme FBPase and the PPP enzyme 6PGDH. In addition, structures are also known from different *Trypanosoma* species for other enzymes of glycolysis and gluconeogenesis, such as PFK, PGK and ENO; GK of glycerol metabolism; PEPCK and PPDK of auxiliary branches of glycolysis and gluconeogenesis; and the PPP enzyme G6PDH. Also the *T. cruzi* GlcK crystal structure is available, but the precise role of the latter enzyme in glucose phosphorylation, next to HK, in glycolysis and/or PPP of trypanosomatids remains to be established.[42] Within the trypanosomatid family, the degree of amino-acid sequence identity between these core metabolic enzymes is generally high, with almost complete conservation of the residues involved in catalysis and binding of substrates and effectors. Therefore, potent, parasite-enzyme-selective inhibitors are expected to be active across species boundaries. Indeed, several examples are available of compounds developed to inhibit enzymes of *T. brucei* carbon metabolism with little effect on the corresponding human enzymes, that are also active on the corresponding *T. cruzi* and *Leishmania* enzymes. This was, for example, demonstrated for ALD,[43] GAPDH,[44,45] and PFK and PYK.[46] For more details, see Section 15.5.

Barros-Alvarez *et al.* presented a summary of the then all available crystal structures of trypanosomatid glycosomal enzymes involved in various core metabolic processes.[1] Here we will focus specifically on structures of three *Leishmania* enzymes of the pathways of carbon metabolism that are likely drug targets: FBPase, PGAM and PYK. Aspects of these structures that are essentially different from those of the mammalian enzymes and offer possibilities for anti-parasite drug discovery, which will be highlighted.

The crystal structure of *L. major* FBPase has been solved at 2.41 Å resolution.[29] Like in other organisms, it is a homotetrameric enzyme. AMP is an allosteric inhibitor of both human[47,48] and *Leishmania* FBPase,[29] transforming the enzyme from its active R-state conformation into an inactive T-state. However, the two enzymes respond to AMP inhibition considerably differently, with inhibitor constant (K_i) values of 0.1–4 μM and 65 μM, respectively. In both enzymes, the binding site for this effector is located at approximately 32 Å distance from the active site but, as is shown in Figure 15.2A, its architecture is poorly conserved. In particular, the alkaline Lys112 of human FBPase is replaced in *Lm*FBPase by an acidic Asp112, which no longer interacts with AMP and additionally attracts and twists the side chain of Arg140, thereby disrupting its bonds to AMP (Figure 15.2B). The distinctive Asp112 provides an acidic, deeper pocket that seems exploitable for design of compounds that selectively interfere with the functioning of the parasite enzyme.

High-resolution crystal structures are available for PYK of each of the three trypanosomatid parasites including the *Leishmania mexicana* enzyme,[49,50] as well as each of the four tissue-specific human PYK isoenzymes (M1, M2, L and R). Trypanosomatid PYKs display unusual regulated properties. Unlike most eukaryotic PYKs which—with the exception of the mammalian M1 isoenzyme—are allosterically activated by F16BP, trypanosomatid PYKs are relatively unaffected by F16BP. In contrast, F26BP which binds to the homologous effector site, activates trypanosomatid PYKs at submicromolar concentrations.[30,31] Figure 15.2C shows the effector sites of human M2PYK

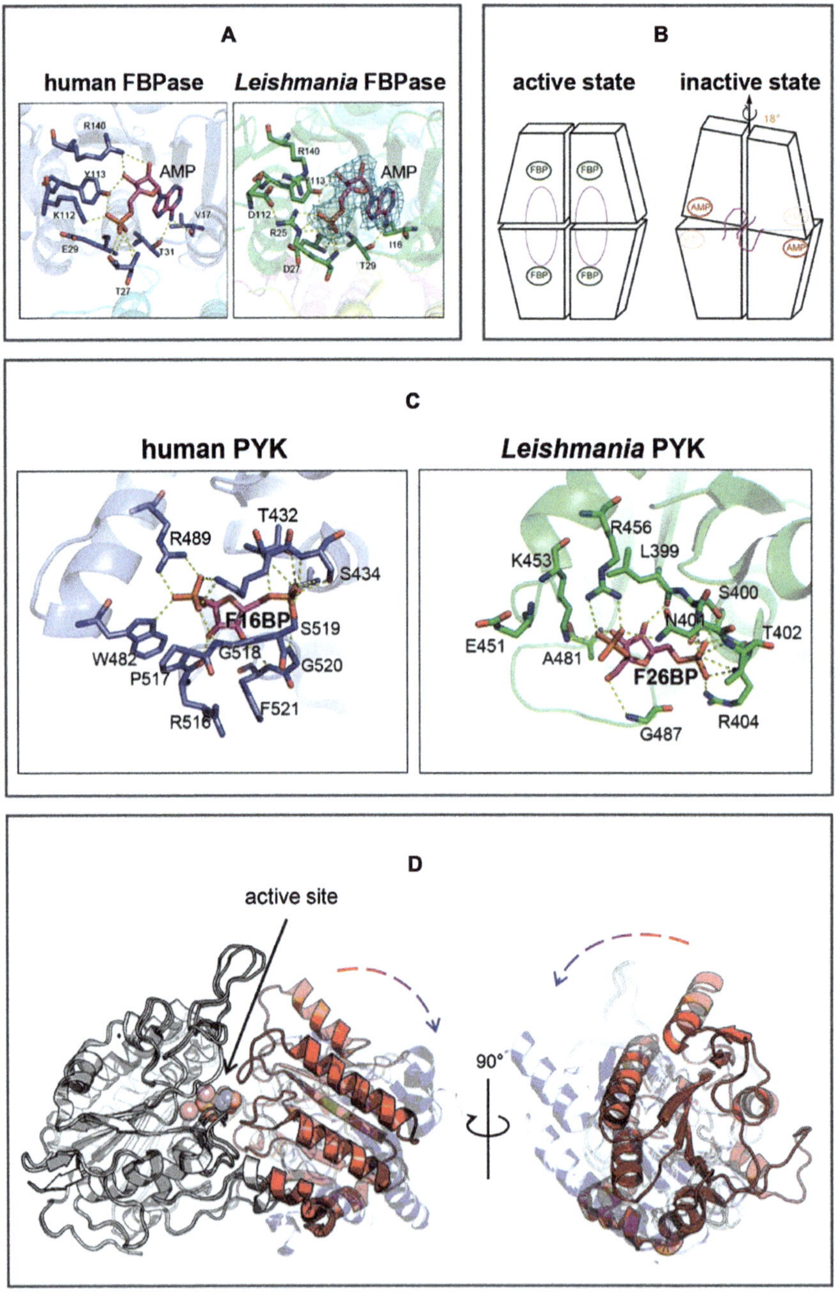

and *Leishmania* PYK, in which the respective activators are bound. As shown in Figure 15.2, many effector-binding residues and the conformation of the site are poorly conserved between the human and *Leishmania* enzymes. These differences provide possibilities for design or selection of compounds that selectively impose an inactive conformation on the parasite enzyme.

Phosphoglycerate mutase (PGAM) has been proposed as a very attractive chemotherapeutic target in trypanosomatids, because these parasites possess an enzyme version, the cofactor-independent PGAM (iPGAM), that is completely different from its human counterpart, the cofactor-dependent PGAM (dPGAM).[51] iPGAMs and dPGAMs are non-homologous enzymes belonging to different superfamilies. iPGAMs are co-factor (2,3-bisphosphoglycerate)-independent, monomeric metalloenzymes with a molecular mass of approximately 60 kDa belonging of the alkaline-phosphatase superfamily; their catalysis of the interconversion of 3-phosphoglycerate and 2-phosphoglycerate involves phospho transfer with a phosphoserine intermediate. dPGAMs are dimeric or tetrameric enzymes with subunits of about 27 kDa; they are part of the acid-phosphatase superfamily and catalyse the phospho transfer *via* a phosphohistidine intermediate without involvement of metal ions. However, based on a low efficacy in high-throughput screening of chemical libraries, the druggability of iPGAMs was reported as unpromising, owing to their small and inaccessible active site.[52] Indeed, the initially solved *L. mexicana* iPGAM structures indicated the presence of a small, buried active

---

Figure 15.2 *Leishmania* carbon-metabolic enzymes displaying features with promise for drug targeting. **A.** Comparison of the AMP binding sites of human and *Leishmania* FBPase. (Left) Conformation of the AMP binding site in human muscle FBPase[75] (blue, PDB code: 3IFA). (Right) Conformation of the AMP binding site in *Lm*FBPase[29] (green). The electron density shown is from $2Fo - Fc$ maps contoured at the 1.0 σ level using a cut-off radius of 1.65 Å. **B.** Schematic models of the allosteric mechanism of *Leishmania* FBPase. The models represent *Leishmania* FBPase in the active (left) and inactive state (right). Each subunit of the homotetrameric enzyme is represented by a trapezoid block. Green and red circles represent active sites and effector sites, respectively. A catalytically important loop is shown in purple. The 18° dimer–dimer rotation upon the binding of the allosteric inhibitor AMP disengages the catalytic loop and disrupts the complete active site. **C.** Comparison of fructose-bisphosphate binding sites of human and *Leishmania* PYK. (Left) Conformation of the human PYK effector site[76] (blue, PDB ID: 4FXF). (Right) Conformation of the *L. major* PYK effector site[50] (green, PDB ID: 3HQP). **D.** Open and closed conformations of iPGAM from trypanosomatids. Shown are front and side views of the superposed iPGAM structures of trypanosomatids. Structures in open (PDB code: 3NVL) and closed (PDB code: 3IGY) conformations are superposed with their phosphatase domains (shown in grey). The rotation of the transferase domain from its closed conformation (red) to an open conformation (blue with transparency) is indicated by dashed arrows. The complete active site that only exists in the closed form is indicated by a black arrow.

site.[53] Nonetheless, recent crystal structure analysis of trypanosomatid iPGAMs and additional biophysical studies revealed that iPGAMs occur in both a closed and open conformation (Figure 15.2D), whereby the region between the two domains forms an accessible, larger and chemically more diverse cavity, indicating that it is druggable.[54,55] The transition between the two forms seems to important for the consecutive processes of substrate binding, catalysis and substrate release. A new HTS assay for iPGAM inhibitors was designed accordingly, in which micromolar-level inhibitors were obtained with considerably higher success than in the initial screens performed under conditions that favoured the closed conformation.[56]

15.5 Drug Discovery

Specific inhibitors have been developed for several enzymes of two pathways of carbon metabolism of trypanosomatids, glycolysis (and gluconeogenesis in case of shared enzymes) and the PPP. A selection of such compounds is presented in Figure 15.3. Some of these compounds kill cultured bloodstream-form parasites at concentrations in the (sub)micromolar range, with little or no effect on growth of cultured mammalian cells. High-throughput screening of a large compound library with *T. brucei* HK yielded a considerable number of enzyme inhibitors.[57] Ten compounds selected for further analysis displayed enzyme inhibition, with IC_{50} values ranging from 41 to 0.05 μM; seven of them inhibited growth of cultured trypanosomes with half maximal effective concentration (EC_{50}) values at low to submicromolar concentrations, with minimal impact on human cell lines. Four of the compounds were also active against promastigotes of *L. major* at low micromolar concentrations.

Bisphosphonates, as metabolically inert mimics of inorganic pyrophosphate, have been developed as selective inhibitors of the PPi-sensitive *T. cruzi* HK.[58,59] The most potent compounds inhibited the purified enzyme with IC_{50} values between 0.81 and 3 μM and were shown to affect glucose consumption in epimastigotes and to kill intracellular amastigotes at an ED_{50} of 2.2 μM, without affecting a human cell line. *L. mexicana* HK is even more susceptible to PPi (K_i 0.035 mM *versus* 0.5 mM for *T. cruzi* HK)[60] and was also inhibited by the bisphosphonates, with IC_{50} values between 2 and 15 μM (AJC and JLC, unpublished). Moreover, the compounds inhibited the consumption of glucose and growth of cultured promastigotes but remain to be tested on intracellular *Leishmania* amastigotes.

Selective inhibitors of *T. brucei* PFK with low- to sub-micromolar IC_{50} values were also obtained by HTS and optimisation of some hits guided by structure–activity relationship (SAR) analysis.[61] Compounds having a characteristic 3,4-dichlorobenzyl motif kill trypanosomes with EC_{50}'s at low micromolar concentrations and lack cytotoxicity in a number of human cell lines. Such compounds have been shown to be equally potent inhibitors of *L. mexicana* PFK but have not yet been tested on growth of this parasite. The *T. brucei* crystal structure indicated the presence of a hydrophobic

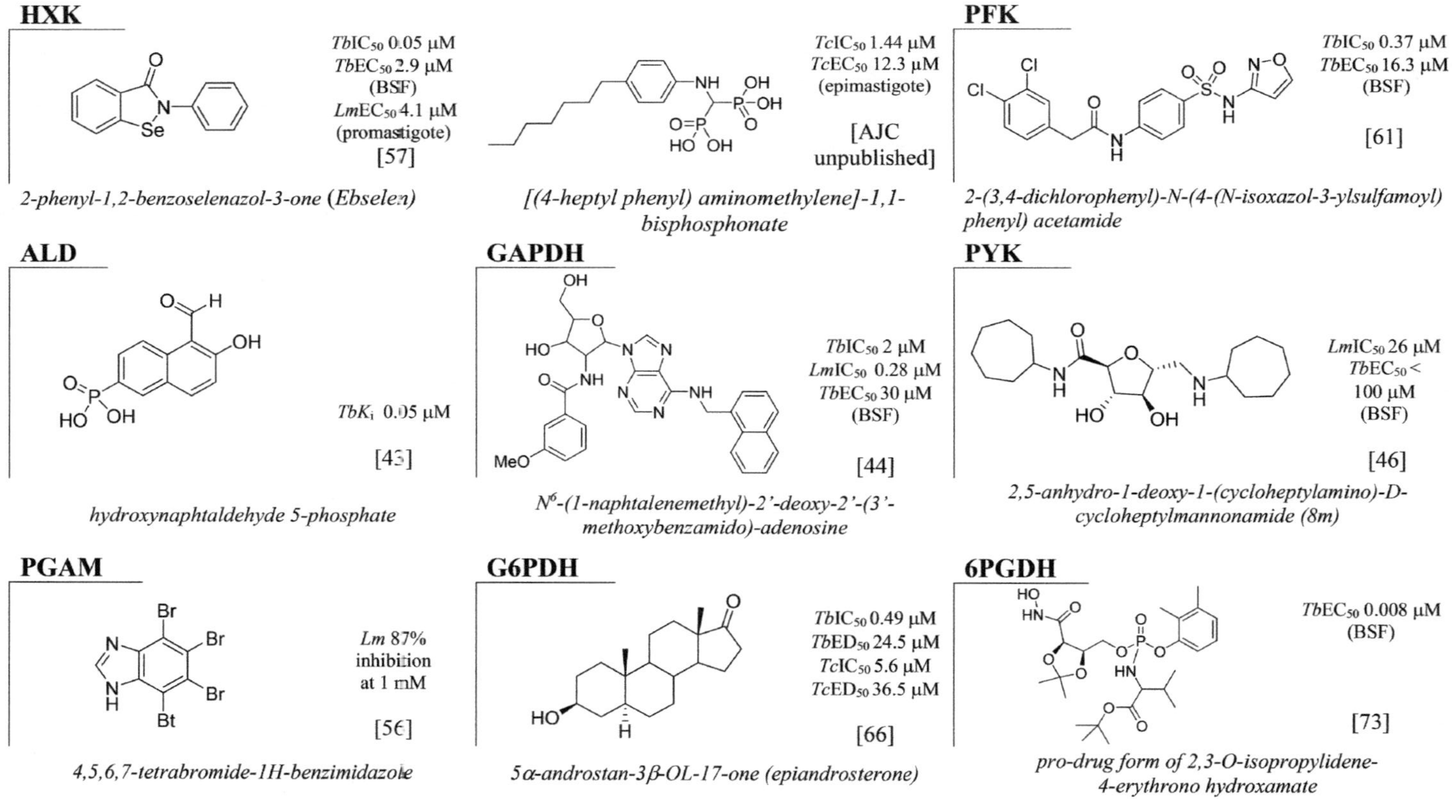

Figure 15.3 Selected inhibitors of some enzymes of carbon metabolism in trypanosomatid parasites. Enzymes are indicated by acronyms: HK, hexokinase; PFK, phosphofructokinase; ALD, aldolase; GAPDH, glyceraldehyde-3-phosphate dehydrogenase; PGAM, phosphoglycerate mutase; PYK, pyruvate kinase; G6PDH, glucose-6-phosphate dihydrogenase; 6PGDH, 6-phosphogluconate dehydrogenase. IC_{50} or K_i values by which the compounds inhibit *T. brucei*, *T. cruzi* and/or *L. mexicana* enzymes and EC_{50} values for growth inhibition of cultured parasites are indicated. References to the original publications are given by numbers in brackets. See the text for a detailed description.

pocket in the active site that is predicted to be occupied by the dichlorobenzyl functionality of the inhibitor.[62] Structure modelling indicated conservation of this pocket in the enzyme of other trypanosomatids, but its absence from mammalian PFKs.

Dihydroxybenzaldehyde phosphate and 2-hydroxybenzaldehyde-5-phosphate derivatives have been developed as quasi-irreversibly binding inhibitors for trypanosomatid ALD with K_i values in the low micromolar range and no effect on mammalian aldolase.[43,63] However, they only killed cultured *T. brucei* when delivered in prodrug form, masking the charged group. Growth of intracellular *L. mexicana* was not affected, probably because the compounds did not reach the parasite upon being unmasked in the cytosol of the host cell.

A considerable number of inhibitors for trypanosomatid GAPDHs have been developed, by structure-based design and screening of libraries composed of synthetic or natural products. A summary of the most parasite-enzyme-selective compounds, in several cases with demonstrated activity on *in vitro* cultured *Trypanosoma* and *Leishmania* species or infected animal models is presented by Barros-Alvarez *et al.*[1] IC_{50} values are typically in the submicromolar or low to medium micromolar range. Among these compounds are adenosine analogues exploiting regions of the cofactor NAD^+-binding site that are not present in the GAPDH of humans or any other species analysed to date.[44,45] Various other successful inhibitors, obtained by design and synthesis or screening, were binding in the active site. Adenosine analogues were also designed to compete with ATP binding at the active site of trypanosomatid PGKs. Compounds were obtained that were active with micromolar IC_{50} values and inhibited growth of *in vitro* cultured *Trypanosoma* parasites, but only with a modest selectivity compared with murine fibroblasts.[64] The best inhibitor obtained in a screen with *L. mexicana* PGAM, as described in Section 15.4, was active at submillimolar concentration,[56] while the approach described for obtaining PYK inhibitors yielded furanose sugar amino amides as mimics for the allosteric effector F2,6BP with IC_{50} values against the *L. mexicana* enzyme as low as 26 μM.[46] The latter compounds exhibited similar values for inhibition of cultured *T. brucei*, but have not yet been tested against *Leishmania* parasites.

With regard to PPP enzymes, steroidal compounds have been obtained that inhibit the enzymes from *T. brucei* and *T. cruzi* in a non-competitive manner at values in the submicromolar–low-micromolar range and kill these parasites efficiently in culture. However, these compounds also inhibit the human G6PDH and have no effects whatsoever on the *Leishmania* spp. enzyme and the growth of the cells.[65–68] The sequence identity between the *Trypanosoma* and *Leishmania* enzymes amounts to about 65%. The nature of this striking difference between the related enzymes is unknown, but insight might be obtained following the recent report of the crystal structure of the ternary complex of *T. cruzi* G6PDH with G6P and NADPH.[69] Other, new non-competitive micromolar inhibitors of *T. cruzi* G6PDH, some with promising trypanocidal activity, have recently been obtained by library screening followed by preliminary SAR.[70] Moreover, several of the compounds showed better selectivity for the parasite *versus* the host enzyme than for the steroidal compounds. They remain to be tested on the *Leishmania* enzymes and cells.

T. brucei 6PGDH is efficiently inhibited by analogues of its substrate, 6-phosphogluconate, and the high-energy 1,2-*cis*-enediolintermediate of its reaction.[71] Notably, very potent and selective hydroxamate derivatives of this intermediate were synthesised with 4-phospho-D-erythronohydroxamate being the most potent one, with a $K_i = 10$ nM for the trypanosome enzyme and a 250-fold selectivity over sheep 6PGDH.[72] Since these compounds were charged, prodrug forms were prepared where the phospho group was masked by producing aryl phosphoramidate derivatives to allow them to cross the trypanosome's membrane. This resulted in compounds with good trypanocidal activity, but unfortunately esterases in the blood probably unmask the prodrugs, reducing the use of these compounds *in vivo*.[73] Therefore, further efforts are required to optimise them for clinical use. Additionally, virtual screening using the *T. brucei* 6PGDH crystal structure yielded three compounds that, when tested, exhibited IC_{50} values in the low micromolar range. Based on these initial hits additional compounds were identified with similar inhibitory activities.[74] All these trypanosome 6PGDH inhibitors deserve further studies, including of their efficacy against the *Leishmania* enzyme and parasites.

The work on trypanosomatid FBPase discussed in Section 15.4 is very recent; hence, no specific inhibitors of this enzyme have been developed as yet.

15.6 Conclusions

Research in recent years, notably by McConville and coworkers,[3,4] has provided new, detailed insights into the carbon metabolic network of the different life-cycle stages of *Leishmania*. From this work, it has become clear that different pathways, or at least parts of them, each play important roles in the human pathogenic stage, the intracellular amastigotes. This renders many of the enzymes involved in these processes candidate drug targets, although genetic and/or chemical validation has been performed only in a few cases. For a considerable number of these enzymes structural and kinetic information is available, or such data are available for the homologous enzymes from *Trypanosoma* species and appear relevant also for *Leishmania*. Compounds have been obtained, by design and synthesis or by screening of chemical compound libraries and subsequent structure-guided optimisation, which efficiently inhibit such enzymes with no or little effect on mammalian enzymes. Many such inhibitors have been shown to affect proliferation of the parasites and, in some cases, their virulence in infected animal models. Future research will aim for further improvement of the potency and pharmacokinetic properties of such compounds, with as the final objective to make them available for treatment of leishmaniasis.

Acknowledgements

The work on trypanosomatid parasites in the research group at the Universidad de los Andes was supported by the "Fondo Nacional de Ciencia, Tecnología e Innovación" (FONACIT) (grant MC-2007001425, awarded to JLC) and the Consejo de Desarrollo Cientifico, Humanistico, Technológico y de las Artes (CDCHTA, grant C-1829-13-03-B, awarded to AJC), and that of the team at the University of Edinburgh by the medical research council (MRC), Biotechnology and Biological Sciences Research Council (BBSRC) and Wellcome Trust. MY was supported by a PhD studentship from the Darwin Trust. We are grateful to Dr Linda Fothergill-Gilmore for stimulating discussions and comments on the manuscript.

References

1. X. Barros-Alvarez, M. Gualdrón-López, H. Acosta, A. J. Cáceres, M. A. Graminha, P. A. Michels, J. L. Concepción and W. Quiñones, *Curr. Med. Chem.*, 2014, **21**, 1679.
2. J. R. Haanstra, E. B. González-Marcano, M. Gualdrón-López and P. A. Michels, *Biochim. Biophys. Acta*, 2016, **1863**, 1038.
3. M. J. McConville and T. Naderer, *Annu. Rev. Microbiol.*, 2011, **65**, 543.
4. M. J. McConville, E. C. Saunders, J. Kloehn and M. J. Dagley, *F1000Research*, 2015, **4**, 938.
5. J. Kovářová and M. P. Barrett, *Trends Parasitol.*, 2016, **32**, 622.
6. P. M. Rainey and N. E. MacKenzie, *Mol. Biochem. Parasitol.*, 1991, **45**, 307.
7. R. J. Burchmore and D. T. Hart, *Mol. Biochem. Parasitol.*, 1995, **74**, 77.
8. D. T. Hart and G. H. Coombs, *Exp. Parasitol.*, 1982, **54**, 397.
9. J. J. Blum, *J. Protozool.*, 1990, **37**, 505.
10. J. J. Blum, *J. Bioenerg. Biomembr.*, 1994, **26**, 147.
11. M. Louassini, M. R. Foulquié, R. Benítez and F. J. Adroher, *Parasitol. Res.*, 1999, **85**, 300.
12. E. C. Saunders, W. W. Ng, J. M. Chambers, M. Ng, T. Naderer, J. O. Krömer, V. A. Likic and M. J. McConville, *J. Biol. Chem.*, 2011, **286**, 27706.
13. D. J. Creek, J. Anderson, M. J. McConville and M. P. Barrett, *Mol. Biochem. Parasitol.*, 2012, **181**, 73.
14. E. C. Saunders, W. W. Ng, J. Kloehn, J. M. Chambers, M. Ng and M. J. McConville, *PLoS Pathog.*, 2014, **10**(1), e1003888.
15. J. E. Vince, D. Tull, S. Landfear and M. J. McConville, *Int. J. Parasitol.*, 2011, **41**, 791.
16. D. Rodríguez-Contreras and S. M. Landfear, *J. Biol. Chem.*, 2006, **281**, 20068.
17. D. Rodriguez-Contreras and N. Hamilton, *J. Biol. Chem.*, 2014, **289**, 32989.
18. A. Garami and T. Ilg, *EMBO J.*, 2001, **20**, 3657.

19. J. E. Ralton, T. Naderer, H. L. Piraino, T. A. Bashtannyk, J. M. Callaghan and M. J. McConville, *J. Biol. Chem.*, 2003, **278**, 40757.
20. T. Naderer, M. A. Ellis, M. F. Sernee, D. P. De Souza, J. Curtis, E. Handman and M. J. McConville, *Proc. Natl. Acad. Sci. U. S. A.*, 2006, **103**, 5502.
21. F. R. Opperdoes and P. Borst, *FEBS Lett.*, 1977, **80**, 360.
22. G. H. Coombs, J. A. Craft and D. T. Hart, *Mol. Biochem. Parasitol.*, 1982, **5**, 199.
23. D. T. Hart and F. R. Opperdoes, *Mol. Biochem. Parasitol.*, 1984, **13**, 159.
24. M. Gualdrón-López, A. Brennand, V. Hannaert, W. Quiñones, A. J. Cáceres, F. Bringaud, J. L. Concepción and P. A. Michels, *Int. J. Parasitol.*, 2012, **42**, 1.
25. C. A. Adjé, F. R. Opperdoes and P. A. Michels, *Mol. Biochem. Parasitol.*, 1997, **90**, 155.
26. M. Gualdrón-López, M. H. Vapola, I. J. Miinalainen, J. K. Hiltunen, P. A. Michels and V. D. Antonenkov, *PLoS One*, 2012, **7**(4), e34530.
27. V. Hannaert, E. Saavedra, F. Duffieux, J. P. Szikora, D. J. Rigden, P. A. Michels and F. R. Opperdoes, *Proc. Natl. Acad. Sci. U. S. A.*, 2003, **100**, 1067.
28. C. López, N. Chevalier, V. Hannaert, D. J. Rigden, P. A. Michels and J. L. Ramirez, *Eur. J. Biochem.*, 2002, **269**, 3978.
29. M. Yuan, M. G. Vásquez-Valdivieso, I. McNae, P. A. M. Michels, L. A. Fothergill-Gilmore and M. D. Walkinshaw, 2016, submitted.
30. E. Van Schaftingen, F. R. Opperdoes and H. G. Hers, *Eur. J. Biochem.*, 1985, **153**, 403.
31. P. A. Michels and D. J. Rigden, *IUBMB Life*, 2006, **58**, 133.
32. D. A. Maugeri, J. J. Cazzulo, R. J. Burchmore, M. P. Barrett and P. O. Ogbunude, *Mol. Biochem. Parasitol.*, 2003, **130**, 117.
33. T. Naderer, E. Wee and M. J. McConville, *Mol. Microbiol.*, 2008, **69**, 858.
34. R. M. de Lederkremer and W. Colli, *Glycobiology*, 1995, **5**, 547.
35. S. Damerow, A. C. Lamerz, T. Haselhorst, J. Führing, P. Zarnovican, M. von Itzstein and F. H. Routier, *J. Biol. Chem.*, 2010, **285**, 878.
36. S. Damerow, C. Hoppe, G. Bandini, P. Zarnovican, F. F. Buettner, M. A. Ferguson and F. H. Routier, *Int. J. Parasitol.*, 2015, **45**, 783.
37. T. Yang and M. Bar-Peled, *Biochem. J.*, 2010, **429**, 533.
38. Á. E. Lobo-Rojas, E. B. González-Marcano, E. A. Valera-Vera, H. R. Acosta, W. A. Quiñones, R. J. Burchmore, J. L. Concepción and A. J. Cáceres, *Parasitol. Int.*, 2016, **65**, 472.
39. E. Suzuki, A. K. Tanaka, M. S. Toledo, H. K. Takahashi and A. H. Straus, *Infect. Immun.*, 2002, **70**, 6592.
40. B. Kleczka, A. C. Lamerz, G. van Zandbergen, A. Wenzel, R. Gerardy-Schahn, M. Wiese and F. H. Routier, *J. Biol. Chem.*, 2007, **282**, 10498.
41. J. Faria, I. Loureiro, N. Santarém, P. Cecílio, S. Macedo-Ribeiro, J. Tavares and A. Cordeiro-da-Silva, *Sci. Rep.*, 2016, **6**, 26937.
42. A. J. Cáceres, W. Quiñones, M. Gualdrón, A. Cordeiro, L. Avilán, P. A. Michels and J. L. Concepción, *Mol. Biochem. Parasitol.*, 2007, **156**, 235.

43. C. Dax, F. Duffieux, N. Chabot, M. Coincon, J. Sygush, P. A. Michels and C. Blonski, *J. Med. Chem.*, 2006, **49**, 1499.
44. A. M. Aronov, S. Suresh, F. S. Buckner, W. C. Van Voorhis, C. L. Verlinde, F. R. Opperdoes, W. G. Hol and M. H. Gelb, *Proc. Natl. Acad. Sci. U. S. A.*, 1999, **96**, 4273.
45. J. C. Bressi, C. L. Verlinde, A. M. Aronov, M. L. Shaw, S. S. Shin, L. N. Nguyen, S. Suresh, F. S. Buckner, W. C. Van Voorhis, I. D. Kuntz, W. G. Hol and M. H. Gelb, *J. Med. Chem.*, 2001, **44**, 2080.
46. M. W. Nowicki, L. B. Tulloch, L. Worrall, I. W. McNae, V. Hannaert, P. A. M. Michels, L. A. Fothergill-Gilmore, M. Walkinshaw and N. J. Turner, *Bioorg. Med. Chem.*, 2008, **16**, 5050.
47. M. Gidh-Jain, Y. Zhang, P. D. Van Poelje, J. Y. Liang, S. Huang, J. Kim, J. T. Elliott, M. D. Erion, S. J. Pilkis, M. Raafat el-Maghrabi and W. N. Lipscomb, *J. Biol. Chem.*, 1994, **269**, 27732.
48. J. Barciszewski, J. Wisniewski, R. Kolodziejczyk, M. Jaskolski, D. Rakus and A. Dzugaj, *Acta Crystallogr., Sect. D: Struct. Biol.*, 2016, **72**, 536.
49. H. P. Morgan, W. Zhong, I. W. McNae, P. A. Michels, L. A. Fothergill-Gilmore and M. D. Walkinshaw, *R. Soc. Open Sci.*, 2014, **1**(1), 140120.
50. H. P. Morgan, I. W. McNae, M. W. Nowicki, V. Hannaert, P. A. Michels, L. A. Fothergill-Gilmore and M. D. Walkinshaw, *J. Biol. Chem.*, 2010, **285**, 12892.
51. N. Chevalier, D. J. Rigden, J. Van Roy, F. R. Opperdoes and P. A. Michels, *Eur. J. Biochem.*, 2000, **267**, 1464.
52. G. J. Crowther, M. L. Booker, M. He, T. Li, S. Raverdy, J. F. Novelli, P. He, N. R. Dale, A. M. Fife, R. H. Barker Jr., M. L. Kramer, W. C. Van Voorhis, C. K. Carlow and M. W. Wang, *PLoS Neglected Trop. Dis.*, 2014, **8**(1), e2628.
53. M. W. Nowicki, B. Kuaprasert, I. W. McNae, H. P. Morgan, M. M. Harding, P. A. Michels, L. A. Fothergill-Gilmore and M. D. Walkinshaw, *J. Mol. Biol.*, 2009, **394**, 535.
54. G. F. Mercaldi, H. M. Pereira, A. T. Cordeiro, P. A. Michels and O. H. Thiemann, *FEBS J.*, 2012, **279**, 2012.
55. E. A. Blackburn, F. A. Fuad, H. P. Morgan, M. W. Nowicki, M. A. Wear, P. A. Michels, L. A. Fothergill-Gilmore and M. D. Walkinshaw, *Biochem. Biophys. Res. Commun.*, 2014, **450**, 936.
56. F. A. A. Fuad, D. R. Houston, P. A. Michels, L. A. Fothergill-Gilmore and M. D. Walkinshaw, *Sains Malays.*, 2016, **45**, 1113.
57. E. R. Sharlow, T. A. Lyda, H. C. Dodson, G. Mustata, M. T. Morris, S. S. Leimgruber, K. H. Lee, Y. Kashiwada, D. Close, J. S. Lazo and J. C. Morris, *PLoS Neglected Trop. Dis.*, 2010, **4**, e659.
58. M. P. Hudock, C. E. Sanz-Rodríguez, Y. Song, J. M. Chan, Y. Zhang, S. Odeh, T. Kosztowski, A. Leon-Rossell, J. L. Concepción, V. Yardley, S. L. Croft, J. A. Urbina and E. Oldfield, *J. Med. Chem.*, 2006, **49**, 215.
59. C. E. Sanz-Rodríguez, J. L. Concepción, S. Pekerar, E. Oldfield and J. A. Urbina, *J. Biol. Chem.*, 2007, **282**, 12377.
60. M. A. Pabón, A. J. Cáceres, M. Gualdrón, W. Quiñones, L. Avilán and J. L. Concepción, *Parasitol. Res.*, 2007, **100**, 803.

61. K. R. Brimacombe, M. J. Walsh, L. Liu, M. G. Vásquez-Valdivieso, H. P. Morgan, I. McNae, L. A. Fothergill-Gilmore, P. A. Michels, D. S. Auld, A. Simeonov, M. D. Walkinshaw, M. Shen and M. B. Boxer, *ACS Med. Chem. Lett.*, 2014, **5**, 12.
62. M. J. Walsh, K. R. Brimacombe, M. G. Vásquez-Valdivieso, D. S. Auld, A. Simeonov, H. P. Morgan, L. A. Fothergill-Gilmore, P. A. Michels, M. D. Walkinshaw, M. Shen and M. B. Boxer, Probe Reports from the NIH Molecular Libraries Program, 2013, http://www.ncbi.nlm.nih.gov/books/NBK154498/.
63. L. Azéma, C. Lherbet, C. Baudoin and C. Blonski, *Bioorg. Med. Chem. Lett.*, 2006, **16**, 3440.
64. J. C. Bressi, J. Choe, M. T. Hough, F. S. Buckner, W. S. Van Voorhis, C. L. Verlinde, W. G. Hol and M. H. Gelb, *J. Med. Chem.*, 2000, **43**, 4135.
65. A. T. Cordeiro, O. H. Thiemann and P. A. M. Michels, *Bioorg. Med. Chem.*, 2009, **17**, 2483.
66. A. T. Cordeiro and O. H. Thiemann, *Bioorg. Med. Chem.*, 2010, **18**, 4762.
67. S. Gupta, A. T. Cordeiro and P. A. Michels, *Mol. Biochem. Parasitol.*, 2011, **176**, 112.
68. S. Gupta, M. Igoillo-Esteve, P. A. Michels and A. Cordeiro, *Mol. Biol. Int.*, 2011, **2011**, 135701.
69. G. F. Mercaldi, A. Dawson, W. N. Hunter and A. T. Cordeiro, *FEBS Lett.*, 2016, **590**, 2776.
70. G. F. Mercaldi, A. T. Ranzani and A. T. Cordeiro, *J. Biomol. Screen.*, 2014, **19**, 1362.
71. C. Dardonville, E. Rinaldi, M. P. Barrett, R. Brun, I. H. Gilbert and S. Hanau, *J. Med. Chem.*, 2004, **47**, 3427.
72. G. F. Ruda, V. P. Alibu, C. Mitsos, O. Bidet, M. Kaiser, R. Brun, M. P. Barrett and I. H. Gilbert, *Chem. Med. Chem.*, 2007, **2**, 1169.
73. G. F. Ruda, P. E. Wong, V. P. Alibu, S. Norval, K. D. Read, M. P. Barrett and I. H. Gilbert, *J. Med. Chem.*, 2010, **53**, 6071.
74. G. F. Ruda, G. Campbell, V. P. Alibu, M. H. Barrett, R. Brenk and I. H. Gilbert, *Bioorg. Med. Chem.*, 2010, **18**, 5056.
75. M. Zarzycki, R. Kołodziejczyk, E. Maciaszczyk-Dziubinska, R. Wysocki, M. Jaskolski and A. Dzugaj, *Acta Crystallogr., Sect. D: Struct. Biol.*, 2011, **67**, 1028.
76. H. P. Morgan, F. J. O'Reilly, M. A. Wear, J. R. O'Neill, L. A. Fothergill-Gilmore, T. Hupp and M. D. Walkinshaw, *Proc. Natl. Acad. Sci. U. S. A.*, 2013, **110**, 5881.

CHAPTER 16

The Redox Metabolism and Oxidative Stress in Leishmania *as a Crossroads for the Lethal Effect of Drugs*

HELENA CASTRO,*[a,b] MARGARIDA DUARTE[a,b] AND ANA M. TOMÁS*[a,b,c]

[a] i3S—Instituto de Investigação e Inovação em Saúde, Universidade do Porto, Portugal; [b] IBMC—Instituto de Biologia Molecular e Celular, Universidade do Porto, Rua Alfredo Allen 208, 4200-135 Porto, Portugal; [c] ICBAS—Instituto de Ciências Biomédicas Abel Salazar, Universidade do Porto, Rua de Jorge Viterbo Ferreira 228, 4050-313 Porto, Portugal

*Email: hcastro@ibmc.up.pt; atomas@ibmc.up.pt

16.1 Reactive Oxygen and Nitrogen Species and *Leishmania*

Reactive oxygen and nitrogen species are terms used to designate a collection of highly reactive oxidants derived from oxygen (ROS) and nitrogen (RNS), which include radicals such as superoxide anion ($O_2^{\bullet-}$) and nitric oxide ($^{\bullet}NO$), and non-radical molecules like hydrogen peroxide (H_2O_2) and peroxynitrite ($ONOO^-$). While low, transient, levels of some reactive species can serve signalling functions and are, therefore, essential for normal physiology, sustained and uncontrolled generation of ROS or RNS overcomes

Drug Discovery Series No. 60
Drug Discovery for Leishmaniasis
Edited by Luis Rivas and Carmen Gil

Published by the Royal Society of Chemistry, www.rsc.org

the cell antioxidant activity and causes oxidative stress and cell death. *Leishmania* are exposed to oxidants, formed both endogenously, as a consequence of the parasite aerobic metabolism, and externally, by the host defensive strategies. Oxidant production can furthermore derive from metabolization of xenobiotic compounds, such as drugs.

16.1.1 Generation of ROS and RNS within *Leishmania*

Different processes and sites may stimulate ROS production within *Leishmania* (Figure 16.1). However, in these parasites, as in many other cells, endogenous ROS release is mainly associated with mitochondria, more concretely, with the respiratory chain. During transference of reducing equivalents along the several intermediates of the chain, electrons may escape, allowing reduction of O_2 to $O_2^{\bullet-}$ and H_2O_2. Although the exact sites for electron leakage are not well established, complexes II and III,[1,2] alternative NADH dehydrogenase[3] and fumarate reductase[4] have been reported to display ROS-generation activity in *Leishmania* or other trypanosomatids. Of these, complex II (or succinate dehydrogenase) may be of particular importance to *Leishmania* given that succinate is the trypanosomatid main respiratory substrate.[5,6] Generation of mitochondrial ROS by these and, perhaps, other enzymes can be part of redox-controlled circuits triggered by specific stimuli, as illustrated by the finding that H_2O_2 fluxes elicited in response to iron deficiency drive differentiation of promastigotes into amastigotes;[7,8] most commonly, however, it occurs hazardously. Either way, under normal growth conditions, endogenous ROS constitute no danger to parasites as they are maintained at low levels,[9] probably through the activity of endogenous scavenging systems. This equilibrium is, however, expected to be affected whenever mitochondrial ROS exceed a certain threshold.[2,10] Excessive ROS generation in *Leishmania* can be elicited by stress signals that directly or indirectly affect respiratory function. For instance, exposure of promastigotes to heat stress causes mitochondrial membrane hyperpolarization and a sharp increase in oxygen consumption, leading to an increment in $O_2^{\bullet-}$ production that is directly responsible for parasite death.[9]

ROS overproduction in *Leishmania* mitochondria also occurs as a consequence of exposure to drugs. Examples of these include therapeutics under development such as the (i) the 8-aminoquinolines sitamaquine[11] and tafenoquine[1] that induce ROS upon interference with complexes II and III activities respectively and (ii) chalcones and aurones that generate ROS by inhibiting mitochondrial fumarate reductase (mFR), as well as the clinically relevant antimonials (Sb^V) [sodium stibogluconate (Pentostam) and meglumine antimoniate (Glucantime)], amphotericin B, pentamidine and miltefosine. Although the mode of action of these four compounds is not well elucidated, their primary targets are different and often unrelated to the respiratory chain enzymes;[12–16] these compounds nevertheless converge in triggering mitochondrial ROS generation which appears to act as the downstream effector mechanism directly responsible for their parasiticidal activity.[17]

Leishmania,[18,19] like *Trypanosoma cruzi*,[20,21] have been suspected to express a nitric oxide synthase (NOS) activity, raising the possibility that these parasites can form RNS. The *Leishmania* activity was furthermore reported to

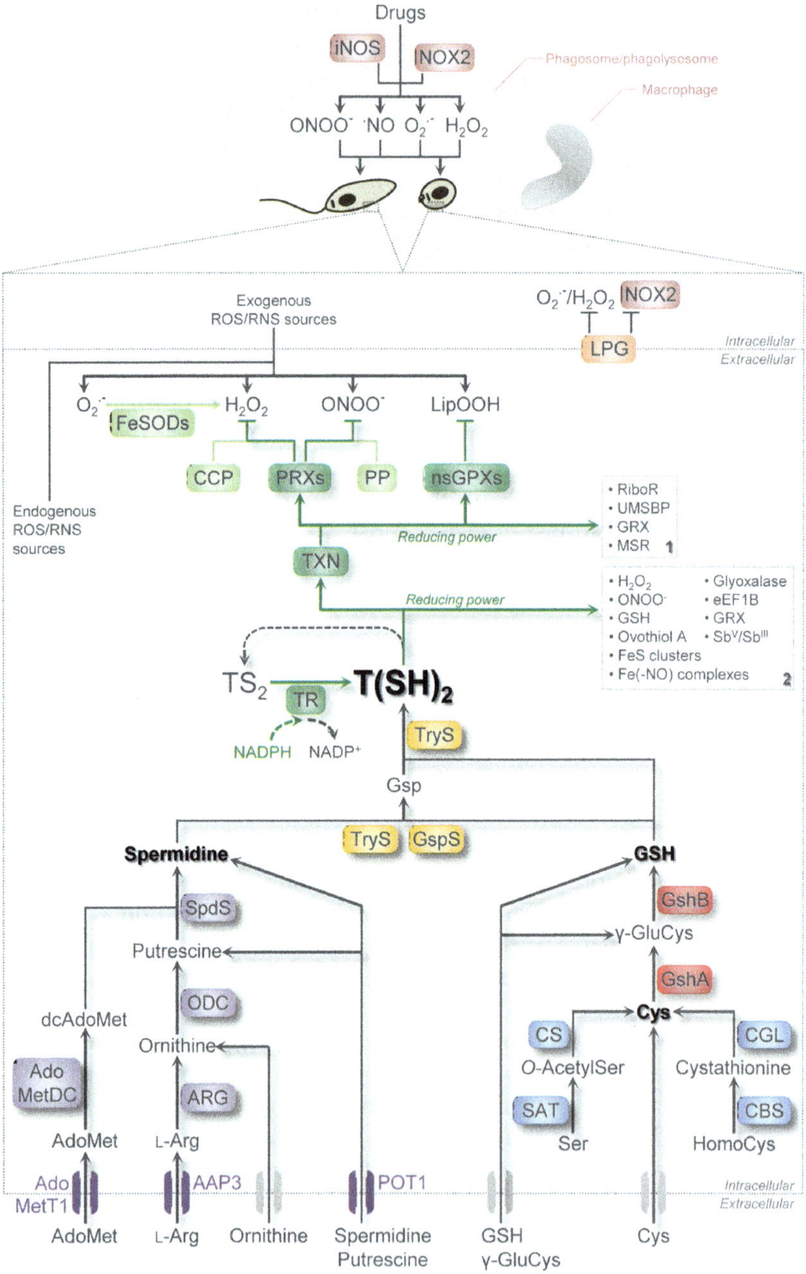

be inhibited by mesoionic 4-phenyl-5-(cinnamoyl)-1,3,4-thiadiazolium-2-phenylamine chloride.[22] Nevertheless, up to now, no enzyme entity displaying $^{\bullet}$NO activity has been identified and the presence of a parasite NOS remains uncertain.

Figure 16.1 The trypanothione system at the cross-roads of *Leishmania* protection against reactive oxygen and nitrogen species. Throughout their life cycle *Leishmania* are exposed to reactive oxygen and nitrogen species (ROS and RNS, respectively), namely hydrogen peroxide (H_2O_2), superoxide anion ($O_2^{\bullet -}$), nitric oxide ($^{\bullet}$NO) and peroxynitrite ($ONOO^-$). These result from the activity of the macrophage enzymes NADPH oxidase (NOX2) and inducible nitric oxide synthase (iNOS) (*pink boxes*), as well as from the reactivity of leishmanicidal drugs. Reactive oxygen species can be further generated endogenously, as a consequence of the aerobic metabolism of the parasites. *Leishmania* have evolved several mechanisms to protect themselves against the oxidative and nitrosative insults of the host. These include components of its glycocalyx (*orange boxes*), namely lipophosphoglycan (LPG), which is involved in inhibition of production and/or scavenging of ROS. Enzymatic antioxidant and antinitrosative defences of *Leishmania* include iron superoxide dismutases (FeSODs) and class I haem peroxidases [pseudoperoxidase (PP) and cytochrome *c* peroxidase (CCP)] (*light green boxes*). The trypanothione [$T(SH)_2$] redox system of *Leishmania* (*dark green boxes*) is also involved in the elimination of ROS and RNS, *via* the activity of two thiol peroxidases, namely non-selenium glutathione peroxidases (nsGPX) and 2-Cys peroxiredoxins (PRXs). These are fuelled by tryparedoxin (TXN), a thioredoxin-like multipurpose oxidoreductase that is directly reduced by $T(SH)_2$. TXN and $T(SH)_2$ can additionally reduce molecules not directly involved in peroxide elimination (white boxes 1 and 2), but with relevance to the overall redox homeostasis of the parasite. The reducing power (*dark green arrows*) required to fuel these $T(SH)_2$-dependent phenomena originates from NADPH. Oxidized trypanothione (TS_2) is regenerated to $T(SH)_2$ by the enzyme trypanothione reductase (TR). Trypanothione is synthesized by trypanothione synthetase (TryS) and glutathionylspermidine synthetase (GspS) from two spermidine and one glutathione (GSH) molecules (*light orange boxes*). Spermidine synthesis, from arginine, involves the activity of four enzymes: arginase (ARG), ornithine decarboxylase (ODC), *S*-adenosylmethionine decarboxylase (AdoMetDC), and spermidine synthase (SpdS) (*purple boxes*). Spermidine can also be taken in from the parasite milieu through the polyamine transporter 1 (POT1). Glutathione is synthesized from cysteine (Cys) by the sequential activity of γ-glutamylcysteine synthetase (GshA), that forms γ-glutamylcysteine (γ-Glu–Cys), and glutathione synthetase (GshB) (*red boxes*). Glutathione and γ-Glu–Cys can also be taken up by *Leishmania*. Cysteine is either acquired from the milieu or synthesized *de novo* through the *cysteine assimilatory pathway*, involving serine acetyltransferase (SAT) and cysteine synthase (CS), or the *reverse trans-sulfuration pathway* via cystathionine-β-synthase (CBS) and cystathionine-γ-lyase (CGL) (*blue boxes*). Transporters not yet identified in *Leishmania* but for the presence of which there is indirect evidence are indicated in *grey boxes* (see main text). Other abbreviations: arginine transporter (AAP3), eukaryotic translation elongation factor 1B (eEF1B), glutaredoxin (GRX), methionine sulfoxide reductase (MSR), ribonucleotide reductase (RiboR), universal minicircle sequence binding protein (UMSBP).

16.1.2 Exogenous Sources of Oxidative and Nitrosative Stress to *Leishmania*

Leishmania infect and replicate in cells—macrophages—that have the inherent capacity to produce large amounts of $O_2^{\bullet-}$, H_2O_2 and $^{\bullet}NO$ as part of their antimicrobial defence; it is therefore of no surprise that to establish themselves in mammalian hosts, parasites have to avoid exogenously-derived ROS and RNS toxicity.

Macrophage $O_2^{\bullet-}$ and $^{\bullet}NO$ are produced by two professional enzymes, NADPH oxidase (NOX2) and inducible nitric oxide synthase (iNOS). The former is a multicomponent protein that assembles on the phagosome membrane during phagocytosis and after activation by soluble inflammatory stimuli,[23] gives rise to $O_2^{\bullet-}$ in the vicinity of the microbe (the so called "respiratory burst"). Although $O_2^{\bullet-}$ can *per se* induce pathogen damage, most phagosome $O_2^{\bullet-}$ is rapidly converted to H_2O_2 and other ROS, including the highly toxic hydroxyl radical ($^{\bullet}OH$), that forms upon reaction of H_2O_2 with transition metals. Inducible nitric oxide synthase, on the other hand, is a cytosolic enzyme that catalyses $^{\bullet}NO$ synthesis once activated by several inflammatory cytokines. Since $^{\bullet}NO$ is very diffusible, it can easily reach microbe-containing vacuoles and react directly with transition metal centres in pathogen proteins, and/or trigger downstream production of oxidants such as $ONOO^-$. Due to its high reactivity, the latter molecule is very hazardous. Studies with animal models of leishmaniases, including mice deficient in NOX2 and iNOS, confirmed that ROS and/or RNS formed in the context of this infection can act as important leishmanicidal mechanisms. Nevertheless, the exact contribution to parasite containment of these reactive molecules depends on the actual *Leishmania* species and strain as illustrated next.

Gathering available information, it has become apparent that the cutaneous leishmaniasis species *Leishmania major* is very susceptible to iNOS-derived oxidants, its replication in the skin of susceptible infected mice being controlled in iNOS-expressing but not in iNOS-knockout animals or animals pharmacologically rendered deficient in $^{\bullet}NO$ production.[24–27] This oxidant was found to contribute to restraining parasite replication during infection, first, by decreasing the overall parasite metabolic activity[28] and, later on when its concentration is much increased due to IFN-γ and TNF-α-induced iNOS activation, by ensuing parasite elimination.[29,30] The effector protagonism of $^{\bullet}NO$ extends to the latent phase of infection, this radical being also required to contain parasite re-activation.[31] Conversely, the multiplication of *L. major* in the skin is not significantly affected by ROS,[32] these molecules being nevertheless reported to prevent visceralization to the spleen.[33] The extent to which *Leishmania donovani* and *Leishmania infantum*, two species giving rise to visceral leishmaniasis, are affected by host-derived RNS is not as clear-cut as in *L. major*, with dissimilar results reported. Illustrating this, macrophage iNOS was found to be essential for control of *L. donovani* in the liver of infected mice by Murray and Nathan[34] but not by White *et al.*[32] who, using a different strain, could only detect a modest protective role of that enzyme in the spleen.

Heterogeneous data was also reported for *L. infantum* by Marques *et al.*[35] who found that •NO contributed to the control of just one of the three strains analysed. Similarly to RNS, NOX2-derived oxidants can either be dispensable,[32,35] play a minor role,[34] or act as a major control mechanism to arrest parasite proliferation in the organs of infected animals.[35]

In short, although *Leishmania* are exposed to host-derived ROS and RNS, not all parasites succumb to the toxic effect of these molecules. At least in part, phenotype differences in animals reflect the inherent capacity of a strain to withstand oxidants, as evidenced in assays where parasites were treated with ROS or RNS donors.[35]

16.2 Elimination of ROS and RNS

The following sections review the redox enzymes/pathways that enable *Leishmania* to maintain homeostasis in face of different ROS and RNS sources (Figure 16.1). It is important to outline in this context that additional mechanisms can contribute to antioxidant protection. For instance, components of the promastigote glycocalyx, namely lipophosphoglycan (LPG), provide an effective first barrier against the host oxidative insult. Lipophosphoglycan can prevent the production of ROS and/or directly scavenge these molecules,[36,37] and has also been reported to suppress iNOS expression and, consequently, •NO generation.[38]

16.2.1 Superoxide Dismutases

The superoxide anion is the precursor of other ROS as well as $ONOO^-$, hence, detoxification of this radical provides an efficient means to prevent redox imbalance and oxidative stress. Although $O_2^{\bullet -}$ can be spontaneously dismutated to H_2O_2 and O_2, this reaction is much more effective when catalyzed by metal-containing oxidoreductases of the superoxide dismutase (SOD) family. *Leishmania* spp. contain in their genomes information for five of these enzymes, SODA, two isoforms of SODB (encoded by two almost identical genes), SODC and an as yet unanalysed SOD (tritrypdb.org/tritrypdb). All characterized enzymes use iron as co-factor (*i.e.* are FeSODs), a factor that might be exploited for drug design as host SODs require manganese or copper or zinc.[39] *Leishmania* SODA is a mitochondrial enzyme,[40] bioinformatics analysis of its *N*-terminal targeting sequence predicting a matrix sub-compartmentalization.[41] SODC was also ascribed to the *Leishmania* mitochondria.[41]

The antioxidant character of mitochondrial SODs was established by genetic studies. Overexpression of SODA was found to increase parasite resistance to the free-radical donors paraquat, nitroprusside[42] and antimycin,[43] while SODC repression by antisense RNA enhanced *Leishmania tropica* sensitivity to this type of compounds,[44] and reduced *L. donovani* survival in mouse macrophages.[44] *Leishmania* mitochondrial SODs have also been involved in resistance to antimonial drugs, with SODA presenting higher activity levels in *Leishmania braziliensis* and *L. infantum* lines resistant to this drug than in sensitive ones.[45] The participation of *Leishmania* mitochondrial SODs in antimony resistance is in agreement with the previously discussed role of mitochondrial ROS acting as the downstream effector mechanism responsible for the leishmanicidal activity of this drug.[17] Unlike SODA and SODC, SODBs are glycosomal. Elimination of one of the SODB isoforms decreased *Leishmania chagasi* survival under conditions of increased endogenous $O_2^{\bullet-}$ levels (generated by the cell-permeable $O_2^{\bullet-}$ donor paraquat) and within macrophages.[46] *Leishmania* SODs have not yet been genetically validated. The finding that the leishmanicidal activity of a family of pyrazole-containing polyamine macrocyclic compounds was associated with FeSOD inhibition[47] nevertheless indicates these antioxidant enzymes to be promising in terms of leishmaniases drug development.

16.2.2 Class I Haem Peroxidases

In recent years, a group of non-thiol enzymes, the haem peroxidases, has emerged as an important player in the peroxide metabolism of these protozoan parasites. The genomes of *Leishmania* harbour several putative haem-containing proteins, most of which remain uncharacterized. Exceptions to this are two class I haem peroxidases of *L. major*, namely a pseudoperoxidase (PP) and cytochrome *c* peroxidase (CCP).

In *Leishmania* spp. PP is located at the plasma membrane of amastigotes and has been reported to be involved in $ONOO^-$ detoxification.[48] Parasites genetically depleted in PP are hampered in their capacity to infect macrophages *in vitro*, especially if these are pre-activated with LPS and IFN-γ. This avirulent phenotype can be rescued by chemical inhibition of nitric oxide synthase. The importance of PP for parasite infectivity was further confirmed in an *in vivo* murine infection model, thus highlighting the potential of this enzyme as a drug target. The unique structural and biochemical features of *Leishmania* PP relative to its mammalian counterparts further encourages its progression for drug development.

The *Leishmania* CCP is not as promising as a drug target as PP. This enzyme, originally categorized as an ascorbate-dependent peroxidase,[49] is presently recognized to use cytochrome *c* as preferred reductant, in other words to function as a cytochrome *c* peroxidase.[50,51] The reducing power used by cytochrome *c* to reduce CCP is originally generated from NADH. *Leishmania* CCP and its cytochrome *c* reductant are co-localized at the intermembrane space of the inner mitochondrial membrane,[52] wherein they act in concert to protect parasites against cell damage induced by oxidative stress.[52,53] The physiological functions of CCP in *Leishmania* further extend to an active role in cell differentiation and virulence as shown in a genetic manipulation study of CCP by Pal *et al.*[54] In this work, the authors found that, contrary to the expectation that a robust antioxidant defence is advantageous for *Leishmania* infectivity, CCP expression inversely correlates with the capacity of parasites to invade and survive within host phagocytic cells. The unexpected behaviours exhibited by CCP-overexpressing and -depleted mutants are consistent with the recent finding that H_2O_2 triggers promastigote differentiation into infective forms.[8] Recent studies have furthermore implicated CCP in drug resistance against antimony and amphotericin.[55,56] From a drug development perspective CCP could be a very attractive target owing to its absence from mammalian hosts. However, based on the advantage that CCP depletion confers during infection, any leishmanicidal strategy based on CCP modulation would have to aim at enzyme activation, rather than its inhibition.

16.2.3 The Trypanothione Redox System

In *Leishmania* spp. detoxification of peroxides and peroxynitrite does not proceed *via* the catalase and selenocysteine-containing glutathione peroxidase enzymes found in higher eukaryotes. Deprived of these classical peroxide-metabolizing enzymes, *Leishmania* have evolved a distinctive antioxidant system almost exclusively dependent on the dithiol trypanothione [N^1,N^8-bis-glutathionylspermidine or $T(SH)_2$].[57] Trypanothione acts as a substitute for glutathione (GSH) in these organisms. It utilizes the reducing power from NADPH to fuel ROS and RNS reduction, among many other cellular functions. Owing to its unique and essential character, the $T(SH)_2$ system offers potential points of attack with specific anti-parasitic drugs. These include the pathways for peroxide-reduction, and for $T(SH)_2$ recycling and biosynthesis (Figure 16.1), as detailed next. The conservation of the $T(SH)_2$ system in other human pathogens of the Trypanosomatidae family (including *Trypanosoma cruzi* and *Trypanosoma brucei*) opens perspectives towards the identification of a unifying leishmanicidal–trypanocidal drug.

16.2.3.1 Utilization of Trypanothione

Trypanothione plays a pivotal role in a number of physiological phenomena that surpass peroxide detoxification and that are critical to the overall redox homeostasis of *Leishmania.*[58,59] While many of these functions are mediated by specialized oxidoreductases remotely related to thioredoxin, known as tryparedoxins (TXNs; reviewed in the next section), others involve the direct interaction of $T(SH)_2$ with redox partners, namely: (i) reduction of glutathione, either spontaneously[60] or *via* a trypanothione–glutathione thioltransferase[61,62] and 2-Cys-glutaredoxins;[63] (ii) reduction of ovothiol A,[64] a mercaptohistidine present in trypanosomatids that acts as a non-enzymatic scavenger of free-radicals (reviewed by Steenkamp)[65] and that catalyses the decomposition of nitrosothiols;[66] (iii) reduction of glutaredoxins, *i.e.* oxidoreductases that contribute to GSH homeostasis and resolution of GSH–protein mixed disulphides; (iv) ligation to iron and iron–nitric oxide complexes, as well as to iron–sulphur cluster protein, free or bound to 1-Cys- and 2-Cys-glutaredoxins;[58] (v) acting as a cofactor of the eukaryotic translation elongation factor 1B (eEF1B)[67,68] and of the glyoxalase system.[69] Notably, most of these $T(SH)_2$-dependent activities were described in *T. brucei* and *T. cruzi* and are assumed to also take place in *Leishmania*.

Importantly, $T(SH)_2$ metabolism is intimately linked to the mode of action of some of the drugs currently used to treat leishmaniases. Particularly in the case of the pentavalent antimonials, $T(SH)_2$ can act by different mechanisms. First, it can directly reduce Sb^V to its trivalent, leishmanicidal, active state (Sb^{III}).[70,71] Second, Sb^{III} impairs the thiol-buffering capacity of *Leishmania*, either through direct inhibition of trypanothione reductase,[13,72,73] or through rapid extrusion of $T(SH)_2$ in the form of thiol–antimony complexes.[13,74–76] On the other hand, $T(SH)_2$ can contribute to *Leishmania* resistance to antimony by promoting the active detoxification of this metal in the form of Sb^{III}–$T(SH)_2$ complexes. Consistent with this, Sb^{III}-resistant strains exhibit elevated levels of $T(SH)_2$[74] or increased biosynthetic capacity of the dithiol.[14,77–79] Another mechanism of Sb^{III} resistance is the upregulation of $T(SH)_2$-dependent peroxidases of the 2-Cys peroxiredoxin family.[80,81] Apart from pentavalent antimonials, other leishmanicidal drugs, namely amphotericin B and miltefosine, are directly or indirectly linked to $T(SH)_2$ metabolism. Specifically, (i) amphotericin B-resistant promastigotes express elevated levels of trypanothione synthetase,[82] the enzyme in charge of $T(SH)_2$ synthesis, and (ii) one of the mechanisms employed by *Leishmania* to tolerate miltefosine pressure is by increasing $T(SH)_2$ metabolism.[83]

For the sake of this review we are particularly interested in the utilisation of $T(SH)_2$ as a supplier of reducing equivalents to the anti-oxidative and anti-nitrosative reactions of *Leishmania*. Trypanothione *per se* is a spontaneous reductant of hydroperoxides[64,84] and peroxynitrite,[85] however its peroxidase activity can be greatly enhanced in the presence of dedicated peroxidases. *Leishmania* harbour two classes of thiol peroxidases that utilise the reducing power of $T(SH)_2$ to catalyse peroxide reduction: (i) two-cysteine peroxiredoxins (2-Cys PRXs),[86–88] and (ii) non-selenium glutathione peroxidases (nsGPXs).[89,90] None of these enzymes is directly reduced by $T(SH)_2$, rather, reducing equivalents flow from the dithiol to the peroxidases *via* a multipurpose oxidoreductase of the thioredoxin family known as tryparedoxin (TXN), which endows these enzymes with tryparedoxin peroxidase (TXNPx) activity.

16.2.3.1.1 Tryparedoxins. TXNs play a prominent role in $T(SH)_2$-dependent peroxide metabolism by acting as redox shuttles between the dithiol and peroxidases of the 2-Cys PRX and nsGPX families. TXNs are related to oxidoreductases of the thioredoxin family, albeit with some distinctive features. In comparison with thioredoxins, TXNs (i) share only 13% sequence identity, (ii) are larger in size, (iii) exhibit a unique active site signature and (iv) preferentially utilize $T(SH)_2$ (instead of thioredoxin reductase) as a source of reducing equivalents. Apart from directly fuelling thiol peroxidases, TXNs can accept other molecules as redox partners and, in this way, participate in various physiologically relevant phenomena, including: (i) DNA synthesis and repair, *via* ribonucleotide reductase;[91] (ii) redox regulation of mitochondrial DNA (kinetoplast) replication, *via* the universal minicircle sequence binding protein (UMSBP);[92,93] (iii) iron–sulphur cluster biosynthesis, *via* GRX;[94,95] (iv) protein repair and methionine homeostasis, *via* methionine sulfoxide reductase (MSR).[96] TXN interactions may extend to a wider collection of molecules, as recently disclosed by proteomics studies.[97,98] Of all these activities, only the reaction of TXN with 2-Cys PRXs, nsGPXs and UMSBP have thus far been confirmed to occur in *Leishmania.*[88,90,93,99,100] The others have been established in *T. brucei*, *T. cruzi* and/or in the model trypanosomatid *Crithidia fasciculata*, but are also expected to take place in *Leishmania* based on the presence of the required protein partners in the parasites' genomes (tritrypdb.org/tritrypdb).

Among the enzymes of the $T(SH)_2$–TXN–peroxidase system, TXN is predicted to be the one exerting more control over the flux of peroxide reduction, a consequence of its slow, rate-limiting recycling by $T(SH)_2$.[100–104] Accordingly, abrogation of TXN is expected to negatively affect the antioxidant capacity of parasites. Such inhibition, should cumulatively impair other critical TXN-dependent reactions (mentioned above), ultimately leading to parasite death. The lethal consequences of TXN abrogation were confirmed by reverse genetics of the cytosolic enzyme of *L. infantum*[105] and of *T. brucei.*[106,107] Importantly, essentiality of cytosolic TXN was extended to the mammalian stage of *Leishmania.*[105] The critical requirement for TXN activity is not verified for the mitochondrion, since the *Leishmania* mitochondrial isoform was found to be redundant during the parasite life cycle and is apparently absent from the genomes of other trypanosomatids.[99] Despite being genetically validated, and kinetically[100,102–104] and structurally characterized,[108–111] TXNs have failed to mobilize researchers to invest in drug screening campaigns, probably because this enzyme is not easy to target with reversible ligands owing to the unstructured nature of its substrate interaction site.[110] Only recently, the first set of small drug-like compounds with TXN-inhibitory *in vitro* activities and trypanocidal effects was reported.[112] Some level of specificity could be achieved, as some of these drugs reacted much more slowly with thioredoxin compared with TXN.[112] The anti-*Leishmania* activity of these TXN inhibitors has not been investigated yet. Notably, the mode of action of these drugs confirmed the early prediction that TXN inhibition could only be attained by irreversible thiol modifiers acting on the surface-exposed cysteine residues of the enzyme.[110] The use of covalent inhibitors is usually discouraged in the pharmaceutical industry due to issues of reactivity and toxicity. This view is nevertheless changing, as some enzyme inhibitors used in the clinic function by covalent modification.[113]

16.2.3.1.2 Two-Cysteine Peroxiredoxins. Two-cysteine peroxiredoxins (2-Cys PRXs) are a group of evolutionarily conserved and ubiquitous peroxidases that make use of two redox-active cysteine residues to catalyse peroxide reduction. *Leishmania* are endowed with one mitochondrial 2-Cys-PRX,[87] plus two to three other isoforms, this number depending on the parasite species (tritrypdb.org/tritrypdb). The latter are encoded within the same chromosomal locus and may specify cytosolic[87] or glycosomal enzymes (Castro H. and Tomás A.M., unpublished results). Common to all these isoforms is their preference to utilize the $T(SH)_2$–TXN redox pair as source of reducing equivalents (Castro H. and Tomás A.M., unpublished results),[88,90,100,114] unlike their eukaryotic counterparts that use thioredoxin. Trypanosomatid 2-Cys PRXs are very efficient at reducing H_2O_2, organic hydroperoxides,[86,88,90] as well as peroxynitrite (Romao S., Tomás A.M. and Radi R., unpublished results). They react poorly with lipid hydroperoxides, being susceptible to inactivation by these compounds.[86,88] There is extensive structural data on these enzymes in *Leishmania* spp.[115–119]

Two-Cys PRXs were the first peroxidases identified in trypanosomatids, their discovery, back in 1997,[103] providing a suitable molecular explanation for the capacity of these parasites, deprived of classical peroxidases, to cope with oxidative stress. This historical context has, ever since, linked trypanosomatid 2-Cys-PRXs with defence against oxidative and nitrosative stress, with very few publications exploring the possibility that these enzymes may be implicated in alternative functions, similarly to orthologous molecules in other organisms.[120] That 2-Cys PRXs can act as antioxidant molecules in *Leishmania* spp. is supported by the observations that parasites overexpressing cytosolic isoforms are more resistant against exogenously-derived hydroperoxides and peroxynitrite.[87,121–123] The increased resistance of these transgenics to peroxides was inevitably used to justify their increased infectivity in *ex vivo*[121–123] and *in vivo* models of infection.[121] The contribution of 2-Cys-PRXs to *Leishmania* virulence was further supported by the fact that their expression is upregulated in highly virulent[124] and in highly metastatic[125] parasite strains. Two-Cys PRXs were also found to be associated with resistance to leishmanicidal antimonial drugs.[80,81,122,126] Interestingly, it has never been genetically demonstrated that the cytosolic 2-Cys PRXs of *Leishmania* are essential for parasite survival and that such essentiality is provided by their peroxidase activity. The best approximation to answering this question was a RNA interference-based study carried out in *T. brucei*,[107] showing that cytosolic 2-Cys PRXs are essential for parasite survival and tolerance to exogenous H_2O_2. The state-of-the-art is different concerning the *Leishmania* mitochondrial 2-Cys PRX. This enzyme was genetically demonstrated to be crucial for the mammalian stage of *Leishmania* spp., its critical activity being that of a chaperone holdase, not of a peroxidase.[127,128]

The well-known functional versatility of 2-Cys PRXs in general[120] is an issue with direct relevance to drug development. In the case of the *Leishmania* mitochondrial isoform it is clear that efficient inhibitors have to be directed against its chaperone activity. The possibility of specifically inhibiting the molecular chaperone function of 2-Cys PRXs was recently validated by Zhao *et al.*,[129] who, using a mammalian enzyme as target, identified compounds capable of disrupting the decameric, chaperone-active quaternary structure of these enzymes. Regarding targeting of the presumably crucial peroxidase activity of the cytosolic enzymes, compounds with peroxidase-inhibitory activity towards 2-Cys-PRXs have been identified in other organisms as irreversible ligands,[130,131] and more recently in *T. brucei*[132] and *Leishmania*[115] as reversible inhibitors. Whether these chaperone- and peroxidase-inhibition strategies offer the possibility of specifically targeting *Leishmania* 2-Cys PRXs without secondary effects on hosts has not yet been explored, even though this may not be easily achieved, based on overall structure preservation of these enzymes across the eukaryotic phylogeny.

16.2.3.1.3 Non-selenium Glutathione Peroxidases. Glutathione peroxidases (GPXs) comprise a group of enzymes that act as pivotal antioxidants by reducing hydroperoxides *via* a selenocysteine (SeCys) residue residing at their active site. In the cases where SeCys is replaced by a Cys residue, the enzymes are referred to as *non-selenium GPXs* (nsGPXs). In *Leishmania* all glutathione peroxidases are of the nsGPXs type. *Leishmania* can potentially encode three nsGPX isoforms (nsGPXA1–3) that differ in possessing or not mitochondrial and/or glycosomal endorsement sequences.[89,90] Except for one report describing biochemical traits of *Leishmania* nsGPXs,[90] little else is known about these molecules. Functional characterization of GPXs is detailed in *T. brucei*, allowing us to speculate about the putative functions of these proteins in *Leishmania* spp. as well as about their potential use as drug targets.

In *T. brucei* the mitochondrial nsGPXA was found by reverse genetics not to be essential during the parasite life cycle, even though its depletion caused a transient growth defect accompanied by cardiolipin peroxidation.[133,134] The reported redundancy of this mitochondrial enzyme is consistent with its apparent absence from some trypanosomatid species (*e.g.*, *L. braziliensis* and *T. cruzi*; tritrypdb.org/tritrypdb). This is not the case for the cytosolic nsGPXA molecule, which was shown to be required for survival of bloodstream *T. brucei* parasites unless these are supplemented with the vitamin E analogue Trolox.[107,133] The critical physiological function of these cytosolic nsGPXAs is not expected to be general antioxidant defence, based on the observations that: (i) the pool of nsGPXAs in *Leishmania* spp. contribute to less than 1% to the overall peroxidatic capacity of the parasite;[90] (ii) in *T. brucei*, knocking down of the nsGPXA pool did not lead to increased parasite susceptibility to exogenous H_2O_2.[107] Rather, these enzymes have more specialized functions, as recently reported for the cytosolic *T. brucei* nsGPXA. The latter was found to protect lysosome and mitochondria membranes from iron-induced lipid peroxidation,[133–135] consistent with the reported ability of nsGPXAs to metabolize lipid hydroperoxides.[133,136] Alternative functions for trypanosomatid nsGPXAs may include hydroperoxide sensing and regulation of redox signalling cascades, as reported in other organisms.[137,138]

The potential essentiality of cytosolic nsGPXA enzymes in *Leishmania* may justify their progress to drug development, as long as requisites of uniqueness and druggability are validated. The first of these conditions is verified, based on the significant divergences in substrate specificity (TXN *versus* GSH) and catalytic mechanism between trypanosomatid nsGPXAs and their closest, classical GPXs, counterparts in mammals.[90,136,139,140] Illustrating this, the peroxidase activity of parasite nsGPXAs depends on two redox-active cysteines that are absent in mammalian counterparts: (i) the peroxidatic Cys that replaces the SeCys residue of classical GPXs, and (ii) the resolving cysteine that generates an intramolecular disulphide bond with the peroxidatic residue, which is subsequently reduced by TXN. The second requisite (druggability) is also confirmed, as supported by the results of structural studies of *T. brucei* and *T. cruzi* nsGPXA enzymes showing potential sites for inhibition with small drug-like compounds,[139–141] and by inhibition assays demonstrating that *T. brucei* nsGPXAs can be inhibited *in vitro* by multi-target trypanocidal quinol compounds.[132]

Leishmania contain a fourth nsGPX-coding gene (*nsGPXB*),[89] which specifies an enzyme with only 30% identity with the nsGPXA1–3 molecules, and for which there is no biochemical and functional data available. In *T. brucei* and *T. cruzi*, nsGPXB molecules were assigned different subcellular localizations, to the mitochondrion[142] and to the endoplasmic reticulum,[143] respectively. The enzymes of both organisms were reported to metabolize fatty acid hydroperoxides *in vitro*, however, with different affinity towards the TXN reductant.[142,143]

16.2.3.2 Recycling of Trypanothione

Trypanothione acts as an electron donor for a multitude of biological reactions, being rapidly wasted and discarded in the form of an oxidized, inactive disulphide (TS_2). The quickest way that parasites have to respond to demands for reduced trypanothione is by regenerating the TS_2 disulphide, a reaction carried out by trypanothione reductase (TR). TR reduces TS_2 at the expense of reducing equivalents derived from NADPH, which, in turn, is produced in the pentose phosphate pathway. As expected from its central role in $T(SH)_2$ metabolism, TR is a critical enzyme for *Leishmania* survival.[144,145] It is also important for antioxidant protection as evidenced by the observation that *Leishmania* mutants with decreased enzyme activity exhibit impaired ability to survive inside macrophages competent to mount an oxidative assault.[144–146]

The nearest mammalian homologue of TR is the enzyme glutathione reductase (GR). Both proteins belong to the large family of FAD-dependent disulphide oxidoreductases, and share a sequence identity of 40% and closely related three-dimensional structures. Despite their overall similarity, TR and GR differ significantly in their active sites, what justifies their mutually exclusive specificity for trypanothione and glutathione disulphides, respectively. The active site of TR is large and with an overall negative charge to accommodate the spermidine moiety of $T(SH)_2$, contrasting with the smaller and positively charged active site of GR.[73,147–150] These differences allow discrimination between TR and GR inhibitors,[151,152] hence making feasible the development of specific anti-parasitic compounds.

TR is the most studied drug target candidate of the $T(SH)_2$ redox metabolism, albeit with no promising *in vivo* inhibitor put forward so far. This preference for TR stems mostly from historical factors, because this was the first enzyme of the $T(SH)_2$-redox system to be identified[153] right after the breakthrough discovery of trypanothione in 1985 by Fairlamb *et al.*[57] Apart from kinetic,[72] structural[73,150] and functional characterization,[144–146] many TR studies have focused on the identification of specific inhibitors. These efforts, in many instances facilitated by the use of high-throughput screening *in vitro*[154] or cell-based assays,[150] in others by the use virtual screening approaches,[155] have altogether delivered a battery of drug-like compounds with diverse chemical scaffolds, modes of action and potent and specific TR inhibitory activity *in vitro.*[156] Some inhibitors, but not all,[157] are active against TRs of all three trypanosomatid parasites, as expected from the overall structural and kinetic resemblance of these enzymes.[158,159] Even though some of these molecules have leishmanicidal activity,[150] none has ever revealed enough potency to progress beyond the early discovery phase. These disappointing *in vivo* performances can be explained in light of the fact that *Leishmania* can survive with only 15% TR activity,[146] a scenario that also extends to *T. brucei.*[160] A rational explanation for this behaviour was recently provided in a kinetic modelling study of the *T. cruzi* $T(SH)_2$ redox system. In their model, Olin-Sandoval *et al.*[161] predicted that TR, owing to its high catalytic efficiency and abundance in cells, must be inhibited by more than 98% in order to decrease $T(SH)_2$ by 50%. To have an efficient anti-parasitic activity, a TR inhibitor must thus be highly specific and potent or be used at very high concentrations. Regardless of how challenging it may be, the quest for TR inhibitors is far from ending, as illustrated by the continuous publications over the years reporting new chemical entities.[156,162–165] The structural information on TR with inhibitors bound into its active site[73,150,166] should contribute to fostering drug discovery against this target enzyme.

16.2.3.3 *Biosynthesis of Trypanothione*

Trypanothione is build up of two molecules of GSH bound to spermidine. Inhibition of uptake or *de novo* biosynthesis of any of these building blocks will unavoidably affect the intracellular availability of $T(SH)_2$, and hence parasite antioxidant defence and overall redox homeostasis. Next, we review the pathways for uptake and biosynthesis of Cys (the precursor of GSH), of GSH and of spermidine in *Leishmania* spp.

16.2.3.3.1 Cysteine Uptake and Biosynthesis. The sulphur-containing amino acid cysteine is the component of $T(SH)_2$ that confers on this dithiol the ability to act as a reductant for numerous biomolecules. Cysteine is incorporated in $T(SH)_2$ as part of its glutathione arms. In *Leishmania* uptake of cysteine from the milieu is insufficient,[167] making the supply of this amino acid largely dependent on its *de novo* biosynthesis *via* two distinct pathways.

One of the routes for cysteine biosynthesis in *Leishmania* is the *cysteine assimilatory pathway*. It consists of a two-step reaction starting from serine and relying on the activity of the enzymes serine acetyltransferase (SAT) and cysteine synthase (CS).[167] The fact that this pathway is absent from mammals and predominantly active in the medically relevant stage of *Leishmania*,[125,168] makes it an attractive target for specific anti-parasitic drugs. One possible strategy towards inhibition of this route may be by blocking the activity of CS with peptides derived from its partner enzyme SAT.[169,170] It is worthy of note that the essentiality of this biosynthetic pathway has never been genetically validated and one possible scenario is that it shares functional redundancy with the alternative *reverse trans-sulfuration* (*RTS*) *pathway* for cysteine biosynthesis. The latter utilizes homocysteine as precursor and the sequential activity of the cystathionine-β-synthase (CBS) and cystathionine-γ-lyase (CGL) enzymes. Even though this pathway is conserved in mammals, it may offer opportunities for specific inhibition based on structural and functional peculiarities of the *Leishmania* CBS enzyme.[167] Moreover, DL-propargylglycine (PAG), an irreversible inhibitor of CGLs, inhibits proliferation of *Leishmania* promastigotes,[171] providing evidence that this biosynthetic route is druggable.

Overall, cysteine biosynthesis in *Leishmania* stands as a potential therapeutic target whose inhibition would disturb the redox homeostasis of parasites. This is supported by the observation that parasites respond to oxidative and nitrosative stress by increasing the expression and activity of some of the enzymes required for the *de novo* synthesis of this amino acid.[168] One chief function of cysteine is to yield glutathione. Whether targeting cysteine biosynthesis in *Leishmania* decreases the pool of glutathione (and, subsequently, trypanothione) remains to be shown. Such a possibility may be verified under conditions where *Leishmania* requirements for glutathione cannot be compensated by scavenging the extracellular environment for exogenous thiol or for its γ-glutamylcysteine precursor.

16.2.3.3.2 Glutathione Uptake and Biosynthesis. Glutathione is a tripeptide made of cysteine, glutamate and glycine, which, in most organisms, acts as a major intracellular antioxidant and has direct participation in vital redox reactions. In *Leishmania* GSH is typically regarded as a mere precursor of trypanothione. However, GSH *per se* may also have a direct role in parasite redox homeostasis, as supported by the presence of high levels of free peptide in these organisms[172,173] and also by the existence of a GSH-dependent reductase implicated in reduction and activation of antimonial prodrugs used to treat leishmaniases.[174,175] *Leishmania* are versatile with respect to GSH supply, relying on both uptake and *de novo* biosynthesis of this molecule.

Acquisition of GSH or of its γ-glutamylcysteine precursor by *Leishmania* from the microenvironment occurs through an as yet unidentified transporter, whose presence can be inferred from the observations that (i) providing extracellular GSH or γ-glutamylcysteine to *Leishmania* increases the levels of intracellular thiols,[172] and that (ii) supplementation with GSH alleviates the cytotoxic effects of intracellular oxidative stress[172] and of antimony.[176] The scavenging of GSH from the milieu may be physiologically relevant, particularly in the intracellular amastigote stage where the concentration of GSH is expected to be within the millimolar range.

Biosynthesis of GSH in *Leishmania* follows a ubiquitous mechanism consisting of two sequential reactions that involve two ATP-dependent enzymes: (i) the rate-limiting conjugation of L-glutamate and L-cysteine to yield γ-glutamylcysteine, catalysed by γ-glutamylcysteine synthetase (GshA), and (ii) the ligation between γ-glutamylcysteine and L-glycine, catalyzed by glutathione synthetase (GshB). *Leishmania* harbours a gene coding for a putative GshB enzyme, which remains uncharacterized. GshB is expected to be a difficult target for drug discovery, based on the fact that the related *T. brucei* molecule shares many structural traits with the human orthologue.[177] Unlike the poorly characterized GshB enzyme, there is more information available on *Leishmania* GshA. Firstly, its expression positively correlates with the intracellular levels of GSH and $T[SH]_2$.[172,178] Secondly, decreased *GshA* expression impairs *Leishmania* resistance to oxidative or nitrosative stress and ability to survive in activated macrophages.[172] Thirdly, and against the expectation that GSH biosynthesis could be compensated for by its uptake, GshA was proven to be essential for *Leishmania* survival.[172] In this reverse genetics study, the authors reported that depletion of GshA could not be metabolically rescued by uptake of exogenous GSH or γ-glutamylcysteine, which led them to suggest that the function of GshA in *Leishmania* could go beyond GSH biosynthesis. Whichever critical function GshA plays in *Leishmania*, its chemical inhibition with L-buthionine sulfoximine (BSO) has a negative effect on promastigote and amastigote growth[179] and tolerance to antimonials.[180]

In summary, the demands of *Leishmania* for GSH can be supported by its uptake and *de novo* biosynthesis. Although such compensatory mechanisms may compromise the design of a drug-based strategy to control parasite growth through GSH (and, consequently, trypanothione) starvation, the fact that GshA was proven to be essential by both genetic and chemical approaches indicates that its inhibition remains a valid therapeutic strategy for leishmaniases. Specific inhibitors for this enzyme are, however, required as BSO also targets mammalian GshA, albeit in a less pronounced way. Such differential BSO toxicity is possibly due to different requirements for and/or bioavailability of GSH between vertebrates and parasites and to specific kinetic traits of the enzymes, as suggested by studies of the closely related *T. brucei* GshA molecule.[181,182]

16.2.3.3.3 Spermidine Uptake and Biosynthesis. Spermidine is the non-thiol component of $T[SH]_2$ that serves as linker between the two GSH arms that compose this molecule. In *Leishmania*, spermidine is the end-product of the polyamine biosynthetic pathway, which initiates with L-arginine and involves the activity of four enzymes: arginase (ARG), ornithine decarboxylase (ODC), *S*-adenosylmethionine decarboxylase (AdoMetDC), and spermidine synthase (SpdS). The importance of spermidine goes beyond supplying building blocks for $T[SH]_2$ biosynthesis. Spermidine is also essential (and sufficient) to support the requirements of promastigotes for polyamines,[183] a family of organic cations implicated in a myriad of physiologically relevant phenomena.

Following the interest in the spermidine biosynthetic pathway for drug development, all its four enzymatic components were functionally analysed. Genetic abrogation of each individual enzyme in *Leishmania* promastigotes was shown to lead to conditionally lethal phenotypes[183–191] that could be rescued by medium supplementation with spermidine or its intermediate precursors.[183,184,186,188–191] The ability of *Leishmania* spp. to circumvent inhibition of spermidine biosynthesis by importing this polyamine through the activity of a dedicated transporter[192,193] may compromise the leishmanicidal efficacy of drugs aimed at targeting such a metabolic route. Nevertheless, mutants genetically deprived of either ARG, ODC or SdpS exhibit attenuated virulence in mice[184,185,187–191,194] indicating that the pool of spermidine in phagolysosomal compartments is insufficient to rescue polyamine auxotrophy and encouraging the quest for inhibitors of the polyamine biosynthetic pathway.

Among the enzymes of the spermidine biosynthetic pathway, ODC and AdoMetDC are perhaps the ones holding more promise as therapeutic targets as they can be inhibited without harm to the hosts, thanks to the different turnover rates exhibited by parasite and mammalian ODCs and AdoMetDCs.[191,195,196] Several trypanosomatidal drugs inhibit ODC and AdoMetDC. This is the case for pentamidine, initially developed as an antitrypanosomal drug and presently considered a second-line treatment for leishmaniases, whose mode of action includes AdoMetDC inhibition (among other mechanisms).[197–203] This is also the case for D,L-α-difluoromethylornithine (DFMO or eflornithine), a therapeutic drug that inhibits ODC and has been on the market for more than two decades to treat sleeping sickness caused by *T. brucei gambiense*. DFMO is also an active inhibitor of *Leishmania* ODC,[204] exhibiting cytotoxic effects against promastigotes,[204–207] and ameliorating, albeit not eliminating, *Leishmania* infections in mice[194,208] and hamsters.[209] *Leishmania* may find ways to overcome DFMO toxicity by inducing ODC expression[210] or by a more complex mechanism involving other proteins.[211]

Another promising target within the *Leishmania* spermidine biosynthetic route is ARG. In this case, parasite-specificity is not a requirement for a drug because the combined inhibition of both host and parasite ARGs may potentiate *Leishmania* elimination,[212–214] as supported by the observations that an active macrophage ARG favours *Leishmania* infections *via* the provision of polyamine precursors and/or through deviation of its arginine substrate from the synthesis of toxic nitric oxide.[212,214–218]

Overall, genetic and chemical evidence indicates that lethal inhibition of the spermidine biosynthetic pathway of *Leishmania* spp. is a realistic, albeit challenging, task. Recently, the structures of some of the enzymes were elucidated in *Leishmania* spp. or in other trypanosomatids.[219–223] This, added to the identification of new small-molecule inhibitors of ARG,[224] ODC,[225,226] AdoMetDC[206] and SpdS,[222] brought new dynamics to this area of research. Since these parasites are adapted to scavenge polyamines from the extracellular milieu, a robust spermidine-targeting strategy to treat leishmaniases would probably have to target both endogenous synthesis and transporter-mediated uptake. Regarding the latter, the hydrophobic character of integral membrane proteins is a clear limitation of drug discovery. If such an enterprise is to be undertaken, it should focus on POT1, the molecule responsible for spermidine internalization in *Leishmania* spp., whose primary structure exhibits some divergence from polyamine permeases from other organisms.[193]

16.2.3.3.4 Conjugation of Glutathione and Spermidine to Yield Trypanothione. Biosynthesis of $T(SH)_2$ occurs in two sequential ATP-dependent conjugations of glutathione to spermidine *via* a glutathionylspermidine (Gsp) intermediate. The first step of synthesis can be catalysed by glutathionylspermidine synthetase (GspS) or by trypanothione synthetase (TryS), whereas the second is driven exclusively by TryS. TryS is ubiquitous along the entire *Leishmania* spp. lineage, but this is not the case for GspS, whose coding sequence is present in some species (*e.g.*, *L. infantum*, *L. donovani* and *Leishmania mexicana*), but either missing or dysfunctional in others (*e.g.*, *L. major*, *L. braziliensis* and *Leishmania tarantolae*) (tritrypdb.org/tritrypdb). For some time it was thought that the presence of GspS could confer advantages on parasites or even allow them to overcome the pharmacological inhibition of TryS. These assumptions were based on the observations that Gsp, the end-product of GspS, may be utilised for the assembly of iron–sulphur proteins[58] or the thiolation of sulfenic acids,[95,227] and also take over some of the redox activities of $T(SH)_2$ *in vitro*,[91,228,229] including recycling by trypanothione reductase.[153,230–232] This ambiguity was recently solved by the demonstration that TryS is *absolutely* critical for survival of the GspS-harbouring *L. infantum* parasite.[233] Importantly, genetic evidence also indicated that GspS can be abrogated from the *L. infantum* genome without any harm to the parasites, thus lending support to the hypothesis that *GspS* is a non-essential gene on its way to extinction in the trypanosomatid lineage.[58,234]

Leishmania TryS presents several advantages as a drug target candidate, apart from its essentiality: (i) it should be difficult for resistance to emerge, because there is no obvious bypass metabolism and also because TryS is encoded by a single-copy gene, for which there are no reports of gene amplification in response to inhibitors; (ii) it is absent from mammals, hence minimising the hazard of drug toxicity (TryS is remotely related to human glutathione synthetase, with a low sequence identity of about 10%[235]); (iii) its mechanisms of action and binding to substrates are well characterised in structural,[235] molecular dynamics[236] and kinetic studies;[234,237,238] (iv) it can be produced in large amounts as a recombinant protein and can be used in convenient automated high-throughput screening assays.[233,234,237,239] Apart from the synthetase activity conferred by its C-terminal sequence, TryS harbours an N-terminal amidase domain capable of hydrolysing glutathione–spermidine conjugates.[234,235,240] At least in the related *T. brucei* parasite, the amidase activity of TryS is negligible and non-essential for parasite survival,[241] hence it is not interesting to follow up in drug screening campaigns.

Over recent years, several efforts have been made towards the discovery of compounds with inhibitory activity against *Leishmania* TryS (Figure 16.2).

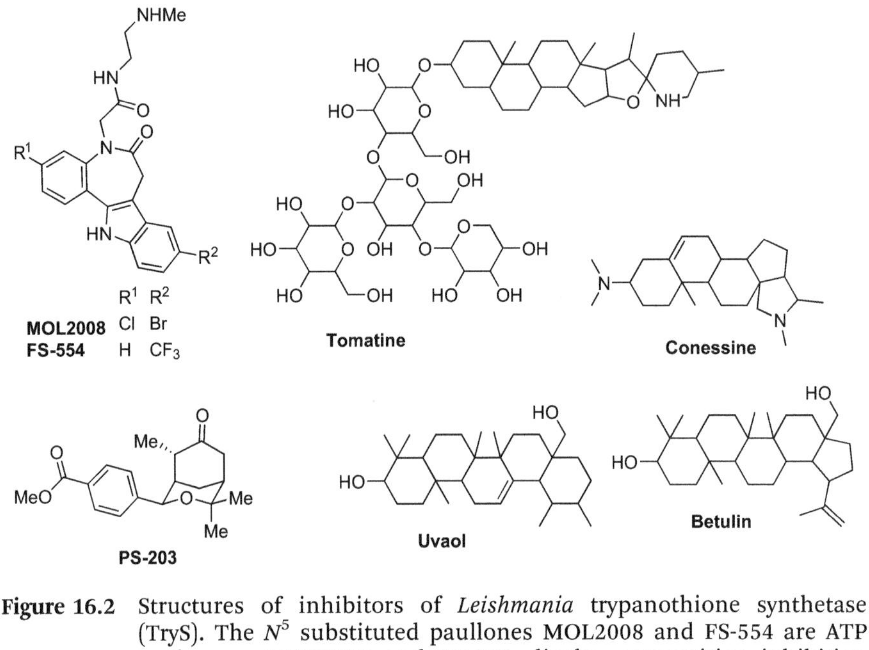

Figure 16.2 Structures of inhibitors of *Leishmania* trypanothione synthetase (TryS). The N^5 substituted paullones MOL2008 and FS-554 are ATP analogues. MOL2008 and FS-554 display competitive inhibition towards *L. infantum* TryS with half maximal inhibitory concentrations (IC_{50}) of 0.15 and 0.35 μM, respectively[233,237] and show mild anti-*Leishmania* activity *in vitro*. The natural compounds tomatine, conessine, uvaol and betulin inhibit *L. donovani* TryS with calculated inhibitory constants (K_i) of 12.54, 3.12, 3.55 and 6.33 μM, respectively.[238] These compounds display competitive inhibition with respect to spermidine and uncompetitive inhibition towards ATP. Tomatine and betulin also present uncompetitive inhibition towards GSH but uvaol and conessine display allosteric inhibition. All four molecules have moderate anti-*Leishmania* activity *in vitro*. PS-203 is an oxabicyclo[3.3.1]nonanone, a class of compounds that interferes with both TryS and TR activities. PS-203 inhibits both enzymes in a competitive way (K_i of 14.2 μM for TR and 26.1 μM for TryS) and is also active *in vitro* towards *L. donovani* amastigotes without affecting macrophages.[243]

Given the conservation of the $T(SH)_2$ redox metabolism along the parasites of the Trypanosomatidae family, the expectation exists that one TryS inhibitor may act as a pharmacological solution to combat *Leishmania*, *T. brucei* and *T. cruzi* infections. This endeavour may, nevertheless, be compromised because, as recently pointed out by Benítez *et al.*,[237] the TryS inhibitory activity of drug-like compounds is remarkably dependent on the parasite species, mirroring the kinetic and structural heterogeneity among the different trypanomatid enzymes. TryS inhibitors thus far identified include molecules that are (i) analogues of Gsp,[242] (ii) simultaneously active against trypanothione reductase,[239–243] or (iii) kinase inhibitors.[233,237] Alternative chemical scaffolds have been identified in low- to high-throughput screening campaigns against TryS from *L. donovani*,[238] *T. brucei*[239,244] or from all three trypanosomatids.[237] Some of these compounds have been proven to be active against *Leishmania* promastigotes,[233,237,238,243] thus providing chemical evidence that TryS is a druggable target in the parasite context. It is worth noting that the anti-parasitic potency of TryS inhibitors is usually weaker than their *in vitro* efficacy,[233,238,239,241,244] a behaviour that may result from cell penetration, stability and bioavailability issues, but also from the parasites' ability to survive with low levels of trypanothione.[239] For a recent review on TryS inhibitors the reader may refer to Leroux and Krauth-Siegel.[156]

16.3 Conclusions

Redox homeostasis in *Leishmania* is largely dependent on the unique molecule $T(SH)_2$. Trypanothione metabolism is crucial for parasite survival and infectivity within their mammalian hosts and is also involved in drug metabolism and resistance. Its inhibition is, therefore, promising from a chemotherapeutic point of view. To adequately prioritize drug targets in redox pathways, validation of the different parasite redox enzymes in the disease-causing stage of *Leishmania* needs to be complete, a venture that might be sped up by the recent adaptation of the CRISPR–Cas9 system to *Leishmania*.[245,246] Up to now several enzymes have been found to be essential to the cell, nevertheless, their drug target potential is not generally ranked as highly as desirable due to issues related to (i) structural characteristics of the enzyme that render inhibitor development particularly challenging (*e.g.*, TXN), (ii) the need to inhibit enzyme activity almost completely (*e.g.*, TR), or (iii) the possibility that *Leishmania* scavenges the synthesised product from the host (*e.g.*, GshA). The current state of research has, nevertheless, identified TryS, the enzyme that conjugates spermidine to GSH to form $T(SH)_2$ as an excellent target, which offers good prospects of success for inhibitor design. Certainly, the simultaneous inhibition of a second redox enzyme or the use of a drug increasing ROS and/or RNS within the parasite would enhance redox imbalance and maximize the chances of exerting successful leishmaniases control.

Acknowledgements

Funding: Norte-01-0145-FEDER-000012—Structured program on bioengineered therapies for infectious diseases and tissue regeneration, supported by Norte Portugal Regional Operational Programme (NORTE 2020), under the PORTUGAL 2020 Partnership Agreement, through the European Regional Development Fund (ERDF). The Fundação para a Ciência e Tecnologia (FCT) provided funding to Helena Castro under the "Investigador FCT" programme.

References

1. L. Carvalho, J. R. Luque-Ortega, J. I. Manzano, S. Castanys, L. Rivas and F. Gamarro, *Antimicrob. Agents Chemother.*, 2010, **54**, 5344.
2. A. Mehta and C. Shaha, *J. Biol. Chem.*, 2004, **279**, 11798.
3. J. Fang and D. S. Beattie, *Mol. Biochem. Parasitol.*, 2003, **127**, 73.
4. J. F. Turrens, *Mol. Biochem. Parasitol.*, 1987, **25**, 55.
5. J. F. Turrens, *Biochem. J.*, 1989, **259**, 363.
6. Z. Verner, P. Cermáková, I. Skodová, B. Kováčová, J. Lukeš and A. Horváth, *Mol. Biochem. Parasitol.*, 2014, **193**, 55.
7. B. Mittra and N. W. Andrews, *Trends Parasitol.*, 2013, **29**, 489.
8. B. Mittra, M. Cortez, A. Haydock, G. Ramasamy, P. Myler and N. W. Andrews, *J. Exp. Med.*, 2013, **210**, 401.
9. J. F. Alzate, A. A. Arias, D. Moreno-Mateos, A. Alvarez-Barrientos and A. Jiménez-Ruiz, *Mol. Biochem. Parasitol.*, 2007, **152**, 192.
10. S. B. Mukherjee, M. Das, G. Sudhandiran and C. Shaha, *J. Biol. Chem.*, 2002, **277**, 24717.
11. L. Carvalho, J. R. Luque-Ortega, C. López-Martín, S. Castanys, L. Rivas and F. Gamarro, *Antimicrob. Agents Chemother.*, 2011, **55**, 4204.
12. Ashutosh, S. Sundar and N. Goyal, *J. Med. Microbiol.*, 2007, **56**, 143.
13. S. Wyllie, M. L. Cunningham and A. H. Fairlamb, *J. Biol. Chem.*, 2004, **279**, 39925.
14. D. Rojo, G. A. Canuto, E. A. Castilho-Martins, M. F. Tavares, C. Barbas, Á. López-Gonzálvez and L. Rivas, *PLoS One*, 2015, **10**, e0130675.
15. T. P. Dorlo, M. Balasegaram, J. H. Beijnen and P. J. de Vries, *J. Antimicrob. Chemother.*, 2012, **67**, 2576.
16. H. Ramos, E. Valdivieso, M. Gamargo, F. Dagger and B. E. Cohen, *J. Membr. Biol.*, 1996, **152**, 65.
17. W. Moreira, P. Leprohon and M. Ouellette, *Cell Death Dis.*, 2011, **2**, e201.
18. M. Genestra, W. J. de Souza, L. Cysne-Finkelstein and L. L. Leon, *Med. Microbiol. Immunol.*, 2003, **192**, 217.
19. M. Genestra, W. J. Souza, D. Guedes-Silva, G. M. Machado, L. Cysne-Finkelstein, R. J. Bezerra, F. Monteiro and L. L. Leon, *Arch. Microbiol.*, 2006, **185**, 348.

20. C. Paveto, C. Pereira, J. Espinosa, A. E. Montagna, M. Farber, M. Esteva, M. M. Flawiá and H. N. Torres, *J. Biol. Chem.*, 1995, **270**, 16576.
21. J. Goldstein, C. Paveto, J. J. López-Costa, C. Pereira, G. Alonso, H. N. Torres and M. M. Flawia, *Biocell*, 2000, **24**, 217.
22. R. J. Soares-Bezerra, E. F. da Silva, A. Echevarria, L. Gomes-da-Silva, L. Cysne-Finkelstein, F. P. Monteiro, L. L. Leon and M. Genestra, *J. Enzyme Inhib. Med. Chem.*, 2008, **23**, 328.
23. P. Nunes, N. Demaurex and M. C. Dinauer, *Traffic*, 2013, **14**, 1118.
24. X. Q. Wei, I. G. Charles, A. Smith, J. Ure, G. J. Feng, F. P. Huang, D. Xu, W. Muller, S. Moncada and F. Y. Liew, *Nature*, 1995, **375**, 408.
25. A. Diefenbach, H. Schindler, N. Donhauser, E. Lorenz, T. Laskay, J. MacMicking, M. Röllinghoff, I. Gresser and C. Bogdan, *Immunity*, 1998, **8**, 77.
26. T. G. Evans, L. Thai, D. L. Granger and J. B. Hibbs Jr., *J. Immunol.*, 1993, **151**, 907.
27. F. Y. Liew, S. Millott, C. Parkinson, R. M. Palmer and S. Moncada, *J. Immunol.*, 1990, **144**, 4794.
28. A. J. Müller, S. Aeschlimann, R. Olekhnovitch, M. Dacher, G. F. Späth and P. Bousso, *Cell Host Microbe*, 2013, **14**, 460.
29. R. Olekhnovitch and P. Bousso, *Trends Parasitol.*, 2015, **31**, 653.
30. R. Olekhnovitch, B. Ryffel, A. J. Müller and P. Bousso, *J. Clin. Invest.*, 2014, **124**, 1711.
31. S. Stenger, N. Donhauser, H. Thüring, M. Röllinghoff and C. Bogdan, *J. Exp. Med.*, 1996, **183**, 1501.
32. J. K. White, P. Mastroeni, J. F. Popoff, C. A. Evans and J. M. Blackwell, *J. Leukoc. Biol.*, 2005, **77**, 311.
33. M. Blos, U. Schleicher, F. J. Soares Rocha, U. Meissner, M. Röllinghoff and C. Bogdan, *Eur. J. Immunol.*, 2003, **33**, 1224.
34. H. W. Murray and C. F. Nathan, *J. Exp. Med.*, 1999, **189**, 741.
35. F. Marques, S. Vale-Costa, T. Cruz, J. M. Marques, T. Silva, J. V. Neves, S. Cortes, A. Fernandes, E. Rocha, R. Appelberg, P. Rodrigues, A. M. Tomás and M. S. Gomes, *Parasites Vectors*, 2015, **8**, 644.
36. G. F. Späth, L. A. Garraway, S. J. Turco and S. M. Beverley, *Proc. Natl. Acad. Sci. U.S.A.*, 2003, **100**, 9536.
37. R. Lodge, T. O. Diallo and A. Descoteaux, *Cell. Microbiol.*, 2006, **8**, 1922.
38. L. Proudfoot, A. V. Nikolaev, G. J. Feng, W. Q. Wei, M. A. Ferguson, J. S. Brimacombe and F. Y. Liew, *Proc. Natl. Acad. Sci. U.S.A.*, 1996, **93**, 10984.
39. I. A. Abreu and D. E. Cabelli, *Biochim. Biophys. Acta*, 2010, **1804**, 263.
40. F. Getachew and L. Gedamu, *Mol. Biochem. Parasitol.*, 2007, **154**, 62.
41. A. Bodyl and P. Mackiewicz, *Parasitology*, 2008, **135**, 1101.
42. W. J. Paramchuk, S. O. Ismail, A. Bhatia and L. Gedamu, *Mol. Biochem. Parasitol.*, 1997, **90**, 203.
43. F. Getachew and L. Gedamu, *Mol. Biochem. Parasitol.*, 2012, **183**, 42.
44. S. Ghosh, S. Goswami and S. Adhya, *Biochem. J.*, 2003, **369**, 447.

45. N. G. Tessarollo, J. M. Andrade, D. S. Moreira and S. M. Murta, *Parasitol. Int.*, 2015, **64**, 125.
46. K. A. Plewes, S. D. Barr and L. Gedamu, *Infect. Immun.*, 2003, **71**, 5910.
47. P. Navarro, M. Sánchez-Moreno, C. Marín, E. García-España, I. Ramírez-Macías, F. Olmo, M. J. Rosales, F. Gómez-Contreras, M. J. Yunta and R. Gutierrez-Sánchez, *Parasitology*, 2014, **141**, 1031.
48. M. Bose, R. Saha, S. Sen Santara, S. Mukherjee, J. Roy and S. Adak, *Free Radical Biol. Med.*, 2012, **53**, 1819.
49. S. Adak and A. K. Datta, *Biochem. J.*, 2005, **390**, 465.
50. V. S. Jasion, J. A. Polanco, Y. T. Meharenna, H. Li and T. L. Poulos, *J. Biol. Chem.*, 2011, **286**, 24608.
51. V. S. Jasion and T. L. Poulos, *Biochemistry*, 2012, **51**, 2453.
52. S. Dolai, R. K. Yadav, S. Pal and S. Adak, *Free Radical Biol. Med.*, 2008, **45**, 1520.
53. S. Dolai, R. K. Yadav, S. Pal and S. Adak, *Eukaryotic Cell*, 2009, **8**, 1721.
54. S. Pal, S. Dolai, R. K. Yadav and S. Adak, *PLoS One*, 2010, **5**, e11271.
55. A. Mukherjee, S. Boisvert, R. L. Monte-Neto, A. C. Coelho, F. Raymond, R. Mukhopadhyay, J. Corbeil and M. Ouellette, *Mol. Microbiol.*, 2013, **88**, 189.
56. A. Kumar, S. Das, B. Purkait, A. H. Sardar, A. K. Ghosh, M. R. Dikhit, K. Abhishek and P. Das, *Antimicrob. Agents Chemother.*, 2014, **58**, 6172.
57. A. H. Fairlamb, P. Blackburn, P. Ulrich, B. T. Chait and A. Cerami, *Science*, 1985, **227**, 1485.
58. B. Manta, M. Comini, A. Medeiros, M. Hugo, M. Trujillo and R. Radi, *Biochim. Biophys. Acta*, 2013, **1830**, 3199.
59. R. L. Krauth-Siegel and M. A. Comini, *Biochim. Biophys. Acta*, 2008, **1780**, 1236.
60. J. M. Kelly, M. C. Taylor, K. Smith, K. J. Hunter and A. H. Fairlamb, *Eur. J. Biochem.*, 1993, **218**, 29.
61. M. Moutiez, M. Aumercier, R. Schöneck, D. Meziane-Cherif, V. Lucas, P. Aumercier, A. Ouaissi, C. Sergheraert and A. Tartar, *Biochem. J.*, 1995, **310**, 433.
62. M. Moutiez, E. Quéméneur, C. Sergheraert, V. Lucas, A. Tartar and E. Davioud-Charvet, *Biochem. J.*, 1997, **322**, 43.
63. M. A. Comini, R. L. Krauth-Siegel and M. Bellanda, *Antioxid. Redox Signal.*, 2013, **19**, 708.
64. M. R. Ariyanayagam and A. H. Fairlamb, *Mol. Biochem. Parasitol.*, 2001, **115**, 189.
65. D. J. Steenkamp, *Antioxid. Redox Signal.*, 2002, **4**, 105.
66. R. N. Vogt and D. J. Steenkamp, *Biochem. J.*, 2003, **371**, 49.
67. T. J. Vickers and A. H. Fairlamb, *J. Biol. Chem.*, 2004, **279**, 27246.
68. T. J. Vickers, S. Wyllie and A. H. Fairlamb, *J. Biol. Chem.*, 2004, **279**, 49003.
69. M. Sousa Silva, A. E. N. Ferreira, R. Gomes, A. M. Tomás, C. Cordeiro, *Drug Discovery in Infectious Disease – Trypanosomatid Diseases*, ed.

T. Jäger, L. Flohé, and O. Koch, Wiley-VCH Verlag GmbH & Co. KGaA, Weinheim, 1st edn, 2013, ch. 8, vol. 4, pp. 153–166.
70. S. Ferreira Cdos, P. S. Martins, C. Demicheli, C. Brochu, M. Ouellette and F. Frézard, *Biometals*, 2003, **16**, 441.
71. S. Yan, F. Li, K. Ding and H. Sun, *J. Biol. Inorg. Chem.*, 2003, **8**, 689.
72. M. L. Cunningham and A. H. Fairlamb, *Eur. J. Biochem.*, 1995, **230**, 460.
73. P. Baiocco, G. Colotti, S. Franceschini and A. Ilari, *J. Med. Chem.*, 2009, **52**, 2603.
74. R. Mukhopadhyay, S. Dey, N. Xu, D. Gage, J. Lightbody, M. Ouellette and B. P. Rosen, *Proc. Natl. Acad. Sci. U.S.A.*, 1996, **93**, 10383.
75. A. Perea, J. I. Manzano, S. Castanys and F. Gamarro, *Antimicrob. Agents Chemother.*, 2016, **60**, 3489.
76. D. Légaré, D. Richard, R. Mukhopadhyay, Y. D. Stierhof, B. P. Rosen, A. Haimeur, B. Papadopoulou and M. Ouellette, *J. Biol. Chem.*, 2001, **276**, 26301.
77. M. Berg, M. Vanaerschot, A. Jankevics, B. Cuypers, I. Maes, S. Mukherjee, B. Khanal, S. Rijal, S. Roy, F. Opperdoes, R. Breitling and J. C. Dujardin, *Mol. Microbiol.*, 2013, **90**, 428.
78. S. Decuypere, S. Rijal, V. Yardley, S. De Doncker, T. Laurent, B. Khanal, F. Chappuis and J. C. Dujardin, *Antimicrob. Agents Chemother.*, 2005, **49**, 4616.
79. S. Decuypere, M. Vanaerschot, K. Brunker, H. Imamura, S. Müller, B. Khanal, S. Rijal, J. C. Dujardin and G. H. Coombs, *PLoS Neglected Trop. Dis.*, 2012, **6**, e1514.
80. S. Wyllie, G. Mandal, N. Singh, S. Sundar, A. H. Fairlamb and M. Chatterjee, *Mol. Biochem. Parasitol.*, 2010, **173**, 162.
81. S. Wyllie, T. J. Vickers and A. H. Fairlamb, *Antimicrob. Agents Chemother.*, 2008, **52**, 1359.
82. A. Equbal, S. S. Suman, S. Anwar, K. P. Singh, A. Zaidi, A. H. Sardar, P. Das and V. Ali, *PLoS One*, 2014, **9**, e97600.
83. A. Kulshrestha, A. Sharma, R. Singh and P. Salotra, *Parasitol. Res.*, 2014, **113**, 1171.
84. E. G. Carnieri, S. N. Moreno and R. Docampo, *Mol. Biochem. Parasitol.*, 1993, **61**, 79.
85. M. Trujillo, H. Budde, M. D. Piñeyro, M. Stehr, C. Robello, L. Flohé and R. Radi, *J. Biol. Chem.*, 2004, **279**, 34175.
86. L. Flohé, H. Budde, K. Bruns, H. Castro, J. Clos, B. Hofmann, S. Kansal-Kalavar, D. Krumme, U. Menge, K. Plank-Schumacher, H. Sztajer, J. Wissing, C. Wylegalla and H. J. Hecht, *Arch. Biochem. Biophys.*, 2002, **397**, 324.
87. H. Castro, C. Sousa, M. Santos, A. Cordeiro-da-Silva, L. Flohé and A. M. Tomás, *Free Radical Biol. Med.*, 2002, **33**, 1552.
88. H. Castro, H. Budde, L. Flohé, B. Hofmann, H. Lünsdorf, J. Wissing and A. M. Tomás, *Free Radical Biol. Med.*, 2002, **33**, 1563.
89. H. Castro and A. M. Tomás, *Antioxid. Redox Signal.*, 2008, **10**, 1593.
90. J. König and A. H. Fairlamb, *FEBS J.*, 2007, **274**, 5643.

91. M. Dormeyer, N. Reckenfelderbäumer, H. Lüdemann and R. L. Krauth-Siegel, *J. Biol. Chem.*, 2001, **276**, 10602.
92. I. Onn, N. Milman-Shtepel and J. Shlomai, *Eukaryotic Cell*, 2004, **3**, 277.
93. R. Singh, B. Purkait, K. Abhishek, S. Saini, S. Das, S. Verma, A. Mandal, A. K. Ghosh, Y. Ansari, A. Kumar, A. H. Sardar, A. Kumar, P. Parrack and P. Das, *Cell Biosci.*, 2016, **6**, 13.
94. M. Filser, M. A. Comini, M. M. Molina-Navarro, N. Dirdjaja, E. Herrero and R. L. Krauth-Siegel, *Biol. Chem.*, 2008, **389**, 21.
95. J. Melchers, N. Dirdjaja, T. Ruppert and R. L. Krauth-Siegel, *J. Biol. Chem.*, 2007, **282**, 8678.
96. D. G. Arias, M. S. Cabeza, E. D. Erben, P. G. Carranza, H. D. Lujan, M. T. Téllez Iñón, A. A. Iglesias and S. A. Guerrero, *Free Radical Biol. Med.*, 2011, **50**, 37.
97. D. G. Arias, M. D. Piñeyro, A. A. Iglesias, S. A. Guerrero and C. Robello, *J. Proteomics*, 2015, **120**, 95.
98. M. D. Piñeyro, A. Parodi-Talice, M. Portela, D. G. Arias, S. A. Guerrero and C. Robello, *J. Proteomics*, 2011, **74**, 1683.
99. H. Castro, S. Romao, S. Carvalho, F. Teixeira, C. Sousa and A. M. Tomás, *PLoS One*, 2010, **5**, e12607.
100. H. Castro, C. Sousa, M. Novais, M. Santos, H. Budde, A. Cordeiro-da-Silva, L. Flohé and A. M. Tomás, *Mol. Biochem. Parasitol.*, 2004, **136**, 137.
101. Z. González-Chávez, V. Olin-Sandoval, J. S. Rodíguez-Zavala, R. Moreno-Sánchez and E. Saavedra, *Biochim. Biophys. Acta*, 2015, **1850**, 263.
102. D. U. Gommel, E. Nogoceke, M. Morr, M. Kiess, H. M. Kalisz and L. Flohé, *Eur. J. Biochem.*, 1997, **248**, 913.
103. E. Nogoceke, D. U. Gommel, M. Kiess, H. M. Kalisz and L. Flohé, *Biol. Chem.*, 1997, **378**, 827.
104. P. Steinert, K. Plank-Schumacher, M. Montemartini, H. J. Hecht and L. Flohé, *Biol. Chem.*, 2000, **381**, 211.
105. S. Romao, H. Castro, C. Sousa, S. Carvalho and A. M. Tomás, *Int. J. Parasitol.*, 2009, **39**, 703.
106. M. A. Comini, R. L. Krauth-Siegel and L. Flohé, *Biochem. J.*, 2007, **402**, 43.
107. S. R. Wilkinson, D. Horn, S. R. Prathalingam and J. M. Kelly, *J. Biol. Chem.*, 2003, **278**, 31640.
108. M. S. Alphey, G. A. Leonard, D. G. Gourley, E. Tetaud, A. H. Fairlamb and W. N. Hunter, *J. Biol. Chem.*, 1999, **274**, 25613.
109. B. Hofmann, H. Budde, K. Bruns, S. A. Guerrero, H. M. Kalisz, U. Menge, M. Montemartini, E. Nogoceke, P. Steinert, J. B. Wissing, L. Flohé and H. J. Hecht, *Biol. Chem.*, 2001, **382**, 459.
110. D. Krumme, H. Budde, H. J. Hecht, U. Menge, U. O. Ohlenschläger, A. Ross, J. Wissing, V. Wray and L. Flohé, *Biochemistry*, 2003, **42**, 14720.
111. M. S. Alphey, M. Gabrielsen, E. Micossi, G. A. Leonard, S. M. McSweeney, R. B. Ravelli, E. Tetaud, A. H. Fairlamb, C. S. Bond and W. N. Hunter, *J. Biol. Chem.*, 2003, **278**, 25919.

112. F. Fueller, B. Jehle, K. Putzker, J. D. Lewis and R. L. Krauth-Siegel, *J. Biol. Chem.*, 2012, **287**, 8792.
113. J. Singh, R. C. Petter, T. A. Baillie and A. Whitty, *Nat. Rev. Drug Discovery*, 2011, **10**, 307.
114. M. P. Levick, E. Tetaud, A. H. Fairlamb and J. M. Blackwell, *Mol. Biochem. Parasitol.*, 1998, **96**, 125.
115. M. Brindisi, S. Brogi, N. Relitti, A. Vallone, S. Butini, S. Gemma, E. Novellino, G. Colotti, G. Angiulli, F. Di Chiaro, A. Fiorillo, A. Ilari and G. Campiani, *Sci. Rep.*, 2015, **5**, 9705.
116. A. Fiorillo, G. Colotti, A. Boffi, P. Baiocco and A. Ilari, *PLoS Neglected Trop. Dis.*, 2012, **6**, e1781.
117. M. A. Morais, P. O. Giuseppe, T. A. Souza, T. G. Alegria, M. A. Oliveira, L. E. Netto and M. T. Murakami, *J. Biol. Chem.*, 2015, **290**, 8582.
118. G. Eslami, F. Frikha, R. Salehi, A. Khamesipour, H. Hejazi and M. A. Nilforoushzadeh, *Mol. Biol. Rep.*, 2011, **38**, 3765.
119. B. K. Singh and V. K. Dubey, *Curr. Pharm. Biotechnol.*, 2009, **10**, 626.
120. A. Perkins, K. J. Nelson, D. Parsonage, L. B. Poole and P. A. Karplus, *Trends Biochem. Sci.*, 2015, **40**, 435.
121. C. A. Henard, E. D. Carlsen, C. Hay, P. E. Kima and L. Soong, *PLoS Neglected Trop. Dis.*, 2014, **8**, e3000.
122. J. P. Iyer, A. Kaprakkaden, M. L. Choudhary and C. Shaha, *Mol. Microbiol.*, 2008, **68**, 372.
123. S. D. Barr and L. Gedamu, *J. Biol. Chem.*, 2003, **278**, 10816.
124. P. Pescher, T. Blisnick, P. Bastin and G. F. Späth, *Cell. Microbiol.*, 2011, **13**, 978.
125. J. Walker, N. Acestor, R. Gongora, M. Quadroni, I. Segura, N. Fasel and N. G. Saravia, *Mol. Biochem. Parasitol.*, 2006, **145**, 254.
126. V. Gómez Pérez, R. García-Hernandez, V. Corpas-López, A. M. Tomás, J. Martín-Sanchez, S. Castanys and F. Gamarro, *Int. J. Parasitol.: Drugs Drug Resist.*, 2016, **6**, 133.
127. H. Castro, F. Teixeira, S. Romao, M. Santos, T. Cruz, M. Flórido, R. Appelberg, P. Oliveira, F. Ferreira-da-Silva and A. M. Tomás, *PLoS Pathog.*, 2011, 7, e1002325.
128. F. Teixeira, H. Castro, T. Cruz, E. Tse, P. Koldewey, D. R. Southworth, A. M. Tomás and U. Jakob, *Proc. Natl. Acad. Sci. U.S.A.*, 2015, **112**, E616.
129. Q. Zhao, Y. Ding, Z. S. Deng, O. Y. Lee, P. Gao, P. Chen, R. J. Rose, H. Zhao, Z. H. Zhang, X. P. Tao, A. J. R. Heck, R. Kao and D. Yang, *Chem. Sci.*, 2015, **6**, 4124.
130. J. D. Haraldsen, G. Liu, C. H. Botting, J. G. Walton, J. Storm, T. J. Phalen, L. Y. Kwok, D. Soldati-Favre, N. H. Heintz, S. Müller, N. J. Westwood and G. E. Ward, *Org. Biomol. Chem.*, 2009, **7**, 3040.
131. C. X. Liu, Q. Q. Yin, H. C. Zhou, Y. L. Wu, J. X. Pu, L. Xia, W. Liu, X. Huang, T. Jiang, M. X. Wu, L. C. He, Y. X. Zhao, X. L. Wang, W. L. Xiao, H. Z. Chen, Q. Zhao, A. W. Zhou, L. S. Wang, H. D. Sun and G. Q. Chen, *Nat. Chem. Biol.*, 2012, **8**, 486.

132. J. König, S. Wyllie, G. Wells, M. F. Stevens, P. G. Wyatt and A. H. Fairlamb, *J. Biol. Chem.*, 2011, **286**, 8523.
133. M. Diechtierow and R. L. Krauth-Siegel, *Free Radical Biol. Med.*, 2011, **51**, 856.
134. C. Hiller, A. Nissen, D. Benítez, M. A. Comini and R. L. Krauth-Siegel, *PLoS Pathog.*, 2014, **10**, e1004075.
135. C. Schaffroth, M. Bogacz, N. Dirdjaja, A. Nißen and R. L. Krauth-Siegel, *Mol. Microbiol.*, 2016, **99**, 172.
136. S. R. Wilkinson, D. J. Meyer, M. C. Taylor, E. V. Bromley, M. A. Miles and J. M. Kelly, *J. Biol. Chem.*, 2002, **277**, 17062.
137. G. Passaia and M. Margis-Pinheiro, *Plant Sci.*, 2015, **234**, 22.
138. R. Brigelius-Flohé and M. Maiorino, *Biochim. Biophys. Acta*, 2013, **1830**, 3289.
139. S. Patel, S. Hussain, R. Harris, S. Sardiwal, J. M. Kelly, S. R. Wilkinson, P. C. Driscoll and S. Djordjevic, *Biochem. J.*, 2010, **425**, 513.
140. M. S. Alphey, J. König and A. H. Fairlamb, *Biochem. J.*, 2008, **414**, 375.
141. J. Melchers, M. Diechtierow, K. Fehér, I. Sinning, I. Tews, R. L. Krauth-Siegel and C. Muhle-Goll, *J. Biol. Chem.*, 2008, **283**, 30401.
142. I. Liu, M. Bogacz, C. Schaffroth, N. Dirdjaja and R. L. Krauth-Siegel, *Mol. Biochem. Parasitol.*, 2016, **207**, 84.
143. S. R. Wilkinson, M. C. Taylor, S. Touitha, I. L. Mauricio, D. J. Meyer and J. M. Kelly, *Biochem. J.*, 2002, **364**, 787.
144. C. Dumas, M. Ouellette, J. Tovar, M. L. Cunningham, A. H. Fairlamb, S. Tamar, M. Olivier and B. Papadopoulou, *EMBO J.*, 1997, **16**, 2590.
145. J. Tovar, S. Wilkinson, J. C. Mottram and A. H. Fairlamb, *Mol. Microbiol.*, 1998, **29**, 653.
146. J. Tovar, M. L. Cunningham, A. C. Smith, S. L. Croft and A. H. Fairlamb, *Proc. Natl. Acad. Sci. U.S.A.*, 1998, **95**, 5311.
147. J. Kuriyan, X. P. Kong, T. S. Krishna, R. M. Sweet, N. J. Murgolo, H. Field, A. Cerami and G. B. Henderson, *Proc. Natl. Acad. Sci. U.S.A.*, 1991, **88**, 8764.
148. Y. Zhang, C. S. Bond, S. Bailey, M. L. Cunningham, A. H. Fairlamb and W. N. Hunter, *Protein Sci.*, 1996, **5**, 52.
149. V. S. Stoll, S. J. Simpson, R. L. Krauth-Siegel, C. T. Walsh and E. F. Pai, *Biochemistry*, 1997, **36**, 6437.
150. P. Baiocco, G. Poce, S. Alfonso, M. Cocozza, G. C. Porretta, G. Colotti, M. Biava, F. Moraca, M. Botta, V. Yardley, A. Fiorillo, A. Lantella, F. Malatesta and A. Ilari, *ChemMedChem*, 2013, **8**, 1175.
151. D. Spinks, E. J. Shanks, L. A. T. Cleghorn, S. McElroy, D. Jones, D. James, A. H. Fairlamb, J. A. Frearson, P. G. Wyatt and I. H. Gilbert, *ChemMedChem*, 2009, **4**, 2060.
152. C. Eberle, J. A. Burkhard, B. Stump, M. Kaiser, R. Brun, R. L. Krauth-Siegel and F. Diederich, *ChemMedChem*, 2009, **4**, 2034.
153. S. L. Shames, A. H. Fairlamb, A. Cerami and C. T. Walsh, *Biochemistry*, 1986, **25**, 3519.

154. C. J. Hamilton, A. Saravanamuthu, I. M. Eggleston and A. H. Fairlamb, *Biochem. J.*, 2003, **369**, 529.
155. R. Perez-Pineiro, A. Burgos, D. C. Jones, L. C. Andrew, H. Rodriguez, M. Suarez, A. H. Fairlamb and D. S. Wishart, *J. Med. Chem.*, 2009, **52**, 1670.
156. A. E. Leroux and R. L. Krauth-Siegel, *Mol. Biochem. Parasitol.*, 2016, **206**, 67.
157. E. Persch, S. Bryson, N. K. Todoroff, C. Eberle, J. Thelemann, N. Dirdjaja, M. Kaiser, M. Weber, H. Derbani, R. Brun, G. Schneider, E. F. Pai, R. L. Krauth-Siegel and F. Diederich, *ChemMedChem*, 2014, **9**, 1880.
158. M. C. Taylor, J. M. Kelly, C. J. Chapman, A. H. Fairlamb and M. A. Miles, *Mol. Biochem. Parasitol.*, 1994, **64**, 293.
159. D. C. Jones, A. Ariza, W. H. Chow, S. L. Oza and A. H. Fairlamb, *Mol. Biochem. Parasitol.*, 2010, **169**, 12.
160. S. Krieger, W. Schwarz, M. R. Ariyanayagam, A. H. Fairlamb, R. L. Krauth-Siegel and C. Clayton, *Mol. Microbiol.*, 2000, **35**, 542.
161. V. Olin-Sandoval, Z. González-Chávez, M. Berzunza-Cruz, I. Martínez, R. Jasso-Chávez, I. Becker, B. Espinoza, R. Moreno-Sánchez and E. Saavedra, *FEBS J.*, 2012, **279**, 1811.
162. R. K. Pandey, P. Verma, D. Sharma, T. K. Bhatt, S. Sundar and V. K. Prajapati, *Biomed. Pharmacother.*, 2016, **83**, 141.
163. R. K. Pandey, B. V. Kumbhar, S. Sundar, A. Kunwar and V. K. Prajapati, *J. Recept. Signal Transduction Res.*, 2016, **5**, 1.
164. R. K. Pandey, B. V. Kumbhar, S. Srivastava, R. Malik, S. Sundar, A. Kunwar and V. K. Prajapati, *J. Biomol. Struct. Dyn.*, 2016, **4**, 1.
165. A. Marcu, U. Schurigt, K. Müller, H. Moll, R. L. Krauth-Siegel and H. Prinz, *Eur. J. Med. Chem.*, 2016, **108**, 436.
166. A. Saravanamuthu, T. J. Vickers, C. S. Bond, M. R. Peterson, W. N. Hunter and A. H. Fairlamb, *J. Biol. Chem.*, 2004, **279**, 29493.
167. R. A. Williams, G. D. Westrop and G. H. Coombs, *Biochem. J.*, 2009, **420**, 451.
168. I. Romero, J. Téllez, A. J. Romanha, M. Steindel and E. C. Grisard, *Antimicrob. Agents Chemother.*, 2015, **59**, 4770.
169. I. Raj, S. Kumar and S. Gourinath, *Acta Crystallogr., Sect. D: Biol. Crystallogr.*, 2012, **68**, 909.
170. P. K. Fyfe, G. D. Westrop, T. Ramos, S. Müller, G. H. Coombs and W. N. Hunter, *Acta Crystallogr., Sect. F: Struct. Biol. Cryst. Commun.*, 2012, **68**, 738.
171. L. Giordana, B. S. Mantilla, M. Santana, A. M. Silber and C. Nowicki, *J. Eukaryot. Microbiol.*, 2014, **61**, 204.
172. A. Mukherjee, G. Roy, C. Guimond and M. Ouellette, *Mol. Microbiol.*, 2009, **74**, 914.
173. A. H. Fairlamb and A. Cerami, *Annu. Rev. Microbiol.*, 1992, **46**, 695.
174. P. K. Fyfe, G. D. Westrop, A. M. Silva, G. H. Coombs and W. N. Hunter, *Proc. Natl. Acad. Sci. U.S.A.*, 2012, **109**, 11693.

175. H. Denton, J. C. McGregor and G. H. Coombs, *Biochem. J.*, 2004, **381**, 405.
176. A. Mehta and C. Shaha, *Free Radical Biol. Med.*, 2006, **40**, 1857.
177. P. K. Fyfe, M. S. Alphey and W. N. Hunter, *Mol. Biochem. Parasitol.*, 2010, **170**, 93.
178. K. Grondin, A. Haimeur, R. Mukhopadhyay, B. P. Rosen and M. Ouellette, *EMBO J.*, 1997, **16**, 3057.
179. P. Kapoor, M. Sachdev and R. Madhubala, *Trop. Med. Int. Health*, 2000, **5**, 438.
180. A. Haimeur, C. Guimond, S. Pilote, R. Mukhopadhyay, B. P. Rosen, R. Poulin and M. Ouellette, *Mol. Microbiol.*, 1999, **34**, 726.
181. D. V. Lueder and M. A. Phillips, *J. Biol. Chem.*, 1996, **271**, 17485.
182. D. L. Brekken and M. A. Phillips, *J. Biol. Chem.*, 1998, **273**, 26317.
183. S. C. Roberts, Y. Jiang, A. Jardim, N. S. Carter, O. Heby and B. Ullman, *Mol. Biochem. Parasitol.*, 2001, **115**, 217.
184. J. M. Boitz, P. A. Yates, C. Kline, U. Gaur, M. E. Wilson, B. Ullman and S. C. Roberts, *Infect. Immun.*, 2009, **77**, 756.
185. U. Gaur, S. C. Roberts, R. P. Dalvi, I. Corraliza, B. Ullman and M. E. Wilson, *J. Immunol.*, 2007, **179**, 8446.
186. Y. Jiang, S. C. Roberts, A. Jardim, N. S. Carter, S. Shih, M. Ariyanayagam, A. H. Fairlamb and B. Ullman, *J. Biol. Chem.*, 1999, **274**, 3781.
187. H. M. Muleme, R. M. Reguera, A. Berard, R. Azinwi, P. Jia, I. B. Okwor, S. Beverley and J. E. Uzonna, *J. Immunol.*, 2009, **183**, 8068.
188. R. M. Reguera, R. Balaña-Fouce, M. Showalter, S. Hickerson and S. M. Beverley, *Mol. Biochem. Parasitol.*, 2009, **165**, 48.
189. S. C. Roberts, M. J. Tancer, M. R. Polinsky, K. M. Gibson, O. Heby and B. Ullman, *J. Biol. Chem.*, 2004, **279**, 23668.
190. C. Gilroy, T. Olenyik, S. C. Roberts and B. Ullman, *Infect. Immun.*, 2011, **79**, 2764.
191. S. C. Roberts, J. Scott, J. E. Gasteier, Y. Jiang, B. Brooks, A. Jardim, N. S. Carter, O. Heby and B. Ullman, *J. Biol. Chem.*, 2002, **277**, 5902.
192. M. Basselin, G. H. Coombs and M. P. Barrett, *Mol. Biochem. Parasitol.*, 2000, **109**, 37.
193. M. P. Hasne and B. Ullman, *J. Biol. Chem.*, 2005, **280**, 15188.
194. T. Olenyik, C. Gilroy and B. Ullman, *Mol. Biochem. Parasitol.*, 2011, **176**, 109.
195. S. Hanson, J. Adelman and B. Ullman, *J. Biol. Chem.*, 1992, **267**, 2350.
196. L. Persson, A. Jeppsson and S. Nasizadeh, *Biochem. Soc. Trans.*, 2003, **31**, 411.
197. C. J. Bacchi, H. C. Nathan, N. Yarlett, B. Goldberg, P. P. McCann, A. J. Bitonti and A. Sjoerdsma, *Antimicrob. Agents Chemother.*, 1992, **36**, 2736.
198. C. J. Bacchi, R. H. Barker, A. Rodriguez, B. Hirth, D. Rattendi, N. Yarlett, C. L. Hendrick and E. Sybertz, *Antimicrob. Agents Chemother.*, 2009, **53**, 3269.

199. R. H. Barker Jr., H. Liu, B. Hirth, C. A. Celatka, R. Fitzpatrick, Y. Xiang, E. K. Willert, M. A. Phillips, M. Kaiser, C. J. Bacchi, A. Rodriguez, N. Yarlett, J. D. Klinger and E. Sybertz, *Antimicrob. Agents Chemother.*, 2009, **53**, 2052.
200. A. J. Bitonti, T. L. Byers, T. L. Bush, P. J. Casara, C. J. Bacchi, A. B. Clarkson Jr., P. P. McCann and A. Sjoerdsma, *Antimicrob. Agents Chemother.*, 1990, **34**, 1485.
201. A. J. Bitonti, J. A. Dumont and P. P. McCann, *Biochem. J.*, 1986, **237**, 685.
202. C. Danzin, P. Marchal and P. Casara, *Biochem. Pharmacol.*, 1990, **40**, 1499.
203. P. C. Ulrich, R. W. Grady and A. Cerami, *Drug Dev. Res.*, 1982, **2**, 219.
204. R. M. Reguera, R. B. Fouce, J. C. Cubría, M. L. Bujidos and D. Ordóñez, *Life Sci.*, 1995, **56**, 223.
205. K. Kaur, K. Emmett, P. P. McCann, A. Sjoerdsma and B. Ullman, *J. Protozool.*, 1986, **33**, 518.
206. R. Mukhopadhyay, P. Kapoor and R. Madhubala, *Pharmacol. Res.*, 1996, **34**, 43.
207. C. P. Sánchez, J. Mucci, N. S. González, A. Ochoa, M. M. Zakin and I. D. Algranati, *Biochem. J.*, 1997, **324**, 847.
208. L. Gradoni, M. A. Iorio, M. Gramiccia and S. Orsini, *Farmaco*, 1989, **44**, 1157.
209. R. Mukhopadhyay and R. Madhubala, *Pharmacol. Res.*, 1993, **28**, 359.
210. T. Coons, S. Hanson, A. J. Bitonti, P. P. Mccann and B. Ullman, *Mol. Biochem. Parasitol.*, 1990, **39**, 77.
211. A. K. Singh, S. Roberts, B. Ullman and R. Madhubala, *J. Proteomics*, 2014, **102**, 44.
212. V. Iniesta, L. C. Gómez-Nieto and I. Corraliza, *J. Exp. Med.*, 2001, **193**, 777.
213. V. Iniesta, L. C. Gómez-Nieto, I. Molano, A. Mohedano, J. Carcelén, C. Mirón, C. Alonso and I. Corraliza, *Parasite Immunol.*, 2002, **24**, 113.
214. P. Kropf, J. M. Fuentes, E. Fahnrich, L. Arpa, S. Herath, V. Weber, G. Soler, A. Celada, M. Modolell and I. Müller, *FASEB J.*, 2005, **19**, 1000.
215. V. Iniesta, J. Carcelén, I. Molano, P. M. Peixoto, E. Redondo, P. Parra, M. Mangas, I. Monroy, M. L. Campo, C. G. Nieto and I. Corraliza, *Infect. Immun.*, 2005, **73**, 6085.
216. A. Taylor-Robinson, *Trends Parasitol.*, 2001, **17**, 262.
217. P. Kropf, S. Herath, V. Weber, M. Modolell and I. Müller, *Parasite Immunol.*, 2003, **25**, 439.
218. N. Wanasen and L. Soong, *Immunol. Res.*, 2008, **41**, 15.
219. E. L. D'Antonio, B. Ullman, S. C. Roberts, U. G. Dixit, M. E. Wilson, Y. Hai and D. W. Christianson, *Arch. Biochem. Biophys.*, 2013, **535**, 163.
220. R. K. Pandey, P. Prajapati, S. Goyal, A. Grover, V. K. Prajapati, *Comb. Chem. High Throughput Screening*, 2016, **19**, 813.
221. N. V. Grishin, A. L. Osterman, H. B. Brooks, M. A. Phillips and E. J. Goldsmith, *Biochemistry*, 1999, **38**, 15174.

222. A. Grover, S. P. Katiyar, S. K. Singh, V. K. Dubey and D. Sundar, *Biochim. Biophys. Acta*, 2012, **1824**, 1476.
223. Y. Amano, I. Namatame, Y. Tateishi, K. Honboh, E. Tanabe, T. Niimi and H. Sakashita, *Acta Crystallogr., Sect. D: Biol. Crystallogr.*, 2015, **71**, 1879.
224. E. Riley, S. C. Roberts and B. Ullman, *Int. J. Parasitol.*, 2011, **41**, 545.
225. M. Das, S. Singh and V. K. Dubey, *Chem. Biol. Drug Des.*, 2016, **87**, 352.
226. R. Mukhopadhyay and R. Madhubala, *Exp. Parasitol.*, 1995, **81**, 39.
227. B. Y. Chiang, T. C. Chen, C. H. Pai, C. C. Chou, H. H. Chen, T. P. Ko, W. H. Hsu, C. Y. Chang, W. F. Wu, A. H. Wang and C. H. Lin, *J. Biol. Chem.*, 2010, **285**, 25345.
228. E. Tetaud and A. H. Fairlamb, *Mol. Biochem. Parasitol.*, 1998, **96**, 111.
229. D. G. Arias, V. E. Marquez, M. L. Chiribao, F. R. Gadelha, C. Robello, A. A. Iglesias and S. A. Guerrero, *Free Radical Biol. Med.*, 2013, **63**, 65.
230. M. C. Jockers-Scherübl, R. H. Schirmer and R. L. Krauth-Siegel, *Eur. J. Biochem.*, 1989, **180**, 267.
231. G. B. Henderson, A. H. Fairlamb and A. Cerami, *Mol. Biochem. Parasitol.*, 1987, **24**, 39.
232. K. Smith, K. Nadeau, M. Bradley, C. Walsh and A. H. Fairlamb, *Protein Sci.*, 1992, **1**, 874.
233. A. F. Sousa, A. G. Gomes-Alves, D. Benítez, M. A. Comini, L. Flohé, T. Jaeger, J. Passos, F. Stuhlmann, A. M. Tomás and H. Castro, *Free Radical Biol. Med.*, 2014, **73**, 229.
234. S. L. Oza, M. P. Shaw, S. Wyllie and A. H. Fairlamb, *Mol. Biochem. Parasitol.*, 2005, **139**, 107.
235. P. K. Fyfe, S. L. Oza, A. H. Fairlamb and W. N. Hunter, *J. Biol. Chem.*, 2008, **283**, 17672.
236. O. Koch, D. Cappel, M. Nocker, T. Jäger, L. Flohé, C. A. Sotriffer and P. M. Selzer, *PLoS One*, 2013, **8**, e56788.
237. D. Benítez, A. Medeiros, L. Fiestas, E. A. Panozzo-Zenere, F. Maiwald, K. C. Prousis, M. Roussaki, T. Calogeropoulou, A. Detsi, T. Jaeger, J. Šarlauskas, L. Peterlin Mašič, C. Kunick, G. R. Labadie, L. Flohé and M. A. Comini, *PLoS Neglected Trop. Dis.*, 2016, **10**, e0004617.
238. P. Saudagar and V. K. Dubey, *Biol. Chem.*, 2011, **392**, 1113.
239. L. S. Torrie, S. Wyllie, D. Spinks, S. L. Oza, S. Thompson, J. R. Harrison, I. H. Gilbert, P. G. Wyatt, A. H. Fairlamb and J. A. Frearson, *J. Biol. Chem.*, 2009, **284**, 36137.
240. S. L. Oza, S. Wyllie and A. H. Fairlamb, *Mol. Biochem. Parasitol.*, 2006, **149**, 117.
241. S. Wyllie, S. L. Oza, S. Patterson, D. Spinks, S. Thompson and A. H. Fairlamb, *Mol. Microbiol.*, 2009, **74**, 529.
242. S. L. Oza, S. Chen, S. Wyllie, J. K. Coward and A. H. Fairlamb, *FEBS J.*, 2008, **275**, 5408.
243. P. Saudagar, P. Saha, A. K. Saikia and V. K. Dubey, *Eur. J. Pharm. Biopharm.*, 2013, **85**, 569.

244. D. Spinks, L. S. Torrie, S. Thompson, J. R. Harrison, J. A. Frearson, K. D. Read, A. H. Fairlamb, P. G. Wyatt and I. H. Gilbert, *ChemMedChem*, 2012, **7**, 95.
245. W. W. Zhang and G. Matlashewski, *MBio*, 2015, **6**, e00861.
246. L. Sollelis, M. Ghorbal, C. R. MacPherson, R. M. Martins, N. Kuk, L. Crobu, P. Bastien, A. Scherf, J. J. Lopez-Rubio and Y. Sterkers, *Cell. Microbiol.*, 2015, **17**, 1405.

CHAPTER 17

DNA Topoisomerases as Promising Targets for Leishmania *Chemotherapy*

ROSA Mª REGUERA, JOSÉ M. ESCUDERO-MARTÍNEZ, BÁRBARA DOMÍNGUEZ-ASENJO, CAMINO GUTIÉRREZ-CORBO AND RAFAEL BALAÑA-FOUCE*

Departamento de Ciencias Biomédicas, Facultad de Veterinaria, Universidad de León, Campus de Vegazana s/n, 24071 León, Spain
*Email: rbalf@unileon.es

17.1 Introduction

Leishmaniasis is a group of diseases prevalent in 98 countries with presence in four continents and approximately 360 million people at risk. The epidemiological burden of these diseases includes 1.6 new million cases each year, over 2-million worldwide-appraised disability-adjusted life-years (DALYs) and 20 000 to 40 000 deaths, most of them in Ethiopia and Sudan. Based on these estimates, leishmaniasis is considered to be the second more neglected tropical disease (NTD) all over the world.[1] The abrupt re-emergence of visceral leishmaniasis (VL)—the most deadly presentation of this disease—in the later decades of the last century has been linked to: (i) immunosuppressant drugs used in organ transplantation;[2] (ii) immunological disturbances due to HIV co-infections,[3] and (iii) climate change due to global warming.[4]

Similarly to other NTDs, first choice drugs against leishmaniasis were developed early in the 20th century. They include pentavalent antimony

Drug Discovery Series No. 60
Drug Discovery for Leishmaniasis
Edited by Luis Rivas and Carmen Gil

Published by the Royal Society of Chemistry, www.rsc.org

(Sb^V) drugs (Glucantime® and Pentostam®) alone or combined with other antibiotic or antifungal drugs, which constitute the keystone of current VL treatment in endemic countries. Little or nothing has changed since these molecules were introduced as anti-leishmanial agents, including the need for these drugs to be repeatedly administered by painful intramuscular injections and the presence of undesirable side effects.[5–7] Additionally, their overuse and treatment discontinuation due to the occurrence of undesirable side effects are an origin of relapses and resistances.[8] Most of the second-line drug arsenal against leishmaniasis is repurposed from other infectious diseases or cancer.[9] In such way, the polyene antifungal drug amphotericin B deoxycholate (AmB) has shown strong anti-leishmanial activity, but its use is limited due to:[6] (i) the high cost of the treatment; (ii) low chemical stability—which compromises the activity of long-storage batches; (iii) remarkable nephrotoxicity[10] and (iv) the need for repeated intravenous administration. These issues make AmB a hard-to-manage medicine where medical facilities are not well established. The liposome formulation of AmB (AmBisome®, Gilead) solves most of the aforementioned problems—nephrotoxicity is ameliorated due to lower dosage, and a single intravenous administration has shown good outcome in several clinical studies in India—but despite negotiations by the WHO with the biopharmaceutical company Gilead, it is still very expensive for low-income countries.[11,12] The introduction of the ether lipid drug miltefosine—a compound initially developed as an anti-tumoural but with potent anti-leishmanial effect—drastically changed the therapeutic perspective on these diseases.[13,14] Miltefosine is an oral drug that allows patient self-medication according to WHO recommendations.[15] However, its proven teratogenicity prevents medical prescription to pregnant women and children. Finally, paromomycin—an aminoglycoside antibiotic—was approved in 2006 in India against VL, constituting one of the drugs for combination therapy.[16] In order to get a definitive remedy against VL by 2020, a massive drug screening programme of small molecules under the supervision of the Drugs for Neglected Diseases initiative organization (DNDi) is on the way, involving public and private funds and institutions. In the meantime, drug combinations using old drugs have been tested in different clinical trials with promising results.[17] In India the use of miltefosine (MTF) + AmBisome and AmBisome + paromomycin (PM), has shown efficacies higher than 90% in clinical trials.[18] For its part, combinations of Sb^V + PM proved efficacious for fighting VL in Africa.[19] However, none of these combinations are suitable for self-medication, entailing mandatory medical intervention. The DNDi expects to find new small molecules that can substitute for those that are currently in use and lack all the problems we have discussed previously. For this purpose, there is an urgent need to find and characterize exclusive and robust parasite targets to attain the DNDi objectives.

Since its discovery, the kinetoplast has been an attractive objective for drug intervention. It is constituted by a unique disc-shaped extranuclear DNA (kinetoplast DNA or kDNA) that represents 10 to 20% of the total DNA in the cell.[20] This unusual organelle is a concatenated DNA network constituted by 5000–10 000 DNA minicircles and 50 copies of DNA maxicircles. Minicircles encode guide RNAs involved in editing mitochondrial nascent RNAs,[21] whereas maxicircles are the equivalent of mitochondrial DNA from other eukaryotes, as they encode mitochondrial proteins, ribosomal and transfer RNAs.[22] Interestingly, replication of kDNA occurs synchronously with that of genomic DNA during cell division. kDNA components must be decatenated from the network before their DNA replication and then catenated again to the nascent mesh when this process is finished.[23] DNA replication from both mini- and maxicircles takes place in theta (θ)-shaped intermediaries, where DNA topoisomerases (Top) play a key role.[24]

Since our group described the huge structural differences between host and trypanosomatid (*Leishmania* spp. and *Trypanosoma* spp.)[25] DNA topoisomerases, many reports have been published regarding the suitability of these enzymes as drug targets.[26–31] These enzymes play pivotal roles in modulating DNA topology during replication, transcription, recombination and repair, and are established sites of action for clinical antibacterial and anti-tumour agents.[32] Top I and II from trypanosomatids exhibit significant structural and biochemical variations from the corresponding human enzymes, and have important functions in organizing the kDNA network unique to these parasites.[33–35]

In the current chapter, we introduce an update about Top enzymes in *Leishmania*, making a special effort to describe the novel structural findings that make them an interesting target. In addition, we also review those compounds targeting Top with proven efficacy against these parasites.

17.2 Supercoiled *vs.* Relaxed DNA. The Need for DNA Topoisomerases

The genetic information of an organism coded into DNA is several millions times longer than the size of the cellular nucleus.[36,37] To pack this information inside a cell it needs to be folded many thousands of times to acquire a more compact configuration than the relaxed DNA molecule. This compacted supercoiled form of DNA may present two principal configurations: (i) interwound supercoiling DNA or plectonemic form (characteristic of prokaryotes) or (ii) solenoidal supercoiling DNA form (typical of eukaryotes), where DNA is condensed around nucleosomes.[38]

Naturally occurring DNA is invariably negatively supercoiled.[39] The term negatively supercoiled DNA refers to underwound DNA, whereas positively supercoiled DNA refers to overwound DNA. Because negatively supercoiled DNA is underwound, it exerts a force that helps to separate the two strands of the helix, which is required during both replication (DNA synthesis) and transcription (RNA synthesis).[40]

The topological status of a DNA molecule is defined by the linking number (Lk). Lk is the number of times a single strand crosses over the other on the projection plane. Lk depends on two other geometric properties, twist (Tw) and writhe (Wr). Tw quantifies the number of times a DNA strand is wound around the axis of the double helix, while Wr indicates the number of times the path of the centreline of the double helix crosses on itself when it is disposed in a plane. Changes in Lk are always natural numbers according to this formula: $Lk = Tw + Wr$. Under physiological conditions relaxed DNA is free of writhe ($Wr = 0$) and has approximately 10.5 bp per turn ($Lk = Tw$). However, when DNA is torsionally stressed, its Lk diminishes, and it becomes negatively supercoiled.[41]

Many essential processes introduce changes into the topological structure of DNA, especially in those processes where separation of DNA strands are needed. Since DNA has double helix structure and lacks the ability to rotate freely inside the cells, when strand separation occurs due to the polymerase–helicase complex, supercoils are generated at the flanking regions. As a result, positive supercoils form ahead of replication forks, and negative supercoils are formed behind them.[42] Without the action of Tops, positive supercoils would accumulate stopping replication and transcription processes almost immediately, whereas negative supercoils may favour the formation of abnormal DNA structures such as D- and R-loops, guanine quartets or Z-DNA structures.[32]

Cells use Top enzymes to change the supercoiled state of a DNA duplex to a more relaxed form.[43,44] For this purpose, Tops catalyse opening or sealing reactions on DNA where a catalytic Tyr residue placed at the active centre of the enzyme plays a role introducing *trans*-esterification reactions to phosphate groups of the DNA backbone.[45] This reaction leads to the formation of a transient enzyme–DNA covalent intermediate that opens a gate that the opposite DNA strand can pass through fuelled by torsional forces. Relaxed DNA is then resealed by means of a second *trans*-esterification reaction.[46]

Top proteins are classified according to the number of DNA strands cleaved by the enzyme. Type I Top (Top I) transiently break one strand of duplex DNA, allowing for single-step changes in the Lk of circular DNAs. Type II enzymes (TopII) cleave both strands of a DNA duplex to generate a gap through which another region of DNA can pass, resulting in Lk changes in steps of two. In addition to relaxation of supercoiled DNA, Top II have two further activities. These enzymes are involved in catenation and decatenation and knotting and unknotting activities of complex DNA networks.[43,44,47,48]

A further subclassification of type I Top enzymes includes three subfamilies: IA, IB, and IC. TopIA was considered to be a Mg^{2+}-dependent prokaryotic-borne enzyme involved in relaxation of positive supercoils through transient covalent binding to the 5′-end of the broken DNA strand followed by passage and rejoining. The Top IA family includes bacterial and archaeal Top I[49,50] and eukaryotic Top III.[51] Type IB (TopIB) are Mg^{2+}-independent enzymes that establish a 3′-phosphotyrosyl bond to one DNA strand,[45] thus nicking the duplex to facilitate the relaxation of both negatively and positively supercoiled DNAs. This allows TopIB to change the Lk in multiples of 1 (n), unlike TopIA, which introduces changes strictly in steps of one in the Lk of a single-stranded circular DNA.[52] TopIB includes eukaryotic and poxvirus Top I, as well as homologous enzymes found in certain bacteria.[53] Finally, type IC includes TopV only, which is found exclusively in a single genus of archaea.[54] Much remains unknown about TopV, but it appears to be functionally similar to Top IB in such a way that the enzyme cleaves the 3′ end of DNA, does not require any cofactors or ATP and relaxes both positive and negative supercoils. Both Top IB and IC use a controlled rotation mechanism for relaxing DNA.[55]

On the basis of sequence and structure, Top II are subclassified as IIA or IIB. TopIIA includes the homodimeric (A2) eukaryotic Top II, and other enzymes from bacteria, virus and plants, including bacterial DNA gyrase. Type IIB are heterotetrameric (A2B2) enzymes that only include one family member (Top VI), which has similar enzymatic characteristics to Top II, but distinct phylogenetic origin.[48]

17.3 DNA Topoisomerases in *Leishmania*

Leishmania parasites display a full set of Top enzymes to handle the topological status of DNA. Localization and putative functioning of these enzymes are described in Figure 17.1: (i) nuclear TopIB and TopIIA enzymes involved in DNA replication and repair that have been fully described as putative drug targets; (ii) mitochondrial TopIA and TopII, which have been described closely linked to kDNA replication; (iii) two putative members of

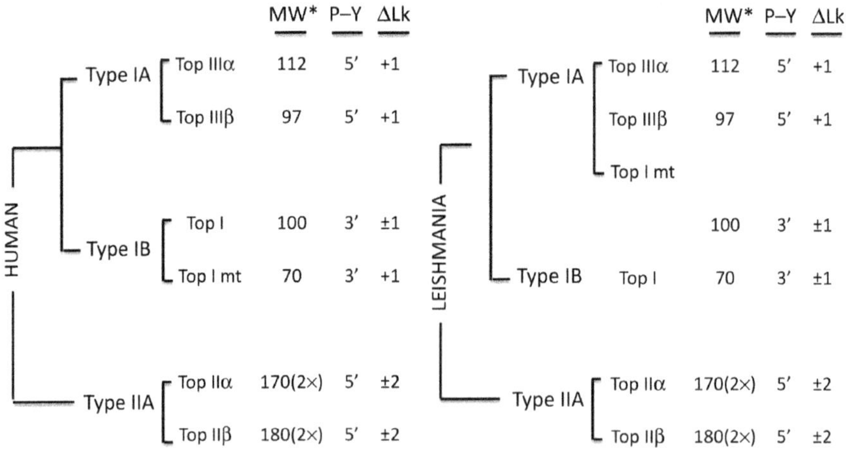

Figure 17.1 Both humans and *Leishmania* parasites have similar Top composition. (Left) Human DNA topoisomerases. (Right) *Leishmania* DNA topoisomerases. MW, Molecular masses in kDa calculated from polypeptide composition. ×2, Dimer. P–Y, Tyrosyl phosphorylated intermediates at the cleaved DNA end. ΔLk, Changes in linking number produced by each catalytic cycle. Based on the scheme published by Pommier and co-workers in 2010.[32]

the TopIA subfamily TopIII (α and β) involved in resolving DNA entanglements that may arise during DNA recombination.

17.3.1 *Leishmania infantum* DNA Topoisomerase IB

Leishmania TopIB is a true Mg^{2+}-independent eukaryotic enzyme that introduces both positive and negative supercoils and establishes temporary covalent bonds to the 3′ end of cut DNA, thus releasing a 5′-OH free end.

Historically, the first report regarding the protein sequence of *Leishmania* TopI was communicated by Broccoli and colleagues,[56] who identified a new gene encoding a 635 amino acid protein in *Leishmania donovani* similar to eukaryotic TopIB, but lacking the phylogenetically conserved "SKxxY" signature containing the catalytically active Tyr residue. These authors argued for the presence of a Ser residue as a putative catalytic amino acid instead of the Tyr described in all the Top enzymes examined so far. Nevertheless, the recombinant polypeptide expressed in *Escherichia coli* was unable to relax supercoiled DNA *in vitro*, which suggested that this open reading frame (ORF) was a non-functional pseudogene. However, it was purposed that the product of a second ORF could be needed to complement the truncated protein, building up a fully active holoenzyme. The solution to this puzzle was given later on by Villa and co-workers,[25] who described a second gene within *L. infantum* chromosome 4 that encoded a protein containing the active site with the lost "SKxxY" signature. Our group expressed both genes successfully using a bi-cistronic yeast expression platform, and the assembled protein was able to relax supercoiled DNA. Soon after, Theresa Shapiro's group confirmed our results in *Trypanosoma brucei* using a classic model of protein purification.[57] As far as we know, this genomic organization in which the genes encoding the protomers of TopIB are placed on different chromosomes, is unique to trypanosomatids. On the basis of the results of these studies, the *Leishmania* TopIB was defined as a heterodimeric enzyme (AB), which consists of a 636 amino acids (73 kDa) large subunit (LdTopIL) associated with a 262 amino acids (28 kDa) small subunit (LdTopIS). These subunits are encoded by two genes located within the *Leishmania* genome on chromosomes 34 and 4, respectively (Figure 17.2A).[25] The protein is only active when the two subunits are co-expressed, hence forming an AB heterodimer. The expression of a synthetic fusion construct of the two genes produced a chimeric polypeptide that displayed a cleavage–religation equilibrium shifted towards cleavage and reduced affinity for DNA.[58] This singular organization has only been described in *T. brucei* and *Leishmania* spp., but not in other unicellular eukaryotes like fungi, yeasts or apicomplexan parasites.[59]

Nevertheless, despite the unusual heterodimeric structure of trypanosomatids' TopIB, it retains all the regions and domains of the canonical human TopIB (hTopIB) required to be fully active and susceptible to inhibitors. From limited proteolysis studies[60] and X-rays for hTopIB–DNA co-crystal structure determination,[61] the primary structure of the human

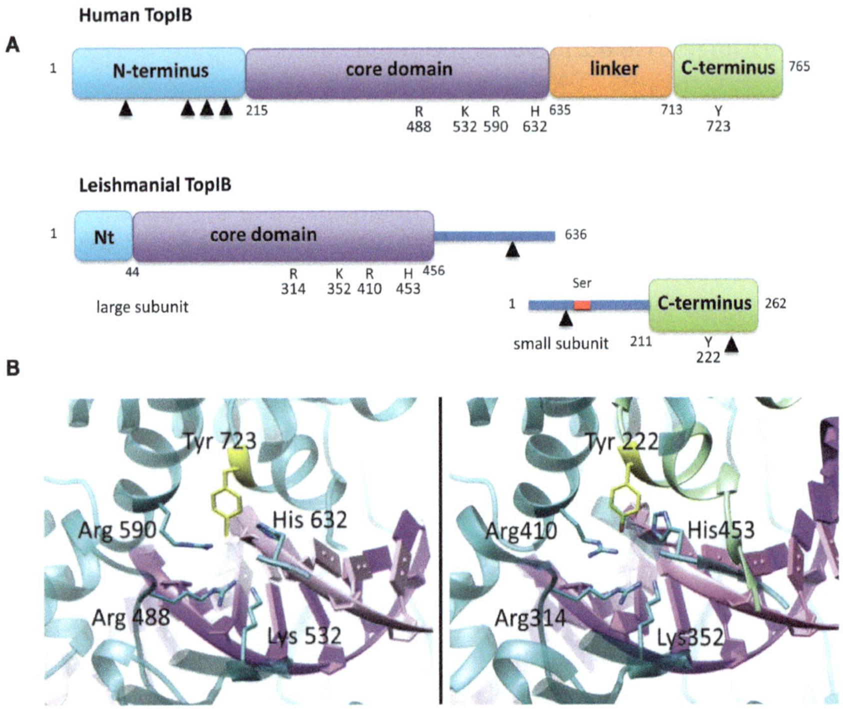

Figure 17.2 (A) Linear schematic representation of human TopIB compared with the subunit holoenzyme from *Leishmania* spp. Structural studies with the human enzyme divide this protein into four structural domains, which can be extended to both subunits of the *Leishmania* enzyme by sequence analogy: (i) non-conserved hydrophilic N-terminal domain; (ii) hydrophobic central core ("core") domain with the catalytic conserved amino acids; (iii) the linker region connecting the core domain to the active site; (iv) the C-terminal domain containing the catalytic Tyr residue involved in transiently cutting one of the DNA strands; (▲) nuclear localisation signal (NLS) motifs; Ser (serine-enriched region needed for nuclear localization in *Leishmania*). (B) Three-dimensional comparison of the amino acids integrating the active centre of human TopIB (left) and *Leishmania* TopIB (right) according to the X-ray structure deduced by Davies and co-workers.[70] Residues belonging to the large subunit are represented in blue, while Tyr222 from the small subunit is shown in yellow.

enzyme was divided into four major regions: (i) the N-terminal domain (amino acids 1–214); (ii) the core domain (amino acids 215–635); (iii) a linker (amino acids 636–712) that connects the core to the catalytic centre and (iv) the C-terminal domain.[62] In order to identify the structural motifs involved in enzymatic activity and susceptibility to inhibitors, we carried out exhaustive structural studies on LTopIB using truncated versions of the wild-type enzyme and a defective-TopIB *Saccharomyces cerevisiae* expression platform. In such a way, all the regions and domains described for hTopIB were also found to be present in the *Leishmania* holoenzyme, but split between the two protomers.[63] Therefore, the large monomer contains the N-terminal domain, shorter in length when compared with the human counterpart and poorly conserved across species. It contains charged amino acids gathered in acidic signatures that drive the enzyme to the nucleus.[62,64] In addition, it was proposed that the N-terminal end serves as a "molecular steer" to guide the assembled heterodimer in binding to DNA within cell nucleus.[65] The central core domain retains high homology with other TopIB and preserves the DNA interacting amino acids Arg314 (Arg488 in human enzyme), Lys352 (Lys532 in human enzyme), Arg410 (Arg590 in the human enzyme) and His453 (His632 in human enzyme). Site-directed mutations of all these amino acids gave rise to non-functional proteins, which confirms the crucial role of these amino acids in Top activity.[66] The large monomer ends in a non-conserved C-terminal domain, which shares part of the linker domain with the N-terminal end of the small subunit. The TopIB small subunit contains a non-conserved N-terminal end needed to interact with the C-terminal extension of the large monomer in order to build up the functional enzyme. It also contains the "SKxxY" signature characteristic of all Top IB enzymes. Within this signature, the Tyr222 plays a role in TopIB activity, as it introduces a nucleophilic attack on a phosphate group of the nucleotide chain and creates a transient phosphodiester bond with the broken DNA strand (Figure 17.2B).[67]

The structure and function of the linker domain in *Leishmania* deserves special mention. It was thought for a long time that the linker region in TopIB enzymes was a simple connector between the core and C-terminal domains. However, it was observed that any change introduced into this region produced impairments in the religation step and loss of susceptibility to Top inhibitors.[68,69] From the studies carried out by Davies and co-workers with LTopIB–DNA co-crystal cleaved with vanadate, mimicking the transition state of the cleavage reaction, it was concluded that LTopIB activity could be partially reconstituted with the amino acids 27–456 region of the large protomer plus amino acids 201–262 from the small subunit. However, it was not possible to obtain any information about the linker domain, since the C-terminal and N-terminal ends of the large and small subunits were removed to prepare the crystal.[70]

Due to the relevant role played by the linker domain in the human enzyme, our group developed a set of double-truncated polypeptides of both large and small subunits using the yeast expression platform. We demonstrated that part of the C-terminal extension of the large protomer and the N-terminal extension of the small subunit must interact in order to build up a fully active and camptothecin (CPT)-sensitive LTopIB. Indeed, the fact that these regions have opposite charge, strongly indicated that their electrostatic affinities were enough to promote assembly into an active dimer.[63,68] However, it is worth remembering that the linker domain is not a simple connector between active amino acids of the core and C-terminal domains. In addition, the linker domain contributes to both DNA binding and sensitivity to TopIB poisons, probably by means of increasing the rigidity of the structure and slowing down the rotation process prior to the religation of the cleaved DNA strand.[71] Our group, using a predictive 3D model combined with a sequential set of internal and lineal truncations on both monomers, were able to identify a conserved heptapeptide RPPVVRS motif (amino acids 175–181) within the N-extension domain of the small subunit, which is spatially close to a Lys-enriched region (amino acids 525–581) of the large subunit. Remarkably, the pentapeptide RPPVV introduces a loop in the secondary structure of the small protomer allowing its interaction with the large subunit. Thus the removal of this motif produced a radical change in protein architecture, resulting in a non-functional enzyme. Furthermore, when the RS motif (residues 180–181) was deleted, the resulting enzyme had lower relaxation activity and less sensitivity to CPT.[72]

Conversely to hTopIB, the N-terminal domain of the large subunit from *Leishmania* TopIB lacks localization signals (NLS) responsible for nuclear localization of the enzyme.[64] However, thanks to a complete set of polypeptides fused to green fluorescent protein (GFP), we were able to identify up to three NLS motifs in both subunits of LdTopIB.[73] NLS1 is located at the C-terminal extension of LdTopIL and consists of a 44-amino acids region enriched in Lys residues (easily identifiable by nuclear importins). NLS2 and NLS3 are located at the N-terminal and C-terminal domains, respectively, of the small LdTopIB subunit. NLS2 is a well-identified 10-amino-acid signature preceded by a cluster enriched in Ser residues that is phylogenetically conserved in trypanosomatids. This Ser-enriched motif seems to have some relevance in Top localization since the phosphorylation of at least one Ser residue is required for it to be driven to the nucleus.[74] More interestingly, NLS3 is placed at the small subunit within a 28-amino-acid region that includes the catalytic Tyr222. The amino acid sequence of NLS3 has no similarity to any other canonical NLS signals described so far. However, it resembles well the PY-NLSs signals proposed by Lee and co-workers[75] in human proteins, where proximal Pro and Tyr residues are recognized by a β-importin to drive this protein into the nucleus. Furthermore, regardless of previous results,[65] it was demonstrated that both subunits can reach the nuclear compartment independently.[73] Although some authors pointed to dual nucleus–kinetoplast LdTopIB localization, no specific mitochondrial localization signals have been identified in any of the subunits of the dimer. In addition, none of the chimeras fused to GFP co-localized with kDNA, thus proving that none of LdTopIB subunits are bound to DNA minicircles, which rules out the hypothesis of a TopIB operating in the kinetoplast.[65]

17.3.2 Inhibitors of DNA Topoisomerase IB

DNA topoisomerases play key roles in cellular processes affecting the topology and organization of intracellular DNA. These enzymes have a great clinical interest, as they are molecular targets of clinically used antitumour[32,76–78] and anti-microbial[32,79] drugs. With the exception of yeasts,[80]

TopIB is essential for eukaryotic organisms. Double gene replacement of the gene encoding LdTopIS, results in a non-viable phenotype in *Leishmania major.*[81] Similar results were found in other trypanosomatids where TopIB genes were silenced by RNA interference (RNAi), which resulted in a drastic reduction in DNA and RNA synthesis *in vitro.*[82] These findings and the structural differences from host orthologues have indicated these enzymes to be putative targets for drug discovery. Anti TopIB compounds fall into two main classes: (i) poisons, such as CPT, which is able to trap the transitory cleavage complex established between the enzyme and DNA forming a stable adduct; (ii) inhibitors, compounds that reversibly interfere with the catalytic functions of the enzyme (Figure 17.3).[83]

CPT derivatives. CPT is a natural plant alkaloid first isolated from extracts of the Chinese tree, *Camptotheca acuminate.* It was identified originally as an anti-tumour TopIB inhibitor in the 1980s.[84] Later on, it was clinically tested and approved by the FDA, showing strong anti-tumour activity against

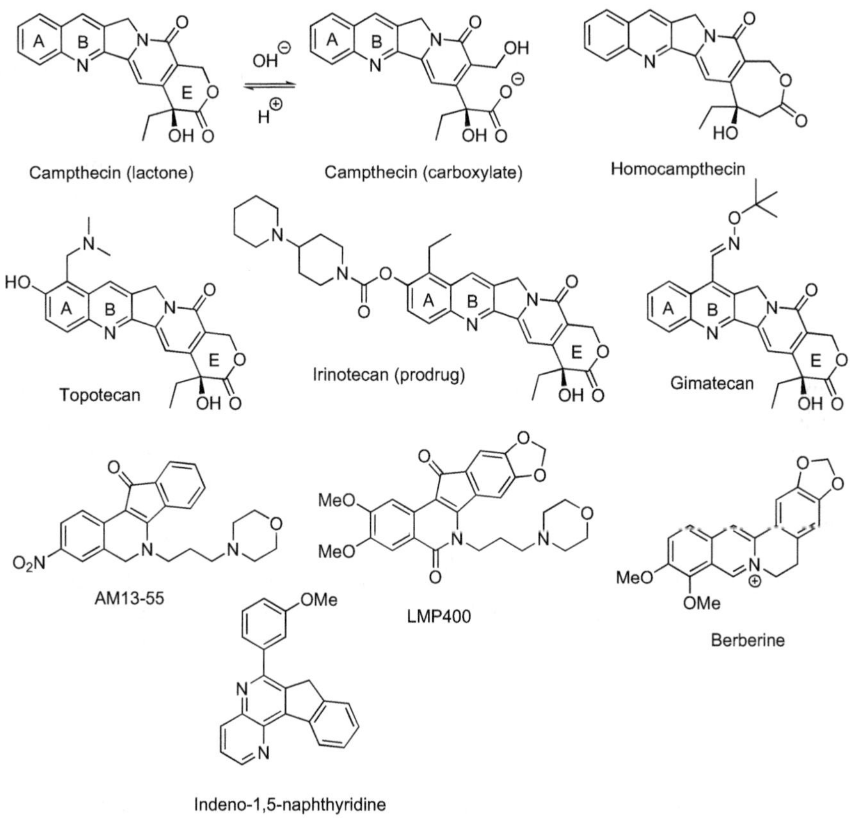

Figure 17.3 Chemical structures of TopIB poisons and inhibitors with clinical relevance. The figure shows the most important chemical families: camptothecins, indenoisoquinolines, berberine and indeno-1,5-naphtyridine.

ovarian, colon and lung cancer in humans. However, its use was discontinued due to undesirable side effects.[83] Over the years, numerous CPT analogues have been developed in order to increase its potency, and more importantly, to extend its stability in the organism. Some of them are still being used in clinical practice and others are under evaluation.[83,85,86] The water-soluble derivatives topotecan (Hycamtin) and irinotecan (CPT-II; Camptosar) are prodrugs used in human medicine since 1997 for the treatment of solid tumours and currently still in use more than 30 years after CPT was isolated for the first time.[87]

The CPT pharmacophore is a pentacyclic ring system with a terminal α-hydroxy-δ-lactone ring that confers a planar configuration needed to intercalate into stacked DNA base pairs.[88] The E-terminal ring contains an asymmetric carbon at position 20, with the S configuration being pharmacologically active.[89] However, the Achilles' heel of this molecule is its poor stability in neutral or basic pH aqueous fluids. They open the α-hydroxy-δ-lactone E-ring giving rise to an open carboxylate, with lower potency to inhibit TopIB.[90] To overcome this instability several modifications have been introduced in the CPT pharmacophore through; (i) enlarging by one carbon atom the E-ring (this limits E-ring opening but also prohibits its reclosure (homocamptothecins); (ii) stabilizing the E-ring transforming it from a six- to a five-membered ring.[32]

CPT is a non-competitive inhibitor of TopIB. It binds reversibly to the DNA–TopIB complex, but is not able to interact with each of these two elements separately. CPT derivatives easily penetrate the interior of target cells and interact with TopIB within the nucleus minutes after exposure. The main inhibitory mechanism of CPT is the stabilization of covalent Top IB–DNA complexes formed during the catalytic cycle, specifically inhibiting the reaction of strand relegation.[91] The three-dimensional structure of the ternary complex has shown that topotecan mimics a DNA base pair and binds at the site of DNA cleavage by intercalating between the upstream (−1) and downstream (+1) base pairs, thus preventing religation of the cleaved strand.[92] Under normal conditions the single strand breaks are transient and in low steady-state concentrations, since the speed of religation is faster than that of cutting (Figure 17.4).[44,93] CPT selectively kills cells during the S-phase of the cell cycle by collision of the advancing replication fork with the cleavage complexes stabilised between TopIB and DNA.[89] This leads to three important events responsible for cell death:[94] (i) double-stranded breaks formation in the hairpin;[87] (ii) irreversible stopping of replication fork advance[89] and (iii) formation of irreversible DNA–TopIB adducts.[95]

CPT and its derivatives topotecan and SN-38 are strong LdTopIB inhibitors of DNA relaxation in the nanomolar range, leading to a strong leishmanicidal effect. Remarkably, these drugs inhibited the *Leishmania* enzyme at lower concentrations than the human enzyme. Gimatecan[96]—an orphan drug approved against glial tumours—was the most active compound preventing the relaxation of supercoiled DNA at submicromolar concentrations. Cleavage equilibrium assays using LTopIB showed that gimatecan changed

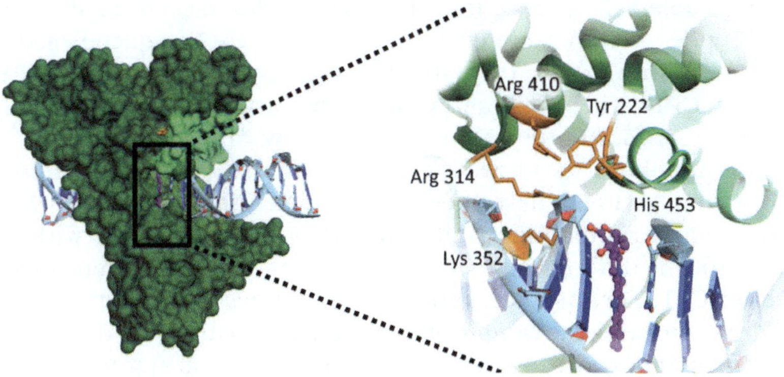

Figure 17.4 Mechanism of interaction of CPT with the covalent DNA–TopIB complex: the drug is interspersed between the base pairs of the DNA, interacting with both the protein and the DNA. The image on the left shows the surface map of *Leishmania* TopIB (green) in ternary complex with a DNA fragment (blue) and CPT. Docking of camptothecin (purple) into the active site of *Leishmania* TopIB. CPT intercalates between the base pairs at position +1 and −1 of DNA (blue). The amino acids (orange) belonging to the large subunit are Arg314, Lys352, Arg410 and His453, while that belonging to the small subunit is Tyr 222.

the equilibrium towards cleavage at much lower concentrations than other CPT derivatives, this effect persisting over time.[97]

All the CPT derivatives assayed are able to poison *L. infantum* DNA topoisomerase IB, which leads to a strong leishmanicidal (both promastigote and amastigote forms) effect. These compounds were tested in an *in vitro* phenotypic platform of genetically modified *L. infantum* amastigotes infecting mouse splenocytes. This platform permits the rapid analysis of thousands of compounds under high-throughput screening conditions, with more realistic infection conditions of splenic macrophages.[98] In this situation, the amastigote form showed higher sensitivity to TopIB poisons (with high therapeutic selectivity indexes) than free-living promastigotes. Gimatecan and CPT were the most powerful compounds for killing *L. infantum* splenocyte-infecting amastigotes within the nanomolar range. The higher selectivity index of gimatecan makes it a promising drug against this neglected disease.[97]

Very few studies have been conducted to determine the activity of CPT and its analogues in experimental infections. Proulx and co-workers[99] showed that a formulation of liposome-encapsulated CPT produced a significant (43–55%) reduction of the parasitic load in a murine model of VL.

Non-CPT derivatives: Several CPT structurally related compounds have been isolated or synthesized and tested against TopIB. Amongst them, naturally occurring flavones, naphtoquinones, indenoisoquinolines and naphthyridines have shown interesting results against *Leishmania*.

Indenoisoquinolines are a group of synthetic drugs that were initially developed as anti-tumour compounds with an improved ability to stabilize cleavage complexes.[100] These drugs have some advantageous characteristics against TopIB:[101] (i) they are chemically more stable in aqueous solutions that CPT derivatives; (ii) they stabilise cleavage complexes at different sites from CPTs; (iii) their anti-proliferative activity is generally higher than that of CPTs; (iv) they are not secreted by ABC transporters.[102] Two indenoisoquinolines synthesized by Cushman's group, indotecan (LMP400) and AM13-55, have shown remarkable inhibitory activity against LTopIB. Both compounds induced cleavage complexes with DNA with significant differences in their relative pattern and intensities with respect to CPT. For instance, indenoisoquinolines tend to trap the TopI sites with C(−1) whereas camptothecins trap sites with T(−1) and G(+1).[103] These drugs showed strong anti-leishmanial effect on naturally infected splenocytes with amastigotes of *L. infantum* with selectivity indexes between nine and twelve times higher than those observed for miltefosine.[104] Furthermore, LMP400 was able to reduce the parasitic load from spleen and liver in an experimental model of visceral leishmaniasis in mice at a similar rate to that observed by other authors with paromomycin alone or in combination with Glucantime[105] or liposomal formulations of CPT.[99]

Some naturally occurring molecules have been shown to be TopIB poisons and have anti-leishmanial effects. The bisnapthaquinones lapachol, β-lapachone and dyospirin are catalytic inhibitors of the *Leishmania* enzyme and require concentrations ten times higher to create a similar effect on the mammalian homologous enzyme.[106] Furthermore, the pentacyclic triterpenoid dihydrobetulinic acid[107] or the naturally occurring flavones (baicalein, luteolin and quercetin) are potent LdTopIB poisons *in vitro*[108,109] (see Chapter 9 of this volume). Berberine is another inhibitor of class I polyheterociclic TopIB, which, due to its structure, is able to intercalate the antitumour drug to benzo[*a*] acridine.[110] Berberine and some of its analogues have been studied *in vitro* against *L. donovani* promastigotes and bloodstream forms of *T. brucei* and *Trypanosoma congolense*, showing efficiency in the micromolar range. The results in animal studies with these compounds have shown efficacy in models of infection in hamster.[111]

Recently, our group in collaboration with Dr Palacios (University of País Vasco, Spain) have tested a series of newly synthesised quasi-flat condensed naphthyridine heterocyclic compounds, which exhibited selective good inhibition of LTopIB activity that correlated-well with their anti-leishmanial effect, thus indicating again that this enzyme is a putative target of these compounds.[112]

Some new inhibitors are coming from the marine invertebrates. Naturally occurring 2-alkynoic fatty acids isolated from sponges have been shown to be promising selective catalytic inhibitors of LTopIB.[113] It has been clearly shown that the inhibitory capacity of these compounds is dependent on the degree of unsaturation, as well as on the length of the acid carbon chain. These compounds probably do not bind to the DNA–LTopIB binary complex, but may be interacting directly with the TopIB enzyme.[114]

17.4 Other DNA Topoisomerases in *Leishmania*

17.4.1 Type II DNA Topoisomerases

Leishmania spp. have a full set of type II topoisomerases (Top II) required for the regulation of DNA topology.[115] Top II are multimeric enzymes that introduce transient phosphotyrosine bonds at the 5′ end of both broken strands of the DNA helix. After cutting, the ends of the DNA are separated and another DNA duplex is passed through the gap, which is finally sealed.[43,44,47,116–118] Unlike TopI, this family of enzymes have additional activities that include relaxation of supercoiled DNA, catenation–decatenation and knotting–unknotting activities required for effective chromosomal segregation. To develop these complexes features, TopII enzymes require ATP and Mg^{2+} for full activity.[119]

Similarly to other eukaryotic organisms[120] Top II enzymes participate in the segregation and maintenance of the chromosome structure in *Leishmania*[121] and are involved in chromosome condensation and decondensation events.[122] Furthermore, trypanosomatids need an extra set of enzymes to perform the segregation, replication, remodelling and maintenance of extrachromosomal kDNA.[123–126]

Leishmania contains a single orthologue of nuclear TopIIα on chromosome 28.[127] Each subunit (A2) of TopIIα is formed by the fusion of domains corresponding to the GyrB/ParE of bacterial DNA gyrase at the N-terminal-end, and GyrA/ParC of TopIV subunits at the C-terminal-end.[44] In addition, three structural domains: ATPase, B′ and A′ are connected to each other by a flexible hinge region.[128] The ATPase domain is found at the N-terminal end of each monomer. ATP promotes dimerization of N-terminal ATPase domains, which control the open–close movements of the protein during DNA release.[129,130] The ATPase domain is followed in turn by the B′ domain and the N-terminal region A′ that contains the catalytic Tyr residue involved in the trans-esterification reaction of DNA.[131–133] Finally, the C-terminal end contains NLSs, but is dispensable for topoisomerization.[134] Since Mg^{2+} ions are involved in TopII activity, a conserved Mg^{2+}-interaction domain called Toprim, is involved in coordinating two Mg^{2+} ions per TopII subunit.[135,136]

Kulikowicz and Shapiro silenced the expression of both nuclear TopIIα and β in *T. brucei*. TopIIα silencing produced nuclear abnormalities and cell growth arrest at early stages of the cell cycle, indicating that this protein is an attractive therapeutic target. However, TopIIβ silencing produced no evident phenotype, indicating that this gene encodes an inactive protein.[127]

Mitochondrial TopII (mtTopII) is the key enzyme involved in disaggregating the kDNA network during cell division in *Leishmania*. A single copy of a mitochondrial TopII orthologue appears in the Genome Database of Trytrips.[35,137,138] The presence of mtTopII in *Leishmania* is mandatory for decatenating the interlocked minicircles during cell division. mtTopII activity releases free 1 kb minicircles from the kinetoplast network before replication as a rolling-circle starts. After all minicircles have been individually replicated they will be segregated into two sets before catenation and condensed again in the two dividing cells. Silencing of the mtTopII gene by RNAi caused progressive loss of the kDNA network, a process called diskynetoplasty.[139]

17.4.2 Type IA DNA Topoisomerases

The subfamily of TopIA enzymes is represented by three type IA topoisomerase-encoding genes in the *L. major* Friedlin database: one on chromosome 21 (annotated as true TopIA) and two putative TopIII ORFs on chromosomes 28 and 36, which correspond to TopIII. TopIA is a DNA-relaxing enzyme,[140] and TopIII enzymes are more specialized to catenating–decatenating double-stranded DNA chains.[48,141,142] The *L. donovani* TopIA gene encodes a protein of 90 kDa that contains an N-terminal domain, which has a mitochondrial localization signal that drives the protein to kDNA. Similarly to TbTopIA, the phylogenetically conserved core region contains all the conserved motifs of the protein with two DNA binding topofolds (amino acids 186–660). Finally, the protein ends in a variable C-terminal domain.[143,144]

Immunolocalization assays in *T. brucei* demonstrated that TopIA operates in the region of the mitochondrion close to the flagellar basal body, associated with other enzymes involved in DNA replication, including mtTopII.[24] The results of these studies indicated that the position of TopIA changes from one pole to the other pole of the kDNA disc when replication is taking place. RNAi-silenced mtTbTopIA showed growth arrest and dyskenoplasty, with no changes in nuclear DNA. In addition, a bulk accumulation of late minicircle θ forms, but not of daughter catenates, indicates that mtTbTopIA plays a role in the resolution on the entwined parental strands during the resolution step of minicircle replication.[143]

Finally, there are TopIII are TopIA subfamily proteins that decatenate rather than relax supercoiled DNA. The *L. donovani* genome database contains two genes encoding a TopIIIα (a 2844-bp long ORF that encodes a 104-kDa polypeptide) and a TopIIIβ isoform (a 2601-bp long ORF that encodes a 95-kDa protein). The deduced amino acid sequence of LdTopIIIβ has the active Tyr at position 327, which is located within a highly conserved GYISYPRTES signature.[145] Null-TopIIIα mutants in *T. brucei* have revealed the critical role of this enzyme in homologous recombination processes and antigenic switching.[146]

17.4.3 Inhibitors of DNA Topoisomerase II

Eukaryotic TopII was proposed as a promising therapeutic target for antiparasitic drugs in the mid 1990s (Figure 17.5).[33] One of the first groups of compounds that showed therapeutic potential against trypanosomatids were derived from the anti-tumoural 9-anilinoacridine scaffold and had antileishmanial activity due to TopII inhibition.[147,148] Fluoroquinolone

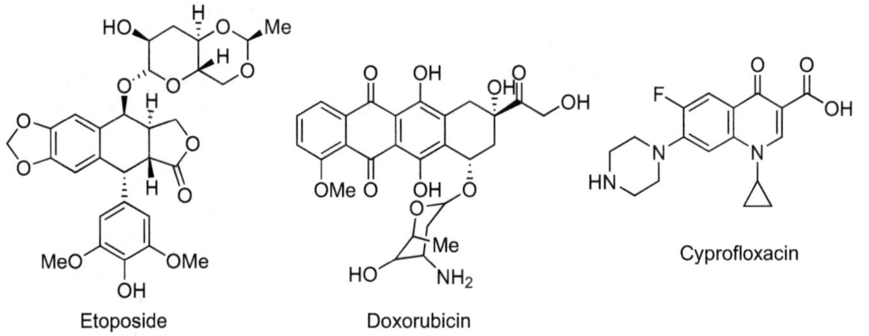

Figure 17.5 Chemical structures of TopII inhibitors: etoposide, doxorubicin and cyprofloxacin.

antibiotics, cyprofloxacin and enoxacin, which inhibited TopII in the parasite, had strong leishmanicidal effects against *Leishmania panamensis* amastigotes.[149,150] The anthracycline anti-tumour antibiotic doxorubicin, derived from the natural alkaloid daunomycin, is a DNA-intercalating agent that stabilises cleavage complex between TopII and DNA at submicromolar concentrations. These compounds, along with aclarubicin, prevented *Leishmania* growth *in vitro* and triggered a programmed cell death process.[151,152] The pharmacological delivery of doxorubicin by nanoparticles in several preclinical trials of visceral leishmaniasis has been recently reviewed.[153] The semisynthetic derivatives of the root extracts of plants of the genus *Podophyllum*, epipodophyllotoxins, stabilise double-stranded cleavage mediated by TopII and inhibit DNA relegation.[154] The two major compounds from this family, etoposide and teniposide, are in clinical use against different cancers[155] and have been proven to be effective against different parasites.[59] However, etoposide showed only moderate activity against trypanosomatids with a poor selectivity index regarding host cells.[151]

17.5 Conclusions

DNA topoisomerases are well-known targets against several tumour processes, the relevance of which in NTDs caused by unicellular parasites is under investigation. *Leishmania*, like other trypanosomatids, has a complete set of six classes of Top enzymes defined by their structure, localization and mechanism of action. The presence of extranuclear kDNA in trypanosomatids that must be replicated and transcribed by specific mitochondrial TopIA and TopII enzymes (encoded by nuclear genes) has been considered particularly attractive for drug development. In addition, some of these proteins exhibit an unusual structure. *Leishmania* TopIB is structurally dissimilar to the rest of the proteins so far described, but it conserves the ability to pass a single DNA strand (relaxase activity) to relieve torsional stresses. Other TopIA and TopII enzymes have decatenase rather than relaxase activity and work in concert with helicases to resolve non-crossover products by dissolution of recombination intermediates.

Despite many Top inhibitors having been tested against tumour cells, there is very poor quantitative structure–activity relationship (QSAR) information on the effects of these compounds on *Leishmania*. Most of the research done to target *Leishmania* Top has been using repositioned drugs previously tested against other pathogens or tumour cells. The lack of information on 3D X-ray analysis of *Leishmania* Top crystals has not permitted the "*ad hoc*" virtual screening of potential inhibitors in a structure-based drug discovery strategy. The crystalline structure of this protein should provide information of crucial importance on those structural determinants involved in the assembly of the active TopIB holoenzyme, as well as in the enzyme kinetic machinery and interaction with inhibitors in the case of the other Top enzymes, which will play a key role in future drug discovery.

Acknowledgements

This research was supported by Ministerio de Economía y Competitividad (AGL2010-16078/GAN, AGL2016-79813-C2-1-R and CYTED 214RT0482) and the Instituto de Salud Carlos III (PI12/00104).

References

1. J. J. Alvar, I. D. Vélez, C. Bern, M. Herrero, P. Desjeux, J. Cano, J. Jannin, M. den Boer and WHO Leishmaniasis Control Team, *PLoS One*, 2012, **7**, e35671.
2. J. van Griensven, E. Carrillo, R. López-Vélez, L. Lynen and J. Moreno, *Clin. Microbiol. Infect.*, 2014, **20**, 286.
3. J. Alvar, P. Aparicio, A. Aseffa, M. Den Boer, C. Cañavate, J. P. Dedet, L. Gradoni, R. Ter Horst, R. López-Vélez and J. Moreno, *Microbiol. Rev.*, 2008, **21**, 334.
4. I. S. Koltas, F. Eroglu, D. Alabaz and S. Uzun, *Trans. R. Soc. Trop. Med. Hyg.*, 2014, **108**, 154.
5. R. Balaña-Fouce, J. C. Cubría, R. M. Reguera and D. Ordóñez, *Gen. Pharmacol.*, 1998, **30**, 435.
6. J. H. No, *Acta Trop.*, 2016, **155**, 113.
7. A. S. Nagle, S. Khare, A. B. Kumar, F. Supek, A. Buchynskyy, C. J. Mathison, N. K. Chennamaneni, N. Pendem, F. S. Buckner, M. H. Gelb and V. Molteni, *Chem. Rev.*, 2014, **114**, 11305.
8. S. L. Croft, S. Sundar and A. H. Farilamb, *Clin. Microbiol. Rev.*, 2006, **19**, 111.
9. L. G. Ferreira and A. D. Andricopulo, *Drug Discovery Today*, 2016, **21**, 1699.
10. C. P. Thakur, R. K. Singh, S. M. Hassan, R. Kumar, S. Narain and A. Kumar, *Trans. R. Soc. Trop. Med. Hyg.*, 1999, **93**, 319.

11. S. Sundar, J. Chakravarty, D. Agarwal, M. Rai and H. W. Murray, *N. Engl. J. Med.*, 2010, **362**, 504.
12. F. Meheus, M. Balasegaram, P. Olliaro, S. Sundar, S. Rijal, M. A. Faiz and M. Boelaert, *PLoS Neglected Trop. Dis.*, 2010, **4**, e818.
13. J. J. Berman, *Expert Opin. Drug Metab. Toxicol.*, 2008, **4**, 1209.
14. S. K. Bhattacharya, P. K. Sinha, S. Sundar, C. P. Thakur, T. K. Jha, K. Pandey, V. R. Das, N. Kumar, C. Lal, N. Verma, V. P. Singh, A. Ranjan, R. B. Verma, G. Anders, H. Sindermann and N. K. Ganguly, *J. Infect. Dis.*, 2007, **196**, 591.
15. B. Monge-Maillo and R. López-Vélez, *Clin. Infect. Dis.*, 2015, **60**, 1398.
16. S. Sundar, T. K. Jha, C. P. Thakur, P. K. Sinha and S. K. Bhattacharya, *N. Engl. J. Med.*, 2007, **356**, 2571.
17. J. van Griensven, M. Balasegaram, F. Meheus, J. Alvar, L. Lynen and M. Boelaert, *Lancet Infect. Dis.*, 2010, **10**, 184.
18. S. Sundar, M. Rai, J. Chakravarty, D. Agarwal, N. Agrawal, M. Vaillant, P. Olliaro and H. W. Murray, *Clin. Infect. Dis.*, 2008, **47**, 1000.
19. A. Musa, E. Khalil, A. Hailu, J. Olobo, M. Balasegaram, R. Omollo, T. Edwards, J. Rashid, J. Mbui, B. Musa, A. A. Abuzaid, O. Ahmed, A. Fadlalla, A. El-Hassan, M. Mueller, G. Mucee, S. Njoroge, V. Manduku, G. Mutuma, L. Apadet, H. Lodenyo, D. Mutea, G. Kirigi, S. Yifru, G. Mengistu, Z. Hurissa, W. Hailu, T. Weldegebreal, H. Tafes, Y. Mekonnen, E. Makonnen, S. Ndegwa, P. Sagaki, R. Kimutai, J. Kesusu, R. Owiti, S. Ellis and M. Wasunna, *PLoS Neglected Trop. Dis.*, 2012, **6**, e1674.
20. T. A. Shapiro and P. T. Englund, *Annu. Rev. Microbiol.*, 1995, **49**, 117.
21. K. Stuart and A. K. Panigrahi, *Mol. Microbiol.*, 2002, **45**, 591.
22. J. Shlomai, *Curr. Mol. Med.*, 2004, **4**, 623.
23. B. Liu, Y. Liu, S. A. Molyka, E. E. C. Agbo and P. T. Englund, *Trends Parasitol.*, 2005, **21**, 363.
24. R. E. Jensen and P. T. Englund, *Annu. Rev. Microbiol.*, 2012, **66**, 473.
25. H. Villa, A. R. Otero-Marcos, R. M. Reguera, R. Balaña-Fouce, C. García-Estrada, Y. Pérez-Pertejo, B. L. Tekwani, P. J. Myler, K. D. Stuart, M. A. Bjornsti and D. Ordoñez, *J. Biol. Chem.*, 2003, **278**, 3521.
26. R. P. Bakshi and T. A. Shapiro, *Mini Rev. Med. Chem.*, 2003, **3**, 597.
27. B. B. Das, A. Dasgupta and T. Sengupta, *Trends Parasitol.*, 2004, **8**, 381.
28. R. Balaña-Fouce, C. M. Redondo, Y. Pérez-Pertejo, R. Diaz-González and R. M. Reguera, *Drug Discov. Today*, 2006, **11**, 733.
29. R. M. Reguera, C. M. Redondo, R. Gutiérrez de Prado, Y. Pérez-Pertejo and R. Balaña-Fouce, *Biochim. Biophys. Acta*, 2006, **1759**, 117.
30. B. B. Das, A. Ganguly and H. K. Majumder, *Adv. Exp. Med. Biol.*, 2008, **625**, 103.
31. I. D'Annessa, S. Castelli and A. Desideri, *Mini-Rev. Med. Chem.*, 2015, **15**, 203.
32. Y. Pommier, E. Leo, H. Zhang and C. Marchand, *Chem. Biol.*, 2010, **17**, 421.

33. C. Burri, A. L. Bodley and T. A. Shapiro, *Parasitol. Today*, 1996, **12**, 226.
34. B. B. Das, T. Sengupta, A. Ganguly and H. K. Majumder, *Mol. Microbiol.*, 2006, **62**, 917.
35. R. Balaña-Fouce, R. Alvarez-Velilla, C. Fernández-Prada, C. García-Estrada and R. M. Reguera, *Int. J. Parasitol. Drugs Drug Resist.*, 2014, **4**, 326.
36. A. Annunziato, *Nat. Educ.*, 2008, **1**, 26.
37. A. Travers and G. Muskhelishvili, *FEBS J.*, 2015, **282**, 2279.
38. A. V. Vologodskii and N. R. Cozzarelli, *Annu. Rev. Biophys. Biomol. Struct.*, 1994, **23**, 609.
39. J. C. Wang, *J. Cell. Sci. Suppl.*, 1984, **1**, 21.
40. G. Witz and A. Stasiak, *Nucleic Acids Res.*, 2010, **38**, 2119.
41. J. E. Deweese, M. A. Osheroff and N. Osheroff, *Biochem. Mol. Biol. Educ.*, 2008, **37**, 2.
42. L. F. Liu and J. C. Wang, *Proc. Natl. Acad. Sci. U.S.A.*, 1987, **84**, 7024.
43. J. C. Wang, *Ann. Rev. Biochem.*, 1996, **65**, 635.
44. J. J. Champoux, *Annu. Rev. Biochem.*, 2001, **70**, 369.
45. J. J. Champoux, *J. Biol. Chem.*, 1981, **256**, 4805.
46. J. C. Wang, *Nat. Rev. Mol. Cell. Biol.*, 2002, **3**, 430.
47. J. C. Wang, *Q. Rev. Biophys.*, 1998, **31**, 107.
48. P. Forterre, S. Gribaldo, D. Gadelle and M. C. Serre, *Biochimie*, 2007, **89**, 427.
49. J. C. Wang, *J. Mol. Biol.*, 1971, **55**, 523.
50. A. I. Slesarev, D. A. Zaitzev, V. M. Kopylov, K. O. Stetter and S. A. Kozyavkin, *J. Biol. Chem.*, 1991, **266**, 12321.
51. D. Raoult, S. Audic, C. Robert, C. Abergel, P. Renesto, H. Ogata, B. La Scola, M. Suzan and J. M. Claverie, *Science*, 2004, **306**, 1344.
52. J. B. Leppard and J. J. Champoux, *Chromosoma*, 2005, **114**, 75.
53. B. O. Krogh and S. Shuman, *Proc. Natl. Acad. Sci. U.S.A.*, 2002, **99**, 1853.
54. A. I. Slesarev, K. O. Stetter, J. A. Lake, M. Gellert, R. Krah and S. A. Kozyavkin, *Nature*, 1993, **364**, 735.
55. B. Taneja, B. Schnurr, A. Slesarev, J. F. Marko and A. Mondragón, *Proc. Natl. Acad. Sci. U.S.A.*, 2007, **104**, 14670.
56. S. Broccoli, J. F. Marquis, B. Papadopoulou, M. Olivier and M. Drolet, *Nucleic Acids Res.*, 1999, **27**, 2745.
57. A. L. Bodley, A. K. Chakraborty, S. Xie, C. Burri and T. A. Shapiro, *Proc. Natl. Acad. Sci. U.S.A.*, 2003, **100**, 7539.
58. B. B. Das, S. Bose Dasgupta, A. Ganguly, S. Mazumder, A. Roy and H. K. Majumder, *FEBS J.*, 2007, **274**, 150.
59. C. García-Estrada, C. F. Prada, C. Fernández-Rubio, F. Rojo-Vázquez and R. Balaña-Fouce, *Proc. Biol. Sci.*, 2010, **277**, 1777.
60. L. Stewart, G. C. Ireton and J. J. Champoux, *J. Biol. Mol.*, 1997, **269**, 355.
61. M. R. Redinbo, L. Stewart, P. Kuhn, J. J. Champoux and W. G. Hol, *Science*, 1998, **279**, 1504.
62. L. Stewart, G. C. Ireton and J. J. Champoux, *J. Biol. Chem.*, 1996, **271**, 7602.

63. R. Diaz-Gonzalez, Y. Perez-Pertejo, D. Ordóñez, R. Balaña-Fouce and R. M. Reguera, *PLoS One*, 2007, **2**, e1177.
64. J. Alsner, J. Q. Svejstrup, E. Kjeldsen, B. S. Sorensen and O. Westergaard, *J. Biol. Chem.*, 1992, **267**, 12408.
65. S. BoseDasgupta, A. Ganguly, B. B. Das, A. Roy, N. V. Khalkho and H. K. Majumder, *Mol. Microbiol.*, 2008, **67**, 31.
66. R. Diaz-Gonzalez, Y. Perez-Pertejo, Y. Pommier, R. Balaña-Fouce and R. M. Reguera, *Biochem. Pharmacol.*, 2008, **76**, 608.
67. R. Diaz-Gonzalez, Y. Perez-Pertejo, C. M. Redondo, Y. Pommier, R. Balaña-Fouce and R. M. Reguera, *Biochimie*, 2007, **89**, 1517.
68. G. C. Ireton, L. Stewart, L. H. Parker and J. J. Champoux, *J. Biol. Chem.*, 2000, **275**, 25820.
69. P. Fiorani, A. Bruselles, M. Falconi, G. Chillemi, A. Desideri and P. Benedetti, *J. Biol. Chem.*, 2003, **278**, 43268.
70. D. R. Davies, A. Mushtaq, H. Interthal, J. J. Champoux and W. G. Hol, *J. Mol. Biol.*, 2006, **357**, 1202.
71. L. Stewart, G. C. Ireton and J. J. Champoux, *J. Biol. Chem.*, 1999, **274**, 32950.
72. C. F. Prada, R. Álvarez-Velilla, R. Diaz-González, C. Prieto, Y. Pérez-Pertejo, R. Balaña-Fouce and R. M. Reguera, *Biochim. Biophys. Acta*, 2012, **1820**, 2062.
73. C. F. Prada, R. Álvarez-Velilla, R. Diaz-González, Y. Pérez-Pertejo, R. Balaña-Fouce and R. M. Reguera, *PLoS One*, 2013, **8**, e73565.
74. A. C. Teng, N. A. Al-Montashiri, B. L. Cheng, P. Lou, P. Ozmizrak, H. H. Chen and A. F. Stewart, *PLoS One*, 2011, **6**, e24100.
75. B. J. Lee, A. E. Cansizoglu, K. E. Süel, T. H. Louis, Z. Zhang and Y. M. Chook, *Cell*, 2006, **126**, 543.
76. H. J. Broxterman and N. Georgopapadakou, *Drug Resist.*, 2001, **4**, 197.
77. J. L. Nitiss, *Curr. Opin. Investig. Drugs*, 2002, **3**, 1512.
78. Y. Pommier, *Curr. Med. Chem. Anticancer Agents*, 2004, **5**, 429.
79. D. T. Bearden and L. H. Danziger, *Pharmacotherapy*, 2001, **21**, 224S.
80. C. Thrash, K. Voelkel, S. DiNardo and R. Sternglanz, *J. Biol. Chem.*, 1984, **259**, 1375.
81. R. P. Bakshi and T. A. Shapiro, *Mol. Biochem. Parasitol.*, 2004, **136**, 249.
82. R. Balaña-Fouce, C. García-Estrada, Y. Pérez-Pertejo and R. M. Reguera, *BMC Microbiol.*, 2008, **8**, 113.
83. Y. Pommier, *Nat. Rev. Cancer*, 2006, **6**, 789.
84. Y. H. Hsiang, R. Hertzberg, S. Hecht and L. F. Liu, *J. Biol. Chem.*, 1985, **260**, 14873.
85. H. Ulukan and P. W. Swaan, *Drugs*, 2002, **62**, 2039.
86. Q. Y. Li, Y. G. Zu, R. Z. Shi and L. P. Yao, *Curr. Med. Chem.*, 2006, **13**, 2021.
87. Y. Xu and C. Her, *Biomolecules*, 2015, **5**, 1652.
88. M. C. Wani, P. E. Ronman, J. T. Lindley and M. E. Wall, *J. Med. Chem.*, 1980, **23**, 554.
89. Y. H. Hsiang, M. G. Lihou and L. F. Liu, *Cancer Res.*, 1989, **49**, 5077.

90. R. P. Hertzberg, M. J. Caranfa, K. G. Holden, D. R. Jakas, G. Gallagher, M. R. Mattern, S. M. Mong, J. O. Bartus, R. K. Johnson and W. D. Kingsbury, *J. Med. Chem.*, 1989, **32**, 715.
91. J. Q. Svejstrup, K. Christiansen, I. I. Gromova, A. H. Andersen and O. Westergaard, *J. Mol. Biol.*, 1991, **222**, 669.
92. B. L. Staker, K. Hjerrild, M. D. Feese, C. A. Behnke, A. B. Burgin and L. Stewart, *Proc. Natl. Acad. Sci. U.S.A.*, 2002, **99**, 15387.
93. B. L. Staker, D. F. Michael, C. Mark, Y. Pommier, D. Zembower, L. Stewart and B. A. Burgin, *J. Med. Chem.*, 2005, **48**, 2336.
94. L. F. Liu, S. D. Desai, T. K. Li, Y. Mao, M. Sun and S. P. Sim, *Ann. N.Y. Acad. Sci.*, 2000, **922**, 1.
95. C. P. Lin, Y. Ban, Y. L. Lyu and L. F. Liu, *J. Biol. Chem.*, 2009, **284**, 28084.
96. J. Hu, P. Y. Wen, L. E. Abrey, C. E. Fadul, J. Drappatz, N. Salem, J. G. Supko and F. Hochberg, *J. Neurooncol.*, 2013, **111**, 347.
97. C. F. Prada, R. Álvarez-Velilla, R. Balaña-Fouce, C. Prieto, E. Calvo-Álvarez, J. M. Escudero-Martínez, J. M. Requena, C. Ordóñez, A. Desideri, Y. Pérez-Pertejo and R. M. Reguera, *Biochem. Pharmacol.*, 2013, **85**, 1433.
98. E. Calvo-Álvarez, K. Stamatakis, C. Punzón, R. Álvarez-Velilla, A. Tejería, J. M. Escudero-Martínez, Y. Pérez-Pertejo, M. Fresno, R. Balaña-Fouce and R. M. Reguera, *PLoS Negl. Trop. Dis.*, 2015, **9**, e0003666.
99. E. M. Proulx, A. Desormeaux, F. J. Marquis, M. Olivier and G. M. Bergeron, *Antimicrob. Agents Chemother.*, 2001, **45**, 2623.
100. S. Kummar, A. Chen, M. Gutierrez, T. D. Pfister, L. Wang, C. Redon, W. M. Bonner, W. Yutzy, Y. Zhang, R. J. Kinders, J. Ji, D. Allen, J. M. Covey, J. L. Eiseman, J. L. Holleran, J. H. Beumer, L. Rubinstein, J. Collins, J. Tomaszewski, R. Parchment, Y. Pommier and J. H. Doroshow, *Cancer Chemother. Pharmacol.*, 2016, **78**, 73.
101. S. Antony, M. Jayaraman, G. Laco, G. Kohlhagen, K. W. Kohn, M. Cushman and Y. Pommier, *Cancer Res.*, 2003, **63**, 7428.
102. Y. Pommier and M. Cushman, *Mol. Cancer Ther.*, 2009, **5**, 1008.
103. Y. Pommier and C. Marchand, *Nat. Rev. Drug Discov.*, 2012, **11**, 25.
104. R. Balaña-Fouce, C. F. Prada, J. M. Requena, M. Cushman, Y. Pommier, R. Álvarez-Velilla, J. M. Escudero-Martínez, E. Calvo-Álvarez, Y. Pérez-Pertejo and R. M. Reguera, *Antimicrob. Agents Chemother.*, 2012, **56**, 5264.
105. J. P. Gangneux, A. Sulahian, Y. J. Garin and F. Derouin, *J. Antimicrob. Chemother.*, 1997, **40**, 287.
106. S. Ray, B. Hazra, B. Mittra, A. Das and H. K. Majumder, *Mol. Pharmacol.*, 1998, **54**, 994.
107. A. R. Chowdhury, S. Mandal, A. Goswami, M. Ghosh, L. Mandal, D. Chakraborty, A. Ganguly, G. Tripathi, S. Mukhopadhyay, S. Bandyopadhyay and H. K. Majumder, *Mol. Med.*, 2003, **9**, 26.
108. B. Mittra, A. Saha, A. R. Chowdhury, C. Pal, S. Mandal, S. Mukhopadhyay, S. Bandyopadhyay and H. K. Majumder, *Mol. Med.*, 2000, **6**, 527.

109. B. B. Das, N. Sen, A. Roy, S. B. Dasgupta, A. Ganguly, B. C. Mohanta, B. Dinda and H. K. Majumder, *Nucl. Acids Res.*, 2006, **34**, 1121.
110. J. L. Vennerstrom, J. K. Lovelace, V. B. Waits, W. L. Hanson and D. L. Klayman, *Antimicrob. Agents Chemother.*, 1990, **34**, 918.
111. F. J. Marquis, D. Makhey, J. E. LaVoie and M. Olivier, *J. Parasitol.*, 2003, **89**, 1048.
112. A. Tejería, Y. Pérez-Pertejo, R. M. Reguera, R. Balaña-Fouce, C. Alonso, M. Fuertes, M. G. González Rubiales and F. Palacios, *Eur. J. Med. Chem.*, 2016, **124**, 740.
113. N. M. Carballeira, *Chem. Phys. Lipids*, 2013, **172–173**, 58.
114. N. M. Carballeira, N. Montano, R. Alvarez-Velilla, C. F. Prada, R. M. Reguera and R. Balaña-Fouce, *Mar. Drugs*, 2013, **11**, 3661.
115. G. R. Buck and E. L. Zechiedrich, *J. Mol. Biol.*, 2004, **340**, 933.
116. J. M. Berger and J. C. Wang, *Curr. Opin. Struct. Biol.*, 1996, **6**, 84.
117. P. O. Brown and N. R. Cozzarelli, *Proc. Natl. Acad. Sci. U.S.A.*, 1981, **78**, 843.
118. L. F. Liu, T. C. Rowe, L. Yang, K. M. Tewey and G. L. Chen, *J. Biol. Chem.*, 1983, **258**, 15365.
119. A. Bergerat, D. Gadelle and P. Forterre, *J. Biol. Chem.*, 1994, **269**, 27663.
120. D. Rose, W. Thomas and C. Holm, *Cell*, 1990, **60**, 1009.
121. S. M. Gasser, T. Laroche, J. Falzaet, E. B. D. Tour and U. K. Laemmli, *J. Mol. Biol.*, 1986, **188**, 613.
122. Y. Adachi, M. Luke and U. K. Laemmli, *Cell*, 1991, **64**, 137.
123. T. A. Shapiro, *Mol. Cell. Biol.*, 1994, **14**, 3660.
124. T. A. Shapiro and A. F. Showalter, *Mol. Cell. Biol.*, 1994, **14**, 5891.
125. T. A. Shapiro, V. A. Klein and P. T. Englund, *J. Biol. Chem.*, 1989, **264**, 4173.
126. M. E. Lindsay, E. Gluenz, K. Gull and P. T. Englund, *Mol. Microbiol.*, 2008, **70**, 1465.
127. T. Kulikowicz and T. A. Shapiro, *J. Biol. Chem.*, 2006, **281**, 3048.
128. T. Sengupta, M. Mukherjee, C. Mandal, A. Das and H. K. Majumder, *Nucl. Acids Res.*, 2003, **31**, 5305.
129. T. Sengupta, M. Mukherjee, A. Das, C. Mandal, R. Das, T. Mukherjee and H. K. Majumder, *Biochem. J.*, 2005, **390**, 419.
130. T. Sengupta, M. Mukherjee, R. Das, A. Das and H. K. Majumder, *Nucl. Acids Res.*, 2003, **33**, 2364.
131. H. Wei, A. J. Ruttenburg, S. K. Bechis and G. L. Verdine, *J. Biol. Chem.*, 2005, **280**, 37041.
132. L. Aravind, D. D. Leipe and E. V. Koonin, *Nucl. Acids Res.*, 1998, **26**, 4205.
133. L. Bjergbaek, P. Kingma, I. S. Nielsen, I. Wang, O. Westergaard, N. Osheroff and A. M. Andersen, *J. Biol. Chem.*, 2000, **275**, 13041.
134. R. J. Reece and A. Maxwell, *Nucl. Acids Res.*, 1991, **19**, 1399.
135. A. D. Bates and A. Maxwell, *Biochemistry*, 2007, **46**, 7929.
136. C. Sissi and M. Palumbo, *Nucl. Acids Res.*, 2009, **37**, 702.

137. A. Das, A. Dasgupta, S. Sharma, M. Ghosh, T. Sengupta, S. Bandopadhyay and H. K. Majumder, *Nucl. Acids Res.*, 2001, **29**, 1844.
138. T. Hanke, M. J. Ramiro, S. Trigueros, J. Roca and V. Larraga, *Nucl. Acids Res.*, 2003, **31**, 4917.
139. T. Kulikowicz and T. A. Shapiro, *J. Biol. Chem.*, 2006, **281**, 3048.
140. D. R. Byrd and S. W. Matson, *Mol. Microbiol.*, 1997, **25**, 1011.
141. Z. Li, A. Mondragón, H. Hiasa, K. J. Marians and R. J. DiGate, *Mol. Microbiol.*, 2000, **35**, 888.
142. T. Viard and C. B. de la Tour, *Biochimie*, 2007, **89**, 456.
143. J. R. Scocca and T. A. Shapiro, *Mol. Microbiol.*, 2008, **67**, 820.
144. M. Duguet, M. C. Serre and C. Bouthier de La Tour, *J. Mol. Biol.*, 2006, **359**, 805.
145. B. Banerjee, N. Sen and H. K. Majumder, *Enzyme Res.*, 2011, **2011**, 230542.
146. H. S. Kim and G. A. Cross, *PLoS Pathog.*, 2010, **6**, e1000992.
147. K. A. Werbovetz, P. G. Spoors, R. D. Pearson and T. L. Macdonald, *Mol. Biochem. Parasitol.*, 1994, **65**, 1.
148. S. A. Gamage, D. P. Figgitt, S. J. Wojcik, R. K. Ralph, A. Ransijn, J. Mauel, V. Yardley, D. Snowdon, S. L. Croft and W. A. Denny, *J. Med. Chem.*, 1997, **40**, 2634.
149. E. Nenortas, C. Burri and T. A. Shapiro, *Antimicrob. Agents Chemother.*, 1999, **43**, 2066.
150. T. M. Cortázar, G. H. Coombs and J. Walker, *Exp. Parasitol.*, 2007, **116**, 475.
151. A. Deterding, F. A. Dungey, K. A. Thompson and D. Steverding, *Acta Trop.*, 2005, **93**, 311.
152. G. Singh and C. S. Dey, *Acta Trop.*, 2007, **103**, 172.
153. V. Gutiérrez, A. B. Seabra, R. M. Reguera, J. Khandare and M. Calderón, *Chem. Soc. Rev.*, 2016, **45**, 152.
154. J. Yang, A. Bogni, E. G. Schuetz, M. Ratain, M. E. Dolan, H. McLeod, L. Gong, C. Thorn, M. V. Relling, T. E. Klein and R. B. Altman, *Pharmacogenet. Genomics*, 2009, **19**, 552.
155. M. Pendleton, R. H. Lindsey, Jr., C. A. Felix, D. Grimwade and N. Osheroff, *Ann. N.Y. Acad. Sci.*, 2014, **1310**, 98.

CHAPTER 18

Molecular Basis of Drug Resistance in Leishmania

ANNELIES MONDELAERS, SARAH HENDRICKX, GUY CALJON* AND LOUIS MAES*

Laboratory for Microbiology, Parasitology and Hygiene (LMPH), University of Antwerp, Universiteitsplein 1, B-2610 Wilrijk (Antwerp), Belgium
*Email: guy.caljon@uantwerpen.be; louis.maes@uantwerpen.be

18.1 Emerging Drug Resistance

For several decades, pentavalent antimonials have been the treatment of choice to treat visceral leishmaniasis (VL).[1] Starting from the 1970s, treatment failures have been increasing in the highly endemic Indian subcontinent, with clinical isolates from post-treatment relapses displaying reduced drug susceptibility *in vitro*.[2] The spread of antimony (Sb) resistance has enforced alternative therapies in these endemic areas, using drugs such as amphotericin B (AmB), miltefosine (MIL) and paromomycin (PMM). Of great concern is that resistant isolates to these drugs have already been reported as well,[3–5] imposing the need for cautious use of the limited number of available anti-leishmanial drugs. To manage and counteract further emergence and spread of drug resistance, a clear understanding of the resistance mechanisms involved remains fundamental. As the development of a drug-resistant phenotype is often multifactorial,[6] large-scale evaluation of clinical isolates with the ambition to identify molecular resistance markers has already proven to be very challenging and complex.[3,7–10] Moreover, this approach requires large and diverse sets of resistant clinical isolates, which for the more recently introduced drugs are not available. Considering the

Drug Discovery Series No. 60
Drug Discovery for Leishmaniasis
Edited by Luis Rivas and Carmen Gil

Published by the Royal Society of Chemistry, www.rsc.org

need for a more proactive approach to combat development of drug resistance, laboratory-generated resistant mutants may provide essential novel insights into potential resistance mechanisms and drug-susceptibility monitoring.

18.2 Methods to Study Resistance

Drug resistance in *Leishmania* can be studied in several ways. A first approach involves comparative evaluation of isolates derived from infected patients with either cure, treatment relapse or treatment failure. However, matched pairs of strains (isolated pre- and post-treatment) are generally not available, precluding proper interpretation of shifting drug-susceptibility. Although unavoidable, the sometimes long adaptation process from initial isolation at the level of the patient until evaluation in the laboratory may further hamper the acquisition of representative clinical isolates, partly related to changes in virulence and genomic profile.[11,12] Another approach is to experimentally select for less susceptible parasites or artificially induce mutations under laboratory conditions. This way, matched pairs (susceptible parent and derived resistant strains) can be obtained with similar genetic and phenotypic backgrounds, also enabling proactive research for (novel) drugs for which resistant clinical isolates are not available. Laboratory selection of drug resistance has generally been performed on axenic promastigotes by gradually increasing the drug concentration in the culture medium.[13–17] However, a stage-dependent outcome of the resistance selection method was recently demonstrated for paromomycin,[15] highlighting the necessity to consider the more relevant intracellular amastigote stage for selection of resistance, whenever possible. Selection on intracellular amastigotes can be achieved either *in vitro* by repeated drug exposure of infected host cells,[4] or *in vivo* by conducting recurrent treatment–relapse cycles in the Syrian golden hamster model.[18] Proactive research on drug resistance before it has emerged in the field may be supportive in (i) elucidation of the putative mechanism-of-action and associated resistance mechanisms; (ii) suggesting alternative therapeutic strategies to overcome drug resistance, and (iii) inclusion of resistant strains in current screening programs for novel drugs to evaluate potential cross-resistance. Whether these laboratory-generated strains will be fully predictive for clinical resistance is still open to debate and can only be confirmed once large-scale comparison with resistant clinical isolates becomes possible.

18.3 Molecular Mechanisms of Resistance and Current Limitations

Identification of the molecular mechanisms of resistance is often hampered by limited knowledge of the mechanism-of-action, which still often relies on hypotheses formulated over the past years. For example and despite their long term implementation, the mechanism-of-action of antimonials is even currently not yet fully understood. Based on ongoing intensive research on clinical isolates and the characterization of laboratory-derived resistant mutants, significant advances are being made for the most widely used anti-leishmanial reference drugs, *i.e.* Sb, MIL, PMM and AmB.

18.3.1 Antimonials

It has been established that pentavalent antimonials [Sb(v)] behave as prodrugs that need to be reduced into their trivalent form [Sb(III)] in order to acquire their full anti-leishmanial potential. Several aspects, including intracellular accumulation, activation and target inhibition, are important for their action and therefore provide an opportunity for the development of Sb resistance.

The results of biochemical assays indicate that the uptake of Sb varies according to oxidation state, given that uptake rates of Sb(III) and Sb(v) do not alter if they are co-administered. Sb(III) is taken up by the parasite through an aquaglyceroporin (AQP1), a transporter for water and uncharged polar solutes that plays an important role in osmoregulation.[19] This uptake can be inhibited by the related metal arsenite [As(III)]. In contrast, Sb(v) uptake is not sensitive to inhibition by As(III) and the results of inhibition experiments with gluconate indicate that it might rely on recognition of the gluconate moieties of stibogluconate.[20] Another hypothesis raised is that Sb(v) enters the cell through a phosphate transporter, which is a mechanism also found in *Escherichia coli* and *Saccharomyces*.[21] Results of experiments with verapamil, a general efflux pump inhibitor, indicate that outward transport from the parasite plays a role in the intracellular drug accumulation.[22] Several intracellular ABCC proteins (MRPA/ABCC3, ABCC4, ABCC5, and PRP1/ABCC7), a subfamily of ATP-binding cassette (ABC) proteins, can confer resistance to Sb by sequestering Sb(III) in complex with trypanothione (TSH) or glutathione (GSH) in vesicles close to the flagellar pocket.[23,24] Overexpression of PRP1/ABCC7 conferred cross-resistance to pentamidine and Sb(III) but not Sb(v).[25] MRPA/ABCC3 was shown to be able to sequester both Sb(v) and Sb(III).[26]

The mechanism of reduction is still enigmatic and could occur in the amastigote as well as in host cells, such as the macrophage. Some reports describe a higher susceptibility of axenic amastigotes as compared with promastigotes, despite the fact that both stages can take up the pentavalent prodrug and are sensitive to Sb(III), indicating that reduction of Sb occurs inside the parasite and relies on stage-specific pathways.[27,28] Other studies could not detect Sb(v) activity against axenic amastigotes in a clinically relevant concentration range and found a higher susceptibility of intracellular amastigotes indicating intra-macrophage reduction.[29] These conflicting findings might result from inter-strain variation, differences in the axenization and hence the amastigote phenotype and assay conditions. Thiol-dependent reductase I (TDR1), a parasite enzyme involved in redox regulation, was found to be able to catalyze the reduction of Sb(v) in the presence of GSH *in vitro*.[30] Results from uptake studies have indicated that amastigotes take up Sb(v) approximately five times more efficiently than promastigotes,[28] indicating that stage differences can relate to rates of uptake and reduction. Nevertheless, it has been shown that Sb(v) reduction can also take place inside the macrophage,[31] although the conversion is probably not very efficient.[32]

Sb(v) in complex with gluconate has been described as affecting glucose catabolism and fatty acid beta-oxidation in the parasite.[33] Once converted from the Sb(v) metalloid into the active Sb(III) state, it is known to influence the parasite redox metabolism and antioxidant defenses resulting in an apoptosis-like cell death with concomitant DNA fragmentation and exposure of phosphatidylserine.[34–36] The major targets of Sb(III)/As(III) in the parasite are the inhibition of TSH-reductase, GSH-reductase and GSH-synthase.[37,38] *In vitro* treatment of *Leishmania* with Sb(III) also results in efflux of TSH and GSH, which are important antioxidants to protect the parasites against oxidative stress.[38] Also in macrophages, Sb(III) has a similar activity on antioxidant efflux, decreasing intracellular free GSH levels resulting in higher relative levels of GSH-disulphide.[32] In addition, antimonials can inhibit protein tyrosine phosphatases [with a prominent activity against Src homology PTPase1 (SHP-1)] resulting in tyrosine phosphorylation of cellular proteins and a modified cytokine responsiveness, and can activate phosphoinositide 3-kinase (PI3K), protein kinase C (PKC) and mitogen-activated protein kinases (MAPKs), leading to elevated levels of reactive oxygen species (ROS) and NO with anti-parasitic activity.[39,40] It is clear from these reports that antimonials exert a broad spectrum of activities, either directly against the parasite or indirectly through immune activation. It has been shown in experimental VL infections that Sb treatment heavily depends on a functional host immune system with functional T cells and Th1 responsiveness with a pro-inflammatory cytokine profile (IL-12, IFN-γ, TNF-α) in order to control hepatic parasite burdens prior to and following treatment.[41–44] Also, IL-4 plays important roles, with Sb being less effective against *Leishmania donovani* in the absence of this cytokine.[45] IL-4$^{-/-}$ mice surprisingly had lower IFN-γ levels following drug treatment, indicating an important role for IL-4 in modulation of IFN-γ production as a determining factor for treatment outcome.

Several studies have demonstrated the possibility of experimentally alter the susceptibility of parasites towards Sb. For instance, deletion of the *AQP1* gene or selection of mutants with telomeric deletions on chromosome 34 affecting *AQP1* result in resistance against Sb(III)[46,47] whereas overexpression of this transporter dramatically increases susceptibility.[48] *In vitro*-selected resistant *Leishmania* strains were shown to have amplified extra-chromosomal circular DNA forms or episomes, the H circles, harboring the ABC transporter MRPA/ABCC3.[49,50] Overexpression of the MRPA efflux transporter is responsible for higher levels of sequestration of Sb(III) into vesicles, conferring a higher resistance.[23,24] Laboratory selection of As(III)- and Sb(III)-resistant parasite lines also revealed changes of intracellular thiol

(GSH and TSH) levels that could be responsible for increased efflux of the Sb(III)–thiol complexes and higher levels of protection against Sb(III).[51,52] Several enzymes involved in the regulation of the biosynthesis of TSH or its precursors spermidine and GSH determine drug susceptibility, with elevated expression of the ornithine decarboxylase (ODC) and γ-glutamyl cysteine synthetase (γ-GSH) enzymes being associated with increased resistance.[53–55] Also, changes in saturation levels of glycophospholipids with an altered membrane fluidity and potential changes in drug uptake and/or sequestration have been found in parasite lines resistant to Sb(III) alone or to Sb(III) in combination with other drugs (AmB or PMM).[56,57] In addition, changes in the proline metabolism were noted that could account for conferring the observed higher levels of resistance to ROS, given that this amino acid has been described as acting as a radical scavenger.[58] Results from comparative proteomic analysis of several *in vitro*-selected Sb-resistant and -susceptible *Leishmania* species lines also indicate other potential components directly or indirectly implicated in resistance, with differential levels of proteins involved in antioxidant defense, general stress response, metabolism and surface and cytoskeletal organization.[59–63] These observations indicate that a range of adaptations can contribute to elevated resistance levels, but these artificially induced changes may not necessarily be representative of the natural situation.

From the Indian Subcontinent (ISC) where Sb resistance is highly prevalent, whole-genome sequencing of 204 *L. donovani* clinical isolates provided a comprehensive insight into the mechanisms of resistance occurring naturally in that area.[64] A high proportion (about 25%) of the sequenced clinical isolates belonged to a genetically distinct population. These parasites were frequently highly resistant to Sb and were all characterized by an insertion of two base pairs in the two AQP1 alleles, resulting in a frameshift and premature stop codon.[64] The affected C-terminal part constitutes a significant proportion of the AQP1 protein, probably heavily affecting the formation and function of the tetrameric transmembrane channel, hence hampering Sb(III) uptake. Evidence for the spreading of this mutated allele conferring increased Sb resistance into the parasite population was provided by the finding of heterozygous hybrids. Beside mutations in the AQP1 locus, reduced levels of transcription of the wild-type gene have also been associated with increased resistance.[65–67] Other genomic features in the clinical isolates from the ISC are the increased copy number of two multigenic intrachromosomal duplicated regions, MAPK1 and the H-locus.[64,68] The H-locus is a 13.5-kb region on chromosome 23 that includes the MRPA/ABCC3-encoding gene, which is implicated in Sb(III) efflux and resistance. Also, circular forms of the H locus carrying the MRPA gene were found in clinical *L. donovani* isolates from relapse patients.[68,69] The MAPK locus is a 15.8-kb segment on chromosome 36 with four genes, of which some have been implicated in virulence,[70] but a link with drug resistance has not yet been described. Results of proteomics studies revealed a relationship between resistance of clinical *L. donovani* isolates and overexpression of the cysteine–leucine-rich protein (CLrP), a DNA-binding glycoprotein present in the membrane and nucleus.[60,62] Results of other studies in *L. donovani* have indicated elevated expression of the MRPA efflux transporter and/or increased expression of the ODC, γ-GSH and TR genes that affect intracellular thiol levels.[66,69,71] In *Leishmania guyanensis*, γ-GSH was overexpressed in isolates from relapsed patients, which is anticipated to affect intracellular thiol levels. Also in Sb(v)-resistant *Leishmania braziliensis* clinical isolates, elevated transcriptional levels of ODC and TR were recorded.[72] Interestingly, the presence of the double-stranded RNA virus Leishmaniavirus 1 (LRV1) in *L. braziliensis* was found to be associated with an increased risk of Sb(v) treatment failure,[73] found not to relate to an intrinsically decreased drug susceptibility of the parasite. A similar role of the LRV1 virus was described in *L. guyanensis*,[74] where it was found to affect pentamidine treatment outcome, further corroborating the hypothesis that the virus influences the host–parasite interplay. It is obvious that the immune status of the host is an important determinant of treatment outcome, as has also been shown in animal models. HIV infection and a range of other medical conditions (related to transplantation medicine, rheumatology, hematology and oncology) resulting in an immunocompromised state, coincide with hampered immunological control and a higher risk of reactivation after treatment.[75,76]

18.3.2 Miltefosine

Miltefosine (MIL) belongs to the class of alkyl-lysophospholipids (ALP), and although various hypotheses have been formulated over the years, the molecular basis of its leishmanicidal activity is still not yet fully understood. Due to its amphiphilic properties, MIL can freely incorporate into phospholipid monolayers.[77] Two entry pathways play a role in the intracellular drug accumulation, the endocytic pathway and entry through flippase activity. The latter relies on a transporter complex composed of two different membrane proteins: the aminophospholipid miltefosine transporter (MT) (a member of the P4 subfamily of P-type ATPases) and its regulatory beta-subunit ROS3 (a member of the lem3/CDC50-like protein family).[78,79] Two inverse mechanisms contribute to export MIL from the cell, *i.e.* exocytosis and protein-dependent floppase activity relying on members of the ABC-transporter family.[80,81]

The activity of MIL results in mitochondrial dysfunction[82,83] and an apoptotic-like cell death.[84–87] The induced mitochondrial dysfunction is associated with a dose-dependent inhibition of cytochrome-*c*-oxidase that is essential in the electron transport chain. ROS accumulation, loss of mitochondrial membrane potential, release of cytochrome *c* into the cytosol and activation of cellular proteases have been documented as outcomes of MIL treatment.[88,89] MIL was also found to modify the fatty acids and sterol metabolism with decreased phosphatidylcholine and increased phosphatidylethanolamine content, changes in the ether-lipid and alkyl-lipid metabolism and in the biosynthesis of alkyl-anchored glycoproteins and glycolipids.[90–92] Perturbation of the ether-lipid metabolism in *Leishmania* parasites was linked to inhibition of glycosomal alkyl-specific acyl-CoA acyltransferase.

In addition to a direct anti-parasitic activity, MIL was also reported to induce pro-inflammatory responses and to promote a shift towards a Th1-biased immune response.[93–96] MIL was found to elevate the levels of inflammatory cytokines such as tumor necrosis factor-α (TNF-α), interleukin-12 (IL-12) and ROS. MIL also triggers an increased expression of pattern-recognition molecules, such as Toll-like receptor (TLR)-4 and TLR-9, and expression of MHC I in mononuclear cells in favor of increased anti-parasitic responsiveness.[97,98]

The increasingly frequent MIL-treatment relapses in the Indian subcontinent[99–102] could not be directly linked to altered drug susceptibility using the standard *in vitro* susceptibility assays, indicating that other parasitological, host and drug factors should be taken into consideration.[99,103,104] *Leishmania infantum* strains with a MIL-resistant phenotype have already been isolated from HIV-co-infected patients in Europe[3] while also *Leishmania chagasi* isolates obtained from VL-patients in Brazil showed variable intrinsic resistance to MIL.[105] MIL-resistance was also explored in experimentally *in vitro*-selected resistant strains with the common feature of decreased intracellular MIL accumulation linked to a defective MT/ROS3 translocation machinery.[78,79] Expression studies revealed that both MT and ROS3 act as limiting factors for translocase activity affecting MIL-susceptibility.[78,106,107] The acquisition of loss-of-function mutations within *MT* or *ROS3* therefore has a great effect on drug susceptibility. The intracellular cytoplasmic loops of the MT were found to be subjected to the highest mutational pressure in *Leishmania amazonensis* and *L. infantum* promastigotes.[108,109] In *L. donovani* clinical isolates with variable response to MIL and in *in vitro* MIL-resistant strains, decreased mRNA expression of both *MT* and *ROS3* genes was reported, indicating that not only loss-of-function mutations but also transporter expression levels play an important determining role in MIL-resistance levels.[103,110–114]

Overexpression of ATP-binding cassette (ABC) transporters is another mechanism that could cause a decline in intracellular MIL-accumulation, potentially expediting the emergence of drug resistance.[115] It was shown that different cell lines overexpressing P-glycoprotein multidrug resistant (MDR) 1 (ABCB subfamily) exhibit resistance to MIL.[116] The results of studies of promastigotes revealed that the overexpression of some other ABC transporters (ABCA7, ABCG5 and ABCG2, ABCG4 and ABCG6) caused acquisition of MIL-resistance by inducing a higher efflux of MIL.[81,113,117] However, the actual implication of this efflux mechanism in clinical resistance remains to be demonstrated.

Next to mechanisms affecting drug uptake and efflux, other adaptations may also contribute to MIL-resistance. For example, MIL-resistant *L. donovani* promastigotes were found to differentially transcribe genes involved in a broad range of processes, such as DNA repair and replication, protein translation and folding, lipid biosynthesis, energy metabolism, antioxidant defense, stress response, membrane structure and amino acid and folate metabolism.[113,114] Whole-genome sequencing revealed mutations in some of these differentially expressed genes[114] while gene expression analysis clearly indicated an enhanced antioxidant defense mechanism to circumvent the cytotoxic effects of MIL.[113,118,119] Proteomic profiling of MIL-resistant clinical isolates revealed differential expression of heat-shock proteins and stress-induced proteins, histone H2A and proteins related to redox homeostasis.[105,120] Metabolomics studies in *L. donovani* additionally revealed alterations in the lipid metabolism of MIL-resistant parasites, with adaptive involvement of the Kennedy pathway responsible for synthesis of phosphatidylcholine and phosphodimethylethanolamine.[112,121] Alterations in C-24-sterol alkylation and fatty acid saturation and elongation may affect membrane fluidity and cause perturbation of drug–membrane interactions.[122] Another biological feature contributing to lower membrane fluidity is the presence of elevated cholesterol levels.[122] MIL-resistant parasites also presented higher levels of amino acids and components of the polyamine biosynthetic pathway,[123,124] anticipated to result in MIL-resistant parasites with a higher tolerance to oxidative stress.

18.3.3 Paromomycin

Paromomycin (PMM), an aminoglycoside antibiotic with a proven efficacy against *Leishmania*, associates with the parasite's negatively charged glycocalyx and is subsequently taken up *via* endocytosis.[16] As PMM-resistant clinical isolates are very scarce so far,[4] not much is known about its mode-of-action nor the molecular basis of resistance. In bacteria, PMM inhibits protein synthesis by binding to a specific region in the bacterial 16s rRNA of cytosolic ribosomes, which is the codon–anticodon recognition site, thereby causing single-base-pair mismatches.[125,126] Although PMM's mechanism-of-action against *Leishmania* is still unclear, a similar action mechanism may be expected as there are a lot of structural similarities between *Leishmania* and eubacterial mitochondrial ribosomes.[127] Once PMM has entered the parasite, it will bind to the 30S ribosomal subunit, thereby impairing protein synthesis.[128] PMM is also reported to alter membrane fluidity and permeability[16,129,130] and known to cause alterations in mitochondrial activity, leading to respiratory dysfunction and associated alterations in the mitochondrial membrane potential.[16,130,131] Based on the results from the few available studies on promastigote-selected resistant strains, some putative resistance mechanisms have been proposed, such as decreased drug uptake,[14,16,131] increased membrane fluidity, leading to a decreased PMM-accumulation or increased expression of several proteins associated with drug resistance (MDR1, MRPA and protein phosphatase 2A).[132] However, given the stage-dependent outcome of drug resistance selection,[15] these findings definitly deserve validation on amastigote-selected resistant strains and/or clinical isolates.

18.3.4 Amphotericin B

Amphotericin B (AmB) is well-known for its broad antifungal properties and is extensively used for systemic fungal infections, for example those caused by *Candida albicans* and *Aspergillus fumigatus*, and has also been proven to be highly effective against leishmaniasis. Since its discovery, extensive studies have been completed to unravel its precise mode of action and it is currently accepted that AmB exerts its action by binding to sterols, especially ergosterol, in the cell membrane, forming channels across the cell membrane, causing leakage of ions and other molecules from the cell, leading to cell death.[133] Binding of AmB to sterols in host–cell membranes would also reduce binding of parasites to the cell membrane and hence diminish parasite uptake.[134] Finally, the presence of intracellular AmB probably causes oxidative-stress-related cell death[5] and mediates the host's immune responses.[135]

In endemic regions with widespread Sb resistance, AmB has gradually replaced Sb(v) as preferred first-line therapy. However, the current extensive use of AmB may now be threatened by the emergence of acquired AmB-resistance. Based on research on experimentally induced strains and on AmB-resistant *L. donovani* field isolates, three mechanisms by which *Leishmania* could resist AmB-pressure have been described.[5] First, AmB-resistant strains could replace ergosterol in their cell membrane by the ergosterol precursor cholesta-5,7,24-trien-3β-ol,[136–138] either leading to increased cell membrane fluidity reducing membrane depolarization, or decreased AmB-binding capacity with subsequent reduced AmB-uptake.[137] Additionally, resistant parasites can efficiently upregulate their efflux machinery and further reduce intracellular AmB, resulting in a diminished oxidative stress response. A last mechanism by which AmB-resistant parasites could resist AmB-pressure is the overexpression of effective ROS-scavenging machinery.[5]

18.4 Conclusions

The emerging resistance observed to all current anti-leishmanial reference drugs is quite alarming given the low number of alternatives in the current drug discovery pipeline and explains why studying resistance mechanisms in experimentally-derived resistant isolates will remain pivotal for the near future. Along with close monitoring of the still existing monotherapy-based treatment strategies, the shift towards combination therapy definitely deserves in-depth consideration to delay the onset and spread of drug resistance.

Acknowledgements

The research of the authors was funded by the Research Fund Flanders (FWO: project G051812N and postdoctoral scholarship 12I0317N for SH), the

Agency for Innovation by Science and Technology in Flanders (IWT: project 121474 for AM), the research fund of the University of Antwerp (TT-ZAPBOF 33049 for GC) and additionally supported by the Drugs for Neglected Diseases initiative (DNDi, Geneva, Switzerland). The LMPH is a partner of the 'Antwerp Drug Discovery Network' (ADDN, www.addn.be) and the Excellence Centre 'Infla-Med' (www.uantwerpen.be/infla-med).

References

1. S. Sundar and J. Chakravarty, *Expert Opin. Pharmacother.*, 2015, **16**, 237.
2. N. Singh, *Indian J. Med. Res.*, 2006, **123**, 411.
3. S. Cojean, S. Houze, D. Haouchine, F. Huteau, S. Lariven, V. Hubert, F. Michard, C. Bories, F. Pratlong, B. J. Le, P. M. Loiseau and S. Matheron, *Emerging Infect. Dis.*, 2012, **18**, 704.
4. S. Hendrickx, G. Boulet, A. Mondelaers, J. C. Dujardin, S. Rijal, L. Lachaud, P. Cos, P. Delputte and L. Maes, *Parasitol. Res.*, 2014, **113**, 1875.
5. B. Purkait, A. Kumar, N. Nandi, A. H. Sardar, S. Das, S. Kumar, K. Pandey, V. Ravidas, M. Kumar, T. De, D. Singh and P. Das, *Antimicrob. Agents Chemother.*, 2012, **56**, 1031.
6. H. C. Maltezou, *J. Biomed. Biotechnol.*, 2010, **2010**, 617521.
7. E. Kazemi-Rad, M. Mohebali, M. B. Khadem-Erfan, M. Saffari, R. Raoofian, H. Hajjaran, R. Hadighi, A. Khamesipour, S. Rezaie, H. Abedkhojasteh and M. Heidari, *Exp. Parasitol.*, 2013, **135**, 344.
8. A. Nuhs, C. Schafer, D. Zander, L. Trube, N. P. Tejera, S. Schmidt, J. Arevalo, V. Adaui, L. Maes, J. C. Dujardin and J. Clos, *Int. J. Parasitol. Drugs Drug Resist.*, 2014, **4**, 37.
9. M. Vanaerschot, S. Decuypere, T. Downing, H. Imamura, O. Stark, D. S. De, S. Roy, B. Ostyn, L. Maes, B. Khanal, M. Boelaert, G. Schonian, M. Berriman, F. Chappuis, J. C. Dujardin, S. Sundar and S. Rijal, *J. Infect. Dis.*, 2012, **206**, 752.
10. Ashutosh, S. Sundar and N. Goyal, *J. Med. Microbiol.*, 2007, **56**, 143.
11. D. Moreira, N. Santarem, I. Loureiro, J. Tavares, A. M. Silva, A. M. Amorim, A. Ouaissi, A. Cordeiro-da-Silva and R. Silvestre, *PLoS Neglected Trop. Dis.*, 2012, **6**, e1469.
12. A. M. Silva, A. Cordeiro-da-Silva and G. H. Coombs, *PLoS Neglected Trop. Dis.*, 2011, **5**, e1451.
13. K. Seifert, S. Matu, F. Javier Perez-Victoria, S. Castanys, F. Gamarro and S. L. Croft, *Int. J. Antimicrob. Agents*, 2003, **22**, 380.
14. M. Maarouf, M. T. Adeline, M. Solignac, D. Vautrin and M. Robert-Gero, *Parasite*, 1998, **5**, 167.
15. S. Hendrickx, R. A. Inocencio da Luz, V. Bhandari, K. Kuypers, C. D. Shaw, J. Lonchamp, P. Salotra, K. Carter, S. Sundar, S. Rijal, J. C. Dujardin, P. Cos and L. Maes, *PLoS Neglected Trop. Dis.*, 2012, **6**, e1664.

16. A. Jhingran, B. Chawla, S. Saxena, M. P. Barrett and R. Madhubala, *Mol. Biochem. Parasitol.*, 2009, **164**, 111.
17. D. Sereno, E. Guilvard, S. Maquaire, M. Cavaleyra, P. Holzmuller, A. Ouaissi and J. L. Lemesre, *Acta Trop.*, 2001, **80**, 195.
18. S. Hendrickx, A. Mondelaers, E. Eberhardt, P. Delputte, P. Cos and L. Maes, *Antimicrob. Agents Chemother.*, 2015, **59**, 4714.
19. M. Plourde, J. M. Ubeda, G. Mandal, R. L. Monte-Neto, R. Mukhopadhyay and M. Ouellette, *Mol. Biochem. Parasitol.*, 2015, **201**, 108.
20. C. Brochu, J. Wang, G. Roy, N. Messier, X. Y. Wang, N. G. Saravia and M. Ouellette, *Antimicrob. Agents Chemother.*, 2003, **47**, 3073.
21. B. P. Rosen, *Comp. Biochem. Physiol. A Mol. Integr. Physiol.*, 2002, **133**, 689.
22. R. Valiathan, M. L. Dubey, R. C. Mahajan and N. Malla, *Exp. Parasitol.*, 2006, **114**, 103.
23. D. Legare, D. Richard, R. Mukhopadhyay, Y. D. Stierhof, B. P. Rosen, A. Haimeur, B. Papadopoulou and M. Ouellette, *J. Biol. Chem.*, 2001, **276**, 26301.
24. P. Leprohon, D. Legare and M. Ouellette, *Antimicrob. Agents Chemother.*, 2009, **53**, 2646.
25. A. C. Coelho, S. M. Beverley and P. C. Cotrim, *Mol. Biochem. Parasitol.*, 2003, **130**, 83.
26. K. El Fadili, N. Messier, P. Leprohon, G. Roy, C. Guimond, N. Trudel, N. G. Saravia, B. Papadopoulou, D. Legare and M. Ouellette, *Antimicrob Agents Chemother.*, 2005, **49**, 1988.
27. M. Ephros, A. Bitnun, P. Shaked, E. Waldman and D. Zilberstein, *Antimicrob. Agents Chemother.*, 1999, **43**, 278.
28. P. Shaked-Mishan, N. Ulrich, M. Ephros and D. Zilberstein, *J. Biol. Chem.*, 2001, **276**, 3971.
29. M. Vermeersch, R. I. da Luz, K. Tote, J. P. Timmermans, P. Cos and L. Maes, *Antimicrob. Agents Chemother.*, 2009, **53**, 3855.
30. P. K. Fyfe, G. D. Westrop, A. M. Silva, G. H. Coombs and W. N. Hunter, *Proc. Natl. Acad. Sci. U. S. A.*, 2012, **109**, 11693.
31. C. Hansen, E. W. Hansen, H. R. Hansen, B. Gammelgaard and S. Sturup, *Biol. Trace Elem. Res.*, 2011, **144**, 234.
32. S. Wyllie and A. H. Fairlamb, *Biochem. Pharmacol.*, 2006, **71**, 257.
33. J. D. Berman, J. V. Gallalee and J. M. Best, *Biochem. Pharmacol.*, 1987, **36**, 197.
34. D. Sereno, P. Holzmuller, I. Mangot, G. Cuny, A. Ouaissi and J. L. Lemesre, *Antimicrob. Agents Chemother.*, 2001, **45**, 2064.
35. G. Sudhandiran and C. Shaha, *J. Biol. Chem.*, 2003, **278**, 25120.
36. N. Lee, S. Bertholet, A. Debrabant, J. Muller, R. Duncan and H. L. Nakhasi, *Cell Death Differ.*, 2002, **9**, 53.
37. M. L. Cunningham, M. J. Zvelebil and A. H. Fairlamb, *Eur. J. Biochem.*, 1994, **221**, 285.

38. S. Wyllie, M. L. Cunningham and A. H. Fairlamb, *J. Biol. Chem.*, 2004, **279**, 39925.
39. M. K. Pathak and T. Yi, *J. Immunol.*, 2001, **167**, 3391.
40. J. Mookerjee Basu, A. Mookerjee, P. Sen, S. Bhaumik, P. Sen, S. Banerjee, K. Naskar, S. K. Choudhuri, B. Saha, S. Raha and S. Roy, *Antimicrob. Agents Chemother.*, 2006, **50**, 1788.
41. H. W. Murray, J. D. Berman and S. D. Wright, *J. Infect. Dis.*, 1988, **157**, 973.
42. H. W. Murray, C. Montelibano, R. Peterson and J. P. Sypek, *J. Infect. Dis.*, 2000, **182**, 1497.
43. H. W. Murray, M. J. Oca, A. M. Granger and R. D. Schreiber, *J. Clin. Invest.*, 1989, **83**, 1253.
44. H. W. Murray, A. Jungbluth, E. Ritter, C. Montelibano and M. W. Marino, *Infect. Immun.*, 2000, **68**, 6289.
45. J. Alexander, K. C. Carter, N. Al-Fasi, A. Satoskar and F. Brombacher, *Eur. J. Immunol.*, 2000, **30**, 2935.
46. B. Gourbal, N. Sonuc, H. Bhattacharjee, D. Legare, S. Sundar, M. Ouellette, B. P. Rosen and R. Mukhopadhyay, *J. Biol. Chem.*, 2004, **279**, 31010.
47. A. Mukherjee, S. Boisvert, R. L. Monte-Neto, A. C. Coelho, F. Raymond, R. Mukhopadhyay, J. Corbeil and M. Ouellette, *Mol. Microbiol.*, 2013, **88**, 189.
48. N. Marquis, B. Gourbal, B. P. Rosen, R. Mukhopadhyay and M. Ouellette, *Mol. Microbiol.*, 2005, **57**, 1690.
49. M. Ouellette and P. Borst, *Res. Microbiol.*, 1991, **142**, 737.
50. P. Leprohon, D. Legare, F. Raymond, E. Madore, G. Hardiman, J. Corbeil and M. Ouellette, *Nucleic Acids Res.*, 2009, **37**, 1387.
51. R. Mukhopadhyay, S. Dey, N. Xu, D. Gage, J. Lightbody, M. Ouellette and B. P. Rosen, *Proc. Natl. Acad. Sci. U. S. A.*, 1996, **93**, 10383.
52. D. Legare, B. Papadopoulou, G. Roy, R. Mukhopadhyay, A. Haimeur, S. Dey, K. Grondin, C. Brochu, B. P. Rosen and M. Ouellette, *Exp. Parasitol.*, 1997, **87**, 275.
53. C. Guimond, N. Trudel, C. Brochu, N. Marquis, A. El Fadili, R. Peytavi, G. Briand, D. Richard, N. Messier, B. Papadopoulou, J. Corbeil, M. G. Bergeron, D. Legare and M. Ouellette, *Nucleic Acids Res.*, 2003, **31**, 5886.
54. A. Haimeur, C. Guimond, S. Pilote, R. Mukhopadhyay, B. P. Rosen, R. Poulin and M. Ouellette, *Mol. Microbiol.*, 1999, **34**, 726.
55. K. Grondin, A. Haimeur, R. Mukhopadhyay, B. P. Rosen and M. Ouellette, *EMBO J.*, 1997, **16**, 3057.
56. M. Berg, R. Garcia-Hernandez, B. Cuypers, M. Vanaerschot, J. I. Manzano, J. A. Poveda, J. A. Ferragut, S. Castanys, J. C. Dujardin and F. Gamarro, *Antimicrob. Agents Chemother.*, 2015, **59**, 2242.
57. M. Berg, M. Vanaerschot, A. Jankevics, B. Cuypers, I. Maes, S. Mukherjee, B. Khanal, S. Rijal, S. Roy, F. Opperdoes, R. Breitling and J. C. Dujardin, *Mol. Microbiol.*, 2013, **90**, 428.

58. S. Kaul, S. S. Sharma and I. K. Mehta, *Amino Acids*, 2008, **34**, 315.
59. M. C. Brotherton, S. Bourassa, P. Leprohon, D. Legare, G. G. Poirier, A. Droit and M. Ouellette, *PLoS One*, 2013, **8**, e81899.
60. S. Das, P. Shah, R. Tandon, N. K. Yadav, A. A. Sahasrabuddhe, S. Sundar, M. I. Siddiqi and A. Dube, *PLoS Neglected Trop. Dis.*, 2015, **9**, e0003992.
61. K. El Fadili, J. Drummelsmith, G. Roy, A. Jardim and M. Ouellette, *Exp. Parasitol.*, 2009, **123**, 51.
62. A. Kumar, B. Sisodia, P. Misra, S. Sundar, A. K. Shasany and A. Dube, *Br. J. Clin. Pharmacol.*, 2010, **70**, 609.
63. F. S. Matrangolo, D. B. Liarte, L. C. Andrade, M. F. de Melo, J. M. Andrade, R. F. Ferreira, A. S. Santiago, C. P. Pirovani, R. A. Silva-Pereira and S. M. Murta, *Mol. Biochem. Parasitol.*, 2013, **190**, 63.
64. H. Imamura, T. Downing, F. Van den Broeck, M. J. Sanders, S. Rijal, S. Sundar, A. Mannaert, M. Vanaerschot, M. Berg, G. De Muylder, F. Dumetz, B. Cuypers, I. Maes, M. Domagalska, S. Decuypere, K. Rai, S. Uranw, N. R. Bhattarai, B. Khanal, V. K. Prajapati, S. Sharma, O. Stark, G. Schonian, H. P. De Koning, L. Settimo, B. Vanhollebeke, S. Roy, B. Ostyn, M. Boelaert, L. Maes, M. Berriman, J. C. Dujardin and J. A. Cotton, *eLife*, 2016, **5**, e12613.
65. S. Decuypere, S. Rijal, V. Yardley, S. De Doncker, T. Laurent, B. Khanal, F. Chappuis and J. C. Dujardin, *Antimicrob. Agents Chemother.*, 2005, **49**, 4616.
66. S. Rai, Bhaskar, S. K. Goel, U. Nath Dwivedi, S. Sundar and N. Goyal, *PLoS One*, 2013, **8**, e74862.
67. V. Gomez Perez, R. Garcia-Hernandez, V. Corpas-Lopez, A. M. Tomas, J. Martin-Sanchez, S. Castanys and F. Gamarro, *Int. J. Parasitol. Drugs Drug Resist.*, 2016, **6**, 133.
68. T. Downing, H. Imamura, S. Decuypere, T. G. Clark, G. H. Coombs, J. A. Cotton, J. D. Hilley, S. de Doncker, I. Maes, J. C. Mottram, M. A. Quail, S. Rijal, M. Sanders, G. Schonian, O. Stark, S. Sundar, M. Vanaerschot, C. Hertz-Fowler, J. C. Dujardin and M. Berriman, *Genome Res.*, 2011, **21**, 2143.
69. A. Mukherjee, P. K. Padmanabhan, S. Singh, G. Roy, I. Girard, M. Chatterjee, M. Ouellette and R. Madhubala, *J. Antimicrob. Chemother.*, 2007, **59**, 204.
70. M. Wiese, *EMBO J.*, 1998, **17**, 2619.
71. N. Singh, M. Chatterjee and S. Sundar, *Parasites Vectors*, 2014, **7**, 596.
72. V. Adaui, K. Schnorbusch, M. Zimic, A. Gutierrez, S. Decuypere, M. Vanaerschot, S. De Doncker, I. Maes, A. Llanos-Cuentas, F. Chappuis, J. Arevalo and J. C. Dujardin, *Parasitology*, 2011, **138**, 183.
73. V. Adaui, L. F. Lye, N. S. Akopyants, M. Zimic, A. Llanos-Cuentas, L. Garcia, I. Maes, S. De Doncker, D. E. Dobson, J. Arevalo, J. C. Dujardin and S. M. Beverley, *J. Infect. Dis.*, 2016, **213**, 112.
74. E. Bourreau, M. Ginouves, G. Prevot, M. A. Hartley, J. P. Gangneux, F. Robert-Gangneux, J. Dufour, D. Sainte-Marie, A. Bertolotti,

F. Pratlong, R. Martin, F. Schutz, P. Couppie, N. Fasel and C. Ronet, *J. Infect. Dis.*, 2016, **213**, 105.
75. J. van Griensven, E. Carrillo, R. Lopez-Velez, L. Lynen and J. Moreno, *Clin. Microbiol. Infect.*, 2014, **20**, 286.
76. J. Alvar, P. Aparicio, A. Aseffa, M. Den Boer, C. Canavate, J. P. Dedet, L. Gradoni, R. Ter Horst, R. Lopez-Velez and J. Moreno, *Clin. Microbiol. Rev.*, 2008, **21**, 334, table of contents.
77. C. C. Geilen, T. Wieder, A. Haase, W. Reutter, D. M. Morre and D. J. Morre, *Biochim. Biophys. Acta*, 1994, **1211**, 14.
78. F. J. Perez-Victoria, F. Gamarro, M. Ouellette and S. Castanys, *J. Biol. Chem.*, 2003, **278**, 49965.
79. F. J. Perez-Victoria, M. P. Sanchez-Canete, S. Castanys and F. Gamarro, *J. Biol. Chem.*, 2006, **281**, 23766.
80. J. M. Perez-Victoria, F. J. Perez-Victoria, A. Parodi-Talice, I. A. Jimenez, A. G. Ravelo, S. Castanys and F. Gamarro, *Antimicrob. Agents Chemother.*, 2001, **45**, 2468.
81. E. Castanys-Munoz, N. Alder-Baerens, T. Pomorski, F. Gamarro and S. Castanys, *Mol. Microbiol.*, 2007, **64**, 1141.
82. R. M. Santa-Rita, A. Henriques-Pons, H. S. Barbosa and S. L. de Castro, *J. Antimicrob. Chemother.*, 2004, **54**, 704.
83. N. K. Verma, G. Singh and C. S. Dey, *Exp. Parasitol.*, 2007, **116**, 1.
84. C. Paris, P. M. Loiseau, C. Bories and J. Breard, *Antimicrob. Agents Chemother.*, 2004, **48**, 852.
85. S. Khademvatan, M. J. Gharavi, F. Rahim and J. Saki, *Korean J. Parasitol.*, 2011, **49**, 17.
86. S. Khademvatan, M. J. Gharavi and J. Saki, *Braz. J. Infect. Dis.*, 2011, **15**, 442.
87. N. K. Verma and C. S. Dey, *Antimicrob. Agents Chemother.*, 2004, **48**, 3010.
88. J. R. Luque-Ortega and L. Rivas, *Antimicrob. Agents Chemother.*, 2007, **51**, 1327.
89. F. Getachew and L. Gedamu, *Mol. Biochem. Parasitol.*, 2012, **183**, 42.
90. H. Lux, D. T. Hart, P. J. Parker and T. Klenner, *Adv. Exp. Med. Biol.*, 1996, **416**, 201.
91. M. Rakotomanga, S. Blanc, K. Gaudin, P. Chaminade and P. M. Loiseau, *Antimicrob. Agents Chemother.*, 2007, **51**, 1425.
92. L. Imbert, R. G. Ramos, D. Libong, S. Abreu, P. M. Loiseau and P. Chaminade, *Anal. Bioanal. Chem.*, 2012, **402**, 1169.
93. P. Wadhone, M. Maiti, R. Agarwal, V. Kamat, S. Martin and B. Saha, *J. Immunol.*, 2009, **182**, 7146.
94. A. K. Mukherjee, G. Gupta, A. Adhikari, S. Majumder, S. Kar Mahapatra, S. Bhattacharyya Majumdar and S. Majumdar, *Int. Immunopharmacol.*, 2012, **12**, 565.
95. C. H. Hochhuth, K. Vehmeyer, H. Eibl and C. Unger, *Cell Immunol.*, 1992, **141**, 161.
96. M. Ghosh, K. Roy and S. Roy, *J. Antimicrob. Chemother.*, 2013, **68**, 2834.

97. C. B. Ponte, E. A. Alves, R. N. Sampaio, A. A. Urdapilleta, S. Kuckelhaus Cdos, M. I. Muniz-Junqueira and S. A. Kuckelhaus, *Int. Immunopharmacol.*, 2012, **13**, 114.
98. I. Eue, *J. Exp. Ther. Oncol.*, 2002, **2**, 333.
99. S. Rijal, B. Ostyn, S. Uranw, K. Rai, N. R. Bhattarai, T. P. Dorlo, J. H. Beijnen, M. Vanaerschot, S. Decuypere, S. S. Dhakal, M. L. Das, P. Karki, R. Singh, M. Boelaert and J. C. Dujardin, *Clin. Infect. Dis.*, 2013, **56**, 1530.
100. S. Sundar, A. Singh, M. Rai, V. K. Prajapati, A. K. Singh, B. Ostyn, M. Boelaert, J. C. Dujardin and J. Chakravarty, *Clin. Infect. Dis.*, 2012, **55**, 543.
101. V. Ramesh, R. Singh, K. Avishek, A. Verma, D. K. Deep, S. Verma and P. Salotra, *PLoS Neglected Trop. Dis.*, 2015, **9**, e0004093.
102. B. D. Pandey, K. Pandey, O. Kaneko, T. Yanagi and K. Hirayama, *Am. J. Trop. Med. Hyg.*, 2009, **80**, 580.
103. V. Bhandari, A. Kulshrestha, D. K. Deep, O. Stark, V. K. Prajapati, V. Ramesh, S. Sundar, G. Schonian, J. C. Dujardin and P. Salotra, *PLoS Neglected Trop. Dis.*, 2012, **6**, e1657.
104. K. Rai, B. Cuypers, N. R. Bhattarai, S. Uranw, M. Berg, B. Ostyn, J. C. Dujardin, S. Rijal and M. Vanaerschot, *mBio*, 2013, **4**, e00611.
105. J. B. Carnielli, H. M. de Andrade, S. F. Pires, A. D. Chapeaurouge, J. Perales, R. Monti-Rocha, S. F. Carvalho, L. P. Ribeiro, R. Dietze, S. G. Figueiredo and E. M. Lemos, *J. Proteomics*, 2014, **108**, 198.
106. M. P. Sanchez-Canete, L. Carvalho, F. J. Perez-Victoria, F. Gamarro and S. Castanys, *Antimicrob. Agents Chemother.*, 2009, **53**, 1305.
107. A. Mondelaers, M. P. Sanchez-Canete, S. Hendrickx, E. Eberhardt, R. Garcia-Hernandez, L. Lachaud, J. Cotton, M. Sanders, B. Cuypers, H. Imamura, J. C. Dujardin, P. Delputte, P. Cos, G. Caljon, F. Gamarro, S. Castanys and L. Maes, *PLoS One*, 2016, **11**, e0154101.
108. A. C. Coelho, C. T. Trinconi, C. H. Costa and S. R. Uliana, *PLoS Neglected Trop. Dis.*, 2014, **8**, e2999.
109. M. C. Laffitte, P. Leprohon, D. Legare and M. Ouellette, *Parasitol. Res.*, 2016, **115**, 3699.
110. D. Kumar, A. Kulshrestha, R. Singh and P. Salotra, *Antimicrob. Agents Chemother.*, 2009, **53**, 835.
111. J. Mishra, R. Madhubala and S. Singh, *Parasitol. Res.*, 2013, **112**, 1001.
112. C. D. Shaw, J. Lonchamp, T. Downing, H. Imamura, T. M. Freeman, J. A. Cotton, M. Sanders, G. Blackburn, J. C. Dujardin, S. Rijal, B. Khanal, C. J. Illingworth, G. H. Coombs and K. C. Carter, *Mol. Microbiol.*, 2016, **99**, 1134.
113. A. Kulshrestha, V. Sharma, R. Singh and P. Salotra, *Parasitol. Res.*, 2014, **113**, 1171.
114. P. Vacchina, B. Norris-Mullins, M. A. Abengozar, C. G. Viamontes, J. Sarro, M. T. Stephens, M. E. Pfrender, L. Rivas and M. A. Morales, *Antimicrob. Agents Chemother.*, 2016, **60**, 4089.

115. M. Ouellette, D. Legare, A. Haimeur, K. Grondin, G. Roy, C. Brochu and B. Papadopoulou, *Drug Resist. Updates*, 1998, **1**, 43.
116. M. Rybczynska, R. Liu, P. Lu, F. J. Sharom, E. Steinfels, A. D. Pietro, M. Spitaler, H. Grunicke and J. Hofmann, *Br. J. Cancer*, 2001, **84**, 1405.
117. E. Castanys-Munoz, J. M. Perez-Victoria, F. Gamarro and S. Castanys, *Antimicrob. Agents Chemother.*, 2008, **52**, 3573.
118. M. Das, P. Saudagar, S. Sundar and V. K. Dubey, *FEBS J.*, 2013, **280**, 4807–4815.
119. J. Mishra and S. Singh, *Exp. Parasitol.*, 2013, **135**, 397.
120. B. Vergnes, B. Gourbal, I. Girard, S. Sundar, J. Drummelsmith and M. Ouellette, *Mol. Cell Proteomics*, 2007, **6**, 88.
121. E. P. Kennedy and S. B. Weiss, *J. Biol. Chem.*, 1956, **222**, 193.
122. M. Rakotomanga, M. Saint-Pierre-Chazalet and P. M. Loiseau, *Antimicrob. Agents Chemother.*, 2005, **49**, 2677.
123. G. A. Canuto, E. A. Castilho-Martins, M. F. Tavares, L. Rivas, C. Barbas and A. Lopez-Gonzalvez, *Anal. Bioanal. Chem.*, 2014, **406**, 3459.
124. E. Inbar, D. Schlisselberg, M. Suter Grotemeyer, D. Rentsch and D. Zilberstein, *Biochem. J.*, 2013, **449**, 555.
125. B. D. Davis, *Microbiol. Rev.*, 1987, **51**, 341.
126. S. N. Hobbie, S. K. Kalapala, S. Akshay, C. Bruell, S. Schmidt, S. Dabow, A. Vasella, P. Sander and E. C. Bottger, *Nucleic Acids Res.*, 2007, **35**, 6086.
127. S. N. Hobbie, M. Kaiser, S. Schmidt, D. Shcherbakov, T. Janusic, R. Brun and E. C. Bottger, *PloS Neglected Trop. Dis.*, 2011, **5**, e1161.
128. R. N. Davidson, M. den Boer and K. Ritmeijer, *Trans. R. Soc. Trop. Med. Hyg.*, 2009, **103**, 653.
129. M. Maarouf, F. Lawrence, S. Brown and M. Robert-Gero, *Parasitol. Res.*, 1997, **83**, 198.
130. B. Chawla, A. Jhingran, A. Panigrahi, K. D. Stuart and R. Madhubala, *PLoS One.*, 2011, **6**, e26660.
131. M. Maarouf, K. Y. de, S. Brown, P. X. Petit and M. Robert-Gero, *Exp. Cell Res.*, 1997, **232**, 339–348.
132. V. Bhandari, S. Sundar, J. C. Dujardin and P. Salotra, *Antimicrob. Agents Chemother.*, 2014, **58**, 2580.
133. A. Lemke, A. F. Kiderlen and O. Kayser, *Appl. Microbiol. Biotechnol.*, 2005, **68**, 151.
134. Y. D. Paila, B. Saha and A. Chattopadhyay, *Biochem. Biophys. Res. Commun.*, 2010, **399**, 429.
135. A. K. Mukherjee, G. Gupta, S. Bhattacharjee, S. K. Guha, S. Majumder, A. Adhikari, P. Bhattachrya, S. B. Majumdar and S. Majumdar, *J. Infect.*, 2010, **61**, 173.
136. S. L. Croft, S. Sundar and A. H. Fairlamb, *Clin. Microbiol. Rev.*, 2006, **19**, 111.
137. N. Mbongo, P. M. Loiseau, M. A. Billion and M. Robert-Gero, *Antimicrob. Agents Chemother.*, 1998, **42**, 352.
138. M. Pourshafie, S. Morand, A. Virion, M. Rakotomanga, C. Dupuy and P. M. Loiseau, *Antimicrob. Agents Chemother.*, 2004, **48**, 2409.

CHAPTER 19

The Macrophage–Parasite Interface as a Chemotherapeutic Target in Leishmaniasis

ALBERT DESCOTEAUX

INRS-Institut Armand-Frappier, 531 boul. des Prairies, Laval, QC H7V 1B7, Canada
Email: albert.descoteaux@iaf.inrs.ca

19.1 Introduction

The protozoan parasite *Leishmania* causes leishmaniasis, a spectrum of human diseases ranging from a confined cutaneous lesion to a progressive visceral infection that can be fatal if left untreated. *Leishmania* is endemic in 98 countries, where it remains a serious health problem, with more than 350 million people at risk.[1] The World Health Organization classifies leishmaniasis as one of the neglected tropical diseases, a group of transmissible diseases found mostly in tropical and subtropical climates in 149 countries and which affect over one billion people, costing billions of dollars annually to these developing economies. Neglected tropical diseases affect populations living in poverty, without adequate sanitation and in close contact with infectious vectors and domestic animals and livestock.

Promastigote forms of *Leishmania* are inoculated into the mammalian host by infected sand flies and are phagocytosed by macrophages. There, they differentiate into amastigotes to replicate inside phagolysosomes or parasitophorous vacuoles (PVs).[2] No effective and safe vaccines to protect against *Leishmania* infections are available,[3] and the current arsenal of

Drug Discovery Series No. 60
Drug Discovery for Leishmaniasis
Edited by Luis Rivas and Carmen Gil

Published by the Royal Society of Chemistry, www.rsc.org

anti-leishmanial compounds is limited to four main drugs: pentavalent antimonials, miltefosine, amphotericin B and paromomycin.[4] Unfortunately, these drugs are difficult to administer, expensive and becoming ineffective due to the spread of drug resistance.[5]

There is thus a pressing need for the development of novel approaches to prevent and treat leishmaniases. Hence, recent initiatives have relied on either target-based screenings of small molecules libraries to selectively inhibit proteins essential for parasite survival, or on high-throughput phenotypic screening of compound libraries.[6] Whereas both approaches may lead to the identification of novel molecules with potent anti-leishmanial activity,[7,8] the development of resistance to these compounds remains an issue that can not be readily put aside. In this context, targeting host processes exploited by intracellular *Leishmania* amastigotes may represent an interesting avenue for the discovery of anti-leishmanial compounds less susceptible to the development of resistance. One of those host processes is the biogenesis, maintenance and expansion of the parasitophorous vacuole (PV), which represents the most intimate interface between the host and the parasite, where the parasite lives and replicates within its host.

The biology of *Leishmania* parasites is quite diverse, not only in terms of pathology and host-specificity. Indeed, it is essential to consider that the intracellular habitat and lifestyle of *Leishmania* parasites vary according to the species.[2,9] Hence, Old World *Leishmania* species (*Leishmania major*, *Leishmania donovani* and *Leishmania tropica*) reside in small, tight-fitting PVs that undergo fission shortly after parasite replication, therefore rarely containing more than a single amastigote. On the other hand, infection by New World species (*Leishmania mexicana*, *Leishmania amazonensis* and *Leishmania pifanoi*) is characterized by the formation of spacious, communal vacuoles that can harbour numerous parasites. Our understanding of the biology of individual and communal PVs as well as the mechanisms governing their development is very limited. It will thus be crucial to extensively study this interface between the host and the parasite to identify potential targets for the development of novel anti-leishmanial compounds.

Leishmania is particularly adept at transforming the macrophage into a hospitable host cell. Hence, promastigotes inhibit the phagolysosomal biogenesis process to create a PV in which they can establish infection, avoid the microbicidal activity of macrophages, and differentiate into amastigotes. The abundant virulence surface glycolipid lipophosphoglycan (LPG) plays a central role in this inhibition.[10] Interestingly, insertion of LPG into the phagosome membrane destabilizes lipid microdomains and causes exclusion of the membrane fusion regulator synaptotagmin V (Syt V) from the phagosome.[11] This discovery indicated that *Leishmania* may use a glycolipid to tamper with the host membrane-fusion machinery to create hospitable PVs. Further studies revealed that *Leishmania* also targets the host-cell membrane-fusion machinery by cleaving key molecules that modulate phagosome functions, including antigen crosspresentation and autophagy and cytokine secretion. The GPI-anchored virulence factor metalloprotease

GP63 is responsible for this cleavage.[12] Disruption of the phagosome membrane-fusion machinery thus appears as an efficient strategy to create an intracellular niche favourable to the establishment of infection and to the evasion of the immune system. This will be further discussed below.

19.2 Phagocytosis and Phagolysosome Biogenesis

Leishmania parasites are taken up by macrophages through phagocytosis, an evolutionarily conserved process by which cells internalize large particulate materials from their milieu and sequester them in vacuoles known as phagosomes. In mammals, phagocytosis plays a central role in linking innate and adaptive immunity, *via* the destruction of invading microorganisms and the processing of their antigens.[13,14] Following internalization, pathogens end up in a vacuole, the phagosome, which rapidly matures into a highly microbicidal compartment, the phagolysosome. The nascent phagosome is formed primarily by invagination of the plasma membrane. However, internalization of multiple or large targets by professional phagocytes requires large amounts of membrane, which is provided to the phagosome by exocytosis of endomembrane from intracellular compartments. Three compartments have been shown to contribute to focal exocytosis during phagocytosis: recycling endosomes, late endosomes–lysosomes and the endoplasmic reticulum (ER).[15] To acquire microbicidal properties, the newly formed phagosome engages in a maturation process involving highly regulated sequential interactions with endocytic organelles.[16] These interactions lead to the (i) assembly of the NADPH oxidase complex, which is responsible for the intraphagosomal production of reactive oxygen species; (ii) recruitment of the vacuolar-ATPase, which leads to the acidification of the phagosome, and (iii) acquisition of an array of hydrolases, culminating in the generation of a microbicidal compartment. Phagolysosome biogenesis is an important means of controlling microbial growth. Yet, several pathogenic microorganisms have evolved structures and mechanisms which allow them to subvert phagosome maturation to create a hospitable intracellular habitat, thereby avoiding an encounter with the macrophage microbicidal machinery, including exposure to reactive oxygen species and acidification.[17,18]

19.3 The Membrane Fusion Machinery and the Phagosome

Ingestion and destruction of pathogenic microorganisms largely depend on the host cell membrane-fusion machinery and intracellular vesicular trafficking.[15,19,20] The vast majority of fusion events that regulate these processes are mediated by a conserved group of membrane-anchored proteins called soluble NSF attachment protein receptors (SNAREs). During fusion between vesicles and target organelles, vesicular SNAREs and target-membrane SNAREs interact and form stable SNARE complexes that pull the two membranes tightly together and exert the force required for triggering their fusion.[21] C2-domain-containing Ca^{2+} sensors, such as members of the synaptotagmin (Syt) family, control the activity of these SNARE complexes and mediate vesicle fusion. Previous studies have clearly established that SNARE-mediated membrane-fusion events play an important role in phagocytosis and phagosome maturation, by regulating interactions between the phagosome and endocytic organelles.[15,19,20] Hence, the early endosomal SNAREs VAMP3 and syntaxin-13 are recruited to the young phagosome, whereas VAMP7 and syntaxin-7 remain associated with the phagosome to regulate interactions with late endosomes–lysosomes.[22,23] The late-endosomal SNARE VAMP8 is recruited to maturing phagosomes and controls assembly of the NADPH oxidase NOX2,[24] whereas Syt V controls phagosomal recruitment of the vacuolar-ATPase.[11] Consistent with the concept that the ER contributes membrane during phagocytosis,[25] the ER SNAREs syntaxin-18 and Sec22b are part of the phagosome proteome and regulate phagolysosome biogenesis.[26] Several other SNAREs are present in the phagosome proteome,[13] but their respective roles in phagosome maturation and function as well as the downstream immune responses remain largely unexplored.

19.4 *Leishmania* and Phagosome Remodelling

Leishmania promastigotes create a PV through the inhibition of phagolysosome biogenesis by preventing interaction of the phagosome with late endosomes and lysosomes.[10,27] This inhibition is mediated by LPG, a virulence glycolipid that consists of a polymer of repeating Galβ1,4Manα1-PO_4 units attached to the promastigote surface *via* a GPI anchor.[28] LPG-mediated phagosome remodelling is characterized by the inhibition of NADPH oxidase complex assembly at the phagosome membrane.[29] Although the exact mechanism remains to be elucidated, the current model of LPG-mediated phagosome remodelling is that LPG inserts into lipid microdomains *via* its GPI anchor, thereby allowing the negatively charged Gal(β1,4)Man-PO_4 polymer to directly interfere with the clustering of molecules into these microdomains.[30] One consequence of LPG-mediated microdomain disorganization is the exclusion of the SNARE regulator Syt V from the phagosome.[11] This in turn abrogates recruitment of the vacuolar-ATPase and impedes phagosome acidification. Collectively, these studies established the important contribution of LPG-mediated manipulation of the host cell membrane fusion machinery in the remodelling of the *Leishmania*-harbouring vacuole.

Pathogens have evolved a variety of strategies to manipulate SNAREs to cause disease.[31] Studies aimed at further investigating the effects of *Leishmania* infection on the host-cell membrane-fusion machinery led to the discovery that this parasite uses GP63, an abundant surface GPI-anchored zinc-metalloprotease, to cleave specific host cell SNAREs and Syts involved in phagolysosome biogenesis.[24,32] These observations indicated the existence of a previously undescribed strategy used by intracellular pathogens for remodelling phagosomes. GP63 is a virulence factor[33] that cleaves several host molecules to modulate macrophage responses to infection. The mechanism by which GP63 accesses and cleaves host-cell substrates remains to be elucidated. The results of previous studies indicated that GP63 is released from the parasite surface, exits the phagosome, and colocalizes with macrophage lipid microdomains.[34] Association with lipid microdomains plays an important role in the ability of GP63 to access host-cell substrates. Interestingly, various SNAREs and Syts associate to diverse degrees with these structures,[35] which may facilitate their cleavage by GP63. Redistribution of GP63 in small vesicular structures present throughout the cell is consistent with the ability of this protease to degrade host molecules outside the parasitophorous vacuoles. Consistent with the fact that GP63 is predominantly expressed and active in promastigotes, there is no cleavage of SNAREs or Syts in macrophages infected with amastigotes.

Leishmania-harbouring PVs interact continuously with host-cell vesicles and the endocytic pathway. Given the importance of SNAREs in mediating trafficking to and from the PVs, it is expected that cleavage of SNAREs affects phagosomal functionality, including antigen crosspresentation.[36] Crosspresentation of microbial peptides on major histocompatibility I molecules is essential for activating $CD8^+$ T cells, which play an important role during cutaneous leishmaniasis. Cleavage of the SNARE VAMP8 by GP63 is sufficient for *Leishmania* to impair processes involved in phagolysosome biogenesis that are required for crosspresentation.[24] Cleavage of Syt XI by GP63 may contribute to parasite survival by impeding recruitment of phagosomal effectors that contribute to *Leishmania* killing.

We have only just begun to uncover how *Leishmania* infection alters the macrophage membrane-fusion machinery and to appreciate the intricacies of this host–pathogen dialog. Clearly, it is imperative to perform a thorough and systematic examination of the membrane-fusion machinery in *Leishmania*-infected cells as the initial step to further investigate PV biogenesis and function. Studying the effects of *Leishmania* infection on host cell functions may lead to the discovery of novel strategies used by *Leishmania* and other intracellular pathogens to evade and subvert host defence mechanisms. This knowledge may then be used to develop novel approaches to treat leishmaniasis.

19.5 The Parasitophorous Vacuole as a Potential Target for Drug Development

It has been proposed that expansion of the PVs hosting parasites of the *L. mexicana* complex is a virulence strategy that enables parasite survival through dilution of the microbicidal effectors to which the parasites are exposed.[37] Disrupting this expansion process may thus represent an interesting approach to interfere with the course of infection. However, little is known concerning the host and parasite factors that contribute to the biogenesis and expansion of *Leishmania*-harbouring communal PVs. Shortly after phagocytosis, small individual phagosomes containing *L. amazonensis* promastigotes fuse extensively with late endosomes–lysosomes,[38] leading to the formation of large PVs containing several parasites. Homotypic fusion between *L. amazonensis*-harbouring PVs is also observed, but its contribution to PV enlargement is unclear.[9] Hijacking components of the host-cell membrane-fusion machinery, including endocytic SNAREs, is thus an important strategy for the biogenesis and expansion of those communal PVs.[19] In addition, communal PVs interact in a sustained way with the host cell's ER, which provides additional membrane needed for expansion. ER-specific SNAREs, including sec22b, D12, syntaxin 5 and syntaxin 18, are present on *L. amazonensis*-harbouring PVs; consistently, knock-down of those ER–Golgi SNAREs restricted PV expansion and was detrimental to parasite replication[39,40] (Figure 19.1). Contribution of ER membrane is thus essential for the biogenesis and expansion of communal PVs, indicating that ER-specific SNAREs could constitute an attractive potential target to treat leishmaniasis. This issue was directly addressed using Retro-2, a compound that blocks retrograde trafficking from endosomes to the ER by disrupting the activity of syntaxin 5. Hence, Retro-2 prevented PV expansion and *L. amazonensis* replication *in vitro* and significantly reduced parasite burden in an *L. amazonensis* experimental model of infection.[41] Similar results were

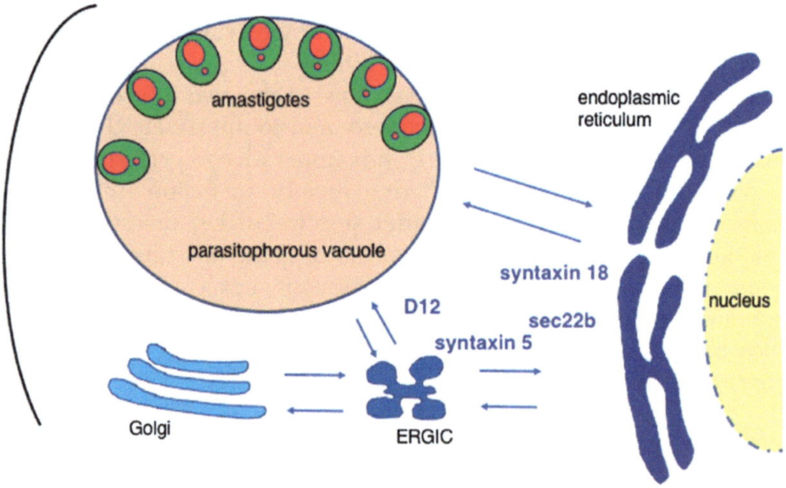

Figure 19.1 The *Leishmania*-harbouring parasitophorous vacuole interacts with various cellular compartments including the ER and the ERGIC. In the case of *L. mexicana* and *L. amazonensis*, these interactions enable the parasitophorous vacuole to acquire the membrane required for expansion and are regulated by ER and ER-to-Golgi intermediate compartment (ERGIC) SNAREs, including sec22b, D12, syntaxin 5 and syntaxin 18. Knockdown of those SNAREs or pharmacological targeting of syntaxin 5 resulted in inhibition of parasitophorous vacuole expansion and parasite replication.

obtained with *L. donovani* and *Leishmania infantum*, both *in vitro* and *in vivo*,[42] indicating that targeting the retrograde trafficking to the ER is a promising avenue for treating infections by *Leishmania* species that live in either communal of individual PVs. Importantly, Retro-2 inhibits parasite replication in axenic cultures, indicating that it may target parasite SNAREs. It may thus be necessary to develop Retro-2 analogues that specifically target mammalian cells, as the possibility exists that *Leishmania* can develop resistance to this compound.

19.6 Conclusions

In conclusion, the host-cell processes involved the biogenesis and maintenance of the intracellular habitat of *Leishmania* represent attractive targets for the development of anti-leishmanial compounds. Current and future efforts must be directed towards identifying key host molecules, such as SNAREs, associated with PV biogenesis and function.

References

1. J. Alvar, I. D. Velez, C. Bern, M. Herrero, P. Desjeux, J. Cano, J. Jannin and M. den Boer, *PLoS One*, 2012, **7**, e35671.
2. J. Alexander and D. G. Russell, *Adv. Parasitol.*, 1992, **31**, 175.
3. R. Kumar and C. Engwerda, *Clin. Transl. Immunology*, 2014, **3**, e13.
4. J. H. No, *Acta Trop.*, 2016, **155**, 113.
5. K. Katsuno, J. N. Burrows, K. Duncan, R. Hooft van Huijsduijnen, T. Kaneko, K. Kita, C. E. Mowbray, D. Schmatz, P. Warner and B. T. Slingsby, *Nat. Rev. Drug Discovery*, 2015, **14**, 751.
6. R. M. Reguera, E. Calvo-Alvarez, R. Alvarez-Velilla and R. Balaña-Fouce, *Int. J. Parasitol. Drugs Drug Resist.*, 2014, **4**, 355.
7. I. Peña, M. Pilar Manzano, J. Cantizani, A. Kessler, J. Alonso-Padilla, A. I. Bardera, E. Alvarez, G. Colmenarejo, I. Cotillo, I. Roquero, F. de Dios-Anton, V. Barroso, A. Rodriguez, D. W. Gray, M. Navarro, V. Kumar, A. Sherstnev, D. H. Drewry, J. R. Brown, J. M. Fiandor and J. Julio Martin, *Sci. Rep.*, 2015, **5**, 8771.
8. S. Khare, A. S. Nagle, A. Biggart, Y. H. Lai, F. Liang, L. C. Davis, S. W. Barnes, C. J. Mathison, E. Myburgh, M. Y. Gao, J. R. Gillespie, X. Liu, J. L. Tan, M. Stinson, I. C. Rivera, J. Ballard, V. Yeh, T. Groessl, G. Federe, H. X. Koh, J. D. Venable, B. Bursulaya, M. Shapiro, P. K. Mishra, G. Spraggon, A. Brock, J. C. Mottram, F. S. Buckner, S. P. Rao, B. G. Wen, J. R. Walker, T. Tuntland, V. Molteni, R. J. Glynne and F. Supek, *Nature*, 2016, **537**, 229.
9. F. Real and R. A. Mortara, *PLoS Neglected Trop. Dis.*, 2012, **6**, e1518.
10. M. Desjardins and A. Descoteaux, *J. Exp. Med.*, 1997, **185**, 2061.
11. A. F. Vinet, M. Fukuda, S. J. Turco and A. Descoteaux, *PLoS Pathog.*, 2009, **5**, e1000628.
12. C. Matte and A. Descoteaux, *PLoS Pathog.*, 2016, **12**, e1005962.

13. J. Boulais, M. Trost, C. R. Landry, R. Dieckmann, E. D. Levy, T. Soldati, S. W. Michnick, P. Thibault and M. Desjardins, *Mol. Syst. Biol.*, 2010, **6**, 423.
14. R. J. Botelho and S. Grinstein, *Curr. Biol.*, 2011, **21**, R533.
15. R. Levin, S. Grinstein and J. Canton, *Immunol. Rev.*, 2016, **273**, 156.
16. I. Jutras and M. Desjardins, *Annu. Rev. Cell. Dev. Biol.*, 2005, **21**, 511.
17. E. D. Case, J. A. Smith, T. A. Ficht, J. E. Samuel and P. de Figueiredo, *Traffic*, 2016, **17**, 461.
18. E. Di Russo Case and J. E. Samuel, *Microbiol. Spectr.*, 2016, **4**, VMBF-0014–2015.
19. J. L. Stow, A. P. Manderson and R. Z. Murray, *Nat. Rev. Immunol.*, 2006, **6**, 919.
20. K. K. Huynh, J. G. Kay, J. L. Stow and S. Grinstein, *Physiology*, 2007, **22**, 366.
21. R. Jahn and R. H. Scheller, *Nat. Rev. Mol. Cell Biol.*, 2006, **7**, 631.
22. R. F. Collins, A. D. Schreiber, S. Grinstein and W. S. Trimble, *J. Immunol.*, 2002, **169**, 3250.
23. V. Braun, V. Fraisier, G. Raposo, I. Hurbain, J. B. Sibarita, P. Chavrier, T. Galli and F. Niedergang, *EMBO J.*, 2004, **23**, 4166.
24. D. Matheoud, N. Moradin, A. Bellemare-Pelletier, M. T. Shio, W. J. Hong, M. Olivier, E. Gagnon, M. Desjardins and A. Descoteaux, *Cell Host Microbe*, 2013, **14**, 15.
25. E. Gagnon, S. Duclos, C. Rondeau, E. Chevet, P. H. Cameron, O. Steele-Mortimer, J. Paiement, J. J. Bergeron and M. Desjardins, *Cell*, 2002, **110**, 119.
26. I. Cebrian, G. Visentin, N. Blanchard, M. Jouve, A. Bobard, C. Moita, J. Enninga, L. F. Moita, S. Amigorena and A. Savina, *Cell*, 2011, **147**, 1355.
27. S. Scianimanico, M. Desrosiers, J.-F. Dermine, S. Méresse, A. Descoteaux and M. Desjardins, *Cell. Microbiol.*, 1999, **1**, 19.
28. S. J. Turco and A. Descoteaux, *Annu. Rev. Microbiol.*, 1992, **46**, 65.
29. R. Lodge, T. O. Diallo and A. Descoteaux, *Cell. Microbiol.*, 2006, **8**, 1922.
30. N. Moradin and A. Descoteaux, *Front. Cell. Infect. Microbiol.*, 2012, **2**, 121.
31. J. Wesolowski and F. Paumet, *Virulence*, 2010, **1**, 319.
32. G. Arango Duque, M. Fukuda, S. J. Turco, S. Stager and A. Descoteaux, *J. Immunol.*, 2014, **193**, 2363.
33. P. B. Joshi, B. L. Kelly, S. Kamhawi, D. L. Sacks and W. R. McMaster, *Mol. Biochem. Parasitol.*, 2002, **120**, 33.
34. M. A. Gomez, I. Contreras, M. Halle, M. L. Tremblay, R. W. McMaster and M. Olivier, *Sci. Signal.*, 2009, **2**, ra58.
35. C. Gil, A. Soler-Jover, J. Blasi and J. Aguilera, *Biochem. Biophys. Res. Commun.*, 2005, **329**, 117.
36. A. Alloatti, F. Kotsias, J. G. Magalhaes and S. Amigorena, *Immunol. Rev.*, 2016, **272**, 97.
37. D. Sacks and A. Sher, *Nat. Immunol.*, 2002, **3**, 1041.
38. N. Courret, C. Frehel, N. Gouhier, M. Pouchelet, E. Prina, P. Roux and J. C. Antoine, *J. Cell Sci.*, 2002, **115**, 2303.

39. B. Ndjamen, B. H. Kang, K. Hatsuzawa and P. E. Kima, *Cell. Microbiol.*, 2010, **12**, 1480.
40. J. Canton, B. Ndjamen, K. Hatsuzawa and P. E. Kima, *Cell. Microbiol.*, 2012, **14**, 937.
41. J. Canton and P. E. Kima, *Am. J. Pathol.*, 2012, **181**, 1348.
42. N. Gupta, R. Noel, A. Goudet, K. Hinsinger, A. Michau, V. Pons, H. Abdelkafi, T. Secher, A. Shima, O. Shtanko, Y. Sakurai, S. Cojean, S. Pomel, V. Lievin-Le Moal, V. Leignel, J. A. Herweg, A. Fischer, L. Johannes, K. Harrison, P. M. Beard, P. Clayette, R. Le Grand, J. O. Rayner, T. Rudel, J. Vacus, P. M. Loiseau, R. A. Davey, E. Oswald, J. C. Cintrat, J. Barbier and D. Gillet, *Chem. Biol. Interact.*, 2017, DOI: 10.1016/j.cbi.2016.10.005.

Subject Index

absorption, disposition, metabolism and elimination (ADME), 27, 40
alkaloids, 185–187
amino acid transport and homeostasis
  arginine transporter AAP3 family, 287–288
  LdAAP24, 286–287
  Leishmania cells, 285–286
  lysine transport, 288
  transporters, 288
amino acid transporters, 292
aminopyrazoles, 34
amphotericin B (AmB), 118, 379
angiotensin converting enzyme (ACE)-related metallopeptidase, 143
anti-leishmanial drug discovery
  leishmaniasis, drugs for, 25–27
  medicinal chemistry
    challenges and strategies, 27–28
      aminopyrazoles, 34
      benzoxaboroles, 32
      drug re-purposing, 30
      nitroimidazoles, 30
      phenotypic drug discovery approaches, 28–34
      proteasome inhibitors, 32
      pyrazolopyrimidines, 32–34
anti-leishmanial potential
  alkaloids, 185–187
  flavonoids, 189–190
  iridoid, 192–196
  lignoid, 191
  naphthoquinones, 191–192
  phenylpropanoids, 188–189
  saponins, 187–188
  terpenoids, 187
anti-leishmanials, 93–96
antimicrobial peptide-based chemotherapy, 248–249
antimicrobial peptide–membrane interaction, 250–251
antimicrobial peptides (AMPs)
  immunomodulation, 268–269
  intracellular targets, 262
  Leishmania, cell death of, 262–263
  mechanism of action
    antimicrobial peptide–membrane interaction, 250–251
    Leishmania, plasma membrane of, 252–254
    leishmanicidal agents, 254–260
    molecular characteristics, 250
  natural history of, 249–250
  plasma membrane permeabilisation, 260–262
  structure–activity relationship, 263–268
  therapies
    enviromental conditions for, 269–270
    intrinsic resistance of *Leishmania*, 269

scale-up production or
in situ expression,
270–271
and synergism, 270
antimonials, 373–376
antimony-based drugs
history and clinical use, 201–203
iridium-based organometallic compounds, 218–219
iron- and ruthenium-based organometallic compounds, 215–217
organoantimonial compounds, 203–208
organobismuth compounds, 208–212
organotellurium(IV) compounds, 215
organotin(IV) compounds, 213–215
palladium, 218–219
rhodium, 218–219
arginine-deprivation response, 291
arginine transporter AAP3 family, 287–288
artificial neural networks (ANNs), 133

Bayesian classifiers (BC), 133
benzoxaboroles, 32

carbon metabolism
drug discovery, 308–311
drug targets of, 303–304
functional and structural properties of, 304–308
network of, 300–303
cell proliferation assay (CPA), 11
chemical genomics, 105–107
cosmid sequencing, 104–105
cutaneous leishmaniasis (CL), 5, 18–19, 63–64
cysteine, 329–330

α-difluoromethylornithine (DFMO), 114
dipeptidylcarboxypeptidase (DCP), 143
disability-adjusted life-years (DALYs), 77–78
distribution, metabolism and pharmacokinetic (DMPK), 184
DNA topoisomerases
in *Leishmania*, 352–353
DNA topoisomerase II, inhibitors of, 362–363
inhibitors of DNA topoisomerase IB, 356–360
Leishmania infantum DNA topoisomerase IB, 353–356
type IA DNA topoisomerases, 362
type II DNA topoisomerases, 361–362
supercoiled *vs.* relaxed DNA, 350–352
drug assay methods, 60
drug delivery systems (DDS), 225–227
drug discovery, 308–311
drug re-purposing, 30
drug resistance
emerging drug resistance, 371–372
molecular mechanisms
amphotericin B (AmB), 379
antimonials, 373–376
miltefosine, 376–378
paromomycin (PMM), 378
study resistance, 372

enzyme-linked immunosorbent assay (ELISA), 14

flagella membrane, 291
flavonoids, 189–190

genomics
classical genomic strategies, 103–104
current limitations and future development, 107
novel gain-of-function genomic approaches
chemical genomics, 105–107
cosmid sequencing, 104–105
Genomics Novartis Foundation (GNF), 97
glucose transport, 288–289
glutathione, 330–331, 333–335

high-performance liquid chromatography (HPLC), 185
high-throughput screening (HTS)
anti-leishmanials, 93–96
new chemical entities (NCEs), 77–79
right screening tactics and assays, 79–80
analysis of screening data, 85–87
host cell, 84–85
parasite, 83–84
progression cascade and selection criteria, 87–92
screening assay format, 85
target *vs.* phenotypic screening, 80–83

immune response modifier (IRM), 18
immunomodulation, 268–269
inositolphosphoceramide species (IPCs), 118
iridium-based organometallic compounds, 218–219
iridoid, 192–196
iron- and ruthenium-based organometallic compounds, 215–217

lateral flow assay or direct agglutination, 14
Leishmania
anti-leishmanial drug development
drug repurposing, 144
ligand-based drug design, 134–137
structure-based drug design, 137–144
anti-leishmanial potential
alkaloids, 185–187
flavonoids, 189–190
iridoid, 192–196
lignoid, 191
naphthoquinones, 191–192
phenylpropanoids, 188–189
saponins, 187–188
terpenoids, 187
bioactive compounds isolation and chemical identification, 184–185
cell death of, 262–263
challenging parasite, 179–181
cycle of, 4
drug discovery
nitroaromatic compounds, 157–158
nitrogen heterocycles, 158–161
quinones, 155–157
gene expression, post-transcriptional regulation of, 242–243
genome, 238
intrinsic resistance of, 269
kinases, 170–171
leishmaniasis drug discovery
folate metabolism, enzymes of, 165–167
polyamine biosynthesis, enzymes of, 162–164

sterol biosynthesis pathway, 167–168
thiol metabolism, enzymes of, 164–165
mitochondrial mRNA editing, 241–242
alternative mRNA editing, 242
editosome, 242–243
natural products
bioactivity-guided fractionation and concentration test, 182–183
chemical tractability, 184
parasite forms, 183
pharmacokinetics and pharmacodynamics (PK–PD), 184
potency, efficacy and selectivity, 183
pre-fractionated samples, 182
solutions and dilution of samples, 183
pentamidine transport, 292–293
perspectives, 243–244
phosphodiesterases, 171
polycistronic transcription, 239–241
proteases, 168–170
scaffolds, 181–182
species and forms of leishmaniasis, 7
spliceosome, 241
target databases, 144–146
targets identification, 131–133
transcriptome, 238–239
trans-splicing mechanisms, 239–241
3′ UTR control and mRNA degradation, 243
Leishmania cells, 285–286
leishmaniasis
antimicrobial peptide-based chemotherapy, 248–249
asymptomatic carriers, 11
control measures
diagnostics and biomarkers, 12–14
East Africa, 15
Latin America, 15
outbreak response, 15–16
South Asia, 14
current drugs for, 25–27
cutaneous leishmaniasis, 18–19
drug access, 19–20
drugs for, 25–27
elimination tool, 20–21
global context of, 8
Leishmania–HIV co-infection, 10
natural history of, 3–10
outbreaks, 11–12
post-kala-azar dermal leishmaniasis (PKDL), 10
visceral leishmaniasis, 16–18
leishmanicidal agents, 254–260
leishmanicidal antimicrobial peptides, 263–268. *See also* antimicrobial peptides (AMPs)
ligand-based drug design (LBDD), 133, 134–137
lignoid, 191
loop mediated isothermal amplification (LAMP), 12
lysine transport, 288

machine learning techniques (ML), 133
macrophage–parasite interface
membrane fusion machinery, 389–390
parasitophorous vacuole, 391–393
phagocytosis and phagolysosome biogenesis, 389
phagosome remodelling, 390–391

membrane fusion machinery, 389–390
memorandum of understanding (MoU), 14
metabolomics, 115–116
  and anti-leishmanial drugs
    amphotericin B, 118
    miltefosine (hexadecyl-phosphocholine), 117–118
    organic antimonials, 117
    resistance and synergy, 118–119
  in *Leishmania* drug development, 119–120
  methods, 116–117
miltefosine, 376–378
miltefosine (hexadecylphosphocholine), 117–118
minimum inhibitory concentration (MIC), 41
models of infection, drug assay methodology
  detection of anti-leishmanial activity, 65–66
    conventional methods, 66–68
    transgenic *Leishmania* and related methods, 68–70
  ex vivo, 63
  in vitro, 60–63
  in vivo
    cutaneous leishmaniasis, 63–64
    pharmacokinetics (PK) and pharmacodynamics (PD), 65
    visceral leishmaniasis, 64–65
  technological innovations and developments, 70–71

naphthoquinones, 191–192
new chemical entities (NCEs), 77–79
next-generation sequencing (NGS), 102
nitroimidazoles, 30
nucleoside transport, 290
nutrient sensing
  arginine-deprivation response, 291
  flagella membrane, 291
  purine starvation, 290
nutrient transport and sensing
  metabolic environment
    insect vector, 283–284
    inside the host, 284–285
nutrient transporters
  amino acid transporters, 292
  pentamidine transport, 292–293

oral administration, 228–230
organoantimonial compounds, 203–208
organoantimony(III), 203
organoantimony(V), 203
organobismuth compounds, 208–212
organotellurium(IV) compounds, 215
organotin(IV) compounds, 213–215

palladium, 218–219
parasitophorous vacuole, 391–393
parenteral administration, 227–228
paromomycin (PMM), 378
pentamidine transport, 292–293
peptide-based therapies, general appraisal of, 249
peripheral blood mononuclear cells (PBMC), 11
phagocytosis, 389
phagolysosome biogenesis, 389
phenotypic drug discovery approaches, 28–34
phenylpropanoids, 188–189
plasma membrane permeabilisation, 260–262

post-kala-azar dermal leishmaniasis (PKDL), 10
Potential Drug Target Database (PDTD), 145
proteasome inhibitors, 32
proteomics
in drug discovery, 115
methods, 112–113
modes-of-action and drug-resistance mechanisms, 113–115
pulmonary administration, 231
purine starvation, 290
purine transport, 290
pyrazolopyrimidines, 32–34
pyrimidine transport, 290

quantitative PCR (qPCR), 11

rapid serologic diagnostic tests (RDTs), 12
redox metabolism and oxidative stress
class I haem peroxidases, 322–323
oxidative and nitrosative stress, exogenous sources of, 320–321
reactive oxygen and nitrogen species, 316–317
ROS and RNS
elimination of, 321
generation of, 317–319
superoxide dismutases, 321–322
trypanothione redox system
biosynthesis of, 329–335
recycling of, 328–329
utilization of, 323–328
rhodium, 218–219

saponins, 187–188
sodium stibogluconate (SSG), 10
soluble *Leishmania* antigen (SLA), 11
spermidine, 331–333, 333–335
structure-based drug design, 137–144
structure-based drug design (SBDD), 133
support vector machines (SVM), 133

target product profile (TPP), 11
terpenoids, 187
Therapeutic Target Database (TTD), 145
transcriptomics
perspectives on, 111–112
transcriptomic toolkit, 108–111
Tropical Diseases of the World Health Organization (TDR–WHO), 42
trypanothione redox system
biosynthesis of
cysteine uptake and, 329–330
glutathione and spermidine, conjugation of, 333–335
glutathione uptake and, 330–331
spermidine uptake and, 331–333
non-selenium glutathione peroxidases, 327–328
recycling of, 328–329
tryparedoxins, 324–325
two-cysteine peroxiredoxins, 325–327
utilization of, 323–324
trypanothione reductase (TR), 139

visceral leishmaniasis (VL), 6, 16–18
clinical efficacy, changing patterns in, 43–46
clinical translation
facilitate clinical development, 41–42
in vitro screening, 38–39
murine models, 39

visceral leishmaniasis (VL) (*continued*)
preclinical animal models for, 39–40
preclinical models, interpretation of, 40–41
Syrian hamster model, 40
discovery and development history, 42–43
dose optimization of, 46–49
translational approach, 49–50

western blot, 14
World Health Organization (WHO), 77

zoonotic VL (ZVL), 12